Culture and Society

Contemporary Foundations of Space and Place
Series Editor: John Agnew

Titles in the Series:

Politics
John Agnew and Virginie Mamadouh

Environment
Kay Anderson and Bruce Braun

Development
Stuart Corbridge

Regions
J. Nicholas Entrikin

Transport
Susan Hanson and Mei-Po Kwan

Culture and Society
Nuala C. Johnson

The City
Jacques Lévy

Economy
Ron Martin

The Rural
Richard Munton

Theory and Methods
Chris Philo

Culture and Society

Critical Essays in Human Geography

Edited by

Nuala C. Johnson

Queen's University Belfast, UK

ASHGATE

Published by
Ashgate Publishing Limited
Gower House
Croft Road
Aldershot
Hampshire GU11 3HR
England

Ashgate Publishing Company
Suite 420
101 Cherry Street
Burlington, VT 05401-4405
USA

Ashgate website: http://www.ashgate.com

British Library Cataloguing in Publication Data
Culture and society : critical essays in human geography. -
(Contemporary foundations of space and place)
1. Cultural geography 2. Human geography 3. Geographical perception
I. Johnson, Nuala Christina, 1962-
304.2

Library of Congress Control Number: 2007941050

ISBN: 978 0 7546 2691 6

Mixed Sources
Product group from well-managed forests and other controlled sources
www.fsc.org Cert no. SGS-COC-2482
© 1996 Forest Stewardship Council
FSC

Printed and bound in Great Britain by
TJ International Ltd, Padstow, Cornwall

Contents

Acknowledgements

The editor and publishers wish to thank the following for permission to use copyright material.

Blackwell Publishing for the essays: D.W. Meinig (1983), 'Geography as an Art', *Transactions of the Institute of British Geographers*, **8**, pp. 314–28. Published by Blackwell Publishing on behalf of The Royal Geographical Society (with The Institute of British Geographers); Pyrs Gruffudd (1994), 'Back to the Land: Historiography, Rurality and the Nation in Interwar Wales', *Transactions of the Institute of British Geographers*, **19**, pp. 61–77; Derek Gregory (1995), 'Between the Book and the Lamp: Imaginative Geographies of Egypt, 1849–50', *Transactions of the Institute of British Geographers*, **20**, pp. 29–57. Published by Blackwell Publishing on behalf of The Royal Geographical Society (with The Institute of British Geographers); David Arnold (2000), '"Illusory Riches": Representations of the Tropical World, 1840–1950', *Singapore Journal of Tropical Geography*, **21**, pp. 6–18. Copyright © 2000 Department of Geography, National University of Singapore, and Blackwell Publishers Ltd; Bronwyn Parry (2002), 'Cultures of Knowledge: Investigating Intellectual Property Rights and Relations in the Pacific', *Antipode*, **34**, pp. 679–706. Copyright © 2002 Editorial Board of Antipode. Published by Blackwell Publishing.

Elsevier for the essays: Denis Cosgrove (1982), 'The Myth and the Stones of Venice: An Historical Geography of a Symbolic Landscape', *Journal of Historical Geography*, **8**, pp. 145–69. Copyright © 1982 Academic Press Inc (London) Ltd: Harcourt Brace Jovanich Publishers. Reprinted 2000. Copyright © 2000 Bell & Howell Information and Learning Company; Setsu Tachibana, Stephen Daniels and Charles Watkins (2004), 'Japanese Gardens in Edwardian Britain: Landscape and Transculturation', *Journal of Historical Geography*, **30**, pp. 364–94; Cheryl McEwan (1996), 'Paradise or Pandemonium? West African Landscapes in the Travel Accounts of Victorian Women', *Journal of Historical Geography*, **22**, pp. 68–83. Copyright © 1996 Academic Press Limited; James S. Duncan (2002), 'Embodying Colonialism? Domination and Resistance in Nineteenth-Century Ceylonese Coffee Plantations', *Journal of Historical Geography*, **28**, pp. 317–38. Copyright © 2002 Elsevier Science Ltd. All rights reserved; David N. Livingstone (2002), 'Race, Space and Moral Climatology: Notes Toward a Genealogy', *Journal of Historical Geography*, **28**, pp. 159–80. Copyright © 2002 Elsevier Science Ltd.

Oxford University Press for the essays: Jackie Huggins, Rita Huggins and Jane M. Jacobs (1995), 'Kooramindanjie: Place and the Postcolonial', *History Workshop Journal* , **39**, pp. 165–181. Copyright © 1995 History Workshop Journal.

Sage Publications for the essays: Tim Cresswell (2003), 'Landscape and the Obliteration of Practice', in Kay Anderson, Mona Domosh, Steve Pile and Nigel Thrift (eds), *Handbook of Cultural Geography*, London: Sage, pp. 269–81; Bruce Braun (2002), 'Colonialism's Afterlife: Vision and Visuality on the Northwest Coast', *Cultural Geographies*, **9**, pp. 202–47. Copyright © 2002 Arnold; Sarah Whatmore (2006), 'Materialist Returns: Practising

Cultural Geography in and for a More-than-Human World', *Cultural Geographies*, **13**, pp. 600–9; Gail Davies (2000), 'Science, 'Observation and Entertainment: Competing Visions of Postwar British Natural History Television, 1946–1967', *Ecumene: A Journal of Cultural Geographies*, **7**, pp. 432–59. Copyright © 2000 Arnold.

Taylor & Francis Ltd for the essays: Catherine Nash (1996), 'Reclaiming Vision: Looking at Landscape and the Body', *Gender, Place and Culture*, **3**, pp. 149–69; Richard H. Schein (1997), 'The Place of Landscape: A Conceptual Framework for Interpreting an American Scene', *Annals of the Association of American Geographers*, **87**, pp. 660–80. Copyright © 1997 Association of American Geographers. Published by Blackwell Publishers; David Ley and Roman Cybriwsky (1974), 'Urban Graffiti as Territorial Markers', *Annals of the Association of American Geographers*, **64**, pp. 491–505. Copyright © 1974 Association of American Geographers; David Harvey (1979), 'Monument and Myth', *Annals of the Association of American Geographers*, **69**, pp. 362–81. Copyright © 1979 Association of American Geographers; Jon Goss (1999), 'Once-upon-a-Time in the Commodity World: An Unofficial Guide to Mall of America', *Annals of the Association of American Geographers*, **89**, pp. 45–75. Copyright © 1999 Association of American Geographers. Published by Blackwell Publishers; Stephen Legg (2003), 'Gendered Politics and Nationalised Homes: Women and the Anti-Colonial Struggle in Delhi, 1930–47', *Gender, Place and Culture*, **10**, pp. 7–27. Copyright © 2003 Taylor & Francis Ltd.

Accademia Venice for the image: J.Tintoretto. Storm rising while the body of St Mark is being transported.

Ashmolean Museum, Oxford for the image: J. Ruskin, View of the Doge's Palace(sea façade), Venice taken from the water (Ashmolean Museum, Oxford).

Lawrence Paul Yuxweluptun for the images: Lawrence Paul Yuxweluptun, Scorched Earth, Clear-cut Logging on Native Sovereign Lands, Shaman Coming to Fix (1991). Acrylic on canvas; no. 36950, National Gallery of Canada, Ottawa. Reproduced with permission of artist; Lawrence Paul Yuxweluptun, Clayoquot (1993). Reproduced with permission of artist.

Museo Correr, Venice for the image: Jacopo de Barbari, plan of Venice in 1500 (Museo Correr, Venice).

Royal BC Museum Corporation for the images: Emily Carr, Tanoo (1913). Oil on canvas. Province of British Columbia Archives, PDP02145. Courtesy of Royal BC Museum, BC Archives; Emily Carr painting in Tanoo, 1913. Province of British Columbia Archives, F00254. Courtesy of Royal BC Museum, BC Archives.

Tate, London for the images: J.M.W. Turner, Venice, the Piazzetta with the ceremony of the Doge marrying the sea (The Tate Gallery, London); J.M. W. Turner, The Sun of Venice going to sea (The Tate Gallery, London); J. Ruskin, The north-west angle of the façade of St. Mark's, Venice (The Tate Gallery).Copyright © 2008 Tate, London.

Every effort has been made to trace all the copyright holders, but if any have been inadvertently overlooked the publishers will be pleased to make the necessary arrangement at the first opportunity.

Series Preface

This series collects together some of the most significant articles from the major fields of human geography published over the past 40 years. During this time something of a renaissance has occurred in the thinking that explores facets of human society using the concepts of space and place (and associated ones such as territory, geopolitics, mobility, diffusion, and locality). This reflects both the rediscovery of cultural difference and local knowledge as claims to universal, objective knowledge have come under critical scrutiny, and the exhaustion of historicist narratives owing to the failure of various projects to end global economic and political inequality. Thinking in terms of fixed territories of statehood has also become problematic owing to increased human mobility in a technologically-driven world in which religious and other, often non-national identities, appear to be in the political ascendancy. The period from 1970 until today has been one of particularly dramatic worldwide geopolitical change. Consider, for example, the impact of the political collapse of the Soviet Union, the economic rise of China, and the spread of globalization around the world. Many of the central tenets of Western social thought on nationalism, the primacy of economic change, and the centrality of states that were established during the Cold War are now subject to revision.

Of course, concepts of space and place as they are applied to human society have old philosophical roots, although these were undoubtedly obscured, particularly in the United States, for many years. Since the seventeenth century 'space' has meant the plane on which events and objects are located, while 'place', the older of the two terms, dates back to the ancient Greeks and refers to the 'lived' or 'occupied' space. What is new in recent years is the varied ways in which these concepts and related ones are now being used. This involves blending older understandings, drawing from the three classic geographic traditions that emphasize the environmental (physical-human), spatial (distributional), and regional (clustering) approaches to geographic sameness and difference, with new sensibilities and concerns about social and political divisions (from social class, ethnicity, disability, and gender to the new political divisions of a post-Cold War world). This sense of older ideas adapting to new circumstances accounts for the seemingly paradoxical title of the series: Contemporary Foundations of Space and Place. If the foundations of recent thinking are rooted in the past, such thinking also mirrors recent imperatives. What is also new is a more self-consciously 'critical' tenor to use of concepts compared to a past when much research was driven less by theoretical or methodological considerations and much more by just an interest in a given phenomenon in itself (landscape features, settlement types, borderlands, etc.) as if the conditions for its study were self-evident.

The ten volumes are divided, using fairly conventional distinctions, between those that are more general or cover large parts of the field (theory and methods, regions) and those that have relatively more constrained empirical subject matter (rural, economic, etc.). Unsurprisingly, there is considerable potential overlap across the volumes. Within the field as a whole, there is no simple division of labor between articles that are just 'theoretical' or 'methodological' and those that employ concepts in more mundanely 'empirical' ways. Nevertheless, there are definitely some articles that have had greater theoretical or methodological influence across the field as a whole. Within their respective areas, however, editors do attempt to cover the main conceptual differences and controversies of the past forty years, paying particular attention to the scope of the influence in question. Each volume is comprised of articles in English. Of course, English dominates global academic production today, but this does not mean that only work written by native English-speakers is significant. In choosing volume editors and volume content, an active effort has been made to ensure the representation of standpoints and perspectives from all over the world. The series as a whole should be particularly helpful to those looking for substantial overviews of the course of human geography between the 1970s and the early 2000s but who lack access to large library collections of academic journals. The individual volumes are crafted by specialists who have expansive rather than narrow conceptions of the purpose of the overall enterprise. The volumes are thus also designed to appeal to those looking for artful reviews of recent developments in different substantive areas of human geography.

JOHN AGNEW
General Editor
UCLA

Introduction

In preparing for a recent visit to the city of Vienna I did three things. I watched, not for the first time, Carol Reed's wonderful film noir movie *The Third Man* (1949), based on Graham Greene's novella of the same name. Second, I revisited a book that had a profound effect on me as a graduate student in Syracuse University in the late 1980s: Carl Schorske's marvellous 1980 work on Vienna. Finally, I re-examined reproductions of the paintings of Gustav Klimt, Oskar Kokoschka and Egon Schiele. It was clear that I had imagined Vienna a long time before I ever set foot in the place and I wanted to refresh those memories through texts and visual images before my flight touched Austrian ground. In so doing I was preparing myself for encountering a culture and society in some ways similar, but in many ways profoundly different, to the one from which I was coming, Ireland.

I was hoping that my experience of Vienna would not be *Lost in Translation* as Tokyo had been in Sofia Coppola's (2003) movie of the same name but there was no guarantee that my preparations would be either relevant or familiarizing. There were barriers and commonalities between Ireland and Austria that would inure and alienate at different junctures. Let me itemize a few of these as a way of introducing how human geographers might interlock with questions of culture and society. At the most basic level, there was the cultural barrier of linguistic difference. I had no German and would have to rely on Austrians conversing with me in English. As a student of nationalism I have long studied the relationships between language and identity politics and my linguistic incompetency, I knew, would shape my encounter with Viennese culture (Johnson, 1992, 1993). Second, Vienna formed the centre of a large and extremely wealthy empire under the Hapsburgs while Ireland had, at best, a very ambivalent relationship with the metropolitan centre of Britain's empire in London. The effect of these contrasting histories is reflected in the scale and physical form of each city. Vienna could boast a palatial architectural heritage, expressive of affluence, privilege and power, quite in contrast with the built environment of Dublin. Large, ornate and at times ostentatious palaces and townhouses of the city's royal and aristocratic forebears have no direct equivalents in the Irish urban scene. The nineteenth-century emergence of a Liberal urban bourgeoisie who radically redefined the urban space and reconjugated ways of living in Vienna through the design of the Ringstrasse has only limited parallels with Ireland. Schorske's (1980) analysis of the links between class and aesthetics in his discussion of the making of the Ringstrasse powerfully elucidated the tensions between liberals and conservatives among Vienna's turn-of-the-century middle class determined to make a material and intellectual mark on the streetscape of their city. This book's close attention to the relationship between urban form and cultural vision among the city's rising bourgeoisie provided me and, I suspect, a host of other geographers with inspiration for a developing interest in public monument and the spatial politics of identity formation. Schorske's sensitivity to the geographies of the everyday and their relationship with the built urban environment offered a template for those concerned with urban cultures, social class and the politics of meaning.

Unlike Dublin Vienna formed an important centre for the visual and musical arts. Boasting a host of well-known painters as well as composers, Vienna has a deserved image as a nucleus of central European intellectual life. Replete with galleries, museums and concert halls, contemporary Vienna celebrates its legacy of artists and musicians, from Schubert to Strauss and from Bosch to Klimt, through a variety of performance spaces. Finally, Vienna's geopolitical position in the centre of Europe guaranteed it an important role historically in the regulation of the continent's political and military life in ways very different to Ireland's position within Europe's international affairs. The key part it played in precipitating the First World War and its position as the birthplace of Hitler speak towards its pivotal geopolitical function. Thus, its geopolitics, which was tied up with its social composition and (multi)cultural identity, marked it apart from Ireland's position on the fringes of western Europe.

Notwithstanding these apparent barriers that I have sketched there are many cultural commonalities between the capital cities of Ireland and Austria. One of the most immediately striking is the importance of religion – Catholicism – in both milieux. The presence of numerous Baroque churches in Vienna's old town point to the long and strong connections between aristocratic and clerical authority in the exercise of power within the Hapsburg dynasty. Richly ornate, heavily gilded and at times breathtakingly ornamented church interiors indicate the solid control that the church hierarchy held over the inhabitants of the city and empire and is a clear manifestation of the social, political and spiritual strength of the church in Vienna's history. The role of Catholicism in Dublin's social and cultural life, although played out differently in the physical landscape of the city, did resonate with familiarity. The fact that I was visiting Vienna in Easter week merely reinforced this sense of the communal. Secondly, there seemed to be certain parallels between the cities as magnets and repellents for literary, intellectual life. If Vienna had its Sigmund Freud and Robert Altenberg, Dublin had its James Joyce and Sean O'Casey. Both cities, in the early decades of the twentieth century, seemed to share a precarious and ambivalent relationship with their independent-minded and experimental thinkers. A complex mix of religious dogma, conservatism and at times outright bigotry made the artists' role in each city difficult and sometimes led to exile. Interestingly, both cities were the settings for iconic films adapted from fiction. John Huston's *The Dead* (1987) and Carol Reed's *The Third Man* are exemplars of their type and each are strongly associated with the city in which they are set, offering a rich insight into the social life of these places at a particular moment in their cultural history.

While my preparation for visiting Vienna was largely textual and visual, my experience of travelling to Vienna was an entirely embodied practice where my whole range of senses were mobilized, both consciously and unconsciously. From the early morning clatter of delivery vans in the old town to the powerful aroma of coffee and cigarettes from the numerous coffee houses, from the sweet flavours of Viennese cakes to the aching muscles of legs having had long days of walking around the city, my body and senses were all engaged in the acquaintance I made with Vienna. My contact was physical as well as social, it was biological as well as intellectual and it was distant as well as intimate. My experience of visiting Austria's capital serves as a metaphor for some of the shifts in emphasis that have characterized geographical investigations of the relationships between culture and society over the past 30 years and are evident in the essays chosen for inclusion in this collection.

The essays that follow spotlight a combination of different approaches which move from engagements that are focused on a highly aestheticized visual take on culture to those which are more performative in their emphasis. These categories, of course, are not mutually exclusive as many have pointed to the highly performative dimension of visual practices (Nash, 2000), but they do, to a degree, signal a shift in direction. In compiling this set of essays I have organized the contributions around four principal themes: landscape, identity politics, imperialism and nature; prefaced by a more general essay on the spaces of interpretation (Part 1). While these themes often overlap in their conceptual and methodological style, they also track some dominant types of work that have inspired geographers in their analysis of the connections between cultures and societies. Although set out chronologically there is no assumed cumulative effect in the ordering or any progressivist agenda underpinning the selection. Each essay is freestanding, chosen on the basis of being an exemplary critical investigation of some aspect of the four themes. As in all such exercises the selection process is in some manner subjective but there was an attempt to use a few criteria to inform the choice. First, I chose essays which seemed to have a major impact on the field. Apart from David Arnold, the authors are all geographers. Second, I chose essays where there was a lucidity of purpose and clarity of execution. Third, I wanted to include historical and contemporary work, as both are vital in this subfield of the discipline. Finally, I wanted the selection to include a diversity of methodological, theoretical and geographical foci.

Part II Landscapes and Cultures

The concept of landscape represents one of the oldest themes in the history of academic geography. Early studies of human–environment relationships and the morphology of landscape focused on treating the past as an assemblage of material artefacts to be excavated, described and used to elucidate the cultural transformations of the land (Sauer, 1963; Schein, 2004). From the 1980s onwards, however, this view of landscape was challenged and new directions for landscape research forged (Cosgrove and Daniels, 1988; Duncan and Duncan, 1988). Associated with what has been broadly described as the 'cultural turn', landscape interpretation charted new territories where there was the explicit application of social and cultural theory in the explanation of landscape change. In particular this body of work centred on social processes and practices to make sense of landscape transformations. Rather than treating the landscape as an object or an image, Denis Cosgrove (1985) redefined landscape as a 'way of seeing' and in so doing opened up a whole field of investigations into the processes of visualizing the landscape. From aesthetic theory and art history to cultural Marxism and feminist studies, landscape interpretation was thoroughly informed by insights derived from work outside the discipline. The selection of essays in this section, therefore, seeks to reveal some of the prototypical work emerging from these newer traditions, while continuing to bear in mind some of the legacies of earlier cultural geographies.

In the first essay I have chosen for this section Denis Cosgrove (Chapter 2) presents an inquiry into the symbolic reading and representation of the landscape of Venice by Renaissance Venetians and in the nineteenth century by English Victorians. Cosgrove charts how a myth of Venice developed through a political and historical discourse promoted by its inhabitants and projected onto the landscape of their city. Drawing on architectural, cartographic and textual sources Cosgrove documents how sixteenth-century Venetians represented their city as the perfect, Utopian space, characterized by liberty and longevity. And this longevity of purpose was to be found in the remarkable visual harmony of the cityscape. This myth was rescripted in the nineteenth century, particularly by the Englishman John Ruskin. It was mobilized for different purposes by Ruskin and its history and meaning reinterpreted. In this fascinating essay Cosgrove highlights with great skill and energy the manner in which a city's form and structure is mythologized through texts and images and he hints in the concluding sections of this essay at the sexualized and gendered focus of the myth which was, and continues, to provide imaginative stimulus to tourists.

This next essay by Donald Meinig (Chapter 3) was delivered as a guest lecture to the Annual Conference of the Institute of British Geographers in 1983. It represents an attempt to position geography more generally, and those geographers interested in landscape and place description more particularly, away from the rationalist, scientific approaches favoured in the natural and social sciences and towards a geography that would lie within an arts and humanist tradition. Rather than focusing on explanation and analytical rigour in our views of landscape, Meinig calls for geographical engagement with interpretive and experiential dimensions of space and landscape. He suggests that disciplines such as literature and history provide fertile ways of writing and interrogating the landscape that could be of benefit to geographers. Geographers could draw on fictional texts as source material for understanding place and peoples' connections to it but also, more provocatively, Meinig suggests that geography might itself create a literature that would merit the attention of other disciplines in the arts and humanities. Using the work of Raymond Williams, Meinig highlights the directions that a literary, artistic geography might take and, over the next 20 years, work would emerge within geography which would draw on the humanities for theoretical inspiration, most particularly in the geographies of fiction.

While themes of sensuality pepper both Cosgrove's and Meinig's discussion of landscape interpretation, Catherine Nash's important essay (Chapter 4) shifts emphasis both by responding to feminist critics who have claimed that the landscape tradition within geography is exclusively masculinist in its approach (Rose, 1993) and by directing our attention to the manner in which visual pleasure and traditions of visual representation can be articulated and portrayed in radically subversive ways. Responding to feminist judgements that all visual pleasure in landscape is mediated through a dominant and singular masculine gaze, Nash suggests that a critical feminist perspective can be brought to bear on the act of viewing and she queries assumptions that visual desire is necessarily oppressive, emanating from a position of domination. Drawing on film theory and work in cultural studies, she interrogates the processes of representation and interpretation by analysing the images of the male body as landscape in the work of two women artists.

Richard Schein's point of departure in his fascinating study (Chapter 5) of an American suburb is to treat landscape as a 'discourse materialized'. In an attempt to highlight the role of individual and collective agency within the context of structural discourses Schein draws our attention to the role of a myriad of individual actions on the development of the landscape and in its alteration. Schein applies his conceptual framework to Ashland Park, a fashionable suburb of Lexington, Kentucky. Using landscape architecture, insurance mapping, zoning, historic preservation and individual consumption patterns, Schein charts how these overlapping and intersecting discourses affect the way that Ashland Park appears and changes. With the American ideal of liberal individualism as the dominant context, Schein illuminates how this can be accommodated within a set of disciplining discourses and practices that, in this case, materialize in the maintenance of a historically sensitive but also changing, desirable American suburb.

The final essay in this section on landscape shifts direction towards an element of the built environment that has played a huge role on both domestic and institutional transformations of landscape – the garden. Instead of featuring the indigenous or native garden, Setsu Tachibana, Stephen Daniels and Charles Watkins (Chapter 6) examine the fashion for Japanese gardens in the development of ideas of landscape, gardening and transculturation in early twentieth-century Britain. They track the development of this fascination with Japan and its gardening styles and methods through a study of two key proponents of the Japanese style and three gardens, which in differing ways have represented the Japanese tradition. Seeing gardens as sites of cultural exchange and contact zones this essay records how the style of japonisme became familiar and desirable in Edwardian Britain as chinoiserie was increasingly rejected. Japanese taste in plants, garden layout and ornament were imported to Britain and translated for the public by key figures such as Josiah Conder in his work of 1893. This essay engagingly illustrates the significance of Japanese gardening in the development of British gardening habits and landscape change. The fashion for Japanese-influenced gardening declined, however, after the First World War as Japan's geopolitical position changed in relation to Britain and its dominions.

Part III Making and Marking Identities

While attention thus far has been explicitly on landscape interpretation, this section of the book draws attention to a body of work that is directed towards an understanding of identity in a variety of different national, regional and local contexts. From racial and religious identity to gender and political associations, this set of essays represents how geographers have dealt with the role space has played in the social and cultural constitution of practices, rituals and performances associated with group identification. Over the past few decades geographers have been fascinated with the analysis of cultural politics and in a particular have shifted focus from institutional and elite identity formation to an analysis of popular ideologies and group identities (Jackson, 1989, Mitchell, 2000). Deploying a range of conceptual perspectives from Marxist to poststructuralist approaches, these essays take seriously how the political is closely bound with social and cultural processes in the making of personal and group identities.

The first essay, written over 30 years ago, represents a landmark in the understanding of urban gang territoriality in the context of the North American inner city. David Ley and Richard Cybriwsky's exploration, in Chapter 7, of the role of graffiti as a boundary marker among Philadelphia's urban gangs demonstrates, with great effect, how demarcating territory and making claim to control it is a highly significant element in identity formation. Well-labelled graffiti in Philadelphia marked the spatial boundaries and social processes involved in the constitution and territorial patterning of youth culture. The authors first analysed the iconography and spatiality of markings made by 'famous' graffiti kings. These markings diffused across the city particularly along arterial links such as railway lines. Teenage gangs, by contrast, produced different images and the spatialities of their graffiti emphasized their claim to local territorial control. In areas of contact between different ethnic or racial groupings, graffiti represented a strong impulse to defend a neighbourhood from external threat and cultural difference. While today we may be very sensitive to the role of territorial markers in the making of identity boundaries, in the 1970s this was innovative. As a resident of Belfast I have noted how graffiti has played a huge role in marking out sectarian boundaries in what has been a heavily divided city and there is a long tradition of geographical study of this city's murals (Boal and Royle, 2006).

Political identities of an entirely different sort inform David Harvey's masterful interpretation (Chapter 8) of the debates surrounding the building of the Basilica of the Sacré-Coeur in northern Paris. Using a Marxist conceptual framework he investigates the historical precedents for the cult of the Sacred Heart in modern French history and in particular the use of this image and this site in a national project with highly contested political interests at stake. From the laying of its foundation stone in 1875 to the completion of the church in 1919 the building and its meaning aroused diverse interpretations. The Basilica drew support from conservative Catholics and monarchists but was opposed by republicans and socialists, and especially those who supported the Paris Commune. Montmartre proved to be a site of great ideological and material import for supporters and opponents of the Commune. This essay represents an early geographical study of the significance of public monuments and architecture in the development of popular politics and would be the foundation for much further research on the links between the public icons and popular memories (Johnson, 1995; 2003).

While Harvey's account dwells on class struggle Pyrs Gruffudd's essay (Chapter 9) concentrates on the development of narratives of national identity. Rather than examining the representations of the nation on the ground, so to speak, Gruffudd examines the development of a particular imagined environment and its presumed moral attributes as an embodiment of national belonging. Specifically, he interprets the role of the 'back to the land' tendencies in interwar Wales. The notions of rural virtue and Utopianism expressed through this movement are considered through the work of the academic geographer, H.J. Fleure and through Plaid Cymru, the Welsh Nationalist Party. By stressing the significance of the history of geographical knowledge in the Welsh case, Gruffudd powerfully demonstrates the important role of the intellectual in the articulation of popular political identities and in this case the significance of an idea of the rural as a defining attribute of Welsh identity.

Switching direction to the role of consumption in the marking of contemporary personal identities, Jon Goss explores, in Chapter 10, through a critical reading of the Mall of America, the role of commodities and their aesthetics in fashioning popular collective consumption identities. As he navigates his way through consumption theory, Goss seeks to develop a framework for understanding consumption that is neither celebratory nor disdainful but instead critical-participatory. An in-depth analysis of this particular mall enables Goss to chart the temporal and spatial journeys that this space invites shoppers to participate in and how the tropes of desire, nostalgia, pleasure and fantasy inform people's experience and understanding of their consumption under late capitalism.

The final essay I have chosen for this section dedicated to marking and making identities moves from the fantasy world of the American mall to the home-space of the women of Delhi in their anti-colonial struggle in the years leading up to Indian independence. Challenging those who conceive of the home as a site for a conservative retreat from political and public life, Stephen Legg (Chapter 11) highlights the ways in which women were active agents from their domestic arenas during the independence struggle, and thus the Indian middle-class home was not inevitably a site for female silence and resubjugation. Using both archival and oral history evidence, and deploying the theoretical insights of subaltern studies, Legg persuasively reveals how, as material and discursive spaces, the homes of Indian women acted as sites in which the nation could be imaginatively conjured and where women participated in a transformative politics. Thereby Legg challenges traditional binary oppositions between public and private space in the exercise of political action.

Part IV Space, Imperialism and Postcolonialism

Historically, geography played a significant, and not uncontroversial, role in the development of geographical knowledge and colonial practices in the world outside Europe. From the days of European overseas exploration geography has been implicated in the description, construction and occupation of non-European lands (Godlewska and Smith, 1994). Whether through cartographic and topographical display or regional and cultural description, geography has played a part in the interpretation of the territories Europeans explored and formally or informally colonized (Bell, Butlin and Heffernan, 1995; Driver, 2001). However, since the influential publication of Edward Said's *Orientalism* (1978) and a variety of subsequent writings by post-colonial, feminist and subaltern analysts (Bhabha, 1990, 1994; Spivak, 1987, 1988) geographers have cast a more critical eye over the discipline's involvement in the representation of and actions towards the non-European 'other' (Gregory, 2002). Taken together the essays in this section provide readers with a flavour of the diversity of geographies that have been written in this genre.

The first essay by Derek Gregory (Chapter 12) takes theoretical inspiration from Said and Michel Foucault and he sketches the role that the physical passage of European travellers through other lands had on the very process of writing about these places. His study centres on Egypt and the imaginative geographies produced by Florence Nightingale and Gustave Flaubert. Together, albeit in different ways, these two travellers objectified Egypt as a text to be read, as a landscape looked at and read about, but not listened to. This practice, Gregory claims, took place across a variety of registers including the library of books already scripting some aspects of Egypt. Finally, the geopolitical relations between Egypt and Europe, and Britain and France in particular, presented a complex context of changing power dynamics that would inform the imaginative geographies produced by each author in each context.

The second essay (Chapter 13) moves continent to Australia and adopts a different methodological procedure and writing style from Gregory's. Written by three authors Jackie Huggins, Rita Huggins and Jane M. Jacobs, it explores the experience of being and writing about *Kooramindanjie* or Carnarvon Gorge, a historical site of Aboriginal settlement and ceremony as well as a contemporary tourist destination. The authors document their experience and impression of this space after a joint visit in April 1993. The Huggins family came from this place whereas Jacobs was approaching it as a relative 'outsider' interested in the impact of tourism on Aboriginal rock art sites. Rather than offering any single representation of Carnarvon Gorge, the authors write in a manner that discloses the tangled history of colonialism that finds expression in a space which is home to some and a tourist venue to others. By treating their experience reflexively and writing in the first person, this journey through an 'Aboriginal' space exposes the diverse ways in which different histories inform, politicize and render ambiguous the 'sense of place' inhereht to this part of Australia.

Women are central to the interpretation of British imperialism in Cheryl McEwan's essay (Chapter 14) on Victorian travel writers to West Africa. She regards these writers as creators of popular geographies of the continent, which spread outside academic and scientific circles, and thus she explores the relationship between gender, travel writing and the regional context in which these narratives were written. Wishing to distance her study from essentialist notions of masculinity and femininity in relation to the experience of empire, she emphasizes the differing manner in which four women writers described, interpreted and understood local cultures and landscapes through their travels in West Africa. The differing spatial context and time of their travel had a profound impact on the issues these women discussed and the images they portrayed in their writing. Although influenced by prevailing imperial tropes about the 'Dark Continent', McEwan illustrates that these writers often challenged such depictions and rendered problematic existing images of West Africa societies as chaotic, disordered and aesthetically unattractive.

The disciplining of bodies in mid-nineteenth coffee plantations is the target of James Duncan's essay on Ceylon (Chapter 15). By treating plantations as colonial laboratories to test and exercise modern technologies of control and domination in the tropics, this essay explores the ways in which British planters and native Tamil workers each developed strategies to exercise or resist this control. Drawing on Henri Lefebvre's idea of abstract space through the bureaucratization of everyday life, Duncan documents British efforts to create abstract, docile bodies on their coffee plantations. Strict work schedules, hygiene regimens, rules of social behaviour towards superiors, for instance, were instituted in efforts to create disciplined and productive native workers. The analysis, however, reveals that such seemingly straightforward bodily regulation was consistently resisted by the Tamils, who developed a variety of overt and covert tactics to resist such colonial power. Such 'weapons of the weak' to use the phrase from James Scott's (1985) influential book were countered by planters through the use of coercion and tightened surveillance. Duncan's exploration of attempts to impose European tactics of space–time discipline literally on the bodies of Tamil workers points to the limitations in practice of British hegemony even at the height of empire.

The final essay in this section (Chapter 16) pauses at a post-colonial moment in the visioning of the 'Northwest Coast' in contemporary Canada. Bruce Braun's exploration of the politics of vision and visuality is rooted in an analysis of the work of two landscape artists – Emily Carr and Laurence Paul Yuxweluptun – and their visual object, the landscapes of the Northwest Coast, as presented at a small exhibition at the Vancouver Art Gallery in 1997. Through the juxtaposition of these artists' representations of this region and an examination of their historical and cultural relationship with the Canadian nation-state, Braun examines the continued influence of colonial ways of seeing this space. In the face of challenges by First Nations, Braun elucidates how this region's placement within an overriding discourse about Canadian national identity must be seen as a provisional achievement. The contrasting ways in which each artist pictures the Northwest Coast point towards the continuing remaking and reimaging of this space within Canadian political and cultural life. It also reveals the continued relevance of colonialism in the making of contemporary identities. In so doing this essay reveals the poignancy of debates about post-coloniality within settler societies.

Part V Cultures of Nature and Science

This final body of work concerns itself broadly with the relationships between space, science and knowledge about nature. While geography has a long genealogy of writing about the connections between nature and human societies that includes environmental determinism and possibilism (Livingstone, 1992), in more recent decades the focus of work in this field has radically changed direction. It has been influenced by at least two strands of inquiry outside of the discipline. First, there are those who draw on a literature from the history and sociology of science where the constructed nature of scientific investigation, procedures, sites and practices are spotlighted (Livingstone, 2003). This body of research has highlighted the social, political and cultural processes undergirding scientific endeavour (Ophir and Shapin, 1991) and stresses the contingent conditions under which particular scientific theories gain warrant and legitimacy (Shapin, 1998). Second, there is a body of work which is also concerned with elements of the social production of nature but which is heavily influenced particularly by Bruno Latour's Actor-Network Theory (Latour, 1987, 1999) and performance theory more generally (Whatmore, 2002). This work challenges the nature–culture dualism by looking at the role of non-human agency in the knowledge-making practices of scientists and societies. These two theoretical concerns do overlap and, as this is a burgeoning branch of geographical inquiry, what follows offers a mere taste of some of the research being carried out in this area.

In the first essay (Chapter 17) Gail Davies examines the early development of the British Broadcasting Corporation's Natural History unit. Drawing on the concept of 'boundary work' she carefully interprets what is involved in the definition and legitimation of specific scientific fields and analyses the heterogeneous sets of processes involved. Through an examination of key actors – broadcasters, ethologists and natural history filmmakers – Davies highlights how their visions intermingled, overlapped and created general templates of what popular natural history programming should present. Taking examples of broadcasts ranging from adventure-based safari hunting expeditions to zoos and, eventually, to programming centred on the observation of animal behaviour, Davies spotlights how these three different formats have represented diverse interpretations of the relationship between television, natural history, science and entertainment. Ultimately, this analysis of the complex and varied connections between expert and popular understandings of nature as articulated through debates and processes surrounding the Natural History unit testifies to the claim that competing visions of nature regularly both define and redefine the boundaries between public and professional understandings of science and nature.

From an altogether different angle, David Arnold's essay (Chapter 18) traces historically the emergence of a European discourse of tropicality. Not only in the domain of literature and the arts, Arnold demonstrates how not only the domain of literature and the arts but also the natural and human sciences (botany, anthropology, geography and zoology) have helped to construct ideas about the tropics and shaped colonial practices in the tropical world. Taking the French geographer Pierre Gourou's 1947 book as a critical case, this essay uncovers how a century of scientific writing, activity and administration of the tropics led to these diverse lands being viewed as a single impoverished and pestilential region. Although the perceived abundance of tropical nature had previously resulted in Edenic visions of these spaces, this version of tropicality was gradually transformed to one which emphasized the primitive, undesirable and uncontrolled nature of tropical flora, fauna, peoples and the concomitant superiority of Northern cultures and civilizations.

In a complimentary analysis, David Livingstone's essay (Chapter 19) investigates the genealogy of an idea and set of practices in Western thought which presented climate in moral categories, with racial consequences. Taking four moments in this genealogical survey: Enlightenment philosophy, early Victorian anthropology, late nineteenth-century tropical medicine and early twentieth-century climatology, Livingstone persuasively examines the links between these different domains of scientific and social scientific knowledge in generating a moral climatology. Although coming from a variety of different intellectual traditions, the actors within these various disciplines all came to the common conclusion that on the basis of global climate the earth could be divided into bipolar categories: superior and subordinate; moderate and intemperate, salubrious and insalubrious. These antithetical distinctions were literally mapped by explorers, military leaders, anthropologists and geographers and gave visual expression to the belief that tropical climates were spaces where inferior races were to be found. Climate could be mobilized from a variety of disciplinary perspectives to produce a singular moral climatology that would be performed again and again across time and space.

Moving firmly into the late twentieth century, Bronwyn Parry's essay (Chapter 20) explores the manner in which Intellectual Property Rights (IPR) – legal instruments used to protect the reproduction of ideas, information and expression that have a commercial value – have extended to include physical materials including biological-genetic matter. IPR laws, developed initially in the USA and Europe, have become universalizing instruments of biological data collection and protection. Using the South and West Pacific as a case study, Parry investigates the local and global impact of corporations and institutions seeking 'patentable' material from these regions under the protection of the IPR legal framework, and the responses and resistance of local institutions and communities to this legally sanctioned acceleration of international bioprospecting. This essay provides geographers with a fresh and challenging agenda for investigating the relationships between law and space in general and in particular the intersections between capitalism, scientific data collecting and hegemonic systems of control and surveillance.

Sarah Whatmore's programmatic statement in Chapter 21 about how cultural geographers might approach questions of science and society in the new century completes this section. Whatmore welcomes a return to the 'material' in the ways that cultural geographers are beginning to explore the relationships between the *geo* (earth) and the *bio* (life). She particularly favours the reconnections that are being forged between questions of science and technology and the lived world of human and non-humans. Drawing inspiration from feminist studies, theories of the corporeal and performance, Whatmore warmly greets the shift from discourse to practice, from meaning to affect and from human to a more-than-human set of concerns in cultural geography and as such is a fitting ending to the essays that I have selected for inclusion in this volume.

Concluding Remarks

I began this introduction with my recent trip to Vienna and I wish to return to some of the themes raised by that story in my concluding comments. Like Vienna our actual encounter with the world is often somewhat different to our theoretical understanding of it and it is these surprises that often animate our continued interest in it. While landscape has played a strong role in the history of geography's encounter with culture and society, more recently this landscape tradition has been queried and the opening essay by Tim Cresswell (Chapter 1) highlights some of the shortcomings of this focus on visioning the landscape. The emerging interest in questions of practice and of the corporeal performance of living in a world made up of diverse convergences of human and non-human materials is providing interesting new research questions.

For many cultural geographers the past continues to intrigue and this volume has included a considerable number of pieces that are historical in emphasis. Geographers' engagement with past landscapes, in the history of scientific cultures and in colonialism's ideologies and practices, continues to fuel important research agendas and to produce detailed book-length studies. Questions of identity are as central in today's society as they were in the past. Nationalism, with its many faces and manifestations, continues to drive individual and group behaviour across a range of places around the globe. Geographers need to continue to understand this phenomenon and its significance in geopolitical, geoenvironmental and geolegal practices. One area of identity that cultural geographers have been relatively silent about has been religion and there seems to me to be plenty of space for engagement with the performances of religious belief and its links not just to political questions but social and ecological ones. The continued interest in questions of imperialism and the impact of colonialism has generated a range of significant research agendas not only for historical geographies but also for an understanding of contemporary political and cultural wars (Gregory, 2002). Theories of performativity and pre-discursive actions/practices have opened up exciting avenues of research, especially in relation to science, biotechnology and the body. Some of the essays included in this volume highlight this perspective. They also raise new methodological questions and, it seems to me at any rate, that ethnography might be a fruitful conduit into these issues. While performativity theory has been applied to contemporary issues, to some effect it would be interesting to explore how well this type of theorizing travels into the past, a question that is just beginning to be asked by cultural geographers.

Bibliography

Bhabha, H.K. (1990), *Nation and Narration,* London: Routledge.

Bhabha, H.K. (1994), *The Location of Culture,* London: Routledge.

Bell, M., Butlin, R.A. and Heffernan, M.J. (eds) (1995), *Geography and Imperialism, 1820–1940,* Manchester: Manchester University Press.

Boal, F. and Royle, S. (eds) (2006), *Enduring City: Belfast in the Twentieth Century*, Belfast: Blackstaff.

Conder, Josiah (1893), *Landscape Gardening in Japan*. Tokyo: Kelly and Walsh.

Cosgrove, D. (1985) 'Prospect, Perspective and the Evolution of the Landscape Idea'. *Transactions Institute of British Geographers N.S.,* **10**, pp. 45–62.

Cosgrove, D. and Daniels, S. (eds) (1988), *The Iconography of Landscape: Essays on the Symbolic Representation, Design and Use of Past Environments*. Cambridge: Cambridge University Press.

Driver, F. (2001). *Geography Militant: Cultures of Exploration and Empire*, Oxford: Blackwell.

Duncan, J.S. and Duncan, N.G. (1988). '(Re)reading the landscape', *Environment and Planning D. Society and Space,* **6**, pp. 117–26.

Godlewska, A. (2000), *Geography Unbound: French Geographic Science from Cassini to Humboldt*, Chicago: University of Chicago Press.

Godlewska, A. and Smith, N. (eds) (1994), *Geography and Empire,* Oxford, UK; Cambridge, MA.: Blackwell.

Gourou, P. (1947), *The Tropical World,* London: Longman.

Gregory, D. (2002), *Colonial Present*, Oxford: Blackwell.

Jackson, P. (1989), *Maps of Meaning*, London: Routledge.

Johnson, N.C. (1992), 'Nation-Building, Language and Education: The Geography of Teacher Recruitment in Ireland 1925–55', *Political Geography Quarterly*, **11**, pp. 170–89.

Johnson, N.C. (1993), 'Building a Nation: An Examination of the Irish Gaeltacht Commission Report of 1926', *Journal of Historical Geography*, **19**, pp. 157–68.

Johnson, N.C. (1995), 'Cast in Stone: Monuments, Geography, and Nationalism', *Environment and Planning D: Society and Space,* **13**, pp. 51–65.

Johnson N.C. (2003), *Ireland, the Great War and the Geography of Remembrance.* Cambridge: Cambridge University Press.

Latour, B. (1987), *Science in Action: How to Follow Scientists and Engineers Through Society*, Cambridge: Harvard University Press.

Latour, B. (1999), *Pandora's Hope: Essays in the Reality of Science Studies*. Cambridge: Harvard University Press.

Livingstone, D.N. (1992), *The Geographical Tradition*. Oxford: Blackwell.

Livingstone, D.N. (1995), 'The Spaces of Knowledge: Contributions towards a Historical Geography of Science', *Society and Space,* **13**, pp. 5–34.

Livingstone, D.N. (2003), *Putting Science in its Place: Geographies of Scientific Knowledge*, Chicago: University of Chicago Press.

Mitchell, D. (2000), *Cultural Geography*, Oxford: Blackwell.

Nash, C. (2000), 'Performativity in Practice: Some Recent Work in Cultural Geography', *Progress in Human Geography,* **24**, pp. 653–64.

Ophir A. and Shapin, S. (eds.) (1991), 'The Place of Knowledge: The Spatial Setting and its Relations to the Production of Knowledge', Special issue of *Science in Context*, **4**.

Rose, G. (1993), *Feminism and Geography: The Limits of Geographical Knowledge,* Cambridge: Polity Press.

Said, E.W. (1978), *Orientalism,* New York: Pantheon Books.

Said, E.W. (1993), *Culture and Imperialism,* New York: Knopf. Distributed by Random House.

Sauer, C.O. (1925), *The Morphology of Landscape*, Berkeley: University of California Publications in Geography, **2**, 2, pp. 19–54. Reprinted in J. Leighly (ed.) (1963), *Land and Life,* Berkeley: University of California Press, pp. 315–50.

Schein, R.H. (2004), 'Cultural Traditions', in J.S. Duncan, N.C. Johnson and R.H. Schein (eds), *A Companion to Cultural Geography*, London: Blackwell, pp. 11–23.

Schorske, C. (1980), *Fin-de-Siècle Vienna: Culture and Politics*, New York: Alfred Knopf.

Scott, J. (1985), *Weapons of the Weak*, New Haven: Yale University Press.

Shapin, S. (1998), 'Placing the View from Nowhere: Historical and Sociological Problems in the Location of Science', *Transactions of the Institute of British Geographers,* **23**, pp. 1–8.

Spivak, G.C. (1987), *In Other Worlds: Essays in Cultural Politics,* New York: Methuen.

Spivak, G. (1988), 'Can the Subaltern Speak?' in C. Nelson and L. Grossberg (eds), *Marxism and the Interpretation of Culture*, London: Macmillan, pp. 271–313.

Whatmore, S. (2002), *Hybrid Geographies: Natures, Cultures, Spaces*. London: Sage.

Part I
Spaces of Interpretation

[1]

Landscape and the Obliteration of Practice

Tim Cresswell

Unlike the other contributors to this section, I do not use 'landscape' as a regular part of my academic vocabulary. Frankly I am not keen on the term. It is a concept too burdened with its own history – too fixated on origins. I have tended to focus on the concepts of place and space. I like place because of its everyday nature. Place, unlike space and landscape, permeates our everyday life and provides meaning in people's lives. Places are quite clearly 'lived'. At the other extreme space has an analytical quality about it. It rubs shoulders easily with social theory and allows conversations to occur with other social science disciplines. Landscape, on the other hand, does not have much space for temporality, for movement and flux and mundane practice. It is too much about the already accomplished and not enough about the processes of everyday life. It is too stuck in the humanities to make it amenable to the kind of critical theory that appeals to me. It seems altogether too quaint. I realize these are personal predilections. It is true that the concept can be made to grow and adapt, to colonize the dynamism of living geography, but I wonder what would be saved in the process. Why not, in other words, abandon the concept altogether and leave it as a dusty anachronism in the glossary of cultural geography? This chapter is about the limits of/to landscape.

Landscape is a very loosely used term. The relative voguishness of geography has ensured that it is no longer predominantly a concern of geography. Part of the problem is that landscape has resonance well beyond geography and even academia: it is part of our everyday vocabulary. There are landscapes of ... just about anything. In popular journalism it is not unusual to see reports on 'the changing leisure landscape' or 'the fantasy landscape'. This is sometimes fancifully changed to shortened forms such as 'mindscape' or 'pleasurescape'. Beyond geography in the academic world, landscape has become a well-worn metaphor. A quick look at a bibliographic database reveals a confusing array of uses for the term. Some are close to geography and have easily intelligible meanings such as 'mapping the criminal landscape', while others appear to use the term as a generic metaphor for a kind of intellectual survey. Examples include 'the legal landscape' and 'white flecks across the landscape of literary history', 'the political landscape of American schools', and 'Germany's changing religious landscape'. Still more take landscape out of the humanities and the social sciences and place it firmly in the lexicon of science, where theoretical physicists use something called 'landscape analysis' to discover the lattice structure of alpha beta transitions in protein folding and talk of 'free energy landscapes'. Urologists meanwhile are busy discussing the 'changing landscape of bladder augmentation'. These uses of the term 'landscape' actually tell us quite a lot about the term. 'The changing landscape of bladder augmentation' suggests an overview, a survey. It also suggests that the author of the paper is in a position to tell us all about it. He is authoritative. An idea such as 'free energy landscapes' points towards the importance of form – of topography. It also suggests we might be able to visualize energy. These are important factors in the notion of landscape we have inherited – an authoritative and surveying position for the viewer and a particular form for that which is being seen. We have become used to the importing of hard science into the social; perhaps there is room here for research into the migration of geography into theoretical physics and urology.

It is not the vagueness of the term that is my principal problem though. My problem with the idea of landscape can be illustrated through the term 'landscapes of practice'. The term is somewhat oxymoronic. Landscape, on the one hand, appears to encapsulate the notion of fixity – of a text already written – of the production of meaning and the creation of dominating power. Landscape is solid. Practice, on the other hand, is about fluidity, flow and repetition. It is about the negotiation between continuity and change. Practice has been seen by social and cultural theorists as an antidote to the representational – as an unexamined component of the everyday. So what could 'landscapes of practice' refer to? In many ways it resembles Raymond Williams' term 'structure of feeling'. Williams developed the notion of 'structure of feeling' in order to simultaneously bring the notion of structure and notion of feeling into question.

> If the social is always past, in the sense that it is always formed, we have indeed to find other terms for the undeniable experience of the present: not only the temporal present, the realisation of this and that instant, but the specificity of present being, the inalienably physical, within which we may indeed discern and acknowledge institutions, formations, positions, but not always as fixed products, defining products. And then if the social is the fixed and explicit – the known relationships, institutions, formations, positions – all that is present and moving, all that escapes or seems to escape from the fixed and the explicit and the known, is grasped and defined as the personal: this, here, now, alive, active, 'subjective'. (1977: 128)

The term 'structure of feeling' simultaneously brings the fixity of structure and the flow of feeling into account. In this formulation structure is not fixed and eternal but a part of a process. Feeling is not just 'personal' but moulded (not determined!) by the social. Structure and feeling do not inhabit two ends of a spectrum but are interconnected and productive of each other. These two terms ('landscapes of practice' and 'structures of feeling') are connected by more than their oxymoronic logic. Williams' language suggests further points of contact. The 'social' is notably 'fixed and explicit' while feeling is 'present and moving'. Williams relies upon an opposition of fixity and flow to make his point about structure and feeling. Structure, the social, institutions and formations have been, in Williams' view, located in the past and seen as finished and coherent – static – while present experience, the lived, the here and now are seen as fleeting and subjective. His argument is that neither of these is the case. It is with this in mind that I explore the possibility that landscape and practice can say something to each other. Necessarily this involves a consideration of the role of vision in landscape studies and the tension between landscapes as seen (and scene) and landscapes as arenas of practice. This chapter, therefore, is structured around two terms and their relation to landscape – vision and practice. The chapter begins with a (brief) rehearsal of the use of the term 'landscape' in cultural geography, pointing out the elaborations, critiques and developments of the central role of vision in the definition of landscape. The tension between vision and practice is developed in the next section with reference to Raymond Williams, Michel de Certeau and J.B. Jackson. This leads into an extended discussion of theories of practice. The theme of landscape as *lived* is central to the final section which focuses on the work of recent archaeologists of landscape and the idea of 'doxic landscapes'.

WAYS OF SEEING LANDSCAPE

Landscape is clearly one of the central themes in contemporary cultural geography. The history of the term hardly needs elaborating given the vigorous reworking of the idea of landscape by what was once known as the 'new cultural geography' (now distinctly middle aged). Views of landscape within cultural geography can broadly be divided into three paradigmatic movements. In the early twentieth century, mostly under the influence of Carl Sauer and the Berkeley School, landscape was seen as a material artefact that was either natural or cultural. Cultural landscapes were made out of natural landscapes through the agency of a mysterious and overarching culture. Thus the landscape of an area was none other than the material expression of the (seemingly unified) group of people who lived in that region. In the 1970s humanistic geographers looked back on Sauer with some fondness (as well as W.G. Hoskin in the United Kingdom) and reformulated his view of landscape. Humanists such as Donald Meinig (1979a), Yi-Fu Tuan (1977) and Edward Relph (1976) argued for a view of landscape that took the human imagination into account. Rather than provide accounts of distinctive regional material landscapes, geographers began to see landscape as residing within the minds and eyes of beholders: landscape as a way of seeing. Finally, in the mid 1980s geographers such as Denis Cosgrove (1984; 1985; 1987), Stephen Daniels (1990; 1993) and Kenneth Olwig (1984), fired by the ascendancy of Marxist and radical accounts of the world, particularly those of Raymond Williams, argued that landscapes were material productions within which were coded particular ideologies. Landscapes were diagrams of power and influence that helped to reproduce the very power structures that produced them in the first place. In Daniels' (1990) terms, landscapes were duplicitous in that they could not be reduced to either their brute materiality or the more ideological notion of 'ways of seeing'.

Central to this potted history has been a critique of thinking of landscapes as purely material topographies. While there is no doubt that Sauer had a broader agenda than the straightforward mapping and description of different landscape morphologies, it is equally clear that the majority of work by his followers has been centred on just such projects. To Sauer landscape was geography's naively given phenomenon. We often forget that Sauer looked to phenomenology to construct his view of the discipline. 'Every field of knowledge is characterized by its declared preoccupation with a certain group of phenomena,' he declared, 'which it undertakes to identify and order according to their relations' (1965: 316). No other discipline could be said to have landscape as its 'object of consciousness', and it was (in Sauer's eyes) the unique and privileged role of geography to describe and understand (which amount to the same thing in much of Sauer's work) this phenomenon:

> The task of geography is conceived as the establishment of a critical system which embraces the phenomenology of landscape, in order to grasp in all its meaning and color the varied terrestrial scene. (1965: 319)

Importantly Sauer was keen to emphasize that landscape was not just a scene for the activities of people but something which included the works of people as a defining element. It was in this way that Sauer rejected the determinism of his predecessors, Semple and Huntingdon, and it is for this reason that Sauer's geography became known as 'cultural geography'. Unfortunately the activities of humans on the landscape became conflated with the action of an overarching and superorganic 'culture' on the land (Duncan, 1980).

Humanistic geographers inherited the concern with the issue of landscape but turned their attention away from the 'morphology' of landscape and towards the 'experience' of landscape. An important link here is J.B. Jackson (in many ways the hero of this story). Jackson followed the Berkeley line in his interest in the cultural landscape and his distrust of formal theory. But Jackson also began to explore the 'symbolic' aspects of landscape:

> [A] landscape is not a natural feature of the environment, but a synthetic space, a man-made system of spaces superimposed on the face of the land functioning and evolving not according to natural laws but to serve a community. A landscape is thus a space deliberately created to speed up or slow down nature. (1984: 8)

Jackson recognized the importance of art and emotion in the study of landscape as he attempted to teach his students to 'read' the landscape as a kind of text full of symbolic clues to the meaning that lies behind the bare morphology. In this sense his work influenced humanistic geographers, as is clear from the long dedication to him in the book *The Interpretation of Ordinary Landscapes* (Meinig, 1979b). The overwhelming theme of Meinig's volume is the way in which culture provides a lens to focus our idea of landscape. The attention throughout is on the way we see the landscape rather than what we see in the landscape. Vision becomes the central way of getting at landscape. Meinig's well-known paper 'The beholding eye' outlines 10 views of the same landscape which depend on the particular way in which individuals with different interests will interpret the same physical features.

> We may certainly agree that we will see many of the same elements – houses, roads, trees, hills – in terms of such denotations as number, form, diversity and color; but such facts only take on meaning through association, they must be fitted together according to some coherent body of ideas. Thus we confront the central problem: any landscape is composed not only of what lies before our eyes but what lies within our heads. (1979a: 34)

The same point is repeated by Yi-Fu Tuan in his essay 'Thought and landscape'. 'Landscape,' he argues, 'is not to be defined by itemizing its parts. The parts are subsidiary clues to an integrated image. Landscape is such an image, a construct of the mind and of feeling' (1979: 89). To many humanists landscape is an image. The focus is turned away from the form of the material landscape and towards the way we see landscape. We might even go further and say that some humanists believe that landscape exists only inside the heads of people as a perspective – a way of ordering the world.

The role of vision was not lost on cultural geographers such as Denis Cosgrove and Stephen Daniels who developed the theme of landscape still further. They agree that landscape is a 'way of seeing' but insist on the relationship between that way of seeing and the material conditions which overdetermine it. Cosgrove was one of the first geographers to suggest that geographers could benefit from the work of cultural studies and began to mix humanism with Marxism. He was worried that humanists had surrendered to 'idealism and subjectivism' in the 'intellectual examination of mind and matter' (1987: 87). The subjective reformulation of the external world (the landscape), Cosgrove tells us, is not the product of an autonomous mind but the product of culture which is itself an element of an overdetermined superstructure. Landscapes as symbolic systems and as ways of seeing needed to be seen in relation to 'social formations'.

The canonical text on landscape in recent cultural geography is Denis Cosgrove's *Social Formation and Symbolic Landscape* (1984). Although the term 'landscape' had been used any number of times before by geographers, it had not been fully defined within a historical context. Cosgrove revealed the lineage of the term in a way that defined some crucial parts of it. He defines landscape as a 'way of seeing – a way in which some Europeans have represented to themselves and to others the world about them and relationships with it, and through which they have commented on social relations' (1984: 1). This first stab at a definition is followed by many others, all centred on a relationship between people and the world revolving around the act of seeing. Eight pages later, landscape is defined as 'the artistic and literary representation of the visible world, the scenery (literally that which is *seen*) which is viewed by the spectator' (1984: 9) which implied 'a particular sensibility, a way of experiencing and expressing feelings towards the external world, natural and man-made, an articulation of a human relationship with it. That sensibility was closely connected to a growing dependency on the faculty of sight as the medium through which truth was to be attained' (1984: 9). Cosgrove clearly agrees that the concept of landscape revolves around vision but, unlike humanists before him, he sought to place this 'way of seeing' in historical and social context. This contextualizing of vision is mirrored by Stephen Daniels in his outline of the 'duplicity' of landscape:

> [L]andscape may be seen as a 'dialectical image', an ambiguous synthesis whose redemptive and manipulative aspects cannot finally be disentangled, which can neither be completely reified as an authentic object in the world nor thoroughly dissolved as an ideological mirage. (1990: 206)

To Daniels the vagueness of landscape makes it an interesting term – one that holds together the material and the ideal, the solid and the superficial. It is still however primarily a visual order whether created through paintings or through country parks.

The other principal way of thinking about landscape in geography has been through the analogy of text. This is principally associated with the work of James Duncan. Duncan (1990) argues against the privileging of vision and suggests that we might think of landscapes as being constructed linguistically – as a signifying system that encourages some readings rather than others. As with a text, the producers write some meanings into a landscape but cannot finally control individual readings.

> The landscape, I would argue, is one of the central elements in a cultural system, for as an ordered assemblage of objects, a text, it acts as a signifying system through which a social system is communicated, reproduced, experienced, and explored. (1990: 17)

In Duncan's work landscape is not the detached object of a powerful gaze but a material symbolic system in which people live. Landscape is a representation of particular ideas produced by people on top of a social hierarchy. But as a text landscape is a space in which meaning and intention get pulled apart. The meaning is unstable because people act in it in ways that cannot be predicted. They read it differently. A key question for Duncan is how landscape acts as a communicative device to produce and reproduce a social and cultural order. Landscapes, he argues, can make subjects and objects appear as fixed and reified – can make what is cultural appear natural. They are the objectifier *par excellence*. They play this role through the use of textual devices such as metonymy, metaphor and allegory. Through these devices powerful people are able to tell morally charged stories about themselves, their relationships to others and their relations with the natural world.

> By becoming part of the everyday, the taken-for-granted, the objective, and the natural, the landscape

> masks the artificial and ideological nature of its form and content. Its history as a social construction is unexamined. It is, therefore, as unwittingly read as it is unwittingly written. (1990: 19)

There is, of course, a problem with pitting text against vision. Texts, after all, have to be read. While it may be possible to read texts haptically, the most obvious way is by looking. Despite this critique Duncan's view of landscape is a more embodied and practised one than Cosgrove's. Cosgrove, adopting the classical definition, resists getting too involved. The landscape is seen from a distant, slightly elevated point. Landscape paintings are seen as part of the control of nature – the taming of chaos. The use of perspective arrests the temporal and makes a moment stand forever. The artist becomes the possessor and the people in the landscape – the world of practice is denied – are frozen out.

VISION AND THE OBLITERATION OF PRACTICE: THREE VIEWS FROM ABOVE

The 'gaze' at the centre of dominant conceptions of landscape has been critiqued for being masculinist. Another effect of this way of seeing is to remove the active subject (whether people in the world or the onlooker) from the world. This effect is explored in similar ways by Raymond Williams, Michel de Certeau and J.B. Jackson. In Raymond Williams' novel *Border Country* (1960) we see the hero Matthew Price returning to the place of his childhood in the Welsh borders after spending many years in England. What follows is an examination of the gap between the idea of the village as 'landscape' and the idea of the village as a lived and felt 'place'. As Matthew realizes he has become an outsider in his own village, he reflects on his change of perspective:

> He realized as he watched what had happened in going away. The valley as landscape had been taken, but its work forgotten. The visitor sees beauty, the inhabitant a place where he works and has his friends. Far away, closing his eyes, he had been seeing this valley, but as the visitor sees it, as the guide book sees it. (1960: 75)

Later in the novel Matthew gets back into the routine of the village: 'It was no longer a landscape or view, but a valley that people were using', yet from the top of a nearby hill 'The patch was not only a place, but people, yet from here it was as if no-one lived there.'

The contemplative gaze obliterates the world of the practical

The world of practice teases apart landscape in its orthodox form. The traditional image of landscape that haunts Matthew Price in *Border Country* is very much like the now poignant and not a little ironic view of New York from the top of the former World Trade Center described by Michel de Certeau in *The Practice of Everyday Life* (1984). From this viewpoint the city was comprehensible and masterable – it appeared laid out before the viewer as though the viewer could command it and take it in – just as the eminent urologist was able to provide a survey of all the latest developments in the field. The grid pattern of upper Manhattan was out there somewhere ordering the world nicely. The exclamation marks of other skyscrapers provided another dimension and sense of perspective to this landscape. When one is 'lifted out of the city's grasp,' de Certeau writes, 'One's body is no longer clasped by the streets that turn and return it according to an anonymous law; nor is it possessed, whether as player or played, by the rumble of so many differences and by the nervousness of New York traffic' (1984: 92). De Certeau does not mention landscape but he does reflect on all the ingredients that make up landscapes – fixity, vision, power.

> The totalizing eye imagined by the painters of earlier times lives on in our achievements. The same scopic drive haunts users of architectural productions by materializing today the utopia that yesterday was only painted. The 1370 foot high tower that serves as the prow for Manhattan continues to construct the fiction that creates readers, makes the complexity of the city readable, and immobilizes its opaque mobility in a transparent text. (1984: 92)

The view from the streets is slightly different. In the turmoil of Fifth Avenue or from the middle of Washington Square it is hard to keep hold of the word 'landscape'. Everything is always changing, in process, becoming. Skateboarders flip their tricks. Young men and women dance in front of the ice cream stand, the trees move in the breeze. The place is not finished, not obviously ordered and not easily framed. It is blurred at the edges. This is the space of practice. The view from on high is, as de Certeau puts it, based on the 'oblivion' and 'misunderstanding of practices' (1984: 93).

We do not have to read de Certeau to come across such observations. We need only to reopen texts we once read avidly in decades past. A third discussion of the obliviation of practice through vision is given by J.B. Jackson. Jackson remains one of the most astute observers of landscape and, indeed, practice:

> I was an early advocate of studying landscapes from the air. At one time I had a large collection of aerial views. But when flying became increasingly unpredictable I gave it up, and I recently drove from New Mexico to Illinois and Iowa in my pickup truck. It was a long trip with many monotonous hours, but I do not regret it. It broke the spell cast by the air-view of the grid system and reminded me that there is still much to be learned at ground level. What goes on within these beautifully abstract rectangles is also worth observing. (1997: 70)

Jackson was disturbed by the distancing involved in an essentially visual definition of landscape. This, to him, ignored the lived character of arrangements of space: 'we are *not* spectators; the human landscape is *not* a work of art. It is a temporary product of much sweat and hardship and earnest thought' (1997: 343). His critique of aestheticized looking permeates his annotations on landscape. In the 1960s he was an early observer of gentrification with its heritage areas, museums and coffee-table books.

> People, the magazines tell us, are once again beginning to appreciate the city. Books of glossy photographs of graffiti, ironwork details, and abstract landscapes are financially successful despite formidable prices. The Museum of Modern Art has its walking tours; restoration of old market districts is in full swing, with the end products economically viable, if often self-consciously precious ...
>
> Building watching, like bird watching, is a rewarding hobby. Both can refresh the person who practices them. Both can drive away apathy and ennui and reawaken a sense of vitality in the watcher by underscoring the beauty to be found in everyday life by one who will search for it. But building watching is a very small and isolated part of the urban experience. It bears the same relationship to living in a city as bird watching does to working a farm or ranch. (1997: 343–4)

Cosgrove, while (ironically) appreciating Jackson's acute powers of observation, is not happy with his definition of landscape because it does not recognize that 'landscape is indeed the view of the outsider, a term of order and control, whether that control is technical, political or intellectual' (1984: 36). He argues that Jackson is not really talking about landscape because he is too inside it. It is true that J.B. Jackson had problems with the term 'landscape'. Although he is very closely associated with a journal of the same name and wrote paper after paper with landscape in the title, when it came down to it he found it too slippery and polysemous. He was too interested in practice and not satisfied with thinking of landscape in a restrictive sense, as a product of vision – as 'a portion of the earth's surface that can be comprehended at a glance' (1997: 306). Jackson, more than anyone, reveals the limits and tensions of the landscape idea. On the one hand he carefully described the origins and history of landscape in terms of vision, and on the other he enacted a critique of the term through an emphasis on everyday routines that produce and reproduce actual living landscapes. He asked those writing about landscape to be careful observers and simultaneously to undermine the priority given to the act of looking. Jackson's landscapes are ones that people inhabit and work in and they are landscapes that people produce through routine practice in an everyday sense.

J.B. Jackson and the practice of vision

Jackson's post-secondary education began in 1928 when he enrolled in the Experimental College of the University of Wisconsin in Madison where he took a broad liberal arts course covering art, literature and society in an interdisciplinary framework. After a year travelling in Europe he went on to Harvard and majored in history and literature. He was particularly inspired by Derwent Whittlesey's course 'Principles of geography'. He moved from there to the School of Architecture at the Massachusetts Institute of Technology in 1932 and withdrew after a year, going to an art school in Vienna where he learned drawing and draughting techniques. Again he left early and bought a motorcycle which he used to tour Europe for the next two years. On return to the United States he moved to New Mexico and became a cowboy. During the war he used his learning to become an intelligence officer – a role that involved the study and interpretations of maps. He saw active service in North Africa, Italy and France. While waiting for the Germans to surrender he occupied himself reading the works of French cultural geographers such as Vidal de la Blache whom he admired. After the war was over he returned to New Mexico and, following a riding accident, decided to match the European geography publications with an American version that was to become *Landscape*. The first issue was published in Spring 1951. In his early essays Jackson outlined the importance and pleasure of careful observation. 'I see things very clearly and I rely on what I see,' he explained, 'And I see things that other people don't see, and I call their attention to it' (Horowitz, 1997: xxiv). By 1956 Jackson was in close contact with the Geography Department at Berkeley, and Sauer (as well as Clarence Glacken, Lewis Mumford, Kevin Lynch and others) began to write for the journal, developing a multidisciplinary and maverick conception of the study of landscape. Jackson continued to criss-cross the United States on his motorbike and in his pickup truck and frequently wrote about the landscape from the perspective of a mobile observer. It is in these essays that the dualism of vision and practice breaks down.

If the equation that links landscape to vision has frequently erased practice, then J.B. Jackson's mobile view of landscape began to show how vision is a practice. J.B Jackson's way of looking is so much less reliant on that distanced gaze from above and so much more practised – more embodied. One of the best essays in this regard is 'The abstract world of the hot rodder'. He introduces us to the world of the pedestrian Sunday. Families would leave their homes on Sunday morning complete with all the accessories of really leisurely leisure. They would take the street car to some 'favorite patch' and walk into the distance 'strolling at a child's pace'. Jackson declares these Sunday excursions a lost treasure which inspired 'schools of painting and writing and music ... Untold minor scientific and artistic accomplishments came from the same abundant source; local botanical and geographical and historical descriptions, small books of nature verse, amateur sketches and compositions.' Despite this romantic notion of a weekend stroll he declares it mere nostalgia to think in those terms in the modern world because the nature of viewing – the way it is embodied – has changed.

> Much more has happened to us than the advent of the automobile; we have learned to see the world differently even on our holidays; we confront the familiar setting in a new manner. Broadly speaking, the former experience of nature was contemplative and static. It came while we strolled (at three miles an hour or less) through country paths with frequent halts for picking flowers, observing wildlife, and admiring the view. Repose and reflection in the midst of undisturbed natural beauty and a glimpse of something remote were what we chiefly prized. I do not wish to decry the worth of these pleasures; none were ever more fruitful in their time; but the layman's former relationship to nature ... was largely determined by a kind of classical perspective and by awe. A genuine sense of worship precluded any desecration, but it also precluded any desire for participation, and intuition that man also belonged. (1997: 202)

Jackson goes on to develop the idea of participation in an 'abstract world' brought on by technologies of speed such as the motorbike and increasingly popular outdoor pursuits such as skiing. Rather then standing back and looking, the hot rodders take part in the landscape. They experience it through embodied practice. A similar perspective is at the heart of his later essay 'The accessible landscape' (1988), in which Jackson describes the view through the window of his favourite pickup truck as he drives through America's vernacular heartland – not the static view from on high but a moving perspective more typical of the everyday practice of American life. Many years earlier in 'Other-directed houses' Jackson celebrated the highway strip as a place of pleasure and diversion for the motorist, pointing out, 16 years before Venturi et al.'s *Learning from Las Vegas*, that the buildings on the strip have to be designed for a kind of vision that travelled at 40 miles per hour – vision as a particular kind of historically specific practice.

Thinking Practice

'Practice' is a term which has begun to make a significant impact on cultural geography. Its origins are in critical theory, performance studies, feminism and post-Marxist social theory. On the whole it has been absent from landscape studies. It is already clear that to talk of 'practice' can mean quite different things to different people. I began this chapter with the observation that the notion of landscapes of practice was based on oxymoronic logic. Landscape, as we have seen, is predominantly thought of as a visual thing – as an image. It is also thought of as material and fixed – a kind of framing. Practice on the other hand is most frequently seen as radically unfixed and unfixable.

Theories of practice from Bourdieu to Judith Butler to Nigel Thrift are indebted to the work of Maurice Merleau-Ponty. He was particularly concerned with what he called the 'body-subject'. The work of Brentano and Husserl had led to the key idea that consciousness is always *consciousness of something*; in other words, there is a directedness to consciousness. The focus of intentionality, post-Husserl, had been the mind. Merleau-Ponty, in his conception of the 'body-subject', wanted to redirect intentionality to the body. Merleau-Ponty argues that 'being in the world' is a bodily thing (rather than mental), a pre-objective view. The fundamental relationship between subject and object exists at a bodily level that pre-exists congnition. In Merleau-Ponty's words:

> The plunge into action is, from the subject's point of view, an original way of relating himself to the object, and is on the same footing as perception. (1962: 110–11)

Bodily practice, then, is the most basic form of intentionality: 'Consciousness is in the first place not a matter of "I think that" but of "I can"' (1962: 137). This phenomenological consideration of the body's movements makes a clear distinction between the world of embodied movement (motility) and objectivity and representation. Basic bodily movement is not 'thought about movement' (1962: 137) but embodied movement that take place in the world of the phenomenal. As examples, Merleau-Ponty points towards fairly complicated sets of bodily practices such as using a needle and thread, or typing on a typewriter. He notes how these operations are not thought about once they are learned. We do not calculate the positions of our hands in space and work out abstractly where they should go next. Rather, our hands act out of habit in a form of bodily consciousness.

> Consciousness is being towards the thing through the intermediary of the body. A movement is learned when the body has understood it, that is, when it has incorporated it into its 'world', and to move one's body is to aim at things through it; it is to allow oneself to respond to their call, which is made upon it independently of any representation. (1962: 138–9)

Independence from representation is the key point here. Merleau-Ponty wants to make sure that we understand that movement is not secondary to consciousness but a more primary form of consciousness.

Merleau-Ponty also considers the relationship between the movement of the body-subject and the world of space and time. His argument is that space and time are not mere backdrops to our movement but are 'inhabited' by movement – that there is a continuum between space/time and the body.

> [It] is clearly in action that the spatiality of our body is brought into being, and an analysis of one's own movement should enable us to arrive at a better understanding of it. By considering the body in movement, we can see better how it inhabits space (and, moreover, time) because movement is not limited to submitting passively to space and time, it actively assumes them, it takes them up in their basic significance which is obscured in the commonplaceness of established situations. (1962: 102)

The body inhabits space and time through motion. 'I am not in space and time,' writes Merleau-Ponty, 'nor do I concieve space and time; I belong to them, my body combines with them and includes them' (1962: 140).

Merleau-Ponty's phenomenology of bodily movement was taken up by David Seamon in the 1970s. Geographers such as Seamon were keen to arrive, through phenomenological enquiry, at the essence of geographical phenomena. Place was the central concept but, in Seamon's case, bodily mobility was a key component of the understanding of place. Like Merleau-Ponty, Seamon fixed on the 'everyday movement in space' – '*any spatial displacement of the body or bodily part initiated by the person himself.* Walking to the mailbox, driving home, going from house to garage, reaching for scissors in a drawer – all these behaviours are examples of movement' (1980: 148).

Through research groups Seamon came to the conclusion, in line with Merleau-Ponty, that most everyday mobile practices take the form of habit. People drive the same route to work and back every day without thinking about it. People who have moved house find themselves going to their old house and only realize it when they arrive at the door. People reach for scissors in the drawer while engaging in conversation. Such practices appear to be below the level of conscious scrutiny. The body-subject knows what it is doing: there is an

> inherent capacity of the body to direct behaviors of the person intelligently, and thus function as a special kind of subject which expresses itself in a preconscious way usually described by such words as 'automatic', 'habitual', 'involuntary', and 'mechanical'. (1980: 155)

Seamon invokes the metaphor of dance (ballet to be specific), in order to describe the sequence of preconscious actions used to complete a particular task such as washing the dishes. He calls such a sequence a body-ballet. When such movements are sustained through a considerable length of time he calls it a 'time–space routine'. This describes the habits of a person as they follow a routine path through the day – driving to work, leaving the kids at school, going to lunch, etc. Seamon also looks beyond the individual body movement to group behaviour. When many time–space routines are combined within a particular place, a 'Place-ballet' emerges which generates, in Seamon's view, a strong sense of place. The mobilities of bodies combine in space and time to produce an existential insideness – a feeling of belonging within the rhythm of life-in-place.

Where am I going with this? Thus far I have developed the idea of landscape as primarily visual and fixed and the idea of practice as about fluidity and embodiment. In other words I have provided a sketch of the contradictions implicit in the term 'landscapes of practice'. The word 'landscape' and the word 'practice' seem to be poles apart. When Raymond Williams developed the idea of a 'structure of feeling' he wanted to draw the two, apparently opposite, terms together – to make structure less fixed and certain and to make feeling less individual and irreducible. To end this chapter I want to do the same with landscape and practice. I want to make landscape seem less fixed, less reliant on the visual, less dependent on authoritative 'framing', and to make practice seem less free-floating and more connected to the forces that shape our lives. To do this I explore some ways forward arising out of recent literature on landscape: ideas of material culture developed by archaeologists and questions of 'moral geographies' developed within geography.

WAYS FORWARD: DOXIC LANDSCAPES

W.J.T. Mitchell, in *Landscape and Power*, makes an argument for landscape as a space in which we lose ourselves – a medium for life. He

> aims to absorb these approaches into a more comprehensive model that would ask not just what landscape 'is' or 'means' but what it *does*, how it works as a cultural practice. Landscape, we suggest, doesn't merely signify or symbolize power relations; it is an instrument of cultural power ... Landscape as a cultural medium thus has a double role with respect to something like ideology; it naturalizes a cultural and social construction, representing an artificial world as if it were something given and inevitable, and it also makes that representation operational by interpellating its beholder in some more or less determinate relation to its givenness as sight and site. (1994: 1–2)

We might think of such a landscape as a 'doxic landscape'. I take the term 'doxa' from Bourdieu's theorizations of practice. I have previously written about doxa in relation to place (Cresswell, 1996). Similar considerations can illuminate the idea of practised landscapes. Doxa refers to the realm of common sense – the taken-for-granted. Doxa is a product of practice. Because people act in certain preconscious ways, any given order tends to get re-established and reproduced owing to the 'naturalisation of its own arbitrariness' (Bourdieu, 1977: 164). In other words a world, or landscape, that is the product of a particular history is made to seem natural and thus becomes an important site for the reproduction of established ways of being. Such a landscape is very much a product and producer of practice. What might the concept of doxic landscapes mean in practice?

Some of the most suggestive work on landscape has come from recent archaeology. Like geography the discipline of archaeology has gone through a reaction against a positivist legacy. Part of this reaction is an increased interest in social theory and cultural geography. Leading exponents of theoretically informed archaeology include Mike Shanks, Chris Tilley and Barbara Bender. Predictably, perhaps, new ways of thinking about landscape have emerged from a critique of the politics of vision and an engagement with the politics of practice.

In Shanks and Tilley's work landscape falls under the more general rubric of 'material culture'. Their argument, crudely put, is that archaeologists studying material culture have simply looked at pottery shards and remnant landscapes with a view to mapping and describing – visualizing if you like. What they have not done is thought about their objects of study in terms of practice. They propose a new way of thinking about material culture that is informed by theories of practice. Material culture is theorized as

> an important element of Bourdieu's doxa, a representation of the given order of the world that constitutes an environment for living; as an effective force for social action; as ideologically informed due to its perceived simple functionality, concreteness and triviality which facilitate naturalisation and misrepresentation. (1987: 113)

Landscape, theorized as material culture in these terms, is not a product of vision in particular but a practised environment. Landscape cannot easily be 'decoded' as there is no code that it can be reduced to. Instead landscape is brought into being through social praxis as an objectified form.

> Material culture is an objectification of social being, a literal reification of that social being in the co-presences and absences embodied in the material form. (1992: 130)

Landscape seen this way is a practised landscape. Practices over time become embedded in the world and leave 'traces of varying degrees of solidity, opacity or permanence' (1992: 131). These traces are 'material culture'. Landscape as material culture is thus implicated in the process of social reproduction which involves the interconnection between materiality, consciousness, action and thought.

Julian Thomas develops this in his analysis of the landscape surrounding Avebury. Thomas applies the kinds of ideas about material culture developed by Shanks and Tilley to landscape. Landscape in the western tradition, he argues, is a product of distance and position which construct 'a particular impression of the world, but are at the same time denied and the view is taken as universal, taking in everything' (1993: 22). The landscape thus becomes a passive object constructed through a process that is denied in the process of construction. Rather than thinking of archaeological landscapes as a distanced 'impression of the world', Thomas sets out to think through how views of landscape would change if they were instead thought of as bound together by the continuous flow of human conduct. Utilizing a phenomenological approach it becomes possible to think of landscapes not as external objects but as sites of dwelling. Thomas reiterates a point made by Ted Relph that 'To think about the world or the entities within it as abstract things is to render them subject to observation, to make them the object of casual curiosity and distance oneself from them' (Thomas, 1993, citing Relph, 1985: 27). Rather than make landscape into an object marked by its existence at a distance, a phenomenological approach focuses on the experiential nature of the world around us. Heidegger's concept of 'dwelling' is the key here:

> Dwelling involves a lack of distance between people and things, a lack of casual curiosity, an engagement which is neither conceptual nor articulated, and which arises through *using* the world rather than through scrutiny. (1993: 28)

Thomas's agenda, in short, is to transform the archaeological landscape from a product of distanced vision to an arena marked by its 'not-to-be-looked-at-ness' (1993: 28). Avebury, then, is understood quite differently when praxis is taken seriously. Thomas thinks through the processes of walking up to the ring of stones: his analysis is less about looking and more about doing, 'How was Avebury used?' not 'What did Avebury look like?'

More recently this idea has been developed on the interface between performance studies and archaeology by Michael Shanks and Mike Pearson (2000). Landscape is at the heart of their concerns. They build on the insights of Merleau-Ponty, focusing on the bodily experience of the land:

> We begin to walk. We feel the ground beneath our feet, the wind in our face. And as we do we leave traces. We are *involved* in the landscape ... We leave the prints of our body, the touch of flesh on metal and stone. We constantly wear things out, with our hands, our feet, our backs, our lips. And we leave the traces of singular actions: the unintentional, the random, the intimate, unplanned touch of history's passing: we break twigs, move pebbles, crush ants ... all the signs that trackers learn to read. (2000: 135)

Landscape becomes a palimpsest – a stratigraphy of practices and texts. In a performance piece called 'The first five miles' Pearson began a walk across Mynedd Bach near the village of Trefanter, south of Aberystwyth, wearing leather boots and gaiters, embroidered waistcoat, frock coat, top hat and lilac gloves and equipped with radio microphone, battery unit, earpiece, receiver and halogen lamp. He was accompanied by an assistant who carried a radio transmitter. As they walked they talked and the transmissions were sent to Aberystwyth and broadcast on Radio Ceredigion. These messages would interrupt a documentary about an attempt to enclose the peat bogs in the area in the 1820s by Augustus Brackenbury and the various attempts to stop him made by local inhabitants in 'The War of the Little Englishman'. Some text was in English and some in Welsh and these were transmitted in stereo to the right and left respectively. As the walk progressed, accompanied by a fractured soundtrack, local memories intervened, telling stories of 'all those things which we might never regard as authentic history but which go to make up the *deep map* of the locale' (Shanks and Pearson, 2000: 144). The point of the performance was to illuminate a particular landscape – to create a work which:

> has none of the dogmatism of the theatrical performance, or architectronics and that distanced aesthetic – framed up, laid out for our pictorial inspection and approval. So that the very *inauthenticity* of the performance allows room for manoeuvre, allows stances, of ownership, identity and interpretation, to be confirmed, challenged, confounded at the same time. (2000: 146)

The integration of landscape with practice inherent in the performances and writings of Shanks and Pearson is also a feature of some recent work on tourism. Tim Edensor's *Tourists at the Taj* (1998) considers the often-looked-at Taj Mahal as a landscape of practice. Four practices he focuses on are walking, gazing, photographing and remembering. Consider walking. Walking is a fairly basic feature of most people's lives – a routine and mundane practice. It is also a practice central to tourism. Most of us, I would guess, have spent time walking around sites while on holiday. Edensor suggests that in tourism 'Landscapes are criss-crossed and imprinted with the bodily presence of the visitor, and symbolic sites are negotiated via various paths' (1998: 105). The movements of tourists, he suggests, become sedimented in the landscape as a kind of 'place-ballet'. Indeed the places of most frequent passage become worn and grooved in the most material sense. The way these practices get to be sedimented is through repetition. Indeed, in the Taj, it seems, most walking is highly organized and choreographed.

> At the Taj, obeying the instructions of guides and tour organisers, most package tourists follow prescribed paths, moving towards certain valorised spaces and features and not others. The performance of these disciplined collective choreographies constitutes a quite precise and predictable 'place-ballet'. Bodies are tutored and disciplined, kept together and directed by assumptions about what is deemed 'appropriate', by group norms, and principally by the orders of the guide. (1998: 107)

Edensor goes on to distinguish these kinds of regulated tourist practices from other ways of walking. Backpackers, apparently, are more likely to wander fairly randomly and unguided, while local visitors make a point of visiting the mosque as part of the itinerary. In each of these cases, though, his point is that the landscape is practised – it comes alive – in both regulated and unregulated ways.

It is notable the degree to which the work I have just summarized looks back to Merleau-Ponty for its inspiration. The habitual and embodied is used to critique a particular form of disembodied gaze. Edensor speaks of 'place-ballets' – a term which must have found its way from David Seamon into tourism studies by some circuitous route. Cultural geographers long ago enacted a trenchant critique of the idealism to phenomenology with its seeming inability to intuit power relations in the process of transcendental reduction. So while a rediscovery of forgotten strains of thought is always welcome, it is necessary to also pay attention to power and its relation to practice.

I think I am in agreement with Don Mitchell here. He has argued, as I do here, that landscape is very much a lived phenomenon. Cultural geography, in its turn to the world of representation, has looked away from material spaces and seen landscape as ideology, as representation. This critique has been framed as a critique of looking at landscapes as merely material forms. Landscape, in the process, has been turned into a rarified realm of art and gardens. By refocusing on the materiality of landscape it is possible for cultural geographers to interpret the role of landscape in the world of practice. As W.J.T. Mitchell has argued, 'an account of landscape has to trace the process by which landscape effaces its own readability and naturalizes itself' (1994: 2).

Recent work has connected the issue of landscape to that of 'moral geographies'. Very simply put, moral geographies concern the normative relationship between space and behaviour. What and who belong where and when? Such considerations necessarily bring together the landscape and body through practice. Cosgrove's landscape was not one to be in but one to be outside of. Landscapes were moral in so far as they could be read as a diagram of a particular ideology or social formation, whether that be feudalism, merchant capitalism or Buddhist kingship. David Matless' work on *Landscape and Englishness* (1998) is quite different. His landscapes are practised landscapes. He asks what practices are appropriate for an English landscape. He connects the way of seeing inherent in landscape to ways of dressing, of bodily comportment, of trespass and of mountain climbing. In this way the English landscape is connected to forms of practical belonging such as citizenship. The English landscape signified proper ways of being through the propaganda of a diverse range of interest groups such as ramblers, scouts, mystics and dancers. The practice of people in the English landscape was not divorced from questions of the visual. As Matless puts it:

> Distinctions of citizenship and anti-citizenship turned on questions of appropriate conduct and aesthetic ability. Landscaped citizenship worked through a mutual constitution of the aesthetic and the social, the eye and the body. The aim of extending visual pleasure to the people was tempered by a desire to control potentially disruptive bodily effects. The education of the eye was to be accompanied by a self-control in the body. (1998: 62–3)

In Matless' moral geography of landscape, landscapes are inhabited, appropriately or otherwise, by people doing things. Like Jackson's landscapes these are not just seen at a distance but are used and lived in. Practice is not obliterated here. Rather landscape is the site of 'dialectical tensions of eyes and bodies, the visceral and the cerebral, pleasure and citizenship, ecstacy and organisation' (1998: 63). Matless and landscape archaeologists such as Shanks and Tilley have succeeded in putting flesh on the bones of J.B. Jackson's practised and everyday landscapes. Along with this return of the practising and mobile body has come a new interest in the everyday and unexceptional. The focus on vision tends to have led cultural geographers towards remarkable and, dare I say it, elite landscapes.

The challenge for cultural geographers of landscape is to produce geographies that are lived, embodied, practised; landscapes which are never finished or complete, not easily framed or read. These geographies should be as much about the everyday and unexceptional as they are about the grand and distinguished. Theories of practice can add theoretical weight to the wonderful insights of a geographer such as J.B. Jackson. What the idea of 'landscapes of practice' allows is an injection of temporality and movement into the static at the same time as practice is contextualized and given a frame.

REFERENCES

Bourdieu, P. (1977) *Outline of a Theory of Practice* Cambridge: Cambridge University Press.

Cosgrove, D. (1984) *Social Formation and Symbolic Landscape*. London: Croom Helm.

Cosgrove, D. (1985) 'Prospect, perspective and the evolution of the landscape idea', *Transactions of the Institute of British Geographers* 10: 45–62.

Cosgrove, D. (1987) 'Place, landscape and the dialectics of cultural geography', *Canadian Geographer* 22: 66–72.

Cresswell, T. (1996) *In Place/Out of Place: Geography, Ideology and Transgression*. Minneapolis: University of Minnesota Press.

Daniels, S. (1990) 'Marxism, culture and the duplicity of landscape', in N. Thrift and R. Peet (eds) *New Models in Geography*, Vol. 2. Boston: Allen and Unwin. pp. 177–220.

Daniels, S. (1993) *Fields of Vision: Landscape Imagery and National Identity in England and the United States*. Cambridge: Polity.

de Certeau, M. (1984) *The Practice of Everyday Life*. Berkeley: University of California Press.

Duncan, J. (1980) 'The superorganic in American cultural geography', *Annals of the Association of American Geographers* 70: 181–98.

Duncan, J.S. (1990) *The City as Text: The Politics of Landscape Interpretation in the Kandyan Kingdom*. Cambridge: Cambridge University Press.

Edensor, T. (1998) *Tourists at the Taj: Performance and Meaning at a Symbolic Site*. London: Routledge.

Horowitz, H. (1997) 'J.B. Jackson and the discovery of the American landscape', in J.B. Jackson *Landscape in Sight: Looking at America*. New Haven: Yale University Press. pp. ix–xxxiv.

Jackson, J.B. (1984) *Discovering the Vernacular Landscape*. New Haven: Yale University Press.

Jackson, J.B. (1997) *Landscape in Sight: Looking at America*. New Haven: Yale University Press.

Matless, D. (1998) *Landscape and Englishness*. London: Reaktion.

Meinig, D. (1979a) 'The beholding eye', in D. Meinig (ed.) *The Interpretation of Ordinary Landscapes*. Oxford: Oxford University Press. pp. 35–50.

Meinig, D. (ed.) (1979b) *The Interpretation of Ordinary Landscapes*. Oxford: Oxford University Press.

Merleau-Ponty, M. (1962) *The Phenomenology of Perception*. London: Routledge and Kegan Paul.

Mitchell, W.J.T. (ed.) (1994) *Landscape and Power*. Chicago: University of Chicago Press.

Olwig, K.R. (1984) *Nature's Ideological Landscape: A Literary and Geographic Perspective on its Development and Preservation on Denmark's Jutland Heath*. London: Allen and Unwin.

Relph, E. (1976) *Place and Placelessness*. London: Pion.

Relph, E. (1985) 'Geographical experiences and being-in-the-world: the phenomenological origins of geography', in D. Seamon and R. Mugerauer (eds), *Dwelling places and Environment: Towards a Phenomenology of Person and World*. New York: Columbia University Press. pp. 24–36.

Sauer, C. (1965) 'The morphology of landscape', in J. Leighy (ed.) *Land and Life*. Berkeley: University of California Press.

Seamon, D. (1980) 'Body-subject, time-space routines, and place-ballets', in A. Buttimer and D. Seamon (eds) *The Human Experience of Space and Place*. London: Croom Helm. pp. 148–65.

Shanks, M. and Pearson, M. (2000) *Theatre/Archaeology*. London: Routledge.

Shanks, M. and Tilley, C. (1987) *Social Theory and Archaeology*. Oxford: Polity.

Shanks, M. and Tilley, C. (1992) *Re-constructing Archaeology: Theory and Practice*, 2nd edn. London: Routledge.

Thomas, J. (1993) 'The politics of vision and the archaeology of landscape', in B. Bender (ed.) *Landscape: Politics and Perspectives*. Oxford: Berg. pp. 19–48.

Tuan, Y.-F. (1977) *Space and Place: The Perspective of Experience*. Minneapolis: University of Minnesota Press.

Tuan, Y.-F. (1979) 'Thought and landscape', in D. Meinig (ed.) *The Interpretation of Ordinary Landscapes*. Oxford: Oxford University Press. pp. 89–102.

Venturi, R., Scott-Brown, D. and Izenour, S. (1972) *Learning from Las Vegas*. Cambridge, MA: MIT Press.

Williams, R. (1960) *Border Country*. London: Chatto and Windus.

Williams, R. (1977) *Marxism and Literature*. Oxford: Oxford University Press.

Part II
Landscapes and Cultures

[2]

The myth and the stones of Venice: an historical geography of a symbolic landscape

Denis Cosgrove

John Ruskin's *Stones of Venice* criticized the Victorian city and Victorian society in the light of a reconstruction of medieval Venice. But Ruskin's reconstruction embodied elements of a long-standing myth propagated by Venetians themselves and inscribed in their organization of urban space and urban landscape. The geographical dimensions of the myth, its changing character through time, its iconographic expressions and its significance for English attitudes to Venice are described and explained. Ruskin used the myth to support his ideology, constructing a homology between architectural development and social structure. But the appeal of Venice transcends historical contingency and may, in part, be understood by reference to psychoanalytic categories, specifically sexual symbolism.

> I stood in Venice, on the Bridge of Sighs:
> A palace and a prison on each hand:
> I saw from out the wave her structures rise
> As from the stroke of the enchanter's wand:
> A thousand years their cloudy wings expand
> Around me, and a dying glory smiles
> O'er far times, when many a subject land
> Look'd to the winged Lion's marble piles,
> Where Venice sat in state, throned on her hundred isles.[1]

For early nineteenth-century romantics Venice was a city of great symbolic force. Like ourselves, they were spellbound by the beauty of the Venetian landscape. Unlike us they were aware of having witnessed the end, in 1789, of the Venetian Republic as a city state which traced its origins to the classical world. Venice was more than a beautiful place; it symbolized loss and decay which appealed to picturesque and romantic tastes. But behind this decadent glamour lay a more lasting complex of ideas and values associated with Venice: a *myth* of Venice, developed in political and historical discourse by Venetians themselves and moulded into the landscape of their city. The myth has distinct utopian characteristics, projecting Venice as a perfect union or conjunction of society and place. Its influence spread far beyond the lagoon and served to sustain Venice as a symbolic landscape of enduring significance for Europeans.

During the period of capitalist transition and formation of modern nation states Venice not only continued to flourish as a prototype of the mercantile city, but held a paradigmatic role in European debates about political constitution, state policy and urban design. The myth of Venice as a special place enjoying time honoured privileges was propagated by Renaissance artists and writers and could still provide material for John Ruskin's critique of social and environmental conditions in the nineteenth century. To fulfil this role the myth had to be reproduced and also modified. Myths, like places, have a history, emerging out of specific conditions and altering as those historical conditions change. Simultaneously and reciprocally, myths serve to inflect the pace and direction of historical changes. Buildings which for Renaissance Venetians were affirmations of the virtues of Venetian life, John Ruskin three centuries later deprecated as symbols of the loss of those same virtues.

It is not possible here to do justice to the complexities of the Venetian myth, nor to its multifarious expressions and changes[2] I shall pick out its salient features as they combined historical conditions and human ideals in a specific geographical setting and shall examine some of the ways in which the myth came to be celebrated in the Venetian townscape. I shall also survey broad changes in the myth and suggest how these changes affected the city's symbolic landscapes. Two stages in the history of the Venetian myth have special geographical significance: in sixteenth-century Venice the utopian characteristics of the myth were dominant and its impact on the landscape most intense. In the mid-nineteenth century, when Venice had become for most people a tourist city, John Ruskin drew upon the myth of Venice to illustrate his theories of landscape and to launch a social and architectural critique of the Victorian city. He was perhaps the last major writer to invoke the myth of Venice as a living force rather than as an archaic tradition. In what follows I hope to illustrate an answer to the question raised some years ago by David Lowenthal. He asked what it was that made for "key symbolic landscapes—landscapes that perennially catch the attention of mankind and seem to stand for, reflect or incorporate, the meaning and purpose of life itself".[3]

The changing myth of Venice

It is not strictly accurate to speak of a *myth* of Venice before the late fifteenth century. During the middle ages historical legends and beliefs abounded concerning the location of Venice but no historical and geographical theory based upon an ideological interpretation of selected facts had been formulated. The dedication of a translation of Plato's *Laws* published in 1450 contained the first historical explanation of Venice's favoured position.[4] Previously Venetian chronicles recorded, and the city celebrated, a number of past events, including the transfer of the body of Mark the Evangelist to Venice in 828, the recognition of Venetian independence and authority at the reconciliation in St Mark's church of Pope and Emperor in 1177, and the annual casting of a gold ring into the Adriatic to commemorate Venice's marriage with the sea (Plates 1 and 3). All these proclaimed the city's special status, and early chronicles refer to Venetian piety, heroism and fortitude and to its citizens' personal self-abnegation in the interests of the commonwealth. Independence from both Empire and Papacy, and suzerainty over the eastern Mediterranean were objective preconditions for the development of Venice as Europe's foremost long-distance trading city, inserted as a non-feudal island between the opposing social formations of western Christendom and the Islamic East.[5] Unlike other Italian cities, medieval Venice had no strong land or agricultural interests. It depended upon independent access to trade. All matters, even ecclesiastical, were firmly subordinated to political and commercial interests.

Out of the chronicles' legends and from their own understanding of the contemporary political organization of the Venetian Republic, humanist scholars in the

Plate 1 (top left). J. Tintoretto, Storm rising while the body of St Mark is being transported (Accademia Venice). The façade on the left is modelled on the east façade of the courtyard of the ducal palace in Venice.

Plate 2 (top right). G. B. Ramusio, Map of Tenochtitlan 1556 (Marciana, Venice).

Plate 3 (bottom). J. M. W. Turner, Venice, the Piazzetta with the ceremony of the Doge marrying the sea (The Tate Gallery, London).

Plate 4. Jacopo de Barbari, Plan of Venice in 1500 (Museo Correr, Venice).

Plate 5 (top left). Gentile Bellini, Miracle of the cross at San Lorenzo (Ditta O. Böhm, Venice).

Plate 6 (top right). Cesare da Conegliano, The Last Supper, detail showing courtyard of Ducal Palace (S.S. Apostoli, Venice).

Plate 7 (bottom left). J. M. W. Turner, The Sun of Venice going to sea (The Tate Gallery, London).

Plate 8 (top right). J. Ruskin, View of the Doge's Palace (sea façade), Venice, taken from the water (Ashmolean Museum, Oxford).

Plate 9 (top left). J. Ruskin, The north-west angle of the façade of St Mark's, Venice (The Tate Gallery, London).

Plate 10 (bottom right). Paolo Veronese, The Battle of Lepanto (Ditta O. Böhm, Venice), Venice, holding the sword of Justice, presents herself as liberator of Europe from the Turk.

Plate 11 (bottom left). John Ruskin, A street in Venice (Southwark Art Galleries and museums, South London Art Gallery).

late fifteenth and sixteenth centuries elaborated a myth of Venice at the very period when Venetian economic power, while still formidable, was threatened by external forces. Turkish pressure in the eastern Mediterranean and Portuguese navigations in the Atlantic constricted Venetian mercantile hegemony. Above all, the Cambrai war[6] which isolated Venice in conflict with the combined powers of western Europe constituted a profound psychological shock for the Venetian patriciate. Their defeat at Agnadello in 1509 brought Imperial troops deep into Venetian *terraferma*, and forced a reassessment of Venice's activities and aspirations that long outlasted the rapid recovery of former possessions. In acquiring Italian territories in the previous century Venice had become a land power seeking, it was feared, to control the peninsula. Constitutionally it remained a city state quite different in its republican values from the signorial communes of Renaissance Italy and the emerging absolute monarchies of western Europe. It was in the context of vulnerability, change and uncertainty that the sixteenth-century myth of Venice was forged.

The humanist historians who articulated the myth of Venice included Sabellico and Giustianini in the closing years of the fifteenth century, but it achieved fullest expression in the writings of Gasparo Contarini, *De Magistratibus et Republica Venetorum* 1543 and in Paolo Paruta's *Della perfezione della vita politica* 1579.[7] To these Venetians their city represented the perfect society, a realized utopia comparable with, but superior in its practical expression to, contemporary utopias conceived in the speculative philosophies of their mainland peers. Venice, they argued, was perfect because the original settlers in the lagoons had grasped the rational structure of creation and had incorporated its rules into their institutions. They were Roman citizens[8] from mainland cities—Padua, Concordia, Aquileia—who had forsaken their homes to preserve liberty. Balance and harmony were products of a *mixed government* which combined monarchy in the person of the Doge, aristocracy in the 300 senators and popular republican rule in the broad membership of the 2,000-strong Great Council. Mixed government had allowed Venice to endure unchanged for a thousand years, sustaining republican liberty while other formerly free cities of Italy had all fallen under signorial despotism. A perfect constitution allowed for individual freedom and was maintained through the individual piety and self-sacrifice of Venetians.

In part the liberty and longevity that proved Venetian perfection was regarded as geographical. The sea had preserved Venice's virginity from barbarian attacks from Attila to Francis I. On the other hand, the islands around the Rialto were located at the centre of the world, the meeting point of commerce and the mediating pivot between Rome and Byzantium, Empire and Papacy. The legend that St Mark himself had introduced Christianity to the lagoons and the miracles associated with his body served to give Venice an apostolic status equal to that of Rome. The enshrinement of the evangelist's body in the Doge's church reinforced the idea of Venice as a sacred centre, an *axis mundi* (Plate 1). Virginal Venice, surrounded by the sea, protected by its saint, could be imagined as a second eternal city. Eternal and perfect, these characteristics spoke to a stasis which denied historical change and legitimated the promise of future permanence.

Bouwsma has shown how in the later sixteenth century Venetian historians like Sarpi, while retaining the central ideas of the myth, accommodated limited scope for development and change to accord with the altered external conditions of the early modern period, especially after the Venetian political crisis of the 1570s.[9] As the decline of the city's economic power and loss of political prestige grew more insistent the ideas of Venetian perfection and permanence were taken up elsewhere in Europe, where social and political thinkers were struggling to develop programmes for the organization of the modern state. Venetian historians were widely translated and read, while travellers of all nations visited and recorded their impressions of the city. Its constitutional organization and personal freedoms excited the imagination of Enlightenment theorists for whom direct lineage from antiquity gave Venice a particular authority.[10] For those attempting to reformulate the relations between church and state, the Papal interdict of 1605 revealed that in the face of counter-reformation ideology Venetians still gave first allegiance to the earthly rather than to the heavenly city, and Venice was admired for its empirical, worldly approach to life and its rejection of authoritarian system-building. Venice was the secular city, sanctified by its constitution, history and landscape whose citizens rejected "a life without particular focus of experience or particular attachment to persons and places; a life that finds no value in the daily emotional and sensory encounters of human existence".[11]

Eighteenth-century Venice offered contradictory images to Europeans, counterposing myth and reality. The image of frugality contrasted with the splendour of the landscape and the luxury of living observed and enjoyed by grand tourists; personal piety coexisted with public secularism; valour in defending freedom with Venice's current neutrality and pacifism; aristocratic rule with the apparent responsibility of the ruling nobility and wide political participation; tight social control and public order with remarkable personal liberty, indeed licence, which made Venice the gaudiest stop on the grand tour. Venice might appear increasingly a gilded anachronism, but Cromwell's parliament sought Venetian advice on the survival of republics, Locke and Voltaire still regarded Venice as "a living demonstration of fundamental truths about life in society", and Venetian diarists and historians informed the discourse of eighteenth-century rationalists.[12] By this time England had come to fulfil something of Venice's role. England was an island mercantile power, a free nation unconquered for 700 years, governed by a secular monarch and parliament with a balance of power preventing autocratic rule. Englishmen displayed precisely the same distaste as Venetians for system and they similarly admired practical empiricism. It is to England and to the response of Englishmen to Venice after the fall of the republic that we should turn if we are to understand Ruskin's use of the Venetian myth. Before doing so we should explore the translation of the Renaissance myth of Venice into the Venetian townscape, for it was the concrete form of Venice which survived its political demise.

The myth of Venice in the Renaissance townscape

Venice's urban landscape presents a remarkable visual harmony. In similarly appealing townscapes, for example that of central Paris, visual unity often derives from an overall design or plan or from a sustained repetitive decorative treatment of street façades and perspectives. In Venice there is neither uniform planning nor a dominating architectural style. The compactness imposed by the site and the counterpoint of enclosed, tortuous streets and canals with architecturally defined open spaces of paving or water contribute to Venice's attraction. They give a sense of mystery and ill-defined expectancy.[13] The ogival arch and the fenestration of Venetian fifteenth-century buildings are unique and arresting features to be sure, but they are by no means ubiquitous. The best-known monuments of Venice, the Doge's Palace and San Marco, defy stylistic classification.[14] They and the spaces around them are results of intervention from many different periods and stylistic traditions. Much of Venice's architectural image is inspired by classical and Renaissance conceptions rather than by those loosely termed gothic. The Marcian Library, the churches of San Giorgio and Il Redentore, the Rialto bridge and the startling mass of Il Salute are examples. It has been suggested that the visual unity of Venice's townscape derives from the quality of light and its reflection upon open stretches of water and from a pervading lightness of architectural form contrasted with the sharp shadows of closed spaces. This may correspond to the Venetian pictorial tradition of *colore* as opposed to *disegno*, of surface decoration rather than architectural mass.[15] Certainly this was one of its attractions for John Ruskin.

The evidence of building

Despite its organic quality, there is evidence of systematic transformation of the Venetian townscape in the sixteenth century. Many of the buildings and spaces which we admire today were regularized in the years between 1480 and 1580 and the concern with urban aesthetics at the time is evidenced by the appointment in 1535 of two officials (*savi*) "a curar et comodar la cita".[16] Apart from private initiatives, particularly large-scale palaces along the Grand Canal, public attention was focused upon three key nodes: the land and water spaces around San Marco; the Rialto Market area; and the Arsenal (Fig. 1).

At the administrative centre of the city, around San Marco, a number of

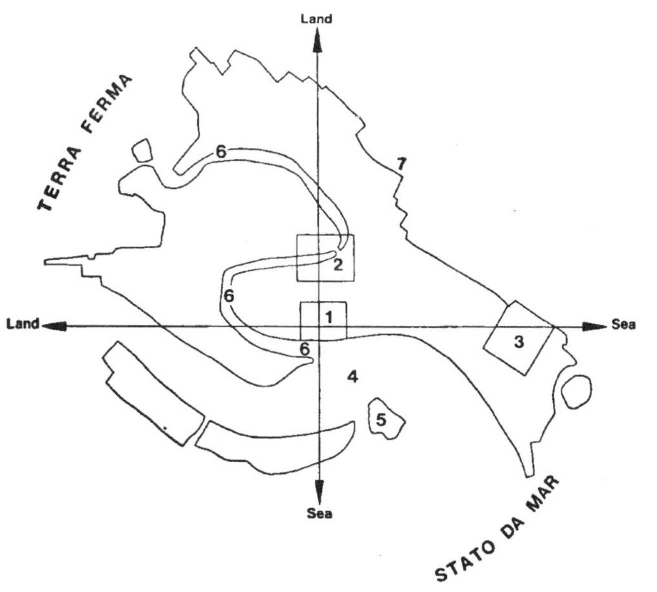

Figure 1. Venice: symbolic nodes and pathways. Numbers refer to: 1, San Marco; 2, Rialto; 3, Arsenal; 4, Basin of San Marco; 5, Island of San Giorgio; 6, Grand Canal; 7, Fondamenta Nuove.

initiatives created a spectacular articulation of open spaces and architectural symbols. The clock tower (*Orologio*) above the entrance to the Merceria was re-designed in the 1490s and the *Procuratie Vecchia* constructed along the northern side of the piazza. This was balanced, and the southern side of the square was defined, by Scamozzi's *Procuratie Nuove*, while J. Sansovino, an official architect designated Procurator of San Marco in 1529, designed the Marcian library, the Mint and the logetta of the campanile, clearing stalls and various other buildings from the area.[17] This work produced two visual axes, one focused on the basilica, and transverse to it a sight-line between the *Orologio* and the island of San Giorgio across the basin of San Marco where Palladio's church was erected in the 1570s. Beside these triumphs of classicism, a strong sense of continuity was endorsed by the rejection of Palladio's proposal to rebuild the fire-damaged Doge's Palace according to a classical design and the decision to restore the Gothic exterior.

By the late sixteenth century the central node of Venice expressed a complex symbolic structure, understandable in terms of the humanist ideas shared by the Venetian patriciate and in terms of the Venetian myth[18] (Fig. 2). San Marco was

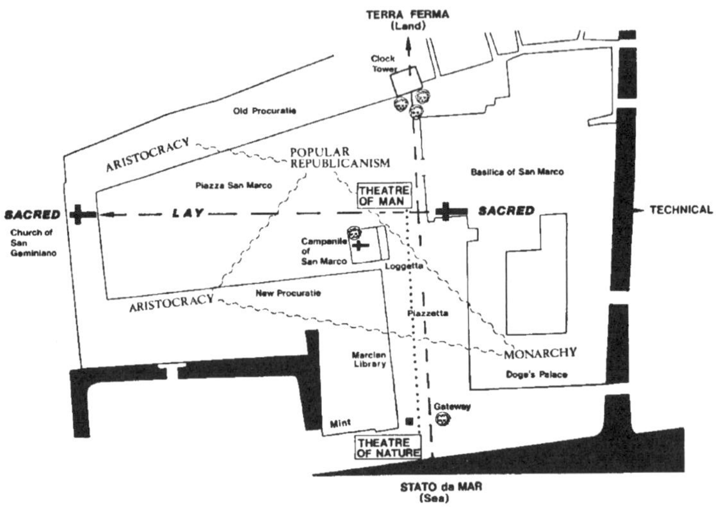

Figure 2. San Marco: the myth of Venice in the Renaissance townscape.

a concrete representation of the perfection of Venetian institutions. The Doge's Palace in its original location and external form represented monarchy, while the aristocracy was housed in the two wings of the *Procuratie* which defined the boundaries of the "place of eloquence, of Minerva": the Piazza San Marco. Here republican freedom was celebrated in the daily discourse of Venetian citizens. It opened towards the sacred legitimation of Venice, the Basilica di San Marco. Opposite the Doge's Palace stood the Marcian Library, the seat of humanist wisdom, lateral to the axis which united the lagoon, the "theatre of nature" gesturing towards Venice's *stato da mar* with the "theatre of man" and, via the symbol of time, the Rialto and ultimately the *terraferma*.[19] Nature's beauty and perfection were thus linked to human wisdom and human perfection just as the sacred and lay aspects of Venice were united in the repeated symbol of St Mark's lion, atop the column at the gateway to the city and presiding over focal points across Venice and its subject territories.

At the Rialto, "the place of function, of Mercury", proposals by Fra Giocondo and Palladio to reorganize the market area according to an absolute geometry of classical space contravened the medieval morphology and were rejected in favour of a piecemeal programme better attuned to the wisdom of commerce than to the humanism which dominated at San Marco.[20] The *Fabbriche Nuove* begun in the 1520s was a speculative venture and da Ponte's bridge favours scenographic impact and commercial use rather than strict adherence to classical regulation. The Arsenal, on the other hand, the industrial centre of Venice, was the "place of technique, of Vulcan". Here were practised the *arte meccaniche*, despised by humanist patricians, and they were to be hidden from public view. The plan of 1539 surrounded the area with a blind wall, three metres high, closing all but one elaborately-decorated entrance guarded by stone lions.

In the initiatives of the sixteenth century we may read the myth of Venice in its townscape (Fig. 3). Constitutional balance, sacred legitimation, natural and human perfection are inscribed in space and architecture, reaching their most complex symbolism in a node of high visibility, the gateway to and heart of the city. The refusal to accept plans for radically restructuring urban space according to the formulae of ideal city theorists suggests that "the preservation of formal continuity was not for Venice simply a question of taste: it speaks a symbolic language, communicating an unequivocal message about the historicity of the republic and its decisively anti-Roman political model".[21]

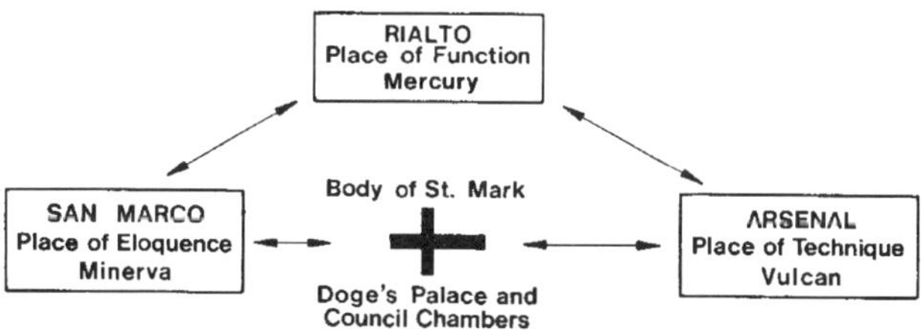

Figure 3. Landscape structure of the Venetian myth.

The evidence of maps

Venetian cartographers provide further evidence of the symbolic significance of the urban landscape and its utopian characteristics. The best known Renaissance map of Venice is that drawn by Jacopo de Barbari at the turn of the sixteenth century (Plate 4). This map is far from being a faithful record of the actual appearance of the city in 1500.[22] The distortion increases dramatically away from its central point in the basin of San Marco and it is revealed as an artistic creation rather than a factual illustration in its selection and exaggeration of building size and height. Schutz has shown how the Barbari map stands in the tradition of *mappae mundi*: synoptic and paradigmatic images of the created world as reflections of divine wisdom. *Mappae mundi* were produced by Venetian cartographers in the fifteenth century, but increasingly they combined an interest in concrete information about the physical world with a moralized geography intended to serve a didactic purpose, as objects of contemplation. Barbari's map illustrates ideas as well as topographical facts and a glance at the marginal information reveals its ideological function. Mercury and Neptune smile on the fortunes of Venice. The subject of the map

> ... is the commonwealth of Venice rather than the physical city: Venice, the premier trading and maritime power of Europe. Her physical features are exhibited as the material manifestations of this state, just as the figures of Mercury and Neptune are the incarnations of its numen. Jacopo's print is a visual metaphor of the Venetian state, like the familiar symbols of Venice, the winged lion of St Mark, and the female figure holding the scales and sword of justice, which represented the divinely protected and justly governed republic.[23]

This map was a model for a series of others serving a similar purpose in exaggerating symbolic elements in the townscape and in isolating the city from the mainland in virginal integrity. Generally they represented the Lido as a circular wall in the sea so that Venice appears like contemporary designs for the ideal city, a self-contained, centralized and circular geographical perfection. One important Venetian cartographer at mid-century, G. B. Ramusio, reproduced in the third volume of his *Delle Navigationi et Viaggi* (1556) idealized representations of Hochelaga and Tenochtitlan—island cities in the recently discovered New World, and Alvise Cornaro proposed to surround Venice with a circular wall of bastions in the lagoon, enclosing newly-drained land for the cultivation of the city's foodstuffs (Plate 2, Fig. 4). In these maps and proposals we see Venice depicted as a practical

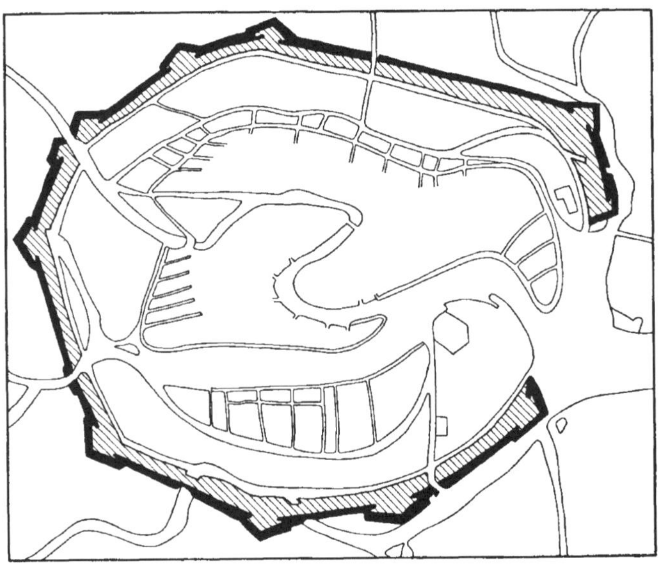

Figure 4. Cornaro's plan for fortifying Venice and reclaiming the lagoon *c.* 1565. Based on a reconstruction by P. Fontana, ASV Casabella M.465.

realization of the Renaissance ideal city. It is interesting to note that these formalized proposals were rejected in favour of a proposal by Cristoforo Sabbadino to link the northern parishes by a single *fondamenta nuove*, leaving the lagoon in its natural state and preserving the "natural" curving outline of the city. The debate between Cornaro and Sabbadino hinged on the issue of how far Venice's perfection was natural and organic or rational and subject to human improvement.[24]

Painting and ceremonial

The townscape paintings of fifteenth century Venice by Carpaccio and Gentile Bellini are well-known.[25] In decorating the *scuole* of the city's guilds and rooms in the Doge's Palace they illustrated legends from the Venetian chronicles, inserting the sense of a special place, blessed by miraculous events into the landscape of the contemporary city (Plates 1 and 5). Venice's high Renaissance artists, Titian, Giorgione and Giovanni Bellini frescoed the external walls of public buildings, adding to the scenógraphic quality of Venice's public face, while Tintoretto, whose St Mark cycle commemorates the central legend of Venetian sanctification (Plate 1), in his vast painting in the Doge's Palace presents paradise as an image of Venice, where God and his angels resemble the Doge and his councillors and the city's official saints achieve prominence before the divinity. More directly, Venetian painters portrayed specific Biblical scenes in the familiar surroundings of Renaissance Venice (Plate 6).

In a landscape charged with emanations from the myth of Venice frequent public ceremonies reinforced the significance of historical legends. The annual ceremony of the Doge's marriage with the sea (Plate 3), his visit to Santa Maria Formosa in memory of the rescue of the twelve Brides of Venice[26] and the reception of the Doge's wife into San Marco had long been holy days in the Venetian calendar and the occasion of elaborate public celebration. Interestingly all have a connection with human marriage, drawing a parallel between female chastity and Venetian virginity. Sixteenth-century celebrations were particularly elaborate and included spectacular humanist displays, notably the construction of floating circular *teatri del mondo* in the basin of San Marco. Designed by intellectuals like Cornaro and Barbaro, these were closely linked in concept to ideal city theory, appealing to the same humanist ideals of cosmological harmonies and microcosmic proportions.[27] Similar iconography was employed in ceremonial arches and architectural *trompes d'oeil* built for the reception into Venice of foreign dignitaries.[28]

These visible expressions of the myth of Venice suggest an ideological programme employing an iconography which integrated historical legends and humanist conceptions of the ideal organization of creation.[29] They declare the virginal perfection of Venice and legitimate the established order and thus the continued rule of the patrician elite. Their reception by ordinary people and their success in sustaining ideological hegemony are difficult to judge, but there was widespread public involvement in ceremonial, and Venice was noted for its relative internal peace and civic harmony. Disputes between members of the ruling class were themselves played out partly in symbolic terms as debates over radical plans for restructuring the Rialto district, the Doge's Palace or the lagoon itself reveal. Opposing sides in these debates were the same as those engaged in the crisis of the *Giovanni* in 1572. Whatever the success of the Venetian myth as an ideological complement to political control, there is little doubt of its importance in Venetian sixteenth-century experience and of its significance in producing a landscape which has given delight to visitors ever since.

The Venetian myth after the fall of Venice

By the time John Ruskin first visited Venice in 1835 he was acquainted with the city through its representation in Romantic art, particularly the poetry of Byron and Samuel Rogers and J. M. W. Turner's paintings. Among Turner's paintings, "The Grand Canal Venice: Shylock", exhibited in 1837, was one of the first of that artist's works to be purchased by the Ruskins.[30] But Ruskin claimed that the vignettes for Rogers' poem *Italy* initially attracted him to Turner's art. In the years following the fall of the Republic Venice had become a landscape of memory viewed primarily as a repository of symbols of its own history.

In *Childe Harold* Byron perceived Italy above all through the medium of its two eternal cities: Rome and Venice. Italy was a sacred landscape, replete with history and art, but it is notable that the idea of classical and enduring perfection in decay which Byron appropriated from Italy as a metaphor for his own feelings was developed from contemplating Venice. Rome perhaps carried too many resonances of rationalism and papism while Venice provided a more secular reference. He regarded the site of Venice as naturally beautiful, remaining so despite the loss of republican liberty:

These days are gone—but Beauty still is here.
States fall, Arts fade—but Nature cannot die.

He transmutes Venice into a work of nature rather than of man, but it is a different conception of nature from that of Renaissance Venetians. This is an evolving, organic nature rather than one ruled by laws of proportion and cosmic harmony, and the reality of the human artifact is uncertain:

. . . a fairy city of the heart
Rising like water columns from the sea.

The themes of the Venetian myth: liberty, duration, privileged centrality and splendour, serve to focus Byron's attention on the significance of Venice's decline. Loss of liberty in a Venice now under Austrian domination symbolized loss and alienation of natural beauty and human freedom in the modern world, a theme echoed in Wordworth's sonnet on Venice.[31] Byron made a direct link with England, "The inviolate island of the sage and free", whose wartime restrictions on personal freedom he deplored. But no theoretical relationship was established between Venetian art and building and its socio-political order, the focus of Ruskin's later study. Both were stressed simultaneously as elements of the uniqueness of Venice whose fate provided a generalized moral for England:

. . . thy lot
Is shameful to the nations—most of all,
Albion to thee: the Ocean Queen should not
Abandon Ocean's children; in the fall
of Venice think of thine, despite thy watery wall.

Byron and Wordsworth used Venice as a vehicle for romantic protest. They drew upon the Venetian myth to emphasize a freedom in nature observable in the past but denied in the contemporary world. Elements of its history and beauties of its townscape were ransacked to reproduce a myth of Venice relevant to its new status and to industrializing England.

Similar ideas informed Samuel Rogers' treatment of Venice, but they were differently employed in his genteel, self-conscious record of a Grand Tour.[32] Rogers devoted seven chapters to Venice, but lacked Byron's moral force. He regarded Venice as a respiring body:

There is a glorious city in the sea
The sea is in the broad, the narrow streets
Ebbing and flowing, and the salt sea-weed
Clings to the marble of her palaces.

The fantasy aspects of Venice were stronger than in Byron's lines. Rogers removed Venice from time and space. We approach it via an "invisible path" and "glide" through its streets "as in a dream". Its history is one of unnamed people escaping a nameless enemy and through the agency of a reified "Enterprise" causing a metropolis to rise in compelling beauty "like an exhalation from the deep". Venice's decline also is mystical: the product of overarching, superhuman and perhaps natural forces of history or destiny. Venice was lost when "the golden stream (of Enterprise) turned to enrich another", but the buildings remained links with the classical world. "From the last glimpse of civilized life—to where Light shone again", it carried the torch of freedom and endurance amidst the rise and fall of empires.

This was the Venice of a mannered romanticism addressed to those "who have learned to live in Past Time", and drained of any moral force. Rogers was writing for an English bourgeois audience; his poem was aimed at the cultivated drawing room where history, literature and painting were called upon to reflect stability, permanence and propriety rather than to challenge assumptions.

This approach to Venice, stripping its myth of contemporary political relevance exemplifies a *Disneyfication* of Venice, a process now complete as any visitor may observe. Within a decade of the appearance of Rogers' poem the first modern-style tourist guide to Venice appeared. *Murray's handbook of northern Italy* was published in 1842, its third edition used by the Ruskins in 1847, and it was soon rivalled by Baedeker.[33] These handbooks, with their starring system for the visit-worthiness of artistic and historical monuments, their concise information on all things necessary to the safety and comfort of a Victorian traveller and their factual summaries of historical events, exemplify the appropriation of Venice for safe, predigested consumption: practical counterparts to Rogers' literary treatment. Venice, like Rome, Florence and Genoa, offered *culture*, its acquisition a mark of social distinction. The culture was largely Classical and Renaissance, consisting of named masters and specific monuments. It allowed an escape from the uncertainties and conflicts of the present into a romantic past. Murray quoted Rogers' lines to introduce Venice and Baedeker recommended such writers as Mrs Oliphant, whose *Makers of Venice* is an example of a genre of historical studies which removed individuals and events as curiosities without context.[34] Victorian visitors were thus encouraged to regard a visit to Venice as a passage through a time-warp into a past from which they could recover a refinement which paid social dividends in the present.

Baedeker recommended his readers to consult John Ruskin's *Stones of Venice*. But he added a strong caveat: "the intelligent reader will temper Mr Ruskin's extreme and sometimes extraordinary statements with his own discretion". The warning was directed less at Ruskin's historical inaccuracies than at his use of picturesque and romantic views of Venice to illustrate, through detailed architectural examination, a powerful polemic aimed directly at the class of people who used Baedeker and Murray, at their England and their Victorian cities. *The stones of Venice* was not intended as a comfortable architectural guide for chaperoned young ladies absorbing culture in the ethereal world of Rogers' marble-banked and shimmering canals. It employed the past to bring into sharp and uncomfortable focus the inadequacies of the present from which Rogers offered an escape.

The Venetian myth in *The stones of Venice*

Ruskin's first extended visit to Venice was in 1845. His letters describe the city through Turnerian eyes, dissolving buildings and water into a vision of light and colour inspired by paintings like "The Sun of Venice going to sea" (Plate 7).[35] He complained bitterly of the damage inflicted by modernization upon Venice's ability to allow one to live in the past:

> There is no single spot, East or West, up or down, where her spirit remains—the modern work has set its plague spot everywhere—the moment you begin to feel, some gaspipe business forces itself upon the eye, and you are thrust into the nineteenth century, until you dream . . . that your very gondola has become a steamer.[36]

Above all, the newly-completed railway causeway to Mestre struck at the roots of Venice's mythical geography, its isolated rising from the sea. Attracted initially to Venice by its light, colour and the reflective quality of its atmosphere, water and buildings which Turner's techniques brilliantly captured, Ruskin had always to temper the argument of the eye with the empirical exactness of science and the force of moral certainty. He was not satisfied by a comfortable immersion in the past but found it necessary to combine visual delight and historical understanding into a lesson for the present. In doing so he drew upon and reinterpreted both the Renaissance and the Romantic versions of the myth of Venice.

The stones of Venice was a product of two extended periods of study in the city in 1849 and 1851. It is the third of Ruskin's major works of art criticism. In earlier publications he had developed a theory of landscape in which divine handiwork was revealed in the forms of natural objects and was demonstrable through close empirical observation of individual instances.[37] To honour their Creator and to achieve true human status, men were obliged to render these forms in their own work. The highest expression of human labour was art, and architecture was that art which bore most directly on the transformation of landscape and nature. Architecture, in Ruskin's view, provides the litmus of a society's adherence to the principles and forms through which divinity was revealed in nature.

> the function of our architecture is, as far as may be . . . to tell us about nature; to possess us with the memories of her quietness; to be solemn and full of tenderness, like her, and rich in portraitures of her.[38]

The Venetian study was intended to illustrate the human consequences of Ruskin's landscape and social theory and the rules for its application to the Victorian city. It combined social, architectural and art history with sustained moral commentary. Although Ruskin frequently digressed from the topic under discussion, he did not deviate from his central aim and purpose.

The theme was introduced with a striking paragraph drawing a parallel between Venice and England:

> Since the dominion of men was first asserted over the ocean, three thrones, of

> mark beyond all others, have been set upon its sands: the thrones of Tyre, Venice and England. Of the first of these great powers only the memory remains; of the Second, the ruin; the Third which inherits their greatness, if it forget their example, may be led through prouder eminence to less pitied destruction.[39]

In summary, Ruskin's argument was that each of the ancient civilizations produced a characteristic architectural form "almost exactly indicative of the degree of intellectual and moral energy of the nations which originated them",[40] and the perfection of each form coincided historically with the culminating point of that people's energy and prosperity. Christianity offered the fullest development of the human spirit and thus of human art. Its civilization therefore gave rise to a new architecture, synthesizing and advancing upon all previous forms. This was *Gothic*, an architecture not only superior to all antecedents, but "demonstrably the best architecture that can exist". It reached its perfection in the thirteenth and fourteenth centuries, but this perfection was its undoing, for it was the nature of Gothic to depend on the free creativity of the human spirit which Christianity offered, and it is in the nature of perfection that it denies further creativity. To reproduce perfect forms is to become a slave to them, alienated from self and the materials of labour. A return to imitating pure forms found in rule-bound Greek and Roman architecture was the inevitable outcome of this decadence. Ruskin regarded classicism as the architecture of slavery and this was his reason for attacking the Renaissance. England, so long a slave to Palladianism, denying free creativity to its factory labour, had to readopt Gothic architecture and its social prerequisites if it were to avoid a fate similar to Venice's. The argument required that Ruskin locate Venice as the central point of Christian civilization, and the myth of Venice allowed for this with its emphasis on personal piety rather than Romanist catechetics and its evangelical foundation.[41] Venice, Ruskin claimed, was the centre of the medieval world, the meeting point of Rome and Byzantium, and the pivot of the two principles that informed that world's moral geography: the spirituality of the desert south and the materialism of the frost-cramped north.[42] Physical geography confirmed its site as special. The tidal range in the lagoons meant that this was the only place where such a city could be built: a lower range would have failed to sluice it, a higher would have rendered internal transport impossible. Venice was geographically ordained:

> One might look over Europe and see how each town takes its natural position —and becomes prosperous if it happens to understand that position and take due advantage of it—and one might say—generally, Genoa grew up in the place for Genoa—and Rotterdam in that for Rotterdam, and Venice in that for Venice . . . It seems that just in the centre of Europe, and at the point where the influence of East and West—of the old world and the new—were to meet, preparation was made for the existence of the city which was to unite the energy of the one with the splendour of the other . . . [43]

The key building in Venice is the fourteenth-century Doge's Palace because it "contains the three elements (of the three antecedent architectures) in exactly equal proportions—the Roman, Lombard and Arab: *It is the central building of the world*"[44] (Plate 8). Centrality is reflected in Venice's position as the greatest of the medieval city-states in material prosperity and in the perfection of its constitution. Ruskin appealed to the keystone of the Renaissance myth of Venice: the integration of civil activities and mixed government. He quoted approvingly Sansovino's summary of Venice's tripartite division of authority.[45] Venice's moral perfection was evidenced in its struggles against the temporal authority of Rome, its public secularism and private piety. Its history was a moral record of the Venetian people. It was not the unfolding of a pre-ordained destiny, as Rogers would have it, nor a chronicle of success and failure in preserving liberty against external enemies, as Byron claimed, but the history of a people making and destroying themselves. They did this, not in confronting material conditions, but in acting in accordance with universal moral principles. This history is recorded in the architecture and townscape of Venice. History and architecture are woven together in *The stones of Venice* through a reappropriation of elements of the Venetian myth so as to develop a homology between history and townscape which serves as an ideological critique of contemporary England.

Ruskin's interpretation of Venetian history

Ruskin offers us an evolutionary history of Venice in three stages: the rise of the city from its fifth-century foundation to 1297, the date of the *serrata*[46]; a period of maturity, of high prosperity and civilization between 1297 and the death of Doge Carlo Zeno in 1418; and a period of decadence, apparent from as early as 1423 when Doge Mocenigo died and fully visible by the sixteenth century. He does not explain why Venice survived four hundred years of this decadence, and he largely ignores political events after 1500.

While Ruskin claimed to reject the popular romantic view of Venetian history, his was hardly a balanced and disinterested account. Its conclusions were anticipated in the first volume, which was written before the detailed study of the 1851 visit.[47] For example, his choice of 1297 as the key date for the opening of Venice's mature period was governed by his belief that the closing of the Venetian oligarchy represented the foundation of the city's constitutional perfection. The combination of paternal authority exercised by a judging but compassionate patriarch with mutual cooperation of all classes under a benevolent elite corresponded with Ruskin's own political beliefs.[48] Contarini had regarded mixed government as immutable from the origins of the republic, but this notion was reformulated by Ruskin to coincide with his social convictions and was given a date of origin to accord with the pattern of architectural development he wished to establish: "Government and cooperation are in all things and eternally the laws of life. Anarchy and competition eternally and in all things, the law of death".[49] It was with the Renaissance, when Venetians were establishing a utopian view of their constitution, that Ruskin argued that the patriciate took on the features of an aristocracy with its "coldness, perfectness of training, incapability of emotion, want of sympathy with the weakness of lower men, blank hopeless haughty self-sufficiency".[50] Ruskin's history incorporated elements of the Venetian myth but transformed them to fulfil his own aims. He denied that constitutional perfection alone could account for Venice's durability because, while the 1297 changes produced a perfect constitution—Ruskin modelled his own utopian Guild of St George upon it—Venetians themselves later abused it. Only in the hands of truly noble men could the Venetian social order follow divine law and provide evidence for the moral status of the fourteenth-century republic.

Ruskin ignored the material history of Venice. He failed to mention Venetian navigation, mercantile relations, the political consequences of its trading hegemony, the *stato da mar* or the *terraferma* and the shift to rentier activity in early modern Venice. He employed various historical legends as evidence for his developing argument, but their selection was governed by his need to establish a relationship between the rise and fall of Venice as a political and economic power and the development and decadence of Venetian art, architecture and townscape.

Ruskin's Venetian townscape

Ruskin approached Venetian architecture through a detailed examination of selected buildings: the Doge's Palace, San Marco Basilica, Torcello cathedral and certain major palaces. From his observations he established generic principles of architectural styles,[51] and provided models for the architecture of the entire city. The Doge's Palace, for instance, was described as the prototype for palaces along the Grand Canal, and the architecture of St Marks "rose in the midst of other work, fanciful and beautiful as itself: . . . every dwelling house in the middle ages was rich with the same ornaments and quaint with the same grotesques which fluted the porches or animated the gargoyles of the cathedral"[52] (Plate 9). Ruskin perceived medieval Venice isolated in the lagoon, framed by the natural backdrop of the Alps, and within it monumental public buildings provided the authority for lesser ones in a hierarchical townscape: "the humblest houses, in the noble times, followed out the system of the larger palaces as far as they could, in their rude materials", just as "the acquiescence in the secure establishment of aristocratic power was an expression, by the people, of respect for the families which had been chiefly instrumental in raising the commonwealth to such a height of prosperity".[53] The Venetian landscape was thus rendered homologous to its political structure, providing the necessary justification for Ruskin's moral and social argument for a divine order which men were called upon to understand and obey.

This of course required Ruskin to establish an historical parallel between Venetian architectural development and the evolution of its political and moral structure. Venetian architecture like its political history was divided into three periods (Fig. 5). First, Venice's rise was characterized by the Byzantine style, most fully realized in the basilica, testament to Venetian piety and the "bible of the city". The second period coincided exactly with that identified as the age of Venice's material and moral supremacy. Venetian Gothic commenced in 1320 with the beginning of work to enlarge the Doge's Palace in order to accommodate the enlarged Grand Council which resulted from the *serrata*. Discussion of the period is founded upon the social and moral principles of Gothic architecture worked out in the best-known chapter of the work, "The nature of Gothic".[54] The new Palace also bore witness to Venetian faith, as Ruskin set out to demonstrate in a detailed iconographic analysis of its capitals. The final period of Venetian building Ruskin called Renaissance and dated from 1423 with the decision to restore the Doge's Palace by reconstructing the Piazetta façade. Although formally a copy of the sea façade and lacking motifs we would call Renaissance, Ruskin refused to dignify it with the name of Gothic because it expressed neither faith nor creative energy. It led, he claimed, ultimately to the vainglorious, faithless extravagances of baroque churches like San Moise and Santa Maria Formosa. Humanism, a respect for the word rather than for the deed, for system rather than experience, a desire for perfection, led to a debased architecture wherein "the lower workman secured method and finish, and lost, in exchange for them, his soul".[55] Renaissance Venetian humanists like Paruta had argued the perfection of Venice precisely on

Figure 5. Periods of Venetian architecture in *The stones of Venice*. Upper to lower: Byzantine, Gothic, Renaissance.

the grounds that it rejected system and placed faith in experience. Ruskin inverted the argument, or rather located it in an earlier period, so directly contradicting the Renaissance myth of Venice.

Byzantine, Gothic and Renaissance architectures mirrored the evolution of Venetian society. Ruskin analysed them visually rather than structurally, basing his discussion on the mass of detailed drawings and notes taken in Venice (Plate 9). He has been accused of failing to treat buildings as structures which occupy and mould space although he devoted much of the first volume to structural problems, insisting on the weight-bearing properties of the cusped arch. Recent study has sought to clear him of this charge, showing that he recognized the articulation of space as the point of origin of architecture while requiring good architecture to go beyond it and to employ decoration and sculpture visually to determine the effect of the whole mass.[56] Because he understood that "a building is, as an optical phenomenon, infinitely and continually variable", changing with atmosphere, light and colour, Venetian architecture was particularly suited to his type of analysis.[57] Venice he treated as *landscape* according to the principles laid out in *Modern painters*, and he was highly responsive to the Venetian tradition of decorating building surfaces in an environment whose reflective qualities are intense. That he should have chosen Venice as the vehicle for his argument had more to do with this visual impact than with its history or historiography, but it is interesting that having so chosen, Ruskin read back from Venetian architecture to political and social values not dissimilar to those which gave meaning to the Renaissance architecture he rejected. The difference in the dates ascribed to these values related to the lessons Ruskin wished to draw from Venetian landscape.

Lessons from the Venetian landscape

Ruskin presented medieval Venice as an ideal society and place because its Gothic architecture revealed faith in divinely instituted natural laws. Its sculpture and decoration followed "the law of organic form" found in the structure, lines and curves of phenomena found in the physical landscape.[58] Gothic submitted to and incorporated this law because it was produced by simple pious men, free to express their inner faith and spirit. Turning to the contemporary English city Ruskin found little to suggest a similar faith: "We have destroyed the goodly architecture of our cities; we have substituted one wholly devoid of beauty or meaning . . ."[59] Gower Street was the antithesis of fourteenth-century Venice, but if the lesson of the latter was heeded then "the London of the nineteenth century may yet become as Venice without her despotism, and Florence without her dis-peace".[60] The historical utopia of Venice was the model for a future utopia in England.

Its realization would be possible only if architecture was a product of proper social relations. Ideally, workers should labour under the severe but just discipline of paternalist masters, but they would not be an alienated proletariat. Proletarian labour, living in the shadow of an aristocratic architecture like that of Versailles, was provoked to social discontent.[61] Gothic architecture induced social harmony not only because it was England's national style[62] but because Ruskin believed that it presupposed a set of social relations that mirrored the order of creation. In his association with the building of the Oxford Museum an early exercise in Venetian Gothic and a trial for many of his ideas, Ruskin supported the freedom of the iconoclastic Irish mason, O'Shea, to carve what he wished on the capitals and window jambs, but complained that his creativity was inadequately disciplined.[63] Elsewhere *The stones of Venice* seemed more effective in popularizing precisely the reproduction of form which Ruskin derided, and had minimal effect on the social relations of labour either in the building industry or outside it.

The success of Ruskin's book in promoting the Gothic Revival is difficult to evaluate. He was one of many advocates, although "The nature of Gothic" seems to have vied with Pugin's *Contrasts* as the theoretical justification for the style.[64] Ruskin did not argue that particular Venetian forms should be adopted in England; he was more interested in Venice as a telling example of the social conditions necessary for a true architecture. But Venetian forms became popular in mid-Victorian buildings. Venetian arches, capitals and polychromatic effects adorned Victorian police stations, public baths, water-works and town halls as well as private houses. It is perhaps too fanciful to argue that in embellishing with Venetian Gothic forms those institutions which characterize the first intervention of the state and its local agencies in the geography of the capitalist city, Victorian architects paid tribute to the ideal of Venetian civic responsibility. Ruskin's own response to the popularity of the style recognized that little of his social and moral argument had been accepted or understood:

> . . . I have had indirect influence on nearly every cheap villa builder between this and Bromley; and there is scarcely a public house near the Crystal Palace but sells its gin and bitters under pseudo-Venetian capitals copied from the church of the Madonna of Health or of Miracles.[65]

The consequence was inevitable.[66] Ruskin had chosen Venice because there he could draw upon the long-standing association of a beautiful city with civic independence, individual piety and personal freedom—in short, the myth of Venice. This he had learned, in bastard form, from Turner, Byron and Rogers. In his book he depicted a mythical perfect landscape and related it to a set of imagined moral and social conditions. These corresponded closely with ideals projected by Venetians themselves in the Renaissance. But in both cases the utopian images lacked a solid basis in history made by real people struggling to change their material conditions. They remained ideological interpretations, and today we regard the splendour of Renaissance Venetian architecture as a prelude to decline rather than as a symbol of the fulfilment of Venice's destiny, and we find Ruskin's didactic architectural analysis naive. Both seem to arrest historical change by constructing out of a social myth the concrete reality of a built environment. But this does not explain why Venice and its architecture and townscape achieved such popularity as forms alone, regardless of their imputed social connections and why they remain so. In broaching that problem we face issues raised by all such symbolic landscapes.

Venice and symbolic landscapes: concluding remarks

The history and meaning of Venice and its landscape have been reinterpreted again and again in the light of perceptions and preoccupations of very different societies. Myth can be built out of myth, but as ideology such reproduction does not occur without connection to material conditions. The sixteenth-century myth of Venice was related to the historical contingencies surrounding the Venetian patriciate, just as Ruskin's use of Venice was a response to his view of Victorian England, which itself conditioned the employment of Venetian architectural motifs in the nineteenth-century city. But across these structural constraints stretches the formal appeal of Venice, its visual impact, the interpretation of land and water, the chiaroscuro and labyrinthine quality of its townscape and the fantasy of colour and decoration which still attract so many visitors. Historically specific analysis does not seem fully to explain this but by extending materialist analysis in perhaps unconventional ways[67] we may make some tentative suggestions for which evidence, although partial and inconclusive, may be found in the material discussed above.

A consistent but subordinate theme which runs throughout this discussion of the Venetian myth has been the sensuality of Venice and its artistic expression. Venice was consistently represented as a female symbol, a virgin undefiled by invaders, pictorially a nubile young woman or an avenging protectress (Plate 10). Even an empirical historian like Fernand Braudel finds it appropriate to employ a sustained metaphor on the "Queen of the Adriatic" when discussing sixteenth-century Venice.[68] The organic, curving outline of the island city is surrounded and penetrated by tidal waters, its streets and canals form a dark and mysterious labyrinth to which we gain access by way of the sparkling and highly-decorated symbolic entrance of the basin of San Marco, the sinuous Grand Canal and the Piazzetta, leading to the womb-like enclosure of the Piazza. Closed circular form and organic nature were stressed in Renaissance planning and cartography and signified in the *imago-mundi* of floating theatres constructed in the lagoon. Similar themes were exploited by Romantic writers and painters and by Ruskin. Even today sensuality remains central to the image of Venice. We need not return to Casanova or Baron Corvo to recognize it. Thomas Mann and Evelyn Waugh in our own century are among those who have employed Venice as the appropriate setting for highly-charged and ambiguous sexual encounters. Films of their work have exploited the Venetian landscape as a visual metaphor for sexuality.[69]

Contemporary writing on aesthetics has sought to escape the weakness of traditional idealist explanations of art's trans-historical and cross-cultural appeal, explanations which claim that it lies in the relevation of universal truth or the employment of "significant form"—a structural category traceable in every great work of art. John Berger has been foremost in relating changes in the character and style of the visual arts to alteration in the social and economic conditions of their production.[70] While acknowledging the significance of Berger's materialist analysis Peter Fuller has pointed to its failure to account for works of art whose appeal transcends the social relations which produced them. But rather than returning to idealist or formalist explanations Fuller pushes materialist analysis further than traditional Marxism by appealing to psychoanalytic concepts. He draws our attention to the significance of those shared infant, childhood and sexual experiences that provide a common set of personal tensions which, he claims, all art seeks in part to resolve. Thus "the aesthetic is an historically specific structuring of certain elements of human experience and potentiality which pertain to that underlying condition of our biological being and which can be occluded or displaced, but never entirely extinguished, by ideology".[71] This claim may provide a foundation for resolving the continued appeal of the Venetian landscape and its symbolism.

Venice as a female symbol seems to incorporate many of those ambiguities that psychoanalysis has revealed in the relationship between the child and its mother and between the adult male and female as sexual partners. The idea of virginity was a central feature of the myth and that of a marriage as yet unconsummated a repeated theme in Venetian ceremonial: the wedding of Venice to the sea, the capture and rescue of the Brides of Venice and the reception of the Doge's wife into the church of San Marco. Venetian patrician society was distinctly masculine. Women's subservient and decorative role as property was expressed in the notorious scale of Venetian dowries and the sumptuous dress styles of Venetian women, including half-metre high shoes. Venice was at once virgin and wife, maternal protectress and a pure form to remain unsullied by alteration of her naturally perfect outline. Sumptuary laws stressed the puritanical aspect of Venetian pietism while abroad Venetian women were long associated with courtesan decadence.

John Ruskin's own sexuality and the nature of his relationship with his mother have been widely discussed.[72] A significant part of his short, unconsummated marriage with Effie was spent in Venice. For Effie it was the happiest part. Her letters record her participation in the city's social life and her acclaim as a beauty by the Venetian and Austrian *élite*.[73] This life, with its hints of sexual possibilities, contrasts strongly with her celibate private life with John Ruskin. He, meanwhile, largely avoided party-going, preferring to spend his days enjoying the sensuous feast of Venetian architecture, including those very buildings he claimed publicly to be decadent.[74] He wrote of Venice with a passion which betrays more than intellectual interest: his paintings reveal an engagement with the mystery and sensuality of the city, and his fascination with the tides suggest that the organic, pulsing, female nature of Venice communicated itself strongly to him. It is as if the Ruskins' marriage was being consummated through Venice's female symbolism while it was denied in their conjugal bed. When Ruskin came to write of Venice it was his own experience as son and husband that allowed him to touch upon a far wider symbolism, as wide even as the heavenly city of St John's *Apocalypse*:

> a city of marble, did I say? nay, rather a golden city, paved with emeralds. For truly, every pinnacle and turret glanced or glowed, overlaid with gold, or bossed with jasper. Beneath, the unsullied sea drew in deep breathing, to and fro, the eddies of green wave. Deep-hearted, majestic, terrible as the sea—the men of Venice moved in sway of power and war; pure as her pillars of alabaster stood her mothers and maidens. . . .[75]

Here possibly is the thread that links the myth of Venice into an historical continuity. If so, it may account for the living attraction of Venice in our own time, but it is a dimension that as historical geographers we have scarcely begun to recognize, let alone found the means to explore and understand.[76]

University of Technology,
Loughborough

Notes

[1] Lord Byron, *Childe Harold's pilgrimage. A romaunt* (London 1853) Canto IV verse i

[2] By far the fullest discussion of the myth of Venice as it was developed in sixteenth-century humanist history is William Bouwsma, *Venice and the defence of republican liberty: Renaissance values in the age of the counter reformation* (Berkeley, California 1968). See also Bouwsma's essay, Venice and the political education of Europe, in J. R. Hale (Ed.), *Renaissance Venice* (London 1974) 445–66. This collection contains a number of other essays on the theme of the Venetian myth. Edward Muir, *Civic ritual in Renaissance Venice* (Princeton, NJ 1981) came to my notice too late to incorporate its findings into this paper but contains a detailed discussion of ritual and ceremonial as expressions of the myth of Venice

[3] David Lowenthal (Ed.), Environmental perception and behaviour. Department of Geography Research Paper **109** (Chicago 1967) 2. The geographical literature on landscape and symbolism has expanded enormously since Lowenthal formulated his question, but it remains an open and challenging one. For a comment on some of this literature and a discussion of its assumptions see my essay, Place, landscape and the dialectics of cultural geography *Canadian Geographer* **XXII** (1978) 66–72; also essays contained in D. W. Meinig (Ed.), *The interpretation of ordinary landscapes* (Oxford 1979)

[4] Bouwsma, *Venice and the defence of republican liberty* 63. A dedication appears in the Greek scholar Trapezuntius' 1452 translation of Plato's *Laws* where he likens Venice's perfection of mixed government directly to the model proposed by Plato. It is thus appropriate that the work be dedicated to the city

[5] In the fifteenth and sixteenth centuries the Italian social formation was by no means dominated by feudalism. The characterization of the medieval city as a non-feudal island in a feudal sea is intended to highlight the role of Venice as a mercantile insertion of first rank within the long-distance trade of a Europe whose formations were dominated still by feudalism. For the economic history of Venice as a trading power see F. C. Lane, *Venice: a maritime republic* (Baltimore 1973)

[6] The war of the League of Cambrai aligned France, Spain, the Empire and Papacy against Venice in 1508. While the League was terminated by Rome's policy change in 1509, the complex ebb and flow of warfare effectively kept north-eastern Italy in turmoil until 1529, when the Peace of Bologna established the *status quo ante* for Venetian territory

[7] Bouwsma, *Venice and the defence of republican liberty* contains a detailed discussion of these works

[8] The ambiguity of ancient Roman citizenship whereby those who possessed it were not necessarily natives of the city, was exploited by Venice to lay claim to the positive features of classical origins while maintaining clear distance between its own perfection and the inadequacies of ancient Rome which had led to its downfall. Thus the claim was not compromised by the distinct anti-Roman aspects of the Venetian myth

[9] The crisis was provoked by the *Giovanni*, a group of nobles intent upon prizing open the increasingly restrictive channels of admission to the Senate and major offices of state. They succeeded in wresting power from the formerly prevailing faction, the *Vecchi*. The dispute also concerned Venice's policy towards landed investment and external relations

[10] Bouwsma, Venice and the political education of Europe *loc. cit.* For a discussion of early visitors' impressions of Venice see Margherita Azzi Visentini, Le testimonianze dei viaggiatori, in *Architettura e Utopia nella Venezia del Cinquecento* (Milano 1980) 71–9

[11] Bouwsma, Venice and the political education of Europe, *loc. cit.* 447

[12] It is significant that Canaletto, painter of Venice in the seventeenth century should have been appointed, under royal patronage, to record the townscape of mercantile London in the years of the Restoration. London is explicitly depicted as a second Venice

[13] Speaking of the poetries of urban space, Armand Frémont, *La région: espace vécu* (Paris 1976) 43 says ". . . la ville de la lagune étend des labyrinthes multiples, ombre et lumière, art et misère, et, surtout, pierre et eau. L'eau enveloppe et pénétre Venise. La ville est accouplement. Elle a toujours exalté les sensualités, y compris les plus scandaleuses, jusqu'à l'inceste". The sexual aspects of the Venetian landscape are explored more fully towards the end of this paper

[14] Deborah Howard, *The architectural history of Venice* (London 1980)

[15] In the literature of painting in sixteenth-century Italy it was argued by Florentine and Roman theorists that the basis of the visual arts lay in the process of translating ideas and conceptions into visual terms, *disegno*. Execution, applying paint to canvas and the techniques of artistic production were regarded as unintellectual and mechanical. Venetians rejected this and concentrated upon the virtuosity of handling their materials, *colore*. This is held to account for the weak development of mannerism in Venetian art. Charles Hope, Made in Venice (review of F. Richardson, "Andrea Schiavone" 1980) *London Review of Books* 3 (1981) 15–16

[16] M. Tafuri, "Sapienza di Stato" e "atti mancati": architettura e tecnica urbana nella Venezia del '500 in *Architettura e Utopia* (1980) 16

[17] Howard, *op. cit.* 144–56

[18] The discussion which follows is derived in large measure from Tafuri, *loc. cit.* The catalogue of which his essay forms a part has a detailed bibliography of studies on urban design in sixteenth-century Venice. For a sixteenth-century townscape description which makes

explicit reference to the perfection of Venice see F. Sansovino, *Venetia, città nobilissima et singolare descritta in XIII Libri* (Venezia 1581)

[19] The *stato da mar* denotes Venice's overseas possessions in the Adriatic and eastern Mediterranean, consisting largely of strategic coastal enclaves and islands. The *terraferma* was the mainland Italian possessions gained by Venice in the fifteenth century and stretching from the Istrian peninsular in the east to Brescia on the Milanese border. Together these two *stati* represented a considerable empire, but one without geographical, religious, ethnic or linguistic homogeneity. Its existence depended heavily upon the centre and accentuated the emphasis placed on the centrality of Venice. The contradictions which the terraferma introduced into Venice's self-image in the sixteenth century as investment began to move into agriculture are discussed in A Tenenti, The sense of space and time in the Venetian world of the fifteenth and sixteenth centuries, in J. R. Hale (Ed.), *op. cit.* 17–46

[20] Tafuri, *op. cit.* 22. Palladio's propsal for the Rialto Bridge is described in his *I quatro libri dell' architettura* (Venezia 1570). The design inspired a number of copies in eighteenth-century England, notably those at Stourhead and at Prior Park, Bath

[21] *Ibid.* 26

[22] Juergen Schultz, Jacopo de'Barbari's view of Venice: map making, city views, and moralized geography before the Year 1500 *The Art Bulletin* LX (1978) 425–74

[23] *Ibid.* 468

[24] Hochelaga is the site of present-day Montreal, discovered by Cartier; Tenochtitlan was Montezuma's capital, site of Mexico City, discovered by Cortez. Both designs are radio-centric, urban views, providing a direct link between Renaissance ideal city theory and European images of the New World. Alvise Cornaro, a Venetian humanist living in Padua, was closely involved in sixteenth-century Venetian drainage schemes in the *terraferma* and wrote extensively on the pleasures and benefits of agrarian life. His work represents one side of the argument within the Venetian patriciate over the future direction of the Venetian economy: whether it should remain committed to maritime commerce, or invest in rentier activity. This debate in part underlay the divisions between the *Giovanni* and *Vecchi* which led to the 1570 political crisis. Cornaro was also a devotee of humanist architecture. His loggia at Padua is the earliest example of Roman Renaissance architecture on the mainland and his plan for Venice integrates his interest in agriculture, defence and ideal city theory. Barbara Mazza, Politicia Lagunare di Venezia nel cinquecento e interventi sul territorio: note di cartografia, in *Architettura e Utopia* 130–43. On the debate between Cornaro and Sabbadino, see Tarufi, *op. cit.* 18

[25] J. Links, *Townscape painting and drawing* (London 1972)

[26] The legend of the Brides of Venice was recorded in the early Venetian chronicles. Twelve young brides of Venetian noblemen were captured by corsairs from the Istrian coast on their wedding night and were subsequently released and reunited with their husbands. The feast of Lady Day was given over to celebration of the event at the church of Santa Maria Formosa

[27] R. Wittkower, *Architectural principles in the age of humanism* (London 1962). For a critique of the symbolic interpretation of ideal city theory see J. R. Hale, *Renaissance fortification: art or engineering?* (London 1977)

[28] Particularly the spectacular ceremonial arch built by Palladio for the entry of Henry III of France at the Lido in 1574. Lina Padoan Urban, Gli spettacoli urbani e l'utopia, in *Architettura e Utopia* 144–66; this article also discusses *teatri del mondo*

[29] Ideological aspects of such iconographic programmes are discussed by Carol Duncan and Alan Wallach, The Museum of Modern Art as late capitalist ritual: an iconographic analysis *Marxist Perspectives* **1** (1978) 28–51. They define such programmes as follows: "These programmes usually rest upon authoritative literary sources—written or orally transmitted myths, litanies, sacred texts, epics. Monumental iconographic programmes frequently evoke a mythic or historical post that informs and justifies the values celebrated in the ceremonial space. As visual commentaries they elucidate the purpose of consecrated ground" (p. 29)

[30] Luke Hermann, *Ruskin and Turner* (London 1968) 22 discusses this disputed claim

[31] ". . . Venice, the eldest child of liberty.
She was a miaden City, bright and free,
No guile seduced, no force could violate,
And, when she took unto herself a mate,
She must espouse the everlasting Sea.

And what if she had seen those glories fade,
Those titles vanish, and that strength decay,
Yet shall some tribute of regret be paid
When her long life hath reached its final day:
Men are we, and must grieve when ever the shade
Of that which once was great, is passed away"

[32] Samuel Rogers, *Italy, a poem* (London 1852). The poem first appeared in 1836

[33] *Handbook for travellers in northern Italy* (London 1842). Ruskin himself had contributed to the third edition some notes on Tuscan cities. In an epilogue to *The stones of Venice* written in the 1870s Ruskin captures something of the spirit of Victorian tourism in Venice, turning it into a bitter comment on the failure of his message in the work: "as far as they found the things I spoke of amusing to themselves, they have deigned for a couple of days or so to look at them,—helped always through the tedium of the business by due quantity of ices at Florian's, music by moonlight on the Grand Canal, paper lamps, and the English papers and magazines at M. Ongario's, with such illumination as those New Lamps contain—Lunar and Gaseous, enabling pursy Britannia to compare, at her ease, her own culminating and co-operate Prosperity and Virtue with the past wickedness and present out-of-pocketness of the umquile Queen of the Sea" **10** 231–2

[34] K. Baedeker, *Italy, handbook for travellers. First part: northern Italy* 10th edn (London 1895) 242

[35] Harold, I. Shapiro (Ed.), *Ruskin in Italy: letters to his parents, 1845* (Oxford 1972) letter 123

[36] *Ibid.* letter 123

[37] I discuss Ruskin's landscape theory in John Ruskin and the geographical imagination *Geographical Review* **69** (1979) 43–62

[38] John Ruskin, *The stones of Venice* 3 vols. **9** (1851–1853) 411. The complete work forms volumes 9, 10 and 11 of E. T. Cook and Alexander Wedderburn, *The works of John Ruskin* (George Allen, London 1904) to which all subsequent references refer. Robert Hewison, *John Ruskin, the argument of the eye* (London 1976) 52 notes that "the connection between architecture and nature in *The stones of Venice* is so close that one comes to be considered as a substitute for the other"

[39] *The stones of Venice* **9** 17

[40] *The stones of Venice*, Explanatory note **11** 356

[41] The pietistic and evangelical character of Venetian religious practice is discussed by Bouwsma, *Venice and the defence of republican liberty* 72ff. Ruskin refers to this particular aspect of Venetian life and, given his background, it was no doubt one of the elements in its history which attracted him. In an appendix to *The stones of Venice* he attacks the Catholic Emancipation Act of 1829. There is a parallel here between Venice's historic struggle with the Roman church and Ruskin's dislike for its form of Christianity

[42] The idea of an environmental foundation for artistic and religious sensibility is a recurrent one in Ruskin's writing. Another example and its elaboration is found in his discussion of cloud study in art. Denis Cosgrove and John E. Thornes, Of truth of clouds: John Ruskin and the moral order in landscape, in Douglas C. D. Pocock (Ed.), *Humanistic geography and literature: essays on the experience of place* (London 1981) 20–46

[43] John Lewis Bradley (Ed.), *Ruskin's letters from Venice 1851–52* (Westport, Conn. 1955) Letter 34. The subject of Venice's tides exercised a particular fascination for Ruskin and his diaries and letters record considerable time devoted to measuring and understanding them

[44] *The stones of Venice* **9** 28, italics added

[45] This is the one reference Ruskin makes to Sansovino's book (see note 18 above). While his reading of Venetian history was wide it is not systematically referenced and appears to have drawn heavily on eighteenth- and nineteenth-century writers

[46] The *serrata del Maggior Consiglio* was the beginning of a process whereby the juridically-defined class of Venetian nobles was separated out and its powers and privileges statutorily identified. Its significance to Venetian constitutional history is a matter of continued debate among historians. See Stanley Chojnacki, In search of the Venetian patriciate: families and factions in the fourteenth century, in J. R. Hale (Ed.), *op. cit.* 47–90, especially 52–8

[47] N. N. Feltes, The stones of Ruskin's Venice: a materialist analysis *Prose Studies* **3** (1980) 54–68. Feltes refers to Ruskin's history of Venice as "fictive", that is, it selects empirical evidence to produce a representation which is ideological and used to describe how certain social groups signify themselves

[48] Ruskin's admiration for Marshal Radetsky, Austrian military commander in charge of the

administration of the Veneto, and an admirer of Effie Ruskin, is based upon the view that Radetsky provided just this kind of authority. Ruskin was unsympathetic to the cause of Risorgimento, applauding the Austrian suppression of the 1848 rising when Venice for a few months regained republican independence

[49] John Ruskin, *The seven lamps of architecture* **7** 207

[50] *The stones of Venice* **11** 74

[51] The method is the same as that applied to the natural landscape, see note 37 above

[52] *The stones of Venice* **10** 118

[53] *The stones of Venice* **10** 302, 340

[54] This chapter, from volume II, was later reprinted separately from the rest of the work, most notably by William Morris in the Kelmscott edition of 1892. It became a manifesto for the art and crafts movement's attack on the factory system and the alienation of workers under industrial capitalism. Morris says in his introduction to it, "the lesson which Ruskin here teaches us, it that art is the expression of man's pleasure in labour" **10** 460. The perception of the medieval period as the high-point of craft labour and of the free guild worker which is expressed in "The nature of Gothic" was popular among English Marxist historians of the 1880s and 1890s. It owes a debt to Ruskin's forceful promotion of medieval craftsmanship, and although he was not alone in holding this view he articulated it with particular force. Raphael Samuels, British Marxists historians, 1880–1980, part one *New Left Review* **120** (1980) 3–96, especially 28

[55] *The stones of Venice* **11** 18

[56] John Unrau, *Looking at architecture with Ruskin* (London 1978). Unrau also gives reference to criticisms of Ruskin's architectural study

[57] *Ibid.* 157

[58] Cosgrove, John Ruskin and the geographical imagination, *loc. cit.* 55–6

[59] *The stones of Venice* **10** 119

[60] *The stones of Venice* **11** 230

[61] This idea anticipates those of the Modern Movement of the 1920s which also saw a role for art as an active social force. Le Corbusier put Ruskin's point forcefully in 1923 in *Towards a new architecture* (London 1946) 14: "It is a question of building which is at the root of the social unrest of today: architecture or revolution"

[62] An argument that had become conventional among promoters of the Gothic revival, particularly from Pugin's *Contrasts* of 1836

[63] The connection between Ruskin and the Oxford Museum is discussed in Peter Ferriday, The Oxford Museum *Architectural Review* **132** (1962) 409–16

[64] It is interesting to note in this connection that one of Ruskin's visitors while in Venice was George Gilbert Scott, among the most influential and prolific architects in the English Gothic movement. Bradley (Ed.), *op. cit.* Letter 40

[65] Letter to *Pall Mall Gazette*, 16 March 1872, in a reply written from Ruskin's Herne Hill home ("this" in the quotation) to a letter criticizing his influence on contemporary architecture **10** 459

[66] Ruskin seems to have recognized this. In the 1874 preface to the third edition of *The stones of Venice* he notes "The relation of the art of Venice to her moral temper, which is the chief subject of the book, and that of the life of the workman to his work, which is the most important practical principle developed in it, have both been ignored, and could not but be so, by modern architectural readers" **9** 14

[67] Sebastiano Timpanaro, *On materialism* (London 1976)

[68] F. Braudel, *The Mediterranean and the Mediterranean world in the age of Philip II* **I** (London 1972) 127–33

[69] Thomas Mann, *Death in Venice* (London 1912); Evelyn Waugh, *Brideshead revisited: the sacred and profane memories of Captain Charles Ryder* (London 1945). In both the feature film of Mann's story and the television production of Waugh's novel explicit use is made of Venetian architecture and street scenes to explore the implicit sexuality in the relationships between the characters. The use of night scenes reinforces the contrast between openness and closure, light and shade in Venetian landscape

[70] John Berger, *Ways of seeing* (Harmondsworth 1972) and *A painter of our time* (London 1976)

[71] Peter Fuller, *Art and psychoanalysis* (London 1980) 193

[72] A summary of the discussion and of the relevant literature may be found in Quentin Bell, *Ruskin* (London 1978) especially 126–61

[73] Mary Lutyens (Ed.) *Young Mrs Ruskin in Venice. Her picture of society and life with John Ruskin, 1849–1852* (New York 1965)

[74] See for example Unrau's comments on Ruskin's response to the church of Il Salute, *op. cit* 48–9

[75] This passage, originally appeared in Ruskin's short guide to Venice *St Mark's Rest* 3 parts (1877–1884). It is quoted in the "Epilogue" to *The stones of Venice* where it is introduced by a reference to Revelations. John's description of the new Jerusalem (Rev. 21: 1, 15, 18, 19, 21) is quoted as the paradigm of the utopian idea by P. W. Porter and F. E. Lukermann, The geography of Utopia, in David Lowenthal and Martyn J. Bowden (Eds), *Geographies of the mind. Essays in historical geosophy on honour of John Kirtland Wright* (New York 1976) 199–200. At the age when he wrote this Ruskin was captivated by the St Ursula legend depicted by the fifteenth-century Venetian painter Carpaccio, closely associating St Ursula with Rose Latouche, the young girl with whom he had fallen hopelessly in love

[76] Thanks are due to Gabriele Zanetto, Stephen Daniels, Hugh Prince and John Lucas for help and advice in the preparation of this article and to Anne Tarver for preparing the figures

[3]

Geography as an art

D. W. MEINIG

Maxwell Professor of Geography, Syracuse University, New York

MS received 1 March 1983

(Guest Lecture delivered at Annual Conference of the Institue of British Geographers, Edinburgh, January 1983).

ABSTRACT. It has been routine to speak of geography as an art as well as a science, but there is little evidence to demonstrate any really firm relationship with the central ground of the arts. Most of the recent emphasis on 'humanistic geography' remains closely associated with science and does not represent any substantial participation in the humanities. A review of recent writings on 'geography and literature' suggests that these two realms of work remain discrete, with geographers as borrowers from rather than creators of literature.

However, there is some evidence, and a growing momentum toward geography as an art in the full sense of highly personal, explicitly emotional works which must be judged as individual creations. Something of the nature of such work with a few examples of its possibilities, including its difficulties and dangers, are discussed, and the work of Henry Glassie and Raymond Williams is cited as especially pertinent for geography. It is concluded that geography will deserve to be called an art only when it has overcome its chronic insecurity, broken through pedantic barriers, and can display a substantial number of geographers as creative artists.

That Geography is an art as well as a science is an old and common assertion with precious little substantiation. I propose to assess this situation, suggest how it might be changed, and why I think it important to do so. Ideally we should have the evidence at hand and review our books in leisurely discussion. As we cannot do that I shall refer to well-known works as far as possible. It will also be to our advantage that several current lines of thought impinge upon my theme, and indeed there are members of this audience who have dealt with some aspects of it better than I can do. At most I can hope to push us rather more firmly on toward what I believe to be a logical conclusion; at the least I can add another voice of encouragement to those already endeavouring in that direction.

Even though I shall confine my remarks to geographical writings in English there will be difficulties of language for several of our key words have such an accretion of meanings that they take up several pages in Raymond Williams' dictionary of that name.[1]

ARTS AND SCIENCES

For example, it might be said that as we organize out thoughts, select our data, choose our words, and make our statements we are, all of us, engaged in the art of geography. But I wish to lift the word above these routine fundamentals—which I think might more exactly be called the craft of geography—and have it refer to that which is implied in its common pairing with 'science'. In my own university, as in many in America, geography is a department within a College of Arts and Sciences. Such a usage represents a modern distinction between different modes of intellectual activity, a divergence which began with the rise of 'rationalism' as a philosophy and reactions to its comprehensive claims although in practice the distinction itself remains complex and uncertain.[2] In a superficial way geography has been one of those fields which has defied that divergence, granting, for example, degrees in arts or in science as appropriate to particular programmes and topics. In a deeper sense, however, that divergence has produced a gulf which cannot be papered over by such administrative strategies. Many would regard it as an abyss whose depths and dangers have been best defined in the famous 'two cultures' of Lord Snow.[3]

GEOGRAPHY AND THE HUMANITIES

My topic can actually be more effectively considered within the common threefold academic division into the sciences, the social sciences, and the humanities. Despite difficulties of definition here too, we might agree that these three terms have served as a general classification for three realms of exploration of the world and ourselves, each with characteristic tools and methods, objectives and results. Logically, geography may be part of all three, but actually, except for certain subfields of physical geography, over the past 50 years or so we have most commonly been classified and self-defined as a social science.

We have just been through a generation of vigorous efforts to reshape geography into a much more formal social science. Whatever the achievements of those efforts, and they were not minor, it is obvious today that such momentum has faltered as geography partakes of the general crisis over the inability of such social science to provide the kinds of answers once so confidently expected of it.[4] And it is a result of this crisis, in reaction to those limitations, that we have witnessed a sudden burgeoning of interest in 'humanistic geography'. Of course such literature has a lineage; one can pick out a few bold threads and trace connections to Carl Sauer and John Kirtland Wright and others, but essentially it *is* a new literature, for while we have long had a 'human geography' we have never before had an explicitly 'humanistic geography' with such a self-conscious drive to connect with that special body of knowledge, reflection, and substance about human experience and human expression, about what it means to be a human being on this earth[5]

'The humanities', like 'the arts', reach across such a variety of fields and philosophies that we ought not expect 'humanistic geography' to have a clear identity or single focus. Actually most of what has been put before us *is* of single type and is humanistic only in a limited sense. In this it reveals its impetus. For in the themes it has seized upon, in its questing, assertive, tentative, unfinished quality, in its lack of substantial product, it demonstrates that, like the rest of social science, human geography has had little effective contact with any humanistic tradition.[6] Thus at the moment our literature brands us as refugees, fleeing a collapsing structure, scattering out to seek shelter and nourishment in hopes of a new start.[7]

'Our roots in the humanities are shallow', writes Yi-fu Tuan in one of his many quiet meditations of the topic.[8] It will be my argument that they are shallow not because the humanities are barren or marginal ground for geography, but because we have not routinely seeded that ground and cultivated and fertilized it, weeding out the worst and nuturing the best, experimenting with new forms, alert for new varieties, all with a convinction that the harvest will yield more than a bit of spice or flavouring, it will provide some of the staples for a healthy geography.

Our roots are not only shallow, they are to be found in only a few of the many fields of the humanities and most thickly in those directly bordering social science. Almost all of this avowedly humanistic geography is analytical in intent, aimed at a systematic examination of subjective meaning and human behaviour. Most of it seems little more than an extension of science. Thus although Professor Tuan in his various writings identifies a broad range of possibilities, he tends to emphasize the humanistic geographer's role as 'an intellectual middleman' who takes 'nuggets of experience as captured in art and decomposes them into simpler themes that can be systematically ordered. Once experience is simplified and given an explicit structure, its components may yield to scientific explanation'.[9] If we were to judge humanistic geography by most of what we find under the label today we would probably concur with Nicholas Entrikin's conclusion that it can be no more than a critique, an effort useful 'to counter the overly objective and abstractive tendencies of some scientific geographers'.[10] And related to these subordinate relationships with science is the fact that to an important degree 'humanism' as a philosophy or ideology has often become narrowed to a form of scientism, it has become manipulative and proscriptive, a rationalism pushed to its limits. There are far too many complexities here to explore in this paper.[11] Let us just be clear that 'humanism' and 'the humanities' are not the same thing. It is quite possible, and some of us would think it entirely appropriate, for a deeply humanistic geographer such as Edward Relph to assess an utterly reject 'humanism', as he has done in his most recent book entitled *Rational Landscapes and Humanistic Geography*.[12]

Unlike geography, history has long had substantial claims to being an art as well as a science (although at the price of chronic internal controversy). Indeed history has sometimes been hailed as the queen of the humanities and no one can deny the central humanistic significance of the historical interpretation of and reflection of human events. But it is clear also that most of our historical geography is more akin to the science than to the art of history. It too tends to be strongly analytical, seeking logical explanations from solid facts rather than less conclusive interpretations of the human condition.

That is not true, however, of all historical geography and there are rich potentials here. Cole Harris has eloquently defined those special qualities of the historical mind: the open creative intelligence, disciplined and rigorous, enthusiastic for hard evidence but wary of elegant explanations and following no model, exercising mature judgement in the attempt to grasp and present some plausible interpretations of the immense complexities of any time and place.[13] He is describing a kind of mind and method very different from those stereotypes of science that have been so insistently urged upon us. He cites Carl Sauer and Fernand Braudel as two of his heroes. Each stimulated by example a whole school of scholarship and they are surely among the very best exhibits of all these traits, of the deeply committed and highly individualistic scholar. Recall the opening phrase of Braudel's monumental work: 'I have loved the Mediterranean with passion . . .'.[14] Such a statement, such an open investment of emotion, aligns the work with art and carries us toward that ground of creative imagination where history merges into 'literature', sustaining an old and influential tradition.[15] The same can hardly be said of geography, or even as yet of historical geography. I shall resist the temptation to focus attention narrowly on the potentials for a truly humanistic historical geography and look more generally at geography and that great central field of creative and imaginative 'literature'.

GEOGRAPHY AND LITERATURE

We may identify several connections with such literature created or advocated by current humanistic geographers. Probably the most common arises from a view of literature as a valuable storehouse of vivid depictions of landscapes and life. At the very least the power of novelists and poets can be a great stimulus to student interest by adding colour and vitality to the drab and lifeless body of regional geography. There can be more to it than that, of course. Most advocates go on to stress the value of literature in sharpening our sensitivities to the human experience with place. The rather sudden emergence of a great concern in geography with environmental perception and behaviour has inevitably prompted forays into the vast realm of literature, for, as Douglas Pocock has remarked, 'literature is the product of perception, or, more simply, *is* perception'.[16] It therefore provides us not only with data it opens up the topic in ways that may be ignored or underplayed in our usual analytical geographic treatments.[17]

Professor Tuan notes that literature not only provides essential clues about human experience with environment, it also contains models for dealing with some of the most difficult problems of descriptive synthesis. And the value of those models rests on the simple but rarely acknowledged fact that the majority of the verities we claim as geographers can be expressed in creative as well as scholarly terms.[18] Thus the possibilities of a formal juxtaposition and mutual enrichment are implied, but we have little evidence for these as yet.

In this recognition of geography and art as complementary approaches to the same general topic it is implicit that artists know what they are talking about. We have a number of studies which have carefully assessed the work of particular authors as to just how revealing their writings are with respect to important geographic themes, such as, to draw upon Graeme Whittington's careful look at the Kincardineshire novels of Lewis Grassic Gibbon: 'the landscape real and perceived', 'the group culture', and 'regional consciousness'.[19] This is to use literature as a document, to be assessed like any other piece of evidence, not by its artistic qualities but by its utility for the scholar. Thus a poor novel may become a good source if the author knew what he was writing about even if by literary standards he failed to write it well.

A logical step beyond all the various supplemental uses is to move toward a closer integration. A bold attempt is the recent work on the *Social Geography of the United States* by your distinguished President, Wreford Watson.[20] The book deals thematically with a number of problems: racial and ethnic tensions, class, poverty, crime, urban and community issues. The chapters are laced with extensive quotations from novelists and other creative writers on American localities, not as a bit of colour to enliven (Professor Watson needs no help in that) but as *primary* evidence central to any penetrating geographic interpretation. For Professor Watson has worked long and hard to get geographers to recognize that writers not only describe the world, they help shape it. Their very portrayals establish powerful images that affect public attitudes about our landscapes and regions. Our creative writers are our mythmakers, therefore, as he has put it, 'the geography of America is the American mind made real in the lanscape'.[21] In this way the connection with art is not supplementary, it is essential; it leads, as he has emphasized, to the very 'soul of geography'.[22]

Another form of integration is to carry our assessments of literature beyond our own utility into the realm of literary criticism. Here the geographer's motivation is not so much to measure a work as a document for geography as to join with other specialists in assessing how a creative writer has worked with ideas and shaped his material so as to convey a set of understandings at various levels of sophistication. An example of this role are the two studies of 'Faulkner's Yoknapatawpha County' by Charles Aiken who was well aware of the efforts of many other labourers in the Faulkner branch of the literary industry but was convinced that a geographer who knew that ground intimately could add something to an understanding of the ways in which Faulkner transmuted fact into fiction.[23] B. P. Birch's recent review of several Wessex novelists is a pertinent British example.[24]

It would have strengthened the case for humanistic geography if the studies of Aiken and Birch had been published in journals of literature rather than in the *Geographical Review* and the *Transactions of the Institute of British Geographers* for one doubts that such periodicals are part of the routine reading of literary scholars. So far in all these discussions about geography and literature we seem to have been talking to ourselves.[25] We have not as yet established geographers as significant participants in the world of literature and thus we do not have much measure of how effective we might be. In those few cases where we do have a critical assessment by a literary figure of our ventures into their territory the response warns against any delusion that we can readily stake out a productive place. Thus the novelist D. M. Thomas, reviewing a recent collection on *Humanistic Geography and Literature,* found a few essays to praise, indeed praise highly, and in the end welcomed the volume as 'a first, clumsy attempt', but he was harshly critical of most of the contents and concluded that 'the thin, jargon-ridden style of most of the essays . . . is a depressing indication of the cultural spirit in our society. Many—thankfully not all—of these humanistic geographers sound de-humanized, or at least not fully human'.[26] Even if we might differ with him in detail it would be hard to deny it in general.

We are still faced with 'two cultures'.[27] It is implicit in all that we have reviewed so far that geography and literature are separate realms of work. Literature remains for us a resource, something we borrow from rather than contribute to, something we use rather than something we create as part of the vocation of geography. So far the connections between the two are few and frail and of uncertain potential. A growing number of geographers are making forays into the heartlands of literature and returning to tell us of the riches they have found, but we have as yet little solid evidence of their own enrichment in the form of major works that have been shaped by the insights of the humanities (although we were offered a tantalizing glimpse of a genuine geographer-poet whose internal integration of these two realms has been consciously disguised from his professional world.[28]) And such connections must serve an exchange rather than a one-sided exploitation. Only when we write books and essays that are so penetrating and powerful in the elucidation of life and landscape, of the grand themes of geography, that people in the humanities cannot afford to ignore us will we have forged enduring links between geography and literature. Without such interaction geography's position remains subordinate and parasitic.

There are hopeful signs but there are formidable barriers, both of our own making. I believe that the limited relationship between geography and literature reflects a strategy rather than necessity, and is the result of a ruling consensus rather than a testing of possibilities. Until now we have tried to model ourselves after the prestigious image of science. But today, along with many other fields, we are caught up in a period of change. The upsurge of interest in humanistic geography is but one of many symptoms. Just where we are heading is not clear but one question looms with inexorable logic: could geography grow, extend and reshape itself so that it could actually stand as one of the arts as well as one of the sciences? Could geographers actually create literature as well as borrow from it? I see nothing in the logic, needs, or possibilities of our field to prohibit it. But what would it require? And how would we recognize it?

THE GEOGRAPHER AS CREATOR OF LITERATURE

We may assume that it would appear in the literature on places, for through all history the literal meaning of 'geography', earth depiction, has by simple extension led to the common understanding that geography deals with the characteristics of places. That should be a comfortable definition for us all, for we can translate it without strain to cover any of our work, from the most particular to the general, the concrete and the theoretical, the practical, regional, and systematic, physical and human, science and art, and all at any scale from the house in its garden to the great globe itself.

As we assess our literature in terms of creative art we can set aside our standard textbooks, for however good a few may be they are explicitly pedagogic in purpose, a tool for use in an assigned task and therefore not 'literature' in this humanistic sense. Regional geographies at a more advanced level are essentially similar, although they may well serve a broader and less captive readership. It is an age-old expectation that geography should provide structured reliable descriptions of areas for reference and study by anyone who wishes to become well-informed about a region, and we can point to a considerable shelf of such books without embarrassment. I think of James Houston's *The Western Mediterranean World* and Dan Stanislawski's book on that small part of it called *The Algarve* as good examples of this kind of avowedly 'descriptive geography'.[29] Such works offer a vast amount of information ordered by area and by topic, integrated by our basic concepts, illuminated by maps, sketches, and photography. No one should underestimate the skill required to write a good book of this kind.[30] Although such works aim at systematic presentation of measurable facts there are no scientific rules governing selection and ordering from the almost limitless body of data available for such regional constructions. Thus, each is idiosyncratic in some degree and this inescapable fact has led many geographers to argue that such regional study can never be the basis for a truly scientific geography.[31]

One response to that has been to disparage regional geography and attempt to get rid of it.[32] Another has been to search for a philosophy and epistemology which would qualify it as a science despite this essential role of the individual writer.[33] Still another is to admit without reluctance that such geography is, ultimately, an art rather than a science. However, as my theme insists such an admission does not by itself carry us very far into the realm of art.

We do have a small body of writing representing a more self-conscious attempt at the art of geography. The most obvious examples are those whose authors have sought to capture what they have termed the 'personality' of an area. Gary Dunbar has traced the distinguished lineage of that term and concept,[34] and the work of two famous geographers can well serve to exhibit this kind of humanistic geography.

One of these is the remarkable set of books and essays by Professor Estyn Evans. He has used the term 'personality' in the title of three of his regional interpretations and his meticulous application of a lifelong concern for 'habitat, heritage, and history' to the study of his homelands of Wales, Ulster, and Ireland display how an avowedly scientific scholar can reach beyond to provide what might well be called wisdom about places.[35] He has made these areas so deeply a part of his life that he has made himself an indelible part of our understanding of them. Who could ever know Mourne Country except through Estyn Evans?[36] The personality of the place has taken on something of the personality of its student. Such translations surely carry us into the realm of creative literature, in a form closely analogous to the art of history.

Throughout his career Carl Sauer argued for the legitimacy and need for geographers to go 'beyond science'.[37] He used the term 'personality' in only one title—for a brief essay on Mexico which he offered as 'an excursion into the oldest tradition of geography', and 'a form of geographic curiosity that is never contained by systems' and seeks 'a quality of understanding' that can best be defined as 'regional interpretation'.[38] During a lifetime of work in the field and the archives, from wide-ranging reading and contact with scholars in a dozen disciplines Sauer created a small shelf of books and essays that bear an unmistakable personal imprint.[39] Deeply grounded in science, yet speculative, reflective, and critical, expressing his strong convictions about some grand themes of man and the earth, it is our most compelling illustration of his view that geography at its best must carry a message for mankind, it becomes a form of moral philosophy.

To call such geography humanistic is not only to classify by its qualities, in this case it is also a testimony of its impact. In 1947 a youngish New England writer, Charles Olson, author of a critically acclaimed study of Melville, came to Berkeley to reconnoitre materials for a large interpretative poetic work he had planned on the American West. Someone, probably George Stewart, that remarkably geographical man of letters in the University English Department, suggested that he talk with Professor Sauer in Geography. He did, and for the next thirty-five years he talked on and on of what an exhilarating, transforming experience it had been. Olson avidly read Sauer and followed many of his leads into archaeology, anthropology, ecology, and economic botany. He now shifted his focus back to a single locality, Gloucester, Massachusetts, which he had known since boyhood days, and began his long, complex *Maximus Poems*, a 'poetic enterprise . . . planned on a scale to challenge comparison with Pound's *Cantos* and Williams' *Paterson*'.[40] It has been said that Olson aspired 'to give in language a *map* . . . of one place', and it is surely a Sauerian map, a radical reading of American history as having a continuity from the pleistocene to the present, with the Europeans as a destructive disruption in the long ecological record. Olson continued a correspondence with Carl Sauer. In the 1950s Olson tried unsuccessfully to get Sauer to lecture at his famous Black Mountain College for writers. Sauer was required reading for all students there. And thus today we can find literary critics talking about 'culture hearths' and 'genetic geography', quoting from 'The Morphology of Landscape', and concluding (as one wrote) that such a geography 'finding its true place near to anthropology and even archaeology, is drawing . . . near also to sculpture, photography, painting, architecture, and literature'.[41] Olson died in 1970 and is now becoming recognized as a major figure in American letters. He had kept in close touch with Sauer all his life and the connection he initiated between literature and geography continues. For in the 1960s another young New England writer, Bob Callahan, came to Berkeley and through Charles Olson met Carl Sauer, with similar catalytic effect. Establishing a small publishing operation, Callahan reissued some of Sauer's books, hailing him as one of the 'three distinctly American intelligences . . . [who] opened up my sense of where the New World might actually be found' and provided 'a refreshed sense of American culture in a new and peculiar and particularly visionary grain'; and thus shortly before his death in 1975 Carl Sauer delivered the manuscript of his last book to Bob Callahan, and thereby *Seventeenth Century North America* carries the imprint of the Turtle Island Foundation, and a cultural geographer has become a cult figure in a small but vigorous literary movement.[42]

I know of no other case in the English-speaking world of geography having such impact on the humanities. But we must note that this relationship, too, is essentially a one-way current, a reverse of the geographers borrowing from literature for special insights into places. I know of no evidence that Carl Sauer was influenced by Charles Olson, and one suspects that he may have found the *Maximus Poems* about as difficult to interpret as most of us do. And so, while recognizing here an unusual connection, we need to explore further in our reconnaissance of geography as an art.

One feels that for all their fine qualities, these books of Evans and Sauer are not quite what the advocates of a more reliable attractive geographic literature have in mind. For although Sauer states that 'really good regional geography is a finely representational art' and encouraged us to try to convey 'the feel of horizon, sky, air, and land' so that our regional studies would not be 'such wooden things',[43] his own work is not strongly evocative in this sense; its power comes from breadth of knowledge and mature reflection rather than from vivid descriptions.

Our need to write far better than we do is a common cry. 'Our work', said the late Professor Gilbert in his Herbertson Memorial Lecture, 'is often so shoddy in craftsmanship that it is quite unreadable. We lack a proper feeling for words' and until we develop 'some literary graces it will not be read by many besides professional geographers'. If geographers 'are to make their regions vivid and alive, they should possess something of the expressive and sensuous imagination of poets and painters; they have all the riches of the English language at their disposal'.[44] That was said 23 years ago but the same general theme of regional geography as 'the highest form of the geographer's art' and requiring 'evocative descriptions' was forcefully put by John Fraser Hart in his Presidential Address to the American geographers just last year.[45] Hart exhorts us to write 'with imagination, flair, style, verve, dash, panache, enthusiasm, vivacity, animation, and perhaps even a bit of flamboyance'.[46] Even allowing for a bit of American extravagance here one feels that we are being pushed well beyond Estyn Evans and Carl Sauer, and we might well pause and ponder the prospect.

Just over 20 years ago in his Presidential Address to this body Clifford Darby gently warned of how difficult this seemingly elementary task can be. 'It is a humiliating experience for a geographer', he said, 'to try to describe even a small tract of country in such a way as to convey a true likeness'.[47] Those wonderful 'riches of the English language' are not lying like nuggets in our streams of consciousness ready for the plucking; for most of us they must be won if at all by hard drilling and repeated refining. To be sure, Hart went on to emphasize that fact, but his exhortation, like the comments of so many others on this theme, seems to focus on style. We are told that we must enliven our descriptions, add colour, drama, or whatever it takes to capture the attention of readers who won't slog through our normal stuff. If we really should believe that our problem is one of style and that it can be solved by fancy dress we shall certainly miss whatever potential there may be in this direction.[48]

If we are at all serious about making geography more fully an art we may indeed take literature, that great body of creative, imaginative writing, as our model, but our attention must be focused first of all not upon styles of writing but upon the motivation of the writer. One does not become an artist by imitation, by learning a craft; one only becomes truly an artist by inner feeling, by conviction, by a compelling need to share those feelings and by developing the ability to translate them to others. The artist must draw upon his feelings and insights and literally *imagine* them into such forms as to make effective connection with the lives of others. He must find his own voice and make it understood; style must arise from purpose and personality.

And here we make the decisive shift from science to art, the shift to the overtly personal statement, to the particular creative act resulting in what Helen Vendler calls 'that single, unduplicatable, compelling literary object.'[49] It is not surprising that our insecure field, so desperate under the academic pressures of its time to be recognized as a science, has dealt very gingerly with this matter of personal creation. We have come to recognize that it is in some degree an inevitable part of our work, but we have made rather frantic efforts to define and limit that ingredient, to keep it under control within some reassuring philosophy and methodology. Like true scientists we are really only comfortable with it as an object of study in itself, as in our expanding literature on perception and sense of place. 'To interpret the subjective meaning places hold for us is clearly not an exercise in alluding to personal experience', writes Edward Gibson; 'humanistic geographers have an obligation to transcend as far as possible a narrowly personal point of view that stands or falls by the mere experience of the observer'.[50] Thus 'mere experience' like 'mere description' is unworthy of our effort; such geographers would have us hover close to the seeming security of science, barely on the borderlands of the humanities.

David Lowenthal and Hugh Prince have argued eloquently and demonstrated persuasively how 'art and imaginative literature provide invaluable routes into realms of environmental meaning' because 'affective response to landscape and place is not simply a decorative façade to life, but forms its core'.[51] Yet even they give no firm sign that geographers might well try to express their own affective responses to that core in their professional work. Prince reminded us years ago that 'it is individual geographers who create the art of geography', but we must remember, he went on to say, 'that what makes them respectable is regard for scholarship'.[52] Hugh Prince is clearly at home in the humanities and one feels he gives us a longer tether than most, yet still holds us in check from the dangers of a too strongly personal statement.

Still, from time to time, a voice may be heard to argue that we ought to break that constraint and move on toward the central ground of the arts. The key to the kind of humanistic geography we need, writes Dennis Jeans in a recent issue of the *Australian Geographer*, requires an 'immersion in the meaning of place' and 'in the end this can only be a personal response . . .

Here we part company with those who insist that a geographical description should be replicable . . . Here we insist on the value of individual experience . . .'. The geographer must 'be aware of the meanings the place has for himself, partly as a technical geographer of some kind, but also as an individual participant in a wider cultural setting'.[53]

And so we are pressured on to introspection as an essential part of a truly humanistic interpretation of places. That in itself may not seen a radical idea. Surely the work of Estyn Evans, for example, has been informed by deep personal reflection.[54] But Jean's remarks emanate from his assessment of the work of such writers as Wordsworth, Zola, Baldwin, Camus, and he clearly infers that like them we should consciously allow our work to be shaped into an explicit message of our personal feelings. And that opens up quite a different prospect.

THE GEOGRAPHER AS ARTIST

Here we look out upon a wide world of creative art, an immensely varied and uncertain ground which if attractive to some will be intimidating or even repellent to others. What might it yield should the geographer try to cultivate some part of it? We do not have much direct evidence to go by.

I know of only one substantial product of this kind: *Fitzgerald*, an intensive treatment of one square mile of Detroit by William Bunge.[55] The author, whose *Theoretical Geography*[56] made him famous as one of our really creative minds, claims the work as science; any reader will quickly realize it is not. It is, rather, a bold illustration and a valuable lesson of what a work of art can be. Bunge, a man of great talent and humane purpose, is a self-styled explorer, an avowed regional geographer immersed in the life of his own community. He is present throughout the work, so deeply, emotionally involved that, as he says, 'the book has had a human power of its own and in a sense, it created the author'.[57] That is a telling clue to art; one can feel the energy surging through it, and the result is unorthodox, unpredictable, full of surprises in topic, argument, format, a truly idiosyncratic creation. It is also a deeply flawed creation. The reviewers were properly devastating of its pretence of being science, for the work is polemic.[58] Written in the highly-charged atmosphere of the late 1960s, directly amidst the excruciating tensions of black and white in battle-scared Detroit, *Fitzgerald* is the shrill political statement of an angry man. Such purpose and passion of course do not deny it as a work of art. But great art is emotion under control, art must derive from an intense experience but it must also be shaped and constructed with skill.[59] Bunge has great energy and deep feelings but he lacks firm discipline and thus has produced an ill-shapen work. His self-righteous rhetoric and deep prejudices vitiate his argument, fail to convince the reader, and waste the worthiness of his cause.

Our most thoughtful guide in these matters, Yi-fu Tuan, has said that to ask humanistic geography actually to create works of art is 'a council of perfection', and therefore unreasonable.[60] But here I cannot follow him. If geography is to become one of the arts we may expect to have all graduations in the quality of that art. Surely we may strive for perfection without the delusion that we can reach it. Who ever heard of a satisfied artist? One who is satisfied with his own creation is dead as a creator. For all its flaws *Fitzgerald* remains a work of art and it is a valuable exhibit of what can happen when we immerse ourselves so deeply in the life of an area and attempt to tell of it with all our energy and emotion. Bunge has cried out what *Fitzgerald* means to him as a microcosm of America, and disappointment in his creation need not denigrate the importance of his attempt.

Fortunately a more compelling exhibit came to hand just as I began to compose this lecture:

a large book, a handsome book called *Passing the Time in Ballymenone.*[61] The author calls it an offering of 'the startling richness of one small spot in Ireland'. It is ostensibly a work in folklore and folklife, but anyone who knows Henry Glassie in book or person will expect something far beyond the vestigial fragments that often appear under such headings.[62] Glassie is thoroughly at home in half a dozen disciplines, including cultural geography (always acknowledging his debt to Fred Kniffen and Estyn Evans), and presents this work as a step 'toward a unified program for the study of human beings'.[63] The book is a remarkable fusion of scholarship, artistry, and emotion. An unknown people in an obscure place are given an intensely real vitality. The ordinary lives and landscapes of individuals and of a whole community are portrayed and analyzed with breathtaking skill in search for the meaning of it all.

Ballymenone is in Fermanagh, on the bitter borderlands of a society that has known the miseries of mankind far longer than Fitzgerald, but there is neither polemic nor sentimentality here. Glassie is a scholar so secure in his science that he can stand firmly on it and, with disciplined emotion, reach beyond: 'rationally', he says,

> study begins in division, in the acceptance of traditional economies [of specialized research in academic disciplines], but if work is good, old categories will slip and shift, and melt away as we find the place where social science and humanities, where art and culture and history, time and space, connect, where theoretical and empirical studies fuse . . . and we will know our work is good when we find in it no roughness, no leaps between bogus 'levels', no gaps between the particular and the universal, the real and the abstract, no disruptive jolts as we swing from fact to theory and back.[64]

That is an artist speaking, keenly sensitive to the structure and shape and tone and finish as being integral to the message of his work.

And so when we join this tall American as he scrambles through a hedge and up a wet slope to his first fortuitous meeting with a man of Ballymenone and accompany him on his journey of discovery of a people and a place and himself, and follow through 700 pages to an end so simple, so poignantly human, as to produce tears, we are vicariously involved in the creation of a work of art. Glassie has immortalized a community; his book is more than an addition to knowledge, it is a sharing of experience and a package of wisdom; we can learn about many topics from it, but the book is more than the sum of its parts and we can admire it as an entity reaffirming the fundamental truth that, ultimately, 'art is not only about something; it is something'.[65]

Passing the Time in Ballymenone is an exhibit, not a model, for art cannot be programmed into objective and method. It is an exhibit of what can come from fully humanistic scholarship that transcends pedantic constrictions. We may assume that the geographer's theme and focus would be somewhat different from that of Glassie but it is the nature of art to be unpredictable and we cannot know just what the result might be.[66]

We may be assured, however, that it is important—and difficult—work to do. 'We learn to see a thing by learning to describe it', writes Raymond Williams, and he goes on to point out that in this sense we can never be describing something that is known, we are discovering it and investing it with meaning.[67] Furthermore, we know, as John Berger says, that 'the relation between what we see and what we know is never settled', for 'seeing establishes our place in the surrounding world; we explain that world with words, but words can never undo the fact that we are surrounded by it'.[68] Here we leave far behind all ideas of 'mere description' and enter into the processes of life itself. In the broadest sense, everyone is involved every day in the ongoing construction of the reality that encompasses. Most of this is, necessarily, conventionalized and inarticulated. To the extent that anyone can convey effectively to others what he sees and feels about a place he is an artist, and great art, says Williams, must come from one who has lived an experience and 'has literally made it part of himself so deeply that his whole energy is available to describe it and transmit it to others'.[69] Thus 'reading the landscape' becomes more than a pleasant pastime, it is a creative act. The interpreter must become so involved that he is deepening and extending his own life (that is what we mean by landscape appreciation as an *enrichment*), and in trying to communicate what he sees, understands, and feels, to define the human significance of a place, he extends this creativity by a major social act, for he is altering the definition of the world for all those who respond to him.[70] Art is by nature idiosyncratic and it may be quite unorthodox, yet it is also social and cultural for every artist is himself shaped by his human context and must make use of established conventions for effective communication. Thus all creative literature is 'social practice', it invokes, or proposes, a relationship with others, it is part of that ongoing dynamic and, as Williams would insist, that dialectic, between society and the individual.[71] Here we deal with truly fundamental matters, and I think that particularly as geographers we might well pay close attention to this versatile philosopher, scholar, critic, and novelist, for Raymond Williams has been a tough, probing, creative thinker about culture and history, art and society, and the individual confronting and coping and trying to shape his world. He speaks so directly to us in so many ways about the interpretation of landscapes and places. Consider his notation of the range of treatment and these things in imaginative literature:

> the presentation of *place* depends on variable conventions from a deliberate unlocation to a simple naming to a brief sketch to variably detailed description, up to the point where, as it is said, the place itself becomes a 'character' or 'the character'. Radically variable assumptions of the relations between people and places, and between 'man' and 'nature', are conveyed in these apparently self-evident ways. Other conventions assume or indicate variable relations between place and societies—'environments'—over a range from the abstraction of place from people, through to perception of people as symptoms of places, to the active apprehension of places as made by people. Descriptions of great houses, or rural landscapes, of cities, or of factories are evident examples of these variable conventions where the 'point of view' may be experienced as an 'aesthetic choice' but where any point of view, including that which excludes persons or converts them into landscape, is social.[72]

All this was richly illustrated in his great interpretive work *The Country and the City.*[73] And the artistry of Williams is beautifully illustrated in his contribution to Ronald Blythe's recent British anthology on *Places* where in a few pages Williams defines and evokes the Black Mountains of his Welsh homeland with consummate skill.[74]

I think we geographers need people like Raymond Williams and Henry Glassie not only to help us see what the humanities can be, what art is, and how all pedantic compartmentalizations of art and science dissolve within the creative individual, we also badly need them to reassure us of the importance, of the absolutely fundamental significance of place—of geography.

CONCLUSION

Thus we turn to the humanities to discover the full possibilities of our birthright. The 'literary imagination is incurably local', said Helen Vendler recently, and she went to remind her literary listeners of the old artistic truth that to immerse oneself in what one most deeply feels is a proven route to the universal.[75] 'We need landscape, a reality out there, as a relief from psychology, from the claustrophobia of the self'.[76] Even though we and the literary critic who wrote that know that the 'reality' of landscape is individual and elusive, his words seem a compelling call, especially in an America mired in the pathology of the self. There is more to life than landscape, but to divorce life from landscape is to invite the kind of alienation all too apparent in the world. All life, all history take place and we need never apologize for being members of a field defined by a concern for places. John Warkentin has noted that 'geography, as F. R. Leavis argues for English literature, has an important dimension as a civilizing subject, and communicates with a wide range of people'.[77] But one feels his description hardly fits the actual form of our modern professional field. J. M. Powell, in a major address, recalls the geographers to their true vocation: 'to act as interpreters of our world; to *bear* witness to its variety and significance, its beauty and horrors' and that 'witness must be by *affirmation* and *protest*, interpreting the world to man, and man to himself'.[78] That is a call to art.

'Art is power. It influences the mind, the nerves, the feelings, the soul. It carries messages of hope, hostility, derision, and moral rebuke. It can fight material and spiritual evils, and transmit the ideals of a community now living, long past, or soon to be born. In a word, Art is deemed universally important', says Jacques Barzun, 'because it helps men to live and to remember'.[79] By limiting ourselves to describing and measuring and analyzing certain aspects of the world as it seems geographers have denied themselves the possibility of probing very deeply into what it all means. Being unable to convey what it means we cannot help shape it toward what it might become. 'Art is that which contributes, however minutely, infinitesimally, to the regeneration of our World'.[80] Modern geography has been so insecure and inward looking that it has closed itself off from that possibility.

This is not a call for geographers to become novelists—although, as the example of Raymond Williams demonstrates, the novel may be an important means of exploration and expression along the way. It is a call only for a greater openness, a clearing away of pedantic barriers, for a toleration of geographical creativity wherever it may lead. Nor should we expect a great surge into new frontiers as a result. It will take courageous pioneers to move in upon the ground of the humanities and for the time being they will have serious handicaps for we do little to equip them with appropriate skills and tools. They will likely always be a very small portion of our profession but they could be among the most important. We shall not have a humanistic geography worthy of the claim until we have some of our most talented and sensitive scholars deeply engaged in the creation of the literature of the humanities. Geography will deserve to be called an art only when a substantial number of geographers become artists.

NOTES

1. WILLIAMS, R. (1979) *Keywords, A vocabulary of culture and society* (New York)
2. BERLIN, I. (1980) 'The divorce between the sciences and the humanities', in *Against the current, Isaiah Berlin; selected writings* (New York) pp. 80–110. Berlin (p. 109) stresses the initiative of Vico in his reaction against the positivist claims of science, by which 'a parting of the ways' developed, characterized by such oppositions as 'the specific and unique versus the repetitive and the universal, the concrete versus the abstract, perpetual movement versus rest, the inner versus the outer, quality versus quantity, culture-bound versus timeless principles, mental strife and self-transformation as a permanent condition of man versus the possibility (and desirability) of peace, order, final harmony and the satisfaction of all rational human wishes.'
3. SNOW, C. P. (1959) *The two cultures and the scientific revolution* (Cambridge)
4. Much the most forceful indictment by a geographer is that of Wilbur Zelinsky (1975) in his Presidential Address 'The Demigod's Dilemma', *Ann. Ass. Am. Geogr.* 65: 123–43; see also POWELL J. M. (1980), 'The haunting of Saloman's house: Geography and the limits of science', *Aust. geogr.* 14: 327–41.
5. Wright urged that geography make specific connection with the humanities and his own creative lines of inquiry laid important foundations; see WRIGHT, J. K. (1966) *Human nature in geography* (Cambridge, Mass.).

6. HARRIS, C. (1978) 'The historical mind and the practice of geography', in LEY, D. and SAMUELS, M. (eds) *Humanistic geography, prospects and problems* (Chicago) pp. 123–37; as Harris states, North American geography in particular 'is not grounded in tradition or culture but rather in a society preoccupied with technique and management', p. 134
7. This is exhibited in much of the landmark volume edited by LEY and SAMUELS (1978) op. cit., the first to bear that title, see also the response to that prospectus by GREGORY, D. (1981) 'Human agency and human geography', *Trans. Inst. Br. Geogr.* N.S. 6: 1–18, and the short 'progress report' LEY, D. (1981) 'Cultural/Humanistic Geography', *Prog. hum. Geogr.* 5: 249–57
8. TUAN, YI-FU (1978) 'Literature and geography: Implications for geographical research', in LEY and SAMUELS (1978) op. cit. (note 6) 194–206, quotation, p. 195
9. TUAN, YI-FU (1976) 'Humanistic geography', *Ann. Ass. Am. Geogr.* 66: 266–76; quotation p.274
10. ENTRIKIN, J. N. (1976) 'Contemporary humanism in geography', *Ann. Ass. Am. Geogr.* 66: 615–32; quotation p. 616
11. Cf. WILLIAMS (1976) op. cit. (note 1) 'Humanity', pp. 121–4
12. RELPH, E. (1981) *Rational landscapes and humanistic geography* (Totowa, N. J.)
13. HARRIS (1978) op. cit. (note 6)
14. BRAUDEL, F. (1972) *The Mediterranean, and the Mediterranean world in the Age of Philip II* (London) Vol. 1, p. 17
15. 'History nears perfection in so far as knowledge and art work in harmony', NEFF, E. (1947) *The poetry of history, the contribution of literature and literary scholarship to the writing of history since Voltaire* (New York), p. 4. The essentially nineteenth century view of this relationship has come under heavy criticism, but Braudel's creative structuring of his work carries it toward the kind of complex, dynamic, contingent interpretation demanded of history by 'modern art' as defined in WHITE, H. V. (1966) 'The burden history', *Hist. Theory.* 5: 111–34. For a compelling personal testimony of the early shaping of an historian as artist (and in a form immediately congenial to the geographer of landscapes and regions) see the statement of the Canadian historian MORTON, W. L. (1970) 'Seeing an unliterary landscape', *Mosaic* 3: 1–10.
16. POCOCK, D. (ed.) (1981) *Humanistic geography and literature, essays on the experience of place* (London) p. 15
17. TUAN (1978) op. cit. (note 8) LOWENTHAL, D. and PRINCE, H. C. (1976) 'Transcendental experience', in WAGNER S., COHEN, S. B., and KAPLAN, B. (eds) *Experiencing the environment* (New York) pp. 117–31; SALTER, C. I. and LLOYD, W. J. (1977) *Landscape in literature, Res. Pap. Coll. Geog. No. 76–3, Ass. Am. Geogr.* (Washington). A few geographers have considered the generally analogous possibilities of the relationships between geography and landscape paintings; e.g. HEATHCOTE, R. L. (1972) 'The artist as geographer: landscape painting as a source for geographical research', *Proc. R. geog. Soc. Aust.* (S. A. Branch) 73: pp. 1–21, and the numerous studies of Ronald Rees, such as REES, R. (1973) 'Geography and landscape painting: an introduction to a neglected field', *Scott. geogr. Mag* 89: 147–58, idem (1976) 'Images of the prairie: landscape painting and perception in the Western Interior of Canada', *Can. Geogr.* 20: 259–78, idem (1978), 'Landscape in art', in BUTZER, K., (ed. *Dimensions in human geography* (Chicago) pp. 48–68
18. TUAN, YI-FU (1978) op. cit. (note 8) Of course many novelists have a natural 'geographic bent', e.g. HUDSON, B. J. (1982) 'The geographical imagination of Arnold Bennett', *Trans. Inst. Br. Geogr.* N.S. 7: pp. 365–79
19. WHITTINGTON, G. (1974) 'The regionalism of Lewis Grassic Gibbon', *Scott, geogr. Mag.* 90: 75–84
20. WATSON, J. W. (1979) *Social geography of the United States* (London)
21. WATSON, J. W. (1970) 'Image geography: the myth of America in the American scene', *Br. J. Advmt Sci* 27: 71–9; also MEINIG, D. W. (1979) 'Symbolic landscapes, some idealizations of American communities', in MEINIG, D. W. (ed.) *The interpretation of ordinary landscapes* (New York) pp. 164–92, and POCOCK, D. C. D. (1979) 'The Novelist's Image of the North', *Trans. Inst. Br. Geogr.* N.S. 4: 62–76
22. WATSON, J. R. (1983) 'The soul of geography', *Trans. Inst. Br. Geogr.* N.S. 8 (4), forthcoming
23. AIKEN, C. S. (1977) 'Faulkner's Yoknapatawpha County: geographical fact into fiction', *Geogrl Rev.* 67: 1–21, AIKEN C. S. (1979) 'Faulkner's Yoknapatawpha County: a place in the American South', *Geogrl Rev.* 69: 331–48, AIKEN, C. S. (1981) 'A geographical approach to William Faulkner's 'The Bear', *Geogrl Rev.* 71: 446–59
24. BIRCH, B. P. (1981) 'Wessex, Hardy and the nature novelist', *Trans. Inst. Br. Geogr.* N.S. 6: 348–58
25. In response to my inquiry Aiken reports that his studies are becoming known to Faulkner specialists and that he is beginning to receive invitations to speak and work with scholars of the humanities. However, I note that the compiler of a survey of Faulkner studies, while valuing Aiken's works, refers to it as 'an essentially non-literary point of view', MCHANEY, T. L. (1978), *Miss. Q.* 31: pp. 435–36. Of course there may be many fruitful interactions between particular geographers, literary scholars, and writers, especially within the same university, but we can hardly claim an effective bond until such relationships become more formal and far reaching, as in the recent publication of a literary scholar's study of Hardy's use of various 'mental geographers' in *Tess* in a geographical periodical: see, BARRELL, J. (1982) 'Geographies of Hardy's Wessex', *J. hist. Geogr.* 8: 347–61
26. THOMAS, D. M. (1981) 'Matters of local interest', *Times Lit. Suppl.*, 13 March 1981
27. *cf.* BERGER, M. (1977) *Real and imagined worlds, the novel and social science* (Cambridge, Mass. and London), esp. ch. 10, 'The challenge of science to literature'.
28. WATSON, J. W. (1983); Watson reports that early in his career he was warned that he must not publish poetry under his own name if he expected to be taken seriously as a professional geographer. I had not seen Jay Appleton's *The poetry of habitat*, Landscape Research Group and Dep. Geogr. Univ. Hull, Misc. Ser. No. 20, 1978, a compelling illustration of this very kind of creative integration which deserves prominent notice in this context, although it cannot by itself alter the general theme and characterization of this essay.
29. HOUSTON, J. (1967) *The Western Mediterranean world, an introduction to its regional landscapes* (New York and Washington); STANISLAWSKI, D. (1963), *Portugal's other kingdom, the Algarve* (Austin)

30. Paterson, J. H. (1975) 'Writing regional geography, problems and progress in the Anglo-American realm', *Prog. Geogr.* 6: 1–26
31. Leighly, J. (1937) 'Some comments on contemporary geographic method', *Ann. Ass. Am. Geogr.* 27: 125–41
32. Gould, P. (1979) 'Geography, 1957–1977: the Augean period', *Ann. Ass. Am. Geogr.* 69: 139–51; Wrigley, E. A. (1970) 'Changes in the philosophy of geography' in Chorley, R. J. and Haggett, P. (eds) *Frontiers in geographical teaching* (London), pp. 13ff.
33. Gregory (1981) op. cit. (note 7)
34. Dunbar, G. (1974) 'Geographical personality', in Walker, H. J. and Haag, W. G. (eds) *Man and cultural heritage, papers in honor of Fred B. Kniffen, Geoscience and Man* (Baton Rouge) 5: 25–33
35. Evans, E. E. (1973a) 'The personality of Wales', *BBC Wales, Annual Radio Lecture* (Cardiff); Evans, E. E. (1973b) 'The personality of Ulster', *Trans. Inst. Br. Geogr.* 51: 1–19; Evans, E. E. (1973c) *The personality of Ireland, habitat, heritage and history* (Cambridge)
36. Evans, E. E. (1967) *Mourne Country, landscape and life in South Down* (2nd rev. edn) (Dundalk)
37. Sauer, C. O. (1925) 'The Morphology of Landscape', *Univ. California Pubs. Geogr.* 2: 19–54, quotation p. 48
38. Sauer, C. O. (1941) 'The personality of Mexico', *Geogrl Rev.* 31: 353–64
39. e.g. Sauer, C. O. (1966a) *The early Spanish Main* (Berkeley and Los Angeles); (1966b) *Northern Mists* (Berkeley and Los Angeles); (1971) *Sixteenth century North America* (Berkeley and Los Angeles), and Leighly, J. (1963) *Land and life, a selection from the writings of Carl Ortwin Sauer* (Berkeley and Los Angeles)
40. Davie, D. (1968) 'Landscape as poetic focus', *Southern Rev.*, N.S. 4: 685–91 quotation p. 687; I am indebted to Cosgrove D. E. (1979) 'John Ruskin and the geographical imagination', *Geogrl Rev.* 69: 43–62, esp. p. 59 for this reference to Davie. Olson, C. (1960) *The Maximus poems* (New York); the essential accompaniment is Butterick, G. F. (1978) *A guide to the maximus poems of Charles Olson* (Berkeley and Los Angeles) which contains a valuable introduction to and chronology of Olson's life, as well as a line-by-line explication. See also Olson, G. (1974) *Additional prose: A bibliography on America proprioception & other notes & essays* (Bolinas, California) pp. 5, 13, 24 for specific references to Sauer
41. Davie (1968) op. cit. (note 40) Paul, S. (1978) *Olson's Push, Origin, Black Mountain, and recent American poetry* (Baton Rouge), esp. p. 168
42. 'Bob Callahan: On Turtle Island' an interview by Michael Helm, published in *City Miner*, (San Francisco). The other two 'distinctly American intelligences' are the historical botanist Edgar Anderson, long a close friend of Sauer, and the poet-ethnologist Jaime de Angulo. The two out-of-print books re-published were *Man in nature*, a children's textbook of 1935, and *Northern Mists*. Sauer, C. O. (1980) *Seventeenth century North America* (Berkeley). I am indebted to Bob Callahan and James J. Parsons for a variety of materials on this relationship.
43. Sauer, C. O. (1956) 'The education of a geographer', *Ann. Ass. Am. Geogr.* 46: 287–99; quotations p. 298
44. Gilbert, E. W. (1960) 'The idea of the region', *Geography* 45: 157–75; quotations pp. 162, 159
45. Hart, J. F. (1982) 'The highest form of the geographer's art', *Ann. Ass. Am. Geogr.* 72: 1–29
46. Ibid. p. 28
47. Darby, H. C. (1962) 'The problem of geographical description', *Trans. Inst. Br. Geogr.* 30: 1–14
48. 'When a geographer asks a student to read literary works in order to improve his style for geographical writing, he surely speaks loosely and out of ignorance', Tuan (1978) op. cit. (note 8). p. 198
49. Vendler, H. (1981) 'Presidential Address 1980', *PMLA* 96: 344–50; quotation p. 348
50. Gibson, E. (1978) 'Understanding the subjective meaning of places', in Ley and Samuels (1978) op cit. (note 6) pp. 138–54: quotation p. 138
51. Lowenthal and Prince (1976) op. cit. (note 17) quotation p. 118
52. Prince, H. C. (1961–2) 'The geographical imagination', *Landscape* 11: 22–5
53. Jeans, D. N. (1979) 'Some literary examples of humanistic descriptions of place', *Aust. geogr.* 14: 207–14
54. However, 'as a professional group geographers are not known for introspection', Tuan (1976) op. cit. (note 8) p. 276
55. Bunge, W. (1971) *Fitzgerald: geography of a revolution* (Cambridge, Mass.)
56. Bunge, W. (1966) *Theoretical geography,* (Lund studies in geography, Lund)
57. Bunge, (1971), op. cit. (note 55) Preface, unpaged
58. The editor of the *Ann. Ass. Am. Geogr.* considered Bunge's *Fitzgerald* sufficiently important and controversial to warrant two reviews, each by a geographer with strong credentials in humanistic geography and in the analysis of blacks in northern American cities. See Lewis, P. F. (1973) and Ley, D. (1973) *Ann. Ass. Am. Geogr.* 63: 131–2 and 63: 133–5 respectively. Lewis's review elicited a pained and angry reply from Bunge. See Bunge, W. (1974) *Ann. Ass. Am. Geogr.* 64: 485–9
59. Fischer, E. (1963) *The necessity of art, a Marxist approach* (London) p. 9
60. Tuan (1976) op. cit. (note 9) p. 273
61. Glassie, H. (1982) *Passing the time in Ballymenone, culture and history of an Ulster Community* (Philadelphia)
62. e.g. Glassie, H. (1968) *Pattern in the material folk culture of the Eastern United States* (Philadelphia); Glassie, H. (1974) 'The variation on concepts within tradition: barn building in Otsego County, New York', in Walker H. J. and Haag, W. G. (1974) op. cit. (note 34) pp. 177–235; Glassie (1975) *Folk housing in Middle Virginia* (Knoxville)
63. Glassie (1982) op. cit. (note 61) p. xiii
64. Ibid. p. xiv
65. Sontag, S. (1961–66) *Against interpretation and other essays* (New York), esp. pp. 3–36; quotation p. 21. 'Unlike history, philosophy or the social sciences, [in literature the] objects of study are actually objects—their existence depends on an exact configuration or form . . . the meaning is determined by the form and the form by the meaning. A literary work must be grasped

whole', McGilchrist, I. (1982) 'A pessimist's solution', in 'Professing literature, A symposium on the study of English', *Times Lit. Supp.* 10 Dec., 1982, p. 1358. Glassie is also in the most obvious sense an artist: the book contains more than a hundred examples of his work with pen and ink and camera.

66. Such idiosyncratic creativity sometimes wells up within a series of books, each volume by a different author but designed to treat a common theme in a generally standard way, wherein a particular author invests himself to such a degree as to transcend the usual product. See Paul Horgan's (1954) magnificent two-volume exposition of *The Great River, The Rio Grande in North American History* (New York/Toronto) in the *Rivers of America* series, and Jacobs, S. W. (1979) *Wayne County: the aesthetic heritage of a rural area* (Lyonds, N.Y.) wherein an architectural historian invited to contribute to a series of small guidebooks on 'Architecture worth saving' in New York State devoted ten years of his life in creating a large and immensely rich work of art.
67. Williams, R. (1961) *The Long Revolution* (London) p. 39
68. Berger, J. (1972) *Ways of Seeing* (London)
69. Williams, R. (1961) *The Long Revolution* (London) p. 50
70. Meinig, D. W. (1979) 'Reading the landscape, an appreciation of W. G. Hoskins and J. B. Jackson' in Meinig, D. W. (ed.) *The interpretation of ordinary landscapes* (New York) pp. 195–244
71. Williams, R. (1977) *Marxism and literature* (Oxford)
72. Ibid. p. 177, quotation by permission of Oxford University Press; For a recent and radical challenge to orthodox modes of interpretation in the humanities by an American literary and historical scholar see Said, E. W. (1982) 'Opponents, audiences, constituencies, and community', *Cultural Inquiry* 9: 1–26
73. Williams, R. (1973) *The country and the city* (New York)
74. Williams, R. (1981) 'Black Mountains', in Blythe, R. (ed.), *Places an anthology of Britain* (Oxford) p. 215–22
75. Vendler (1981) op. cit. (note 49) p. 348; e.g. Spender, S. (1974) *Love-hate relations, a study of Anglo-American sensibilities* (London) pp. 243–44; and Williams, R. (1979) *Politics and letters, interview with New Left Review* (London) pp. 301–2, wherein he discusses his intention of undertaking an experiment in writing a novel about history from a socialist philosophical standpoint: 'I don't know yet whether it can in practice be done. The great problem is obviously to find the necessary unity. I thought of trying to achieve it through continuity of place rather than of people—taking the same place as it is inhabited in different periods . . . by different kinds of people, manifesting different social relations, and exercising different ways of using the land. It happens that the district I know best in Wales would be very suitable for this kind of continuity . . .'.
76. Broyard, A. (1981) 'Disappearing Landscape', *N. Y. Times, Sun. bk. Rev.*, 20 Sept. 1981
77. Warkentin, J. (1978) 'Epilogue', in Gibson, R. (ed.) *European settlement and development in North America, essays on geographical change in honour and memory of Andrew Hill Clark* (Toronto), p. 208
78. Powell (1980) op. cit. (note 4) p. 333
79. Barzun, J. (1973) *The use and abuse of art* (Princeton) p. 21
80. Franck, F. (1981) *Art as a way, a return to spiritual roots* (New York) p. 146

[4]

Reclaiming Vision: looking at landscape and the body

CATHERINE NASH, *University of Wales, Lampeter, UK*

ABSTRACT *This paper responds to feminist critiques of the masculinity of the landscape tradition within geography. It draws upon reassessments of the gaze within film theory, art history and cultural studies as well as within representational practice. It does so in order to reclaim the concept of landscape as a theoretical tool and subject of study for a feminist cultural geography. In theorising a reclamation of looking and landscape through a critical feminist approach, issues of vision and space, gender and representation, politics and pleasure are brought forcefully together through considering images of the male body as landscape by two contemporary women artists. While recognising the politics of representation, the aim is to deconstruct ideas of an unproblematic women's vision and of a singular or essential male or female gaze. Despite the way in which the metaphor of the body/land has been employed to justify both approaches to women and the environment and to legitimate colonisation, this paper suggests that with regard to both the body and landscape we need to look again and reconsider the radical potential of visual pleasure and traditions of visual representation.*

Introduction

Feminist critiques of the concept of landscape within cultural geography have made its use questionable. This paper firstly discusses feminist responses articulated by Gillian Rose and Susan Ford to the problems raised by feminist theory in relation to landscape representation within cultural geography. Gillian Rose's critique of landscape study in cultural geography employed film theory and psychoanalytic accounts of the gaze of the 1970s to discuss the complex feminisation of landscape (Rose, 1992, 1993). My response to this critique similarly draws on contemporary reassessments of visual pleasure and anxiety within film theory and cultural studies as well as within representational practice. By applying these approaches to representational politics, to a discussion of women's images of the male body as landscape, I attempt to reconcile a feminist approach which retains the idea of landscape as a focus of substantive and theoretical concerns, despite feminist critiques of the masculinity of the landscape tradition within geography. This paper seeks to examine the possibility of a feminist politics of visual pleasure which does not entail abandoning a critical perspective on the politics of representation. These images suggest that rather than simply assert the oppressive nature of images of feminised landscapes or of women's bodies as terrain, it is necessary to engage with them to disrupt their authority and exclusive pleasures and open up possibilities for difference, subversion, resistance and reappropriation of visual traditions and visual pleasure (Kotz, 1993). Pleasure in research, writing, or looking at landscape or the body is political but this does not render this representation or vision automatically unacceptable. Working towards a critical perspective on visual pleasure and landscape imagery involves considering issues

Correspondence: Catherine Nash, Department of Geography, University of Wales, Lampeter, Dyfed SA48 7ED, UK.

FIG. 1. Diane Baylis *Abroad* 1992. Published with permission of the artist.

of desire, consent and representation. It raises the issue of whether visual desire is always dependent on a position of domination or pleasure always oppressive. It thus entails reassessing the criteria by which the politics of representations are judged.

In theorising a reclamation of looking and landscape through a critical feminist approach, issues of vision and space, gender and representation, politics and pleasure are brought forcefully together through considering the image of the male body as landscape. Since much of the criticism of landscape has been based on the feminisation of the 'field of vision', these masculine objects of vision provide a focus for considering the radical potential of returning to traditions of visual representation. Two images of the male body by women artists: the photograph *Abroad* (1992, Fig. 1) by Diane Baylis [1] and *Inis t'Oirr/Aran Dance* (1985) a video-installation by Pauline Cummins [2] provide means to address the specific and complex politics and pleasures, potential and problems of the body/landscape metaphor. In addressing two explicit images of the male body by women, this discussion is located in a series of overlapping academic, cultural and social developments. It is framed by recent attention to the representation and production of those identities against which 'marginal' groups have been defined: whiteness, masculinity, heterosexuality. This attention to the constructed nature of the identities of the 'centre' supplements a longer critical study of the representation of the 'marginal'.

Secondly, it draws on reassessments of women's visual pleasure as artists and audiences as well as objects of vision, in film theory, art history and creative practice. This consideration of women's images of men also reflects developments within feminist art history, which has moved from a focus on representations of women and the retrieval of lost histories of women artists, to consider gender difference in representational practice through differing combinations of Marxist, post-structuralist and psychoanalytic approaches (Gouma-Peterson & Mathews, 1987; Pollock, 1995). Finally, it is located also within debates concerning sexuality, pornography, power and ethics, within a diverse range of sexualities, sexual practices and sexual relationships, their representation and censorship (de Lauretis, 1988; Dyer, 1990; Gibson & Gibson, 1993; Harwood *et al.*, 1993). Significantly, in both published work and in representational practice issues of women's visual pleasure and the production of erotic imagery of men and women for women are being addressed (Gamman & Marshment, 1988; Boffin & Fraser, 1991; Lau, 1993, Salaman, 1994). Both *Abroad* and *Inis t'Oirr/Aran Dance* represent the male body in different ways, one through using a Western, pictorial, landscape format, the other referring to the cultural symbolism of an Irish regional landscape. In doing so they appropriate visual traditions and cultural symbols, despite the way these traditions have been implicated in relations of power and oppression. They thus prompt a reassessment of approaches to the politics of representation within geography and more widely.

Valid and valuable engagements with the politics of representation from feminist, antiracist and post-colonial positions, have undermined ideas of cultural production as an autonomous artistic rather than socially grounded practice, yet have tended to render all forms of visual pleasure and representational practice deeply problematic. Vision and visual representation become equated with generalised notions of masculinism, imperialism and oppression, leading to a deep unease about representation and a form of critical iconoclasm (Solomon-Godeau, 1992). In addition the media of representation from maps to photographs have come to be deeply and often implicitly essentially, associated with oppressive power–knowledge relations. While much of this work has been dependent on careful historical research on the use of mapping or picturing in strategies of control and subordination, the simple suggestion that landscape representation is ideological, or vision oppressive, suggests an ahistorical condemnation of the genre, practice or media rather than attention to the particular effects of images in specific and finely differentiated social contexts. This kind of criticism may collapse the differences between different landscape representations and kinds of seeing.

Critical interpretation of an image needs to address its relation to the history of the representational tradition to which it belongs since its meaning cannot be completely disentangled from the history of its genre. It is in part the use of landscape imagery in the history of Western culture in establishing ideas of hierarchical class, gender and racial difference which makes landscape problematic. In particular landscape imagery and the ability to view landscape according to ideas of picturesque taste, helped secure the social and cultural authority and status of white, upper- and middle-class men in Britain in the late eighteenth century (Barrell, 1990). Both images discussed here relate to traditions of representation which have produced and naturalised forms of social organisation through landscape imagery. *Abroad* deploys the format of conventional landscape imagery, and thereby evokes its attendant history. *Inis t'Oirr/Aran Dance* draws heavily on the symbolism of the West of Ireland and may then be considered complicit with essentialist ideas of national, racial, sexual and gender identity that images of the West have been used to support. The Aran of the title and of the knitting tradition refers to the Aran Islands off the coast of County Galway in the west of Ireland, which were the focus of intense national desire and anxiety about racial, moral, linguistic, spiritual and sexual purity in the early twentieth century (Nash, 1993). Yet, through my discussion of these images, I will argue that this history does not fix an image as eternally oppressive by association with images which were used in particular times and places in legitimating and reinforcing social power and control. More effective critiques of particular representational practices can be made through the more demanding and complex task of understanding images through specific contexts of production and reception which include but are not finally determined by the representational tradition to which they belong, evoke or work through. My reading of the radical potential of these two art works, which bring together a sexualised male body and images of landscape, is through their different and historically and geographically specific contexts. These contexts include the deliberate framing of the images through the artist's intention. This intentionality should neither be dismissed, thereby reinforcing the silencing of disruptive voices, nor solely relied upon, but understood as part of the contingent politics of the image. These images seem to unsettle any easy claim of oppression in any act of seeing or visual representation. Visual representation of landscape and the body can clearly operate to reinforce gender, sexual as well as class and racial oppression, but, as I will suggest through these two examples, can do otherwise.

Using a concept of contextual politics entails being attentive to the initial time and place of production; the relationship between this location and the places figured within the image and to the endlessly variable arenas in which the image may be consumed. When considering the circuit of meaning production from producer to text to consumer, it also entails thinking through ideas of scale, from shared contexts to individual responses. Different issues will emerge in addressing the politics of an image at every scale. Rather than interpretation being framed by a concentration on male visual pleasure in images of landscape or the female body, interpretation needs to acknowledge the multiplication of contexts and relations of reception and the diverse and contingent meanings and effects of representations. Acknowledging the contingent politics of representation may avoid the danger of general feminist or antiracist critiques and condemnation of forms of representation being recuperated by the Right to censor representations of resistance and difference (Butler, 1990; Williams, 1993). The 'messy' ambiguity of the meaning of an image, produced through the 'structures of feeling' that occur across multiple relations between authors, texts and readers (Mercer, 1994) make representations themselves and attempts to judge their political effects both more undecidable and more effective.

Representing the Male Body as Landscape

Abroad (Fig. 1) is part of contemporary visual and textual assertion of and investigation of women's images of men and of the relationship between sexuality and representation (Kent & Morreau, 1985; Caught Looking Inc., 1992; Gibson & Gibson, 1993; Salaman, 1994; Smyth, 1994). The photograph was exhibited in and was used as the cover illustration for the catalogue accompanying the Arts Council funded exhibition 'What she wants: women artists look at men' which toured Britain and Northern Ireland in, 1994. The exhibition was a response to the pornography debate and the largely unsatisfactory commercial production of images of male bodies for women. While Diane Bayliss's art work was produced in a context in Britain in which representation of the body has been politicised through pornography debates, the AIDS crisis and the expression of sexual diversity, *Inis t'Oirr/Aran Dance* by Pauline Cummins deals with craft production, sexuality, and cultural traditions in Ireland. The installation is a series of slowly changing slide images of a male torso and images of wool, knitting patterns and an Aran jumper, accompanied by the voice of the artist over its 10-minute duration. The artist narrates her thoughts about knitting, story-telling, landscape and the movement of her hands over the male body while images of the body, first dressed in the Aran jumper, then naked, slowly evolve on the screen. This installation was made and exhibited in the mid-1980s in Southern Ireland when women's reproductive rights were being debated and women were attempting to negotiate personal identity with traditional Catholic and nationalist ideas of Irish femininity (Cummins *et al.*, 1987; Smyth, 1988).

Diane Bayliss's *Abroad* and Pauline Cummins's *Inis t'Oirr/Aran Dance* thus raise issues of pleasure, power and representation and questions of what it means to depict the male body as landscape. Does it simply replicate masculinist vision? Or does it free looking from its automatic implication in patriarchal and phallocentric representations and power? Can it suggest not only other forms of feminine sexuality but other versions of masculine sexuality as passive and desired rather than active and desiring? Can it point to ways of acknowledging the erotics, pleasure and power of landscape imagery while relating to landscape in less oppressive ways? Again the answer is geographically and historically contextual. There is no inherently bad or good looking. This article discusses different ways of considering gender and visual pleasure and their implications for understanding landscape and sexuality within and outside cultural geography. In doing so the different ways that these two images relate to landscape should not be collapsed or simply opposed. *Abroad* and *Inis t'Oirr/Aran Dance* draw on two different but connected understandings of the term 'landscape'. *Abroad* adopts with irony the pictorial conventions of picturesque landscape imagery. Its rectangular format, framing devices, distant horizon and intersecting, receding topographical planes within the image evoke the conventions of late eighteenth and nineteenth-century Western landscape painting. Alternatively *Inis t'Oirr/Aran Dance* refers to another understanding of landscape. Here the male body is represented in conjunction with, and in parts of the installation is indistinct from, a set of meanings and symbols derived from a whole series of written and visual representation of a region, which constructed the west of Ireland as a highly significant and contested national landscape. The installation thus relates to a set of representations which constitute the meaning of the landscape or imaginative geography of the west of Ireland, while *Abroad* engages with a convention of landscape depiction. The relationship between the image of the male body and landscape is also different in both. The association between body and landscape is less overtly defined in *Inis t'Oirr/Aran Dance* than in *Abroad* but the associations work on a number of levels: the clothed body evoking the rural landscape in the patterns of the jumper, the jumper standing in for the body, the knitting evoking the west of Ireland, the male body as fleshy terrain. Yet both play upon ideas of closeness and distance: in *Abroad* the closeness implied in seeing the body as a body, and the distance required to see it as a landscape, and in *Inis t'Oirr/Aran Dance* the intimate tactile presence of the body and implicit presence of desire and sexual pleasure, in spaces and activities defined through ideas of domestic femininity, and in attachment to a national landscape and cultural traditions.

These images of the male body also provide a means to address the representation of men and especially those representations produced by women, since to continue to study the representation of women without considering the representation of men elides the cultural investment expended in display of the male and reinforces the apparent effacement of masculinity as a social construction (Cohen & Hark, 1993). Both images of women and gay men are produced against the structuring norm of heterosexual masculinity (Neale, 1982). Only recently has attention turned to the discussion and analysis of the process of constructing and maintaining this norm (Cook, 1982; Dyer, 1982, 1993; Neale, 1982, 1993; O'Pray, 1982). Film theory has mostly:

> equated the masculinity of the male subject with activity, voyeurism, sadism, fetishism, and story, and the femininity of the female subject passivity, exhibitionism, masochism, narcissism, and spectacle. In this scheme of homologous differences the power stability, and wholeness of masculine subjectivity at the

> expense of femininity seem all too axiomatic and, thus, universal and uncontestable. (Cohen & Hark, 1993, p. 2)

Questioning rigid gendered positions in film means that men can be considered in relation to visual spectacle, masochism, passivity, and masquerade normally linked to the feminine, rather than reinforcing the 'unperturbed monolithic masculinity produced by a decontextualised psychoanalysis' (Cohen & Hark, 1993, p. 3). The male body can be investigated as it signifies gendered, class, racial and generational differences. Images of men represent them 'overtly performing their gender, in neurotic relationship to it, or seeking alternatives to masculinity as their culture defines it'. Yet even dominant images of men only precariously produce and reproduce masculinity. They 'rarely efface the disturbances and slippages that result from putting men on screen' (Cohen & Hark, 1993, p. 3). While the construction of hegemonic masculinity is precarious, so too is the production of alternative versions of masculinity. However, alternative images of men have potential to disrupt the unquestioned equation of vision with power and the binaries of 'normal' gender and sexual identity. The possibility of women's images of the male body, and the male body as landscape, which are not simply an inversion of power, may lead to the possibility of retrieving images, the study of landscape and visual pleasure from inevitable entrapment within the logics of the 'male gaze'. As Salaman suggests,

> [when] women artists attempt to represent men as the beautiful object they will not effect a social reversal of power, but they can affect the scopic regimes, and this can translate into a new knowledge, new 'abstract power' in the debates of pleasure and representation. [...] both begin to articulate ways in which a modernized female observer can look at the male body and play with the legacies of his vision without having to occlude her desiring body and its variety of identifications. (Salaman, 1994, pp. 23–24)

While artistic intention cannot fix the meaning of these images, my interpretation of them derives in part from the way they relate to the intentions of the exhibition organisers and artists. Attending to artistic intention as part of this context of meaning production allows the specific assertion of women's pleasure as a political strategy of these artists to remain highlighted. This does not mean that they cannot also be read as offering pleasures which escape the binaries of gay or straight, masculine or feminine. Reading these images as objects of women's sexual pleasure is not to suggest that they cannot function as a source of men's pleasure or to fix visual pleasure in the male or female body to tightly defined sexual identities. It seems limiting to suggest that women's visual pleasure in the male body is necessarily heterosexual or in the female body is only lesbian. If sexuality cannot be neatly attached to overt gender identity, then by extension neither can pleasure in looking at male or female bodies be simply harnessed to polarised sexual identities. *Inis t'Oirr/Aran Dance* apparently focuses on female heterosexuality, yet this does not mean its pleasures are limited to heterosexual women identifying with the artist. It may offer positions of identification for men and women, gay or straight. This is not to undermine gay and lesbian claims to difference but is an understanding which has emerged through the way in which gay and lesbian representations can work to disrupt binary understandings of gender and sexuality (Harwood *et al.*, 1993). Attention to the text—audience relations entails thinking through the possible meanings and effects of representations for individuals and social groups without suggesting that images will automatically be read in certain ways, or that the possibility of multiple readings and complex viewing practices means there is 'equality and freedom in the regime of representation' (Dyer, 1993, p.2) The concentration here on one author–text–audience relation, that between images of men, produced by women and read by a woman, is strategic. The readings of *Abroad* and *Inis t'Oirr/Aran Dance* which follow suggest the way these images may work in their specific contexts to produce alternative and non-traditional understandings of male and female sexuality through reworking traditional representational formats and cultural symbolism. While this reading can never be final or universal, neither does it rest only in a personal and internalised subjectivity. Rather than deliver a statement of personal positionality and somehow fix myself categorically and my reading as a direct function of this, it is part of my argument that positionality in looking cannot be easily defined. The disputed meaning of these representations and the shifts in spectatorship of an individual viewer point to the context-bound nature of both the politics of representation and social subjectivity.

Landscape and the Problem of Visual Pleasure

Considerations of gender, visual images of women and spectatorship were stimulated in the 1970s by Laura Mulvey's influential article which linked psychoanalytic perspectives to a feminist analysis of images of women within mainstream film (Mulvey, 1975; Marcus, 1991). According to her analysis, women in film have a passive role as objects of male viewing. This understanding of the construction of femininity through women's to-be-looked-at-ness, means that female viewers can either identify with the object of the gaze or adopt a masculinist viewing position. Gillian Rose has powerfully combined this psychoanalytic perspective with a critique of the equation of women with nature and the gendering of knowledge production within geographical epistemology. She suggests that within cultural geography, Nature, landscape, femininity and the unknown are figured as objects of masculine desire and fear, and thus linked to senses of loss and lack across which male subjectivities are produced and managed according to psychoanalysis. For Gillian Rose, studies within the landscape tradition in British cultural geography of the class relations constructed and reproduced with nineteenth-century English, rural landscapes, and virtual silence on issues of gender, signifies not only the omission of gender issues and critical engagement with patriarchal power but a deep ambivalence regarding the object of study. The ambivalent pleasures of landscape within cultural geography draw 'not only on a complex discursive transcoding between Woman and Nature, but also on a specific masculine way of seeing'. For Rose both fieldwork and images of landscape are structured by the 'gendered logic of the gaze' within the dominant visual regime of white, heterosexual, masculinism. This 'masculine gaze':

> sees a feminine body which requires interpreting by the cultured knowledgeable look; something to own, and something to give pleasure. The same sense of visual power as well as pleasure is at work as the eye traverses both field and flesh: the masculine gaze is of knowledge and desire. (Rose, 1993, pp. 98–99)

The pleasure and emotive force which landscapes may provide (Daniels, 1989), according to Rose, is an ambivalent pleasure which disrupts the construction of modern masculinity as scientific, rational and distanced. The pleasure of the text disrupts distanced, rational and scientific knowledge, producing a 'tense oscillation between knowledge and pleasure' (Rose, 1993, p. 101). For her the male gaze moves between scientific viewing and a sexualised aesthetics, between voyeuristic distance and power and narcissistic identification with the image. Images of women or nature suspend fear of lack while distance from them supports masculine self-identification. Thus men's relationship to landscape for Gillian Rose is mediated by desire for integration and fear of engulfment (Kolodny, 1975; Rose, 1993, p. 105).

Yet in drawing upon versions of psychoanalytic theory which problematically suggest a universal schema of development of self and gender identity and rely on ideas of masculine pleasure, desire and seduction, this critique may inadvertently reproduce the tropes of 'normal' sexuality. Desire is based on lack of the phallus rather than on any sense of productive multifarious desire (Kerr & Quintanales, 1982; Kelly, 1984; Best & Kellner, 1991; Besley, 1993; Spargo & Botting, 1993). Arguments based upon Mulvey's initial critique have tended to ignore the issue of women's spectatorship and reproduce the dominance of heterosexism in understanding gender and visual pleasure. If the specific discourses of landscape within geography have led to the conclusion that valid pleasure in landscape 'is for straight men's eyes only' (Rose, 1993, p. 99), lesbian and gay and straight women's images and accounts of looking point to the reclamation of visual pleasure despite its historical connections with heterosexual male power. Important issues of consent, trust and censorship have been brought to prominence by sado-masochism and its imagery (Butler, 1990; Henderson, 1992). My re-examination of the pleasures and politics of the visual responds to the suggestion by Gillian Rose that 'interpretation of geographical knowledge needs the specificities of historical geography often missing from psychoanalytic accounts of the visual' (Rose, 1992, p. 18). As she suggests, the power relations of pleasures in landscape and looking can only be addressed through specific images and contexts. To accept the way her critique works for a particular moment in the history of cultural geography, but to point to its omissions and risks of generalisation, in effect, matches her call for historical and spatial contextualisation of representation. It is this method of contextual interpretation that is applied in discussion of *Abroad* and *Inis t'Oirr/Aran Dance*.

Susan Ford has also responded to the tension between working within a 'geographical tradition formulated through visual means' and feminist critiques of the 'male gaze' and sought to 'explore a tenable viewpoint or viewpoints from where a feminist geographer might look upon a landscape' (Ford, 1991, p. 151). She argues that while vision, distance and power may have been central to the construction of the modern masculine subject, to relegate women to the other senses reinforces the dualisms between mind/body, culture/nature and masculine/feminine. In contrast to the conditions of distance, objectification and control within classical landscape art, acknowledging emotion and celebrating landscapes of intimacy, she suggests, may offer a means to reconcile feminist critiques with personal investment in landscapes. This feminist stance posits a 'transvestite gaze' which can adopt a distanced and thereby masculinist position and an intimate and thereby feminist position (Ford, 1991, p. 153). Yet this movement between a masculinist and feminist way of seeing tends to polarise both forms of looking and tie them to gender identities. It implies that women look intimately at the local, small scale and places tied to bodily experience, maintenance and reproduction. Their look is custodian, reverent and gentle. However socialised this vision may be, to claim a feminist position which can move between this and a masculinist violent gaze tends to reinforce ideas of deep gender differences and gendered difference in looking. I want to consider here the possibilities for multiple and mobile identifications with and ways of seeing landscape, and to unfix both versions of gender and sexual identity from ideas of masculine activity and the 'male gaze' and feminine passivity and reception of this gaze.

In reclaiming pleasure in vision and landscape for a feminist cultural geography, rather than turning to ideas of a feminine and body-centred identification with nature, it may be productive to think through a range of potential identifications with landscape or nature including those of indifference or disinterest. In recognising the constructed and therefore unstable basis of both ideas of identity and landscape, feminists could employ non-essential ideas of landscape, place and nature without being tied to the choice between masculine distance from and feminine closeness to nature. Avoiding this dualism may be more disruptive and enabling than deciding between them. Self-conscious play with identifications and positions of dominance or submission, distance or closeness frees women from the pressure to adhere to ideas of women as both nature and body-centred. Painted, filmed, and written images of women, in indifferent, reverent or dominating relations to landscape, point to varied and non-essentialist, feminine subject positions in relation to nature or landscape, which do rely on the discourses of eco-feminism for their articulation. By suggesting, for example, the validity of a feminine heroic approach to landscape through a distanced and elevated viewing position, the power and naturalness of the masculine heroic is subverted. This is not a liberal feminist wish to claim male privilege but an attempt to pry apart the rigid equations of all looking with masculinity, and thus masculinist power and oppression (Grosz, 1992a). Women looking and representations of women looking may deconstruct essentialist notions of place and gender identity. Yet, this looking operates within and across subjectivities defined through class, race, sexuality and geography, rather than resting on an impossibly asocial concept of nature or landscape that offers a universal source of identification for all women. Thus women's varied relations to these objects of vision are never innocent. Calls for attention to women as active viewers have to confront the politics of these varied visual encounters.

Women Looking

Recent research in both art history and cultural studies is producing a more complex account of the production and consumption of visual imagery than the general contention that images of women are produced for the pleasuring and empowering of men. Male anxieties about women's spectatorship and active production of visual imagery in the history of Western visual culture reveal that despite the role of visual imagery in the construction of femininity, women have also been understood as viewers and potential artistic producers, as well as objects of vision. Within this history considerable efforts have been made to regulate images of women, images by women and women's vision in order to maintain normative gender identities and relations. Studies of nineteenth-century Britain and France show the way in which looking by both men and women was central to the construction of gender, class and national identities. Art exhibitions, museums and 'great exhibitions' produced images of the gendered self and images of the class, gender and racial 'others' through which national, gender and class identities were constructed and made visible, often with different sections aimed at working- and middle-class women and the production of appropriate femininities (Greenhalgh, 1985; Coombes, 1988; Breckenridge, 1989; Mitchell, 1989). The street of the 'modern' city was itself an arena for class- and gender-specific relations of looking and their power relations, but a space also in which these social relations could be disrupted by woman's returning look (Pollock, 1988, 1992, 1995; Bershard, 1992). In relation to the production of imagery, it was considered improper for the male nude to be a women's life class model and ruled out within the art academies (Garb, 1993). Thus the centrality of figure painting to the elevated status of history painting before its demise in the nineteenth century, had an exclusionary force which protected a masculinity based on men as cultural producers of representations of men's or women's bodies. Similarly, from the mid-nineteenth century in Britain the suitability of landscape as a place for middle-class women to paint in terms of their presence in the countryside, or as a subject to depict was seriously debated (Cherry, 1993, p. 118). Concerns over women's vision did not end in attempts at its regulation but extended to the production of images for women's viewing and thereby the construction of class-specific versions of femininity. These often contained images of women viewing children, loved ones, domestic interiors and gardens, thus constituting these as the appropriate objects of a feminine gaze (Cherry, 1993, p. 118). Within the production of popular painting, by the late nineteenth century the function of images of women was less didactic as their primary function returned to providing visual pleasure for bourgeois men, to represent difference and to signify and secure this masculinity. But as Deborah Cherry suggests, this did not reduce the importance of women's vision as 'femininity was constituted as the locus of the network of *women's* gazes: pleasuring looks and appraising glances were directed to women's visual representations of femininity, whether these were constituted in their dress, appearance and "personal style" and/or their high cultural images of women' (Cherry, 1993, p. 119).

This concern regarding women as viewers indicates the potency of women's active viewing. Asserting women's visual pleasure resists both the idea of women as passive objects of the male gaze and hegemonic versions of what is an appropriate feminine viewing position and objects of view (Betterton, 1985; Doane, 1981, 1982, 1988; Moore, 1988). To consider issues of female visual pleasure is not to suggest a way of seeing shared by all women. As Deborah Cherry writes, nineteenth-century women 'were located in and took up viewing positions in a social formation in which femininity was not monolithic but multiply shaped and fractured by colliding determinants such as age, dis/ability, religion, politics, sexuality and location as much as race or class'. Her understanding of nineteenth-century looking as 'a social and cultural activity framed by and taking place within the broader field of power relations which were not polarised but a tense and unstable web of interrelated and productive forces' remains valid for the late twentieth century (Cherry, 1993, p. 116).

Addressing more recent and contemporary images, studies of audiences of a range of genres in Hollywood film from fantasy to film noir point to both constantly shifting and never finally settled identifications and subjectivities in audience responses, which do not conform to the dominant and expected gender identifications. These studies discuss the shifting identifications that often structure an individual response and how apparently limiting gender roles may inspire members of the audience rather than offend (Kaplan, 1978; Dyer, 1993, p. 2). This research indicates the complex relationships between the psychic subject produced through the image and the social subject as audience and consumer of visual images, despite the problematic separation of the psychic and social they seem to imply (Stacey, 1994). Recent studies of mainstream film as well as pornography suggest that looking does not produce static positions of identification, distance, voyeurism, narcissism or fetishism but movement between these possible spectator positions for women and men. The understanding of consumption which these studies use is not one of simple assimilation of the imagery by the viewer but of active manipulation of its components and coding of its meaning (Wicke, 1993, p. 70). This means that a range of subject positions are available to the viewer which are not simply determined by an inherent psycho-sexual structure in the individual. These terms from psychoanalysis can remain useful in considering visual representation if freed from a universalising schema of the development of gender identity based in and reproducing early twentieth-century ideas of femininity, and if the terms describe positions and relations to the viewed which can be adopted by women and men (Williams, 1994a).

Thus to adopt landscape as a focus of research means neither that it can be detached from its history of use within cultural and social domination nor that to do so entails the simple masochistic replication of a coherent and singular male gaze or narcissistic identification with the objectified and feminised object. Rather than argue that the politics of representation or visual pleasure can be assessed by reference to a male or female gaze, it is more useful to think of a multiplicity of shifting viewing positions, gazes or ways of seeing. Thus the discussion of the two art works below, integrates two understandings of meaning in visual representation: established traditions of representing the body and landscape which are linked to but not limited by their shared meaning in particular geographical and historical contexts, and the possible responses by audiences of individuals in a variety of contexts. Attention to context, then, is mobilised by considering scale from general context of production, display and collective meaning to individual interpretation.

Abroad

The context of interpretation here includes the exhibition 'What she wants', which explicitly addressed female desire and visual pleasure. The images suggested alternative responses to a question asked and answered inadequately, from Freud to the contemporary media industry (Pollock, 1995, pp. 11–12). In the words of the catalogue:

> 'What she wants' establishes a site for women to take their pleasure seriously, eschewing market demands for images replete with fake seductiveness, which fail to 'clitillate'. The show attempts to chart current changing masculinities and situates 'want' back into the discourse of straight feminism, beyond the tortured defensiveness of anti-porn feminism and the *laissez faire* transgression of some pro sex representations. Even a couple of years ago such a project would have been constrained by women's fears of colluding with an oppressive system of looking, of indulging in a luxury feminism could ill afford, or of trying to imitate the clichéd codes of porn in a didactic demonstration of revenge. (Smyth, 1994)

Abroad has thus been produced in a context of the politicisation of sexual identity, and the situation of diverse, highly visible and highly contested representations of sexuality (Harwood *et al.*, 1993). By figuring the male body as a landscape, the image and its title draw on an exotic and erotic iconography of landscape, and the historical motif of equating sexual and geographical travel and exploration. The title implies being away from the familiar and known. *Abroad* may be about visual pleasures which are different or foreign, but as with imaginative geographies, also deeply known and present in the familiar. This is an image of the foreign and of the unknown, not because of the absence of women's sexual and visual enjoyment of men but because images of the male body by women until recently have been virtually unknown and rarely seen and women's visual pleasure rarely acknowledged (Walters, 1978; Saunders, 1989). While the title may imply a representational practice outside existing representational systems, importantly *Abroad* is neither fixed within nor escapes from traditions of depicting the body or landscape, but is powerful because it plays upon existing traditions.

The body in the photograph is aestheticised and sexualised through the landscape format. The spatial organisation of the image creates a vanishing point and visual focus centred on the genitals. The body is truncated in order for the allusion to landscape to work and thus there is no returning gaze. While truncation of the body may be criticised as objectifying the man, it also avoids a submissive expression which could reinforce objectification or an aggressively sexual returning look which could make the female viewer again an object of his desire. More significantly the faceless body may allow the representation of female sexual and aesthetic enjoyment of an anonymous male body rather than love of a known man within conventional constructions of femininity. It is both an intimate image and distanced since the viewing position implied is both one of closeness to the body and distance from the landscape. To see the image as a body entails a viewing position whose closeness to the body does not allow the kind of distance and elevation associated with landscape representation and with power. It is this position that allows the body as a landscape image to be composed according to picturesque conventions of framing, receding horizontal topographic planes and central vanishing point. The movement between the kind of spatial ordering and imaginary viewing positions that are required in seeing the image as a body or landscape make both the conventions of landscape and the materiality of the body apparent.

The male body composed according to these landscape conventions is far from the sculptural traditions of representing the male body. The body in *Abroad* is neither marble-like nor tautly poised for action, nor representative of an 'ideal' muscular masculinity. Making the male body a landscape in this way, points to the arbitrary and constructed authority of figuring the female body as nature. It is as easy in this instance to find curves in the male body to correspond to apparent rounded forms in the natural landscape. This frees the body–landscape metaphor from any essential equation between nature and femininity. The politics of imaging either the male or female body as landscape are always contextual, and thus analysis of these issues needs to address the social specificity of the viewer and viewed and with knowledge of the historical alliance of colonialism and representation in construction of racialised and sexualised bodies (Young, 1994) [3]. This awareness does not delegitimate the strategic or pleasurable representation of the male body as landscape in women's images of men. *Abroad* also allows alternative conceptions of masculinity and male sexuality.

Within cultural representations male sexuality is predominantly centred on the genital and equated especially with the penis. Even if it is not directly shown, the penis is the symbol of an aggressive, active male sexuality, figured as a weapon, gun or knife (Dyer, 1993). Ideas of male sexuality as the result of an autonomous drive originating in the penis form the basis of representations of sex as violence. The uncontrollability of this drive then absolves men of the responsibility for their actions (Lehman, 1988). Hegemonic images of insistent aggressive male sexuality put women in their place as objects of vision to be acted upon but also increase the difficulty for men attempting to subvert dominant ideas of masculinity and male superiority.

> Such representations help to preserve the existing power relations of men over women by translating them into sexual relations, rendered both as biologically given and as a source of masculine pleasure. What is perhaps more surprising is that these images should, by and large, be so unattractive, so straight and narrow, so dreary. Men too are fixed in place by this imagery, and if theirs is a place of superiority it is none the less a cramped, sordid, compulsive little place with its hard edged contours and one off climaxes [...]. Here and there we do get a vision of a sexuality that is not nasty and brutish, silly and pathetic, but varied, sensuous, languorous, warm and welcome. We need to see it more often in order to live it more fully. (Dyer, 1993, p. 121)

Unlike mainstream images of men the body in *Abroad* is static, passive, inert and unthreatening (Dyer, 1982). Richard Dyer's own writing and Diane Baylis's *Abroad* are Dyer's 'exceptional cases':

> Male genitals are fragile, squashy, delicate things; even when erect, the penis is spongy, seldom straight, and rounded at the tip, while the testicles are imperfect spheres, always vulnerable, never still. There are very exceptional cases where something of the exquisiteness and softness of the male genitals is symbolised. (Dyer, 1993, p. 112)

Images of the variable form and softness of the genitals displace the representation of the penis as either weapon or anatomical joke (Dyer, 1993). In *Abroad* the penis is flaccid and as such it does not reproduce the tropes of 'normal' masculinity. In this landscape the penis does not break the skyline of the body's gentle contours. If male power is dependent upon the construction of a masculinity which is in turn a product of representation and understanding of the male body as inviolable, impenetrable and whose boundaries are hard-edged and distinct, these representations of the penis as flaccid may disrupt the stability of unequal social relations based on normative and essentialist understandings of femininity and masculinity (Schor, 1988). They also do not easily facilitate forms of criticism based on the presence or absence of the phallus, which however much defined as a symbol of male power, still ultimately depend upon theories of genital difference in which the fear of loss or lack of the penis, structures the formation of male and female subjectivity (Macey, 1992; Grosz, 1992b).

Women's images of men are not automatically unproblematic or redeemed by female sensitivity or goodness: as Linda Williams writes, '[t]hough it is certainly significant that women image-makers are negotiating the line between the erotic and the pornographic, and that male bodies are balancing on that line, this reversal alone may not lead to any startling liberations from the presumed objectification of an all-controlling, and increasingly dubious, "male gaze" ' (Williams, 1994b, p. 8). Yet this one, in this interpretative context, provides alternative versions of male sexuality and feeds women's 'greedy' and deprived sexual vision, offering alternative versions of both male and female sexuality, gay or straight (Williams, 1994b, p. 11). Looking is not always necessarily powerful and oppressive, nor is distanced vision always an appropriation. As studies in post-colonialism and sexual politics have shown, the experience and exploitation of the position of desired object can be empowering and pleasurable. To picture the male body as landscape is not simply to invert customary gendered positions within a scopic regime but to suggest other pleasures and subject positions which are not determined by this regime.

Inis t'Oirr/Aran Dance

Though the metaphor of body as landscape is employed less directly, *Inis t'Oirr/Aran Dance* is more explicit in its expression of female heterosexual desire than *Abroad*. Its references to specific traditions of Irish rural, domestic femininity are central to its subversive force and political strategy but do not limit its focus to women's heterosexual desire. The artist's voice and the images of the body offer two roles, that of being touched or touching, but ones whose relationship may be imagined as same sex as well the active feminine pleasure in the male body most apparent in the installation. Within the specific context of its exhibition within Ireland, Cummins radicalises rather than rejects traditional customs, crafts and landscapes of the nation to express an active and self-determined sexuality. Since the late 1970s in Ireland, unreflective adoption of conventional landscape painting and the iconography of the west of Ireland have come under criticism because of the way they have been used to support notions of an inherent Irish identity and artistic tradition in the visual arts dependent on ideas of racial essentialism (Brett, 1986, 1989; Duddy, 1987; Fowler, 1990; Stinton Cosgrove, 1990). Despite this, Cummins expresses possibilities for alternative understandings of sexuality, gender and these national traditions themselves through, rather than against, the symbols of cultural identity she enlists.

The installation explores the products and results of female desire through describing the Aran jumper and male body. Both are constructed through fantasy and women's craft activity. In addressing her attention to the dressing and undressing of a male body in *Inis t'Oirr/Aran Dance*, Pauline Cummins raises important questions of the politics of representation, desire and visual pleasure for women (Cummins, 1987). The image of the Aran sweater, symbolic of traditional Irish life in the west of Ireland, evocative of the deep significance of the Aran Islands in Irish cultural nationalism, is used in her installation to draw attention to women's activity in knitting and in sexual enjoyment of the male body. Rather than rejecting this craft tradition because of its possible associations with reactionary views of the role of women, Pauline Cummins undermines the customary understanding of the domestic itself. By linking the domestic to autonomous and active female sexuality through an emblem of traditional Irish rural life, she prompts reconceptualisation of the meaning of traditional crafts, the domestic and the feminine. To imagine that women knitting in past and contemporary Ireland could and can be absorbed in fantasies of sexual pleasure is radically disruptive of traditional versions of Irish femininity. The artist describes the sources and structure of the video.

> My interest was aroused by an old Aran sweater I saw, on a tailor's dummy. It clung so tightly, it was like no Aran sweater I'd seen before. It was sensuous and strong. It was voluptuous and warm, gentle and soft. It was very male. It was made by a woman, for a man—it was sexual. The first part of the tape is like an outer layer, a description of knitting, a suggestion of the fantasy imposed on the man by the woman. The second part is a love song—the enjoyment of the male body, by the female. A renaming—an inquiry into what is opposite—and how one creates the 'other'. (Cummins n.d.)

The first part of the sequence begins with images of balls of wool, then the jumper they produce. The second shows the torso of a man wearing it and then the body beneath. She begins her script with the words:

> Begin at the beginning.
> In a sense knitting is like drawing with a long piece of wool, its finger weaving, from the fleece to a warm pullover, a jumper, a jacket, a waterproof coat.
> And where do the patterns come from? From the woman's head?
> 'Did you make that out of your own head now?'

In the installation Pauline Cummins deals with the origin of the garments in images of the landscape and women's imagining of the male body. Her installation weaves the body and landscape together as sources for the women's craft. In taking women's knitting of the traditional Aran jumper as her subject she evokes the traditional image of Irish women as domestic reproducers of the crafts, traditions, language and population of the nation. Knitting often connotes ideas of passivity, privacy and maternal care in clothing the family. But Pauline Cummins emphasises the tactile nature of knitting and the fantasies of touching the male body of its future wearer woven into the garment, a practice that knits thought and knits desire. Her suggestion that knitting is a form of drawing breaks down the distinction between art and craft, active and passive, which has devalued women's cultural production. The knitted is represented as the product of women's active, sexual narratives of desire; it tells stories of thought and touch.

> And if knitting is like drawing with a long piece of wool, women have been drawing for quite some time now, weaving their dreams and fantasies into garments, for their children, for themselves, for their menfolk.
> I'll spin you a yarn.
> I'll weave you a tale.
> And as the lines weave in and out of each other they weave a pattern.

The artist repeats her question of where the complex, geometrical patterns and winding lines in the knitting pattern come from. The answer for her is in the creativity of the women and the way in which the rhythms of soil and water and divisions of land in the west of Ireland are recreated by them in the patterns they knit.

> Ploughed fields.
> Waving sands.
> Weaving water.
> And what are the berries? They look like nipples to me, but they're like berries too or little houses set in walled fields, with the sea on either side.
> Aran...

The patterns of the Aran jumper echo the morphology of the landscape and hug the contours of the male body it covers. For the artist, the knitting is about body and land, terrains of houses and fields as well as the terrain and features of the body. This association and slippage between the form of the male body, the texture of the garment and the topography of the Irish landscape, inverts and subverts the customary metaphor of the landscape as feminine and female body as a landscape. *Inis t'Oirr* translates as golden island and for the artist the male body is a golden, island landscape. Her appropriation of the body/landscape metaphor, like Nuala Ní Dhomhnaill's poem *Oileán/Island*, which figures the male body as a landscape, is profoundly political in a historical context of the feminisation of the Irish nation and landscape. The installation takes the nationalist and often patriarchal symbolism of the west of Ireland landscape to produce a feminist statement which brings together attachment to these specific cultural traditions and a radical sexual politics. Though the body in *Oileán* is not named as male, the poet intended the piece to be understood as about a male body as landscape against the poetic traditions of representing Ireland as a female figure [4].

Oileán

Oileán is ea do chorp
i lár na mara móire.
Tá do ghéaga spréite ar bhraillín
gléigeal os farraige faoleán.

Island

Your body an island
in the great ocean.
Your limbs spread
on a bright sheet
over a sea of gulls.

(Ní Dhomhnaill, 1988)

Rather than reject local and national cultures in favour of an international feminist alliance, Cummins and Ní Dhomhnaill return to these national cultural traditions in order to disrupt the meanings these traditions have been used to construct and support. They indicate a form of national identity which can retain attachment to national traditions without accepting the essentialist and uncorrupted versions of national, gender and sexual identity they imply. As I discussed in relation to *Abroad*, this use of the body/landscape metaphor does not in this case entail the adoption of a masculinist viewing position but destabilises the social and representational trope of the active masculine artist or viewer and the passive feminine object of representation or receiver of the gaze. For the woman artist to depict the male body as object of desire disrupts gendered notions of visual pleasure and representational practice which have governed art criticism. The suggestion that visual pleasure and representation is inherently exploitative and masculinist has led many artists informed by feminism to eschew figurative art in favour of abstraction (Deepwell, 1989; Fortnum & Houghton, 1989). *Inis t'Oirr/Aran Dance* prefigures more recent returns to the negotiation of visual pleasure in representation of desire by women artists.

The second section of the installation suggests a recovery of women's active sexuality through uncovering the male body. In the sequence of frames from the clothed to naked body through the gradual removal of the jumper, the installation follows erotic conventions of undressing the body. In attempting to convey an alternative version of feminine, visual, sexual pleasure the artist draws on familiar tropes of Western erotic representation. She does so in the absence of a body of knowledge or an artistic tradition which recognises and describes female sexuality as a potent initiating force, rather than merely as a response to masculine desire. Her doing so suggests that it is not possible to find a language of representation that is outside and purified of existing pictorial traditions other than in untenable notions of a disembodied aesthetic originality or an asocial artistic source in the female body. This does not negate attempts to appropriate, renew and disrupt the meaning of representational traditions. Pauline Cummins's voice and the cultural referents in the installation locate it within the particular cultural and social context of contemporary Ireland. The costs and achievements of the installation and its particular sexual politics can only be judged in this specific context.

The body which is exposed from neck to pubic shadow, is shown with lines of knitting superimposed on it. These linear traces are of the woman's touch. She narrates the movements of her hands over his body and the enlarged chevrons of the knitting pattern. The body is thus both naked and clothed, exposed and protected by the pattern projected upon the body. The narrative describes the movement of her hands over him and to the genitals, erection and ejaculation. Her voice describes her uncovering of the male body and her tactile and visual pleasure.

> The hidden male body, buried, suppressed.
> Touch the hip, into the waist. Squeeze, Rub up.
> The back. The spine bending, extending, joining the shoulder, broad, wide, thick.
> Arms, hairy, sinuous, strength.
> Thighs, joining, apart. And the butterfly motif.

While producing a version of male sexual arousal which is climactic, her hand movements across his body do not reproduce the focus and limits of male sexuality defined exclusively through the genital. The narrative does not end in ejaculation but describes its aftermath. 'Spilling, subsiding, the swelling, shrinking, sinking, back to size, back to place, hidden, buried, always there'. Through her words the penis is brought back to size and back to place, which undermines images of the phallic. While the installation expresses a woman's enjoyment of the male body, in describing semen as 'hundreds and thousands' Cummins puns on the name for tiny coloured cake decorations and thus, through this woman-centred humour, avoids reifying the male body as object of reverent veneration, and locates her pleasure again within the space of the domestic and her social context rather than suggesting a universal feminine sexuality. She eschews the verbs and metaphors of conventional descriptions of male sexuality and the slow measured pace of her voice does not reproduce senses of urgency or violence in orgasm. The artist's insistence in her narrative that she is describing stimulation by hand rather than intercourse ('tipping the navel, pressed against the belly, *his* belly', 'Rubbing, pressing, raising, standing, ejaculating') emphasises that the man is brought to arousal by her touch, not by his sight of or action on her. That the orgasm is erotic rather than reproductive asserts the validity of sexual pleasure despite and against Catholic denigration of the body; that she is stimulator asserts the validity of active female sexuality against women's idealisation as mothers in Ireland.

Though focusing on a male body and suggesting a masculine sexuality which can be passive and gentle, *Inis t'Oirr/Aran Dance* does so, also, to provide a means of expressing a creative and active female sexuality through images of the male body. The suggestion that the body enjoyed is a known, loved body since the jumper has been knitted for him may connect women's sexuality to the love of a domestic partner. This, however, may make a too literal connection between the first and second part of the installation. That the body is unidentifiable produces a more radical representation of female sexuality than tying it to an acceptable monogamous stability. It is a male body, not necessarily limited to a known, loved individual, which is pleasurable as an object to touch and view. In representation of the sexualised adult body in commercial pornography or art it is often the face which is the point at which the politics of the image can be judged. It is the alluring or resistant expression of the face and the direction of the figure's gaze that often determines the visual relations possible in the representation of the erotic body and is the point of tension in the image. The face must not disrupt the tight and fragile alignment of viewer and viewed. The absence of a returning gaze in the installation again avoids the ever-present threat of the artist and viewer becoming again the object of his vision.

Unlike *Abroad*, this installation operates through two representational media and sensory registers: the spoken work and vision. It also draws heavily on ideas of touch and texture in describing the knitted garment and the male body. To a certain extent this mixing of media decentres vision. It could also suggest a version of feminine sensuality, sexuality and representational form which rejects vision in favour of touch and rejects a subjectivity structured through masculinist separation of subject and object in favour of loss of difference and individualism in a sensuous merging of both. Yet this reading would deny the very visibility of this video-based installation; it would entail ignoring her account of the visual pleasure that the sight of the jumper initially gave her; it would miss her insistent maintenance of distance in her explicit reference to *his* belly and *his* body. As she says, the installation is about 'how one creates the 'other' ', and the formation of a feminist subjectivity through ideas of difference and objectification. It is this difference, separateness and distance within the subject–object relations in the image which allows, produces and feeds desire (Wicke, 1993, p. 79). The image of the male body as landscape does not escape the connotations of objectification or pornographic representation. Yet neither term offers unproblematic criteria of judgement. While studies of pornography have moved from a simple understanding of pornography as always an objectification of women's bodies for men, to try to locate these images inside or outside of this category would mean supporting the class-based distinction between sexually explicit original art work as erotica and mass-produced sexual imagery as pornography. Resisting this distinction avoids the political danger of legitimating certain sexual pleasures and censoring others through definitions of what is or is not pornographic. It may be contentious to abandon objectification as a criterion of critical judgement and to suggest that objectification in representation is not always oppressive, but the images discussed here seem to suggest that a radical politics of visual pleasure may in some cases entail and endorse objectification. The politics of objectification in representing different bodies are not the same since the meaning of gendered, racialised and dis/abled bodies are linked to their history of representation with Western visual culture. The objectification of the male body in *Abroad* and *Inis t'Oirr/Aran Dance* does not simply produce new relations of power and oppression, but undermines the authority of concepts of beauty, sexuality, pleasure and social relations supported by traditional depictions of women and readings of them as sources of male heterosexual pleasure (Nead, 1992).

In *Inis t'Oirr/Aran Dance*, the images on the video do not explicitly show the genitals and ejaculation her voice describes. To describe the ejaculation verbally rather than picture it may be a response to the threat of artistic censorship and an attempt to negotiate the representation of the male body and female sexuality, with distance from phallocentricism. Yet to represent a sexualised male body by a woman artist is significant within the context of contemporary Ireland. The installation deals innovatively and radically with subjects which are problematic in Ireland, within cultural representation and social relations. It untangles knitting from implications of domestic passivity, it engages with images of the west of Ireland despite the way it has been delegitimised within avant-garde Irish art as an artistic subject because of its connotations of unconsidered nationalist nostalgia, and it represents Irish women as sexually active and desiring both now and in the past. Most importantly it weaves together images of the male body and landscape through the jumper. The landscape is sexualised and eroticised through its conjunction with the male body. This masculinising of the landscape thus denaturalises the femininity of the body/landscape metaphor. *Inis t'Oirr/Aran Dance* suggests women's activity as viewers of the body and landscape, recovering craft, female sexuality and landscape for non-essentialist understandings of each. The radical potential of the installation is achieved through using rather than rejecting this culturally specific landscape symbolism.

Conclusion

This paper has not intended to dismiss or diminish the important way in which feminist geography has impacted upon and is contributing to the study of landscape representation in cultural geography. Nor is it to suggest that representations of landscape or of the body can be easily detached from a set of limiting and oppressive meanings and subject positions. The history of Western representation, knowledge production and its effects cannot be erased. Yet the production and circulation of images like *Abroad* and *Inis t'Oirr/Aran Dance* are disruptive of dominant representational systems and gendered and sexual subject positions and offer pleasures which are not limited to male heterosexual appraisal of women. Despite the way in which the metaphor of the body–land has been employed to justify both approaches to women and the environment and to legitimate colonisation, this discussion of the politics of visual pleasure suggests that with regard to both the body and landscape, we need to look again. As the work of many contemporary women artists illustrates, neither landscape nor looking is redundant or inherently oppressive. Neither can their approaches to the relationship between the body and landscape be generalised; the land can be entered into, the body can be earth, landscape can be a motherland, the motherland can be deconstructed, history evident in the landscape can be mourned through ideas of the female body and emotion, the female body can be reclaimed as a landscape or the male body be revisioned as land to change conceptions of both male and female sexuality [5]. *Abroad* and *Inis t'Oirr/Aran Dance* work with and radically deploy traditions of landscape representation and landscape symbolism rather than reject them. These images work precisely because they work through the representational traditions and cultural meanings they draw on and disrupt. Certain forms of visual representation may support patriarchal power relations, but looking is never only or just masculine. To view body as land or land as body has no essential meaning, yet neither can it ever be innocent. Its politics are always contextual; there are different kinds of looking.

Acknowledgements

This paper owes much to the stimulating work of Gillian Rose and Susan Ford and to discussions with Mandy Morris. I am also grateful for comments made by Matthew Gandy, Steve Pile, Miles Ogborn, two anonymous referees and audiences at presentations of these ideas in Oxford and Exeter in 1994 and Chicago, 1995. I am especially indebted to Pauline Cummins and Diane Baylis for sharing their work.

NOTES

[1] Diane Bayliss was born in Canada in 1943 and now lives and works in southern England (Salaman, 1994).

[2] Pauline Cummins was born in Dublin in 1949 and is working in Ireland.

[3] If race has not been a central concern in my reading of these two images, it is because I have felt that neither of them can definitely be fixed as white male bodies and an attempt to decide the racial identity of the men in this way would reproduce those desires to pin down and determine the meaning of racial identity which have been part of racist ideologies.

[4] Comment by N. Ní Dhomhnaill, Poetry Reading, Douglas Hyde Gallery, December 1991.

[5] I am referring here to Catherine Harper, Kathy Prendergast, Dorothy Cross and Alanna O'Kelly whose work has been discussed in Nash (1994, 1995).

REFERENCES

BARRELL, J. (1990) The public prospect and the private view: the politics of taste in eighteenth-century Britain, in: S. PUGH (Ed.) *Reading Landscape: country–city–capital*, pp. 19–40 (Manchester, Manchester University Press).

BERSHARD, D. (1992) Looking, power and sexuality: Degas' *Woman with a Lorgnette*, in: R. KENDALL & G. POLLOCK *Dealing with Degas: representations of women and the politics of vision*, 95–105 (New York, Universe).

BESLEY, C. (1993) Desire in theory: Freud, Lacan, Derrida, *Textual Practice*, 7, pp. 384–411.

BEST, S. & KELLNER, D. (1991) *Postmodern Theory: critical interrogations* (London, Macmillan).

BETTERTON, R. (1985) How do women look? The female nude in the work of Suzanne Valadon, *Feminist Review*, 19, pp. 3–24.

BOFFIN, T. & FRASER, J. (Eds) (1991) *Stolen Glances: lesbians take photographs* (London, Pandora).

BRECKENRIDGE, C.A. (1989) The aesthetics and politics of colonial collecting: India at World Fairs, *Comparative Studies in Society and History*, 31, pp. 195–216.

BRETT, D. (1986) From the local to the global: the place of place in art, *Circa*, 29, pp. 17–21.

BRETT, D. (1989) The land and landscape, *Circa*, 43, pp. 14–18.

BUTLER, J. (1990) The force of fantasy, *Differences*, 2, pp. 105–125.

CAUGHT LOOKING INC. (Ed.) (1992) *Caught Looking: feminism, pornography and power* (East Haven, CT, Long River Books.

CHERRY, D. (1993) *Painting Women: Victorian women artists* (London, Routledge).

COHEN, S. & HARK, I.R. (1993) Introduction, in: *Screening the male: exploring masculinities in Hollywood cinema*, pp. 1–8 (London, Routledge).

COOK, P. (1982) Masculinity in crisis, *Screen*, 23, pp. 39–48.

COOMBES, A. E. (1988) Museums and the formations of national and cultural identities, *Oxford Art Journal*, May, pp. 57–68.

CUMMINS, P. (1987) *Aran Dance*, *Circa*, 36, pp. 34–35.

CUMMINS, P., JONES, B., MURPHY, P. & SMYTH, A. (1987) Image making, image breaking, *Circa*, 32, pp. 13–19.

CUMMINS, P. (n.d.) *Artist's Statement*, supplied by the artist.

DANIELS, S. (1989) Marxism, culture and the duplicity of landscape, in: N. THRIFT & R. PEET (Eds) *New Models in Geography, vol. 2: The Political Economy Perspective*, pp. 196–220 (London, Unwin Hyman).

DEEPWELL, K. (1989) Landscape and feminist art practices, *Women Artists' Slide Library Journal*, 27, pp. 25–26.

DE LAURETIS, T. (1988) Sexual difference and lesbian representation, *Theatre Journal*, 40, pp. 155–177.

DOANE, M.A. (1981) Woman's stake: filming the female body, *October*, 17, pp. 22–32.

DOANE, M.A. (1982) Film and the masquerade—theorising the female spectator, *Screen*, 23, pp. 74–88.

DOANE, M.A. (1988) Masquerade reconsidered: further thoughts on the female spectator, *Discourse*, 11, pp. 42–54.

DUDDY, T. (1987) Irish art criticism—a provincialism of the Right? *Circa*, 35, pp. 14–18.

DYER, R. (1982) Don't look now—the male pin-up, *Screen*, 23, pp. 61–73.

DYER, R. (1990) *Now You See It: studies in gay and lesbian film* (London, Routledge).

DYER, R. (1993) *The Matter of Images* (London, Routledge).

FORD, S. (1991) Landscape revisited: a feminist reappraisal, in: C. PHILO *New Words, New Worlds: reconceptualising social and cultural geography*, pp. 151–155 (Lampeter, Social and Cultural Geography Study Group of the Institute of British Geographers.

FORTNUM, R. & HOUGHTON, G. (1989) Women and contemporary painting, *Women Artists' Slide Library Journal*, 28, pp. 4–5.

FOWLER, J. (1990) Regionalism reconsidered, *Circa*, 50, pp. 22–25.

GAMMAN, L. & MARSHMENT, M. (Eds) (1988) *The Female Gaze: women as viewers of popular culture* (London, Women's Press).

GARB, T. (1993) The forbidden gaze: women artists and the male nude in late nineteenth century France, in: K. ADLER & M. POINTON *The Body Images: the human form in visual culture since the Renaissance*, pp. 33–42 (Cambridge, Cambridge University Press).

GIBSON, P.C. & GIBSON, R. (Eds) (1993) *Dirty Looks; women, pornography and power* (London, British Film Institute).

GOUMA-PETERSON, T. & MATHEWS, P. (1987) The feminist critique of art history, *Art Bulletin*, 69, pp. 326–357.

GREENHALGH, P. (1985) Art, politics and society at the Franco-British Exhibition of 1908, *Art History*, 8, pp. 434–452.

GROSZ, E (1992a) Voyeurism/exhibitionism/the gaze, in: E. WRIGHT (Ed.) *Feminism and Psychoanalysis: a critical dictionary*, pp. 447–450 (Oxford, Blackwell).

GROSZ, E (1992b) Phallus: feminist implications, in: E. WRIGHT (Ed.) *Feminism and Psychoanalysis: a critical dictionary*, pp. 320–323 (Oxford, Blackwell).

HARWOOD, V., OSWELL, D., PARKINSON, K. & WARD, A. (Eds) (1993) *Pleasure Principles* (London, Lawrence & Wishart).

HENDERSON, L. (1992) Lesbian pornography: cultural transgression and sexual demystification, in: S. MUNT *New Lesbian Criticism: literary and cultural readings*, pp. 173–191 (Hemel Hempstead, Harvester Wheatsheaf).

KAPLAN, A. (Ed.) (1978) *Women in Film Noir* (London, British Film Institute).

KELLY, M. (1984) Desiring images/imaging desire, *Wedge*, 6, pp. 4–9.

KENT, S. & MORREAU, J. (Eds) (1985) *Women's Images of Men* (London, Readers and Writers).

KERR, B.T. & QUINTANALES, M.N. (1982) The complexity of desire: conversations on sexuality and differences, *Conditions, 8*, 3, pp. 52–71.

KOLODNY, A. (1975) *The Lay of the Land: metaphor as experience and history in American life and letters* (Chapel Hill, NC, University of North Carolina Press).

KOTZ, L. (1993) Complicity: women artists investigating masculinity, in: P. C. GIBSON & R. GIBSON (Eds) *Dirty Looks; women, pornography and power*, pp. 101–123 (London, British Film Institute).

LAU, G. (1993) Confessions of a complete scopophiliac, in: P. C. GIBSON & R. GIBSON (Eds) *Dirty Looks; women, pornography and power* (London, British Film Institute).

LEHMAN, P. (1988) In the realm of the senses: desire, power and the representation of the male body, *Genders*, 2, pp. 91–110.

MACEY, D. (1992) Phallus: definitions, in: E. WEIGHT (Ed.) *Feminism and Psychoanalysis: a critical dictionary*, pp. 318–320 (Oxford, Blackwell).

MARCUS, L. (1991) Taking a good look, *New Formations*, 15, pp. 101–109.

MERCER, K. (1994) Reading racial fetishism: the photographs of Robert Mapplethorpe, in: K. MERCER, *Welcome to the Jungle: new positions in black cultural studies*, pp. 171–220 (London: Routledge).

MITCHELL, T. (1989) The world as exhibition, *Comparative Studies in Society and History*, 31, pp. 217–236.

MOORE, S. (1988) Here's looking at you kid, in: L. GAMMAN & M. MARSHMENT (Eds) *The Female Gaze: women as viewers of popular culture*, pp. 44–59 (London, Women's Press).

MULVEY, L. (1975) Visual pleasure and narrative cinema, *Screen*, 16, pp. 6–8.

NASH, C. (1993) Embodying the nation: the west of Ireland landscape and Irish identity, in: M. CRONIN & B. O'CONNOR *Tourism and Ireland: a critical analysis*, pp. 86–114 (Cork, Cork University Press).

NASH, C. (1994) Re-mapping the body/land: new cartographies of identity by Irish women artists, in: A. BLUNT & G. ROSE *Writing Women and Space*, pp. 227–250 (New York and London, Guilford Press).

NASH, C. (1995) Landscape, body, nation: cultural geographies of Irish identities, unpublished PhD thesis, University of Nottingham.

NEAD, L. (1992) *The Female Nude: art, obscenity and sexuality* (London, Routledge).

NEALE, S. (1982) 'Chariots of Fire': images of men, *Screen*, 23, pp. 47–53.

NEALE, S. (1993) Masculinity as spectacle: reflections on men and mainstream cinema, in: S. COHEN & I. R. HARK *Screening the Male: exploring masculinities in Hollywood cinema*, pp. 9–20 (London, Routledge).

NÍ DHOMHNAILL, N. (1988) *Selected Poems: Rogha Danta* (Dublin, Raven Arts Press).

O'PRAY, M. (1982) Movies, mania and masculinity, *Screen* 23, pp. 63–71.

POLLOCK, G. (1988) *Vision and Difference: femininity, feminism and the histories of art* (London, Routledge).

POLLOCK, G. (1992) The gaze and the look: women with binoculars—a question of difference, in: G. POLLOCK & R. KENDALL *Dealing with Degas: representations of women and the politics of vision*, pp. 106–130 (New York, Universe).

POLLOCK, G. (1995) The 'View from Elsewhere': extracts from a semi-public correspondence about the politics of feminist spectatorship, in: P. FLORENCE & D. REYNOLDS *Feminist Subjects, Multi-media: cultural methodologies*, pp. 3–38 (Manchester, Manchester University Press).

ROSE, G. (1992) Geography as a science of observation: landscape, the gaze and masculinity, in: G. ROSE & F. DRIVER *Nature and Science: essays on the history of geographical knowledge*, pp. 8–18 (Cheltenham, Historical Geography Research Series).

ROSE, G. (1993) *Feminism and Geography: the limits of geographical knowledge* (Cambridge, Polity Press).

SALAMAN, N. (1994) Regarding male objects, in: N. SALAMAN *What She Wants: women artists look at men*, pp. 13–26 (London, Verso).

SAUNDERS, G. (1989) *The Nude: a new perspective* (London, Herbert Press).

SCHOR, M. (1988) Representations of the penis, *M/E/A/N/I/N/G*, 4, pp. 3–17.

SMYTH, A. (1988) The contemporary women's movement in the Republic of Ireland, *Women's Studies International Forum*, 11, pp. 331–341.

SMYTH, C. (1994) What she wants and what she gets, in: N. SALAMAN *What She Wants: women artists look at men*, pp. 51–61 (London, Verso).

SOLOMON-GODEAU, A. (1992) Image, in: E. WRIGHT (Ed.) *Feminism and Psychoanalysis: a critical dictionary*, pp. 171–173 (Oxford, Blackwell).

SPARGO, T. & F. BOTTING (1993) Re-iterating desire, *Textual Practice*, 7, pp. 379–383.

STACEY, J. (1994) *Star Gazing: Hollywood cinema and female spectatorship* (London, Routledge).

STINTON COSGROVE, M. (1990) Irishness and art criticism, *Circa*, 52, pp. 14–19.

WALTERS, M. (1978) *The Male Nude* (Paddington Press, London).

WICKE, J. (1993) Through a gaze darkly: pornography's academic market, in: P. C. GIBSON & R. GIBSON (Eds) *Dirty Looks; Women, Pornography and Power*, pp. 62–80 (London, British Film Institute).

WILLIAMS, L. (1993) Second thoughts on *Hard Core*: American obscenity laws and the scapegoating of deviance, in: P. C. GIBSON & R. GIBSON (Eds) *Dirty Looks; Women, Pornography and Power*, pp. 46–61 (London, British Film Institute).

WILLIAMS, L. (1994,a) The pornographic subject: feminism and censorship in the 1990s, in: S. LEDGER, J. MCDONAGH & J. SPENCER (Eds) *Political Gender: texts and contexts*, pp. 189–203 (London, Harvester Wheatsheaf).

WILLIAMS, L. (1994,b) Introduction: What do I see? What do I want?, in: N. SALAMAN (Ed.) *What She Wants: women artists look at men*, pp. 3–11 (London, Verso).

YOUNG, L. (1994) Mapping male bodies: thoughts on gendered and racialised looking, in: N. SALAMAN (Ed.) *What She Wants: women artist look at men*, pp. 39–50 (London, Verso).

[5]

The Place of Landscape: A Conceptual Framework for Interpreting an American Scene

Richard H. Schein

Department of Geography, University of Kentucky

This paper presents a conceptual framework for interpreting cultural landscapes in the U.S. through the example of a residential suburb. The particular landscape is presented as the tangible, visible articulation of numerous discourses. Several examples of a "discourse materialized" in the landscape—landscape architecture, historic preservation, neighborhood associations, insurance mapping, zoning, consumption—are employed to demonstrate how the landscape at once constricts and is constructed by individuals who live in a particular place. U.S. cultural landscapes ultimately are viewed as material phenomena, reflective and symbolic of individual activity and cultural ideals, as they simultaneously are central to the constitution and reinforcement of those activities and ideals.
Key Words: American cultural landscape, interpretation, social life.

This paper contributes to a geographic tradition broadly defined, yet coherent through its focus upon the *cultural landscape* as a tangible, visible entity, one that is both reflective and constitutive of society, culture, and identity. I shall first propose a brief conceptual framework for interpreting cultural landscapes in the U.S., then illustrate this framework through the example of a particular landscape that bears the weight of representing cultural landscapes in general across the country. The research is based on the principle that interpreting a cultural landscape is a geographically specific exercise that requires interrogating the role of landscape in social and cultural reproduction, as well as understanding the landscape within wider social and cultural contexts. The paper concludes that the cultural landscape is ultimately one mechanism through which such connections are formulated and maintained.

In order to achieve this end, I will first present a brief (and selective) theoretical genealogy of the landscape idea, especially as it has been used to examine cultural landscapes in the U.S. This will lead to the conceptual framework and its empirical example, Ashland Park, a contemporary neighborhood that originated early in the twentieth century as a suburb of Lexington, Kentucky. Finally, I will draw some conclusions about the interpretation of ordinary landscapes in the U.S. and their role in the everyday experiences of the people who live in and through them.

A Selective Genealogy of "Landscape"

"Landscape" has been a fundamental concept and subject within North American cultural geography throughout this century. Most notably, since Sauer's (1925) influential "Morphology of Landscape," cultural geographers have contributed to a widely subscribed literature concerned with the cultural or humanly created landscape as autobiography (e.g., Lewis 1979, 1983) and as a common and ordinary part of everyday life (e.g., Jackson 1984; Meinig 1979a; Stilgoe 1982). We look at rural landscapes (e.g., Hart 1975; Zube and Zube 1977), urban landscapes (e.g., Knox 1993; Jackson 1972), and suburban landscapes (e.g., Stilgoe 1988). Landscape is seen as symbolic, as representative, and as a representation (e.g., Cosgrove and Daniels 1988; Duncan and Ley 1993; Schein 1993); as duplicitous (e.g., Daniels 1989); and as gendered, class-based, politicized, and central to the (re)production of social life (e.g., Duncan 1990; Mitchell 1994; Anderson and Gale 1992; Zukin 1991). We look at landscapes of production and consumption (e.g., Meyer 1990; Sorkin 1992), as well as at the production of landscape, both as material form and as an epistemological framework (e.g., Roberts and Schein 1993; Cosgrove 1984, 1985). Landscapes are seen as texts, as metaphors, and as part of discourses (e.g., Duncan and Duncan

1988; Barnes and Duncan 1992a). Landscapes can be premodern (e.g., Kniffen 1965), modern (e.g., Relph 1987), or postmodern (e.g., Mills 1988). Landscape can be used to refer to a spatial and visual ensemble, often equated with particular places and times (e.g., Conzen 1990); it can refer to the constituent parts of such an ensemble (e.g., Murray-Wooley and Raitz 1992; Noble 1984). There are even (sometimes acrimonious) debates about the "proper" way to approach a landscape within cultural geography (Price and Lewis 1993a, 1993b; Cosgrove 1993; Duncan 1993; Jackson 1993).

It is the existence of such debates, of course, that suggests that the list above is somewhat disingenuous—included are scholarly works that, despite a common focus upon the cultural landscape are, at first glance, widely divergent and seemingly incommensurate in their theoretical and methodological approaches to the subject. Despite that diversity, and despite recent "civil wars" between the "old" and the "new" cultural geography, all of the works in the above list may be seen as part of a tradition, a landscape genealogy, with at least two common foci. First, all are particularly concerned, first and foremost, with the cultural landscape as a material object—what Lewis (1979) calls the "tangible, visible scene." Second, all are cohered by a loosely defined set of commonly accepted key works, in which individual components alternately serve as either a touchstone of intellectual and scholarly affiliation or as a foil and point of divergent theoretical positioning—a conceptual "other."

While a claim to such a genealogy may be teleological, it nevertheless is useful for erecting a heuristic set of signposts toward the conceptual framework presented in the next section. Such scholarly signposts serve two purposes. First, they are a reminder of past themes in the interpretation of landscape, providing a familiar point of entry or touchstone for landscape scholars with divergent interests—thereby serving as a rapprochement and a reminder to abide by what Daniels (1989) calls the duplicity of landscape. Second, these signposts establish a particular set of interpretive issues as prolegomena to the conceptual framework. In brief, the genealogy develops the idea of the cultural landscape from cultural spoor through unwitting autobiography reflecting "American culture" to the discursive place of landscape in a dynamic sociospatial relationship with the very individuals and societies that create it in the first place. Selective by nature, like all genealogies, in the end, it forms part of a claim to the future of landscape interpretation.

Sauer's "Morphology of Landscape" (1925 [1963:343]) generally stands as the first U.S. theoretical claim for a specifically *cultural* landscape as the product of human activity: "[c]ulture is the agent, the natural area is the medium, the cultural landscape is the result." Perhaps following Sauer, Lewis (1979:12) is widely cited for his suggestion that the "human landscape is our unwitting autobiography." There is a slight change of material landscape emphasis from Sauer to Lewis; the former was primarily concerned with the landscape as the impress of human activity, while the latter taught us to "read" that impress as the *reflection of* "our tastes, our values, our aspirations, and even our fears. . . ." But together, these two essays stand by association for the empiricist tradition in cultural geography. And the actual scholarly practice of Sauer and Lewis notwithstanding (as opposed to their theoretical positioning), these particular statements have stood as important touchstones for a positivist landscape tradition, as foils for critiques of that tradition, and as the object of challenges by what has retrospectively come to be called the "new" cultural geography.

While the Sauerian claim that the cultural landscape stands as the impress of human activity is relatively undisputed, the fundamental question of landscape *authorship* within that claim is problematic. Duncan's (1980) challenge to the traditional assumption of the "cultural" in landscape interpretation pointed to a reliance upon a superorganic, or reified, conception of culture, which assigned it "ontological status and causative power" (181). Culture appeared external to the individual, thus denying human agency or the individual's ability to effect cultural (and thus landscape) change. Related to that focus upon human agency in the cultural landscape is the need to understand temporal change. Despite claims that modern geography privileges the historical actor (Soja 1989), we have not thoroughly theorized temporal relations, the notion of landscape change, or the underlying power relationships implicit in change. Lewis (1979) suggests that the cultural landscape presents us with a series of sedimentary layers of social accretion, each cultural stratum reflecting particular ideological origins, intentions, and contexts; but this is still an undifferentiated, perhaps authorless, wholesale approach to landscape change (resulting in, say, a "French colonial" or an "industrial" landscape). It might be more useful to begin by viewing the landscape as a palimpsest rather than cultural strata—an analogy that at least provides the possibility for erasure and overwriting and the co-existence of several different scripts, implying not just different historical eras, but several historical and contemporary actors as well.

Asserting the role of agency or author in cultural landscape interpretation stems from the emergence of humanistic geography's postempiricist challenge in recent disciplinary history (e.g., Samuels 1979). Similar challenges came from scholars concerned to explicate the structural conditions of the human condition, including Cosgrove's work on the role of symbolic landscapes in particular social formations and his observation that "a cultural concept like landscape does not emerge unprompted from the minds of individuals or human groups" (1984:2).

The eventual attempt to mediate the now-familiar (and usually caricatured) structure-agency opposition that resulted from these parallel challenges can be found within the cultural landscape literature, such as in Daniels's (1989) attempt to transcend the dualism. An important consideration to emerge from these postpositivist turns in geography is the idea that the cultural landscape is more than just the "final product" of human activity that can be read simply as a reflection of those who created it. We must also interrogate our embeddedness in and interaction with the cultural landscape. Corollary to this interrogation is the requirement to see the cultural landscape as not only a "thing" built by human hands—a material palimpsest—but also as a theoretical construct, with certain ontological and epistemological assumptions and ramifications (Cosgrove 1985; Schein 1993).

The cultural landscape, as both a material presence and conceptual framing, serves to discipline interpreting subjects alongside their objectification of landscape's form and meaning. This can be illustrated through the spatial and visual components of the cultural landscape. Like "space", the cultural landscape is produced and is ultimately implicated in the ongoing reproduction of social and cultural life (e.g., Gottdiener 1985; Massey 1984; Soja 1989; Lefebvre 1989). As part of that production, spatial relationships—distributions, partitioning, enclosures, circulation, division—serve as part of the dispersed disciplinary mechanisms of modernity, what Foucault (1980) calls "capillaries of power." As a material object, the cultural landscape is *seen*. As an epistemology, vision also has a set of hidden rules, such as reliance upon linear perspective, pretension to mimesis, and the claim to the objective view. Landscape observers are disciplined by time-honored rules about what constitutes valid evidence and the legitimate object of inquiry (Cosgrove 1985; Crary 1990).

If it is accepted that the landscape in its disciplinary functions is implicated in the ongoing formulation of social life—or, to paraphrase Massey (1984), that *landscape matters*—then it is necessary to ask *how* it matters. In addition to theorizing the place of the cultural landscape in the social relations and spatial arrangements of daily life, this includes understanding how a particular, identifiable cultural landscape in *this* place is related and connected to landscapes and social processes in *other* places. Traditional landscape studies often approached this question by positioning a particular cultural landscape (or landscape artifact) at the end of one-way arrows representing migration or a diffusion pathway (e.g., Kniffen 1965). More useful theorizations can be found in recent works addressing the question of *scale*.

Traditional views of "scale" as a nested spatial hierarchy consisting of fixed, bounded, and reified levels—local, regional, national, global—have been challenged by more fluid approaches to spatial relations, focused upon connections and oppositions, and the processes that construct scale in the first place (Agnew 1993; Brown 1993; Lipietz 1993). Reconceptualizing scale involves a sense of "active process" which is "always in a grand sense open. . . ." What is called for is a sense of "scaling" (Roberts forthcoming). This is achieved in Massey's suggestion of the power geometries inherent in a "progressive sense of place." Places, for Massey, are not scale-bound and reified: "They can be imagined as articulated moments in networks of social relations and understandings" (Massey 1993:66).

So, too, can the cultural landscape be envisioned as an articulated moment in networks that stretch across space. Not only does this work toward situating the landscape, it also privileges geographical connectivity along with temporal change. Landscapes are always in the process of "becoming," no longer reified or concretized—inert and there—but continually under scrutiny, at once manipulable and manipulated, always subject to change, and everywhere implicated in the ongoing formulation of social life.

Toward a Conceptual Framework

In an essay entitled "The Beholding Eye," D.W. Meinig (1979b) suggested that ten different individuals faced with the same visible scene might perceive ten different landscapes, alternately seeing in the view before their eyes landscape as nature, as habitat, as history, and so on. Meinig's point, of course, is that we interpret landscapes through the ideas we bring to the project.[1] But the essay also presents the possibility that the cultural landscape can itself capture different, even competing, sets of meaning, or independent, thematic networks of knowledge—networks presenting the landscape as nature, habitat, or history—and that these really *are* inherent in each cultural landscape. We can position a particular landscape as a *node* at the intersection of any number of these knowledge networks. To use a language linked to more contemporary interpretation and analysis, a particular landscape may articulate a series of relatively independent discourses. These discourses are "shared meanings which are socially constituted, ideologies, sets of 'common sense' assumptions" (Duncan 1990:12). Each discourse is a "social framework of intelligibility, within which all practices are communicated, negotiated, or challenged" (Duncan 1990:16).

Duncan (1990) has demonstrated the potential of discourse analysis in interpreting a heavily coded landscape in the Kandyan Kingdom of Sri Lanka. His work explores "the relationship between a landscape and the competing discourses within the discursive field of which it is a constitutive part" (Duncan 1990:181). Although Duncan's study is convincing, especially regarding the general utility of discourse analysis for interpreting landscapes, it leaves open the question of such analysis in a North American—specifically U.S.—context. For reasons I discuss below, the U.S. landscape poses a particular challenge to the landscape interpreter.

Most U.S. landscapes are created piecemeal within a cultural milieu that idealizes liberal individualism, laissez-faire capitalism, and political democracy. One of the principle tenets in the historical formation of the American cultural system is the establishment and protection of property rights and freehold land tenure. Americans firmly believe in their lawful ability to do what they like with their individual pieces of the planet. At the very least, this mitigates against powerful, centralized, large-scale landscape authorship like that found in the preindustrial nineteenth-century Kandyan Kingdom. While there certainly are identifiable landscape authors, e.g., Robert Moses (Caro 1974), most landscapes in the U.S. are the result of countless individual, independent, self-interested decisions that create, alter, and maintain landscapes, their meaning, and their symbolism.

Each seemingly individual decision behind any particular U.S. landscape is embedded within a discourse. When the action results in a tangible landscape element, or total ensemble, the cultural landscape becomes the *discourse materialized*. Examples of such discourses might include zoning theory and practice, architectural design trends, economic consumption patterns, and others. As a material component of a particular discourse or set of intersecting discourses, "the cultural landscape" at once captures the intent and ideology of the discourse as a whole and is a constitutive part of its ongoing development and reinforcement.

Through its material form as a cultural landscape, each discourse presents competing social and visual disciplines or *strategies* that combine to constrict or limit human action within and interpretation of any particular landscape. Although we may "unpack" or trace any individual discourse, it is the combined effect that creates the cultural landscape as Bordieu's (1977:82,79) *habitus*: history turned into nature through an amnesia of genesis. In our day-to-day lives, lived in ordinary vernacular landscapes, we take the tangible, visible scene for granted, especially as an ensemble. "For most Americans, cultural landscape just *is*" (Lewis 1979:11). This naturalization makes the (seemingly unproblematic) vernacular landscape (of our suburb, our drive to work, our college campus) perhaps even more powerful in its disciplinary capabilities.

More geographically, what Giddens (1990:18–19) calls distanciated relations ultimately play a role in structuring the local landscape as a node at intersecting discourses. Just as the landscape as discourse materialized belies its social and historical construction, the visible form of the local conceals the reach of influences whose origins often are far removed from the immediately local.

Described thus far, any discourse materialized in the cultural landscape would seem to have superorganic qualities—it "appears" out of nowhere, to be captured in the local scene. This raises a question, if not of where the discourse itself arises, at least of the possibility for change in the discourse. The answer lies in the cultural landscape, and in the possibilities for human agency in altering that landscape. At one level, our individual actions within and upon a particular landscape may serve as what de Certeau (1984:xix) calls "tactics"—those actions in the practice of everyday life with which we inhabit the strategies of the dominant discourse. Each tactic has the capability of altering the discourse. More empowering to the individual is the realization that discourses themselves, in all their manifestations—textual, material, ideological—are humanly created, and our myriad individual acts of inhabiting the landscape are part of the ongoing reformulation of those discourses. In any event, the purpose here is not to theorize the concept of discourse per se, but to position the cultural landscape within a discursive framework. The cultural landscape, as discourse materialized, is simultaneously disciplinary in its spatial and visual strategies and empowering in the possibilities inherent for individual human action upon the landscape. The cultural landscape thus is continually implicated in the ongoing reconstitution of a discourse, or set of discourses, about social life, and it is in this sense that it serves as both a disciplinary mechanism and a potentially liberating medium for social change.

The Framework in Place: Ashland Park

Any conceptual framework is best explicated through an empirical example. In this paper a neighborhood called Ashland Park is the example which I have chosen for two reasons. First, I live there and have an intimate, ongoing familiarity with the place.[2] Second, Ashland Park is a suburb, one of those symbolic landscapes representing "part of the iconography of nationhood, part of the shared set of ideas and memories and feelings which bind people together" (Meinig 1979c:165).

The term suburb seems to deserve "key word" status (Williams 1983). In a U.S. context, suburb variously can refer to a political jurisdiction at the edge (and thus outside) of the incorporated urban area, to incorporated residential subdivisions, or to, more generally, the cultural normality of a residential life noted for a gendered division of labor and a separation of home and workplace characteristic of the modern, industrial era. Ashland Park has been all three. As a suburb, it is a place where seemingly innumerable individual human actions combine to create the landscape in toto. It is perhaps in their own homes and on their own properties that most Americans have the ability to effect change in the built environment. Those same principles of freehold land tenure and individualism, the belief in laissez-faire capitalism, and the promise of liberal democracy that are so central to the U.S. experience are writ small in every suburban residential landscape, in both material and ideological form. More than three-quarters of households in the U.S. are in privately owned homes, and many are in places similar to, if not exactly like, Ashland Park.

Admittedly, the middle-class resident of Ashland Park has perhaps a greater ability to modify his or her own immediate landscape than, say, the resident of Bluegrass-Aspendale, a government-subsidized housing project only one mile away. But residents of Bluegrass-Aspendale, despite their renter's status, are not completely disempowered from creating and alterating their immediate landscape (Nast and Wilson 1994). In short, it is in our residential settings that Americans are most likely to have the ability to live both in and through the landscape as active agents of change. And in the end, the intent of this example is not to create a template for interpretation that assumes all cultural landscapes in the U.S. to be the same. The intent rather is to explore the place of landscape in social life as discourse(s) materialized. What better place to start than in a setting familiar to many by first-hand experience or through its idealized qualities as an American scene?

Ashland Park originated as an early twentieth-century middle-class subdivision at the southeastern edge of Lexington (Figure 1), a wealthy market town atop the central-place hierarchy in the agriculturally rich Kentucky Bluegrass (Ulack et al. 1994; Raitz 1980; Wright 1982; Davis 1927; Reisenweber and Hudson 1990). As the city expanded, many inner-Bluegrass "gentleman farms" were subdivided into suburban developments. Ashland Park comprises ninety-five acres carved from the 600-acre former estate of Kentucky favorite-son Congressman Henry Clay. The core of the old family property remains in the middle of the neighborhood. The present great house, built by Clay's son in the 1850s, has been a museum since 1964 (Clark et al. 1996) with its remnant eighteen-acre grounds and gardens serving as a quasi-public park. The surrounding neighbor-

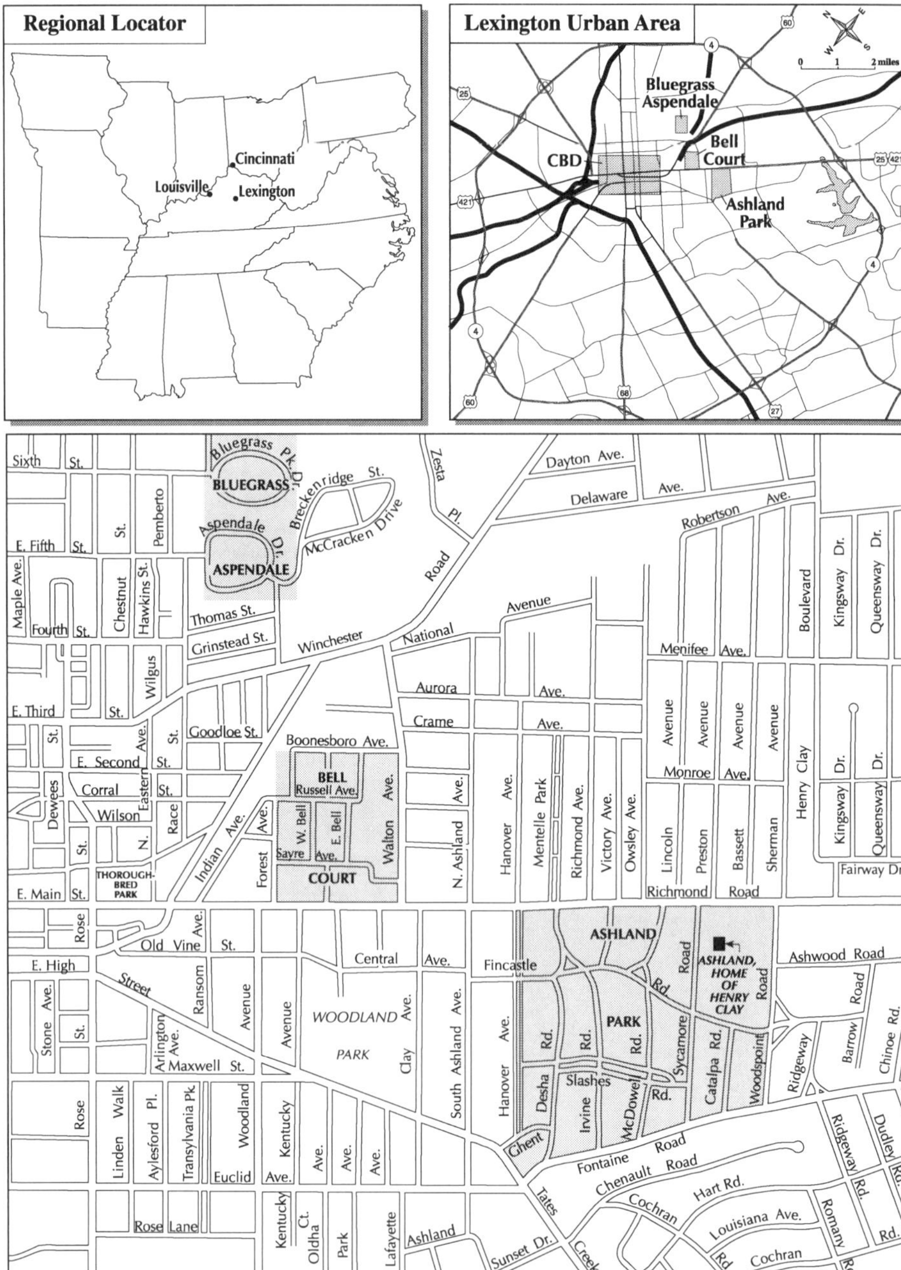

Map prepared by Dick Gilbreath, University of Kentucky Cartography Lab

Figure 1. Ashland Park, a subdivision of Lexington in the Bluegrass region of Central Kentucky, was originally on the edge of the city.

hood was designed by the Olmsted Brothers, a prominent national landscape architecture firm. The majority of Ashland Park's homes are period houses ranging from the 1910s to the interwar era, including California bungalows, American Foursquares, Colonial Revivals, Craftsman houses, and a number of apartments which are large, one-floor flats in two-story buildings. The timing of Ashland Park's development meant that it was accessible via the streetcar that ran along the neighborhood's northern edge, providing access to downtown for resident and servant alike; most houses were built with detached garages. As Lexington's first twentieth-century suburb, Ashland Park is now closer to downtown than to the farthest suburbs as the city's population has exploded since the postwar period from about 50,000 to nearly 250,000 residents today. The patina of seventy-plus years is evident in the mature street trees, individual gardens, numerous house additions in period style, and the continual maintenance of the neighborhood since its inception. Today Ashland Park stands as one of the more recognized city neighborhoods, in which a third generation of solidly middle-class families is being raised by primarily dual-income professional parents. At least one faction of the local growth coalition heralds Ashland Park as the neotraditional prototype for continued suburban expansion into the horse-farm landscape surrounding the city (Figure 2).

The preceding description of Ashland Park and the accompanying figures, while accurate, conjure an image that portrays the place and landscape as primarily the outcome of historical process. But the landscape is more than that. Unlike the drawing rooms displayed in the Ashland museum, the neighborhood is not frozen in time. The Ashland Park landscape is a reflection of the people who created it, but as discourse(s) materialized, it also is implicated in the lives of those people and others in ways not accessible through such a traditional description.

Discourse Materialized: Ashland Park

While any number of discourses may flow through the Ashland Park landscape, I offer here the examples of landscape architecture, insurance mapping, zoning, historic preservation, neighborhood associations, and consumption as brief indications of how relatively discrete discourses "touch ground" and how the actions of individuals in and through that landscape are captured in the discourses themselves.

Landscape Architecture[3]

The very plan of Ashland Park presents the landscape as a spatial and visual discipline with material and symbolic implications. In a strictly Euclidean sense, the neighborhood plan as manifest in the landscape is a form of discipline, directing daily physical activity along streets, sidewalks, and greenspaces. In a more symbolic sense, the formal plan provides a spatial and visual template for the playing out of cultural rules such as those governing public and private space (e.g., Davis 1990); an anglophile residential aesthetic (e.g., Jenkins 1994); the gendered division of labor, the cult of domesticity, and an invented family tradition (e.g., Bondi 1992; Hareven 1991; Marsh 1990; Walker 1981); or, more generally, for what has come to be seen as a problematic "suburban ideal" (e.g., Jackson 1985; Sies 1987).

Ashland Park is today a bounded, identifiable, and recognized "neighborhood" of Lexington, whose origins are traced to its creation as a "whole": a formal subdivision, designed by one firm, at the request of a single client early in this century. House spacing, street orientation, tree and shrub planting, infrastructural support systems, street lighting, street names, and the general overall landscape appearance were, first and foremost, determined by the original neighborhood plan. That plan, the collective work of Olmsted Brothers, the Brookline, Massachusetts landscape architects who had inherited the firm and the tradition from Frederick Law Olmsted, still has a recognized place in the city's historical and contemporary geographies. Ashland Park was not the only Olmsted Brothers project in Lexington. Beginning in 1904, the firm participated in at least twenty-nine local projects, including parks, a city planning survey, institutional designs, residential estates, and subdivisions. John Charles Olmsted, nephew and son-in-law of the firm's founder, visited Lexington in 1904, 1906, and 1916. His visits underscore the incorporation of Lexington into a national (even international) discourse concerning landscape architecture and urban design. Into this discourse, both national and local, the McDowell family inserted a small piece of Lexington that would become Ashland Park.

(a) **Strategically placed signs mark the entrances to the historic district.**

(b) The towering tree canopy is a source of neighborhood pride. Residents are admonished to look after and venerate the trees and to plant replacements in keeping with the original plantings.

(c) Curvilinear streets, narrow driveways, mature trees, and period houses are hallmarks of the neighborhood.

Figure 2. Landscape details in Ashland Park.

(d) Mature, diverse gardens and detached garages provide a patina of age and elegance to the contiguous backyard areas.

(e) A historical marker on the remnant grounds of the Ashland estate reminds visitors of the association with Henry Clay. **(f)** Pockets of greenspace and boulevards are products of the Olmsted Brothers plan.

Figure 2. cont.

Henry Clay McDowell, whose wife was a descendant of Clay, owned the Ashland Estate at the end of the nineteenth century. When he died in 1899, his family began plans to subdivide part of the family property. Those plans included more than twenty-eight subdivision plans by Olmsted Brothers drawn between 1906 and 1917. A 1917 article in the local newspaper touted the McDowells' plans by noting that the Olmsteds were the leading landscape architects in the country; and that same year, the McDowells took an ad in the newspaper promoting the first sale of lots in the subdivision (Froehling 1993). The Ashland Park plan bore all the hallmarks of a standard Olmsted residential design established in Frederick Law Olmsted and Calvert Vaux's plan for Riverside, Illinois in 1869: informal open spaces, gracefully curving roadways, an abundance of trees and shrubs, and a feeling of spaciousness and quiet—a park in the city (Reps 1965:347–48).

The appearance of the Olmsted Brothers firm in Lexington in the early decades of the twentieth century linked the city and Ashland Park to visual and spatial strategies of a landscape architecture/suburban design discourse that seemingly originated and was centered in places far removed from provincial Lexington. That link was not entirely one way, however. It was not simply a matter of an "outside" consultant dictating design to local (Lexington) clients. First, there is evidence that landscape design criteria were not applied whole cloth but were modified by the landscape architects to account for local situations. For example, the 1904 report to the Lexington Park Commissioners by Olmsted Brothers noted the "marked peculiarity of Lexington" where "there is a strikingly great interest in horses held for pleasure. . . ." Accordingly, it was deemed "peculiarly important" that the master plan for Lexington's park system make "provision for pleasure driving and riding much greater than other cities. . . ."

In the case of Ashland Park, the McDowell family members, like the other twenty-eight Olmsted Brothers clients in Lexington, were *active* clients who hired the Olmsted firm and presumably were not "pushovers" for every design option suggested. The design of Ashland Park was not merely one plan dropped onto the 95 acres, but underwent more than two dozen permutations. Changes were made through an interactive process that included the clients, the firm, and the local context. An Olmsted representative visiting Lexington in 1926 reported that the "old Ashland Subdivision" was "built in general per plan *but revised*. . ." (emphasis added). (Smith to McDowell, 30 June 1926, McDowell Correspondence) The representative was in town at the request of William C. McDowell, who was considering engaging Olmsted Brothers to lay out the remainder of his family's Ashland estate. McDowell's addition plans were obviously structured, in part, by the material presence of the first Olmsted Brothers subdivision of Ashland. McDowell ultimately balked at the firm's fees, however, and bought only an outline and some advice for much less money. In short, he modified the general tenor of the national firm's suggestions to better serve his own local interests. In trying to convince McDowell of their plan's value, Olmsted Brothers compared their proposal to similar subdivisions in nearby Louisville, hoping both to sway their potential client with allusions to a city with which he was no doubt familiar, and to solidify the link between the Ashland Park subdivision and a national discourse about residential suburbs.

What emerges in the twenty-two years between John Charles Olmsted's first appearance in Lexington and W. C. McDowell's Ashland Park addition is a story bound together by a discourse about suburban residential design linking a national firm, local clients, and regional comparisons, all of which had human agents as active promoters and creators of real cultural landscapes. As one of more than 5,000 American projects undertaken by the firm between 1857 and 1950, Ashland Park was both influenced by projects that came before it and, through local agency, presumably became a component of Olmsted Brothers' continual development of both their art form and enterprise; for the firm could not have been successful without adapting ideas to suit the clients.

Insurance Mapping

Central to fire insurance underwriting for much of the past century has been the practice of insurance mapping. Fire insurance maps probably originated in London at the end of the eighteenth century, and were in common use in mid-nineteenth-century America "in response to the need for detailed information concerning potential fire risks of individual commercial, residential, and industrial structures" (Ristow 1968:198). In the infancy of fire insurance, underwriters personally inspected properties before writing a policy. As the nation rapidly urbanized and industrialized, especially after the Civil War, personal visits were not always feasible or economical. Insurance maps made risk information available to underwriters who were unable to personally inspect properties. These maps constitute a form of systematic, professionalized knowledge available in standardized form about specific locales and buildings. In effect, insurance maps were a strategic means whereby the building activities of individual citizens could be monitored and directed. In this way, the practice of fire insurance mapping stands as a discourse that has disciplined the residents of Ashland Park through the cultural landscape.

During the era that Ashland Park was being designed and built, the Sanborn Co. dominated the fire insurance mapping business in the U.S. It had already mapped more than five thousand American towns and cities, and the total number would eventually exceed 12,000. Instructions for Sanborn surveyors were specific and strict. A 1936 *Surveyor's Manual* (Sanborn Map Co. 1936) was more than 160 pages long and included thirteen chapters of explicit instructions. The introduction admonished the surveyor that "our customers rely upon the authenticity of our publications. Surveyors must recognize the necessity for the highest degree of accuracy and thoroughness" (1936:5). Sanborn Maps were extremely detailed, usually produced at a scale of around 100 feet to the inch. Early Sanborn atlases noted eight categories of building materials (stone, wood, brick, etc.), used different symbols for roof types, firewalls, windows, stables, steam boilers, fire hydrants, and warehouses, and included classifications according to fire hazard (Ristow 1968). Over the course of almost a hundred years, many more features were added to the key. According to Ristow, "most insurance underwriters, at the beginning of the second decade of the twentieth century, were dependent upon Sanborn maps" (1968:204). Fire insurance maps' large scale and colorful appearance comprehended a wide range of structural information about properties and urban areas, and included virtually all information relevant to efficient and effective insurance underwriting. With a Sanborn map in the files, a company in Philadelphia could issue a policy on a building in Omaha based on map information. More to the point, insurance underwriters anywhere, with access to the Sanborn Map of Lexington, were able to draw Ashland Park into the discourse of the insurance industry, albeit usually at the request of a resident policy holder.

Fire insurance maps of Lexington—eventually including Ashland Park—showed spatial and visual information about the structures that local residents built and lived in (e.g., Sanborn Map Co. 1886, 1896, 1907, 1934, 1958). Beyond their immediate function as "inert records of morphological landscapes," fire insurance maps, like maps generally, also serve(d) as "refractive images contributing to dialogue in a socially constructed world" (Harley 1988:278). As Harley (1988:279) has demonstrated, a map "replicates not just the 'environment' in some abstract sense but equally the territorial imperatives of a particular political system." They do not simply reflect; they also codify, concretize, and naturalize spatial information. Thus, maps ultimately help to structure the social world. That structuring includes links into discourses that transcend the immediately local, and that move beyond the limited confines of fire insurance underwriters' strategies. The Sanborn Map Company noted that their maps served "numerous diversified industries and purposes" (1953:4). The Company boasted "an unparalleled record of the physical growth and current status of municipalities throughout the country," which was also useful to market analysts, various governmental agencies, utility companies, banks, mortgage companies, and life insurance companies who "utilize them in underwriting mortgage loans."

As a surveillance technique of the insurance company, the fire insurance map determined more than what Ashland Park residents paid for fire insurance. The map monitored the ways residents built, what they built, where they built, and when. This is not to suggest that the heavy hand of the insurance industry was the sole determinant of the Ashland Park landscape, but just as many urban houses today employ deadbolt door locks and have smoke detectors and fire extinguishers in part for insurance purposes, the requirements of the insurance underwriters have always been a strong disciplinary component of individual building owners' decisions.

It cannot be said, however, that simply because insurance premiums are high, a particular individual will forgo a favored building style, material, or addition. Ashland Park residents certainly *were* free to alter their houses and garages, and they did. Consequently, fire insurance maps were periodically updated to reflect changes in the material landscape. Chapter 10 of the 1936 manual was titled "Corrections for Pasters," and most insurance maps were multilayered, fragmented accretions of an original field map and subsequent addenda, a paper palimpsest of landscape history. Thus, the fire insurance maps served as a template helping to determine insurance risk, playing directly on the financial decisions and visual qualities of the neighborhood at the same time that individual local decisions were incorporated into subsequent mappings. Changes in the local landscape were recorded and taken into the network of fire insurance knowledge for use in distant places. The Ashland Park neighborhood was constricted by, yet simultaneously and recursively constitutive of, the discourse of insurance mapping that was played out *through* the cultural landscape.

Zoning[4]

On June 27, 1928, the newly formed Lexington Planning Commission held its first meeting. Those present learned that nearby Louisville's temporary zoning ordinance had been approved by the State Court of Appeals. Within two months, the Lexington Planning Commission had proposed a temporary zoning ordinance of its own, written by a Mr. L. Segoe, a consultant from Cincinnati, and published in a Sunday edition (August 19, 1928) of the *Lexington Leader* and *Lexington Herald*. A few days later, the temporary ordinance was passed by the City Commissioners, who, in the following October, traveled to Cincinnati to attend the convention of the United City Planners of Ohio. That same month, the Fayette County Fiscal Court approved the city's official planning body, the Lexington Planning and Zoning Commission.

Convinced of the need for a more permanent zoning ordinance, the city then solicited and received outside bids from firms in Cincinnati, St. Louis, and New York. Segoe's Cincinnati consulting firm was hired, and at his suggestion the Planning and Zoning Commission proposed to the Mayor that it become a member of the National Conference on City Planning. The Mayor approved and sent the Commission secretary to national meetings in Buffalo, New York. Community input into the newly developing zoning ordinance came from the Lexington Real Estate Board, some thirty local real estate agents, developers, builders, and architects. At the invitation of the chair of the Commission, the leaders of numerous civic clubs sent representatives to Commission meetings regarding zoning. Public meetings and hearings were advertised and held, one of which drew approximately 125 interested citizens and property owners. Changes to the draft ordinance were made in response to these public meetings. For example, in April 1930, the Maxwell Realty Company made suggestions that were adopted regarding sewer requirements (including the new sewers in Ashland Park). By the summer of 1930, Lexington had a formal zoning ordinance, prepared by the Planning and Zoning Commission and consultant Segoe, approved by the Mayor, presented to the City Commissioners, and tested in the Fayette County Court.

As this brief story suggests, Lexingtonians during the years 1928–1930 did not independently formulate and enact their ideas about urban zoning wholly in situ or within a cultural vacuum. A number of local actors drew upon people and information within a developing broader network that linked contemporaneous ideas taking place in, at least, Louisville, Cincinnati, St. Louis, and New York and connected through institutional structures such as regional and national conferences of professional city planners.

More generally, Lexington was among the hundreds of American urban places subscribing to an emerging discourse on city planning and zoning regulation. American zoning history usually is traced to New York City in the second decade of the twentieth century. The landmark test case for zoning, however, was *The Village of Euclid v. Ambler Realty Company* (1926). A conservative Supreme Court ruled in 1926 that industry and apartments could be viewed as similar to a "public nuisance" in single-family zones, prompting one scholar to claim that "zoning has provided the device for protecting the homogeneous, single-family suburb from the city" (Babcock 1966:3). The practice was supported enthusiastically by municipalities nationwide, and "zoning swept the country in the twenties" (Weaver and Babcock 1979; Babcock 1966:5).

Like other new zoning ordinances across the country, the Lexington Zoning Ordinance of 1930 partitioned the city and established land use rules, dictating what one could and could not do with individually owned "pieces" of the landscape. The basic unit of control in zoning's so-called "Euclidean envelope" is the zoning "lot," usually a parcel of land under single ownership/control, over which zoning "gives broad powers" (Cook 1980:17–20). As the present-day

Lexington Zoning Ordinance states, the strategic framework of zoning is created "to facilitate orderly and harmonious development in the visual and historic character of Lexington-Fayette Urban County." The zoned landscape is a disciplined landscape, and that discipline extends to the dweller within.

The regulation of space enacted was and is not absolute, however. Within the context of Ashland Park, through their everyday practices, residents made modifications to their property that were not always "up-to-code." Their individual tactics in the city, manifested in the material landscape, challenged the strategic discourse of zoning theory and practice. When cited for violations, many Lexingtonians took their appeals to the local hearing boards, who often granted variances or even permanently altered the planning code itself. They still do. The zoning ordinance is not "set in stone," but is continually under review and challenge as the process of strategies and tactics plays out through the local cultural landscape.

The local adoption of zoning principles joined Lexington and Ashland Park with an ever-developing discourse that included local residents, local government, city planners and zoning officials, outside consultants, the universities that train such professionals, other municipalities, and the courts that test and uphold the ideals inherent in zoning. Those ideals incorporate notions of spatial and visual discipline imposed on residents and landscapes. Like any law in a democracy, however, zoning rules are continually under scrutiny and subject to modification. Individual challenges to zoning rules serve to continually modify and reconstruct the rules, and in fact, are expected to do so, as evidenced by the common inclusion in urban policy and planning texts of numerous "case studies" of past zoning conflicts. When such conflicts are recorded for posterity as court cases, they not only alter the localities in which they occur but enter the zoning discourse to play out in future applications of zoning and planning strategy (e.g., Schneider 1980; Kenny 1992).

Historic Preservation[5]

Historic veneration, preservation, and a quest for architectural authenticity in the U.S. all are supported by a tradition notable for its nineteenth-century patriotic hagiography—the preservation of Mount Vernon, Monticello, Independence Hall—and the twentieth-century reconstruction of colonial Williamsburg (Stipe and Lee 1987; Zelinsky 1988). The discourse of historicity, especially as promoted by the historic preservation movement, is especially evident as it plays out through the Ashland Park landscape. Ashland Park became a federally registered historic district in Lexington in 1986, one of eighteen districts, ninety-six separate sites, and 3,704 buildings within the city so designated. National registry embeds the Ashland Park landscape within three governmental levels of historic regulation: the Lexington-Fayette Historic Commission (local), the Kentucky Heritage Council (state), and the National Park Service (federal). The Federal Register was established through the National Historic Preservation Act of 1966, and now includes more than 750,000 national entries institutionally supported by a National Trust Membership of almost 200,000 people and thirty-five university graduate and technical courses. The Ashland Park "historic" landscape is truly one node in a far-ranging network that includes ideas, ideals, institutions, regulation, and preservation: a disciplining discourse encompassing the local landscape and its residents.

Yet, while many of the general institutional and cultural contexts and connections for Ashland Park's historicity are directed by such extra-local imperatives, the *specific* details of the district's official "history" are unique to the area and constitute one part of the vast, nationwide, even international, historic preservation network. Historic preservation in Ashland Park primarily is grounded in the suburb's period architecture (the district's 505 buildings are nearly all residential and more than 97 percent are built in one of five nationally recognized styles), the "original design by the firm of the Olmsted Brothers," the "mature landscaping," and the remnant eighteen acres and great house of the original Clay estate (Kerr, nomination form, 1985). Among the first on the National Register, Ashland is now a museum that stands as the sociospatial focus of the neighborhood, a rallying point of invented tradition for the city and, especially, nearby residents. It is Lexingtonians who decide just how to venerate Clay and his estate within the historic preservation context. The fact that Clay himself actually died before the house on view was built seems not to detract from the site's status as a historic shrine. Similarly, the manner of historic representation is highly selective and locally determined. The house exhibit is frozen in the late nineteenth-century, and there are no traces of slaves quarters on the grounds, despite written records indicating their remnant presence on the former plantation during the time period represented. Local decisions about what is to be preserved within the larger cultural context and included in networks of knowledge *are* important and are a constant element in larger, extra-local processes.

The structural demands of (one version of) historical accuracy specified within the discourse of historicity are manifest in the landscape, which in turn serves as a local disciplinary mechanism. Local preservation groups, neighborhood associations, and real estate agents draw on the discourse to effect control over the landscape and the neighborhood's overall character. For example, recent attempts to include a teahouse on the Ashland grounds for summer visitors were seen to denigrate both the authenticity of the estate itself and, by association, the entire neighborhood. The resulting controversy prompted a special meeting of the Ashland Neighborhood Association. The locally distributed flyer advertising that meeting informed residents that the attempt to open a cafe was being promoted by the Museum's board of directors in order "to place Ashland on par with comparable national and international historic sites." It is significant, however, that within the national preservation network, there are possibilities for local resistance that have the potential to reach beyond the merely local. According to the same flyer, the board of directors also "invited the input of neighbors and neighborhood associations on this subject, prior to the Foundation reaching a final decision." No doubt the results of this local process will be noted by similar institutions nationwide, just as one public meeting that ensued over the cafe issue at Ashland included reference to historic sites in Tennessee and North Carolina.

The intersection of competing discourses to combine as habitus is illustrated in the potential link between historicity/historic preservation and zoning. Zoning usually applies to land use, or spatial practice, while historic preservation embeds a visual code. The combination provides a discipline for both space and landscape. Through zoning, the ideals circulating within the preservation discourse are made concrete in the form of law, albeit law that, like zoning strategies in general, is manipulable. One illustration of that intersection comes from a letter to the editor of the *Lexington Herald Leader* under the title "Don't Kill Neighborhoods" (Casey 1993). In it, an Ashland Park resident recounted his appearance before the Urban County Government Board of Adjustment in order to "oppose a request to change the building setback line to allow an upper front porch on Sycamore Road to be enclosed" (p. A12). The letter writer acknowledged that such a building action "may seem minor, not affecting the historic or aesthetic character of Ashland Park. However, incremental 'changes' . . . seem to be part of a subtle trend that may undermine the essence and stability of the neighborhood" (p. A12). The cultural landscape—in this case, "front porches . . . oversized garages and large additions" (Casey 1993:A12)—are the material habitus through which neighborhood social structures are disciplined, challenged, and, perhaps, ultimately altered.

The link between historic preservation and zoning generally is forged when a historic district is granted a regular zoning status, usually in the form of an "H-1 Overlay" classification. Although Ashland Park has yet to take that final preservation step to H-1 classification, all the preliminary pieces are in place, and the possibility of zoning reinforcement has been seriously raised. Many Ashland Park residents and local preservationists are keeping a keen eye on the Bell Court historic district, six blocks closer to the city center and twenty years older (see Figure 1). Bell Court is the city's most recent H-1 zone and is locally famous for one homeowning dissenter to the zoning overlay who has challenged the preservationist discourse (Staddon 1993). The homeowner was thrust into the preservation network when he failed to replace his front porch pillars with similar, historically relevant support posts. His response was to paint the porch columns on his highly visible Main Street house as if draped in the American flag, symbolically challenging one set of ideals with another. He also erected a very well-crafted, intricately lettered sign in his frontyard, reading: "If our founding fathers thought taxation without representation was TYRANNY, they should try living in LEXINGTON, KY in a HISTORIC DISTRICT." The Main Street homeowner's tactics for subverting the discipline of historic preservation will be worked out in time. The individual protest and its landscape manifestation will, in their very small way, become a part of the larger preservation discourse, with cumulative structural effects at district, city, state, and national scales, played out in the ongoing and cumulative interaction between structure and agency in and through the cultural landscape.

More immediately, the Bell Court preservation battle has and will continue to affect the Ashland Park landscape. An interview with a neighborhood architect, who also is an officer in the Ashland Neighborhood Association, provides a case in point. The architect has taken the possibility of H-1 zoning to the appropriate neighborhood committee, which balked at the idea, mostly in response to the negative publicity and public opposition galvanized against historic-district status through the case of the "flag pillars" (interview with Pohl, 3 March 1994).

The Neighborhood Association[6]

The strategies of historic authenticity and preservation are reproduced in Ashland Park by the Ashland Neighborhood Association (ANA), whose logo is a black and white drawing of Ashland, the Henry Clay estate. The ANA is part of yet another discourse, as one of more than 160 neighborhood associations in Lexington. Their activities are monitored in the local newspaper and by the Lexington-Fayette Urban County Government. The neighborhood associations often communicate with each other, whether informally through existing social networks, on an ad hoc basis when spatially contiguous associations have shared concerns, or formally, as in their recent coordination through the Fayette County Neighborhood Council, which has proposed additions to the urban area's comprehensive planning document. More generally, the ANA is part of the surge in neighborhood-based organizational activities and information networks that have developed in the U.S., especially since World War II (e.g., Paddison 1985).

The ANA coordinates visual and spatial strategies directly affecting the landscape and residents of Ashland Park. In the absence of an H-1 zoning classification, the ANA has taken a lead in monitoring and setting aesthetic guidelines drawn from diverse sources. For example, the president of the ANA recently appeared before the Lexington-Fayette County Board of Adjustment to oppose a resident's building variance request, citing the need to maintain the historic and aesthetic character of Ashland Park and urging the board to "place more emphasis on the aesthetic implications of their decisions" or else "Ashland Park may no longer be seen as unique or desirable and we all would lose" (Casey 1993:A12). The ANA also is active in seeking out and implementing projects such as matching funds programs to purchase historical plaques and to plant replacement street trees in keeping with the original Olmsted design. The ANA Newsletter, "Ashland News," regularly urges neighborhood residents to consider the material integrity of original design when renovating houses: "We urge you to be very careful not to accept building materials or practices that are not in keeping with the historic nature of the area" (Sept–Oct. issue 1992). It even reminds residents to return garbage and recycling cans to the back of the house in a timely manner after streetside collection. The ANA takes organizational and operational clues from an extra-local discourse that includes other neighborhood associations and their activities across the country.

Yet the ANA's organizational and disciplinary structure is equally constituted by relatively independent local actors. The ANA is democratically structured with a constitution and bylaws, the election of six officers on a regular basis at an annual public meeting in a nearby school, twelve block captains, and a membership open to anyone living in the neighborhood. There is opposition to ANA positions, such as the presently brewing battles over historic preservation and over the implications of proposed "neighborhood watch" committees. ANA disciplinary strategies involving life, landscape, and place do not preclude challenge and alteration, and these tactics affect not only the immediately local, but eventually, in some small way, the entire discourse of neighborhood associations and community preservation activists.

Consumption

The final discourse briefly illustrated here is probably the one most obvious in any American landscape, capturing the theme of consumption in a capitalist economy. Within Ashland Park, the design of the period houses, the construction materials, the paint and double-glazed windows, and even the house numbers were all bought as commodities, and together these elements make a statement about who the residents think they are or want to be (e.g., Sack 1992). It is incorrect to assume that each of these commodities is *only* seen in terms of its economic value and is everywhere and always translatable by the everyday landscape reader into dollar terms. The cultural landscape is not simply epiphenomenal to the economy. People attach cultural meaning to objects separate from the economic value of those objects. In this way, consumption practices, undoubtedly overlain by symbolic associations, play an important role in shaping the contemporary American landscape. The network of knowledge involving American consumption patterns entails a strong structural discipline for the average suburban resident.

The individual Ashland Park resident who goes to the hardware store on a Saturday morning and buys a gallon of "Williamsburg Blue" paint for the family room or an imitation coach lamp for the front porch appears to be making an independent consumption decision about the product, and its placement in the cultural landscape. That decision, however, is in part structured by what is available at the hardware store or by what is advertised on the television. Even garden planting choices—rosebushes, shrubs, or spring annuals—are bought from local nurseries and plant catalogues. All of these choices are made available by the discourse that includes product designers, manufacturers, advertisers, wholesalers, and retailers, and even the similar choices of friends and neighbors, or the recommendations of groups like the ANA. Consumers are disciplined in their choices through the illusion of choices; they choose, however, from what is already in place.

The capitalist discourse of consumption is perhaps the most sensitive of all to individual human action. Retailers monitor inventory, recording popular items and ordering more from the product chain, canceling those that do not sell well. Any successful manufacturer and advertiser is sensitive to changing consumer tastes, no matter how manipulated the cynic might claim those tastes to be. "The Market" is continually in the process of altering the array of products available within the network of consumption. The consumer is therefore disciplined at the same time he or she is changing the discourse itself simply through purchase.

Beyond Ashland Park

The Ashland Park landscape exists as a visible ensemble of houses, streets, trees, gardens, and related elements that can be conceptualized as the tangible intersection of several discourses, including the examples of landscape architecture, insurance mapping, zoning, historic preservation, neighborhood associations, and consumption. While the Ashland Park landscape—or any U.S. landscape—in toto stands as a relatively unchanging, monolithic ensemble, conceptualizing elements within that ensemble—house paint, building materials, garden shrubs—provide access to the discourse(s). Each discourse includes ideals, guiding principles, and often even explicit institutionally mandated instructions that serve to discipline Ashland Park residents. Park residents, in turn, can accept, challenge, modify, and ultimately constitute those ideals through their individual landscape contributions that individually appear as seemingly minute increments of landscape change. Conceptualizing a landscape within a set of discourses also provides a vision of landscape change (over time) attributable, in part, to individual, fragmented activity rather than simply presented as strata of (superorganic) cultural accretion(s). Furthermore, the discourses provide conceptual connections between different places and landscapes. In the case of Ashland Park, for example, these same discourses of landscape architecture, historic preservation, and zoning intersect in other places; yet the possibilities for individual contribution to the landscape and the discourses insure that while there may be *types* of landscapes (e.g., suburbs), no two cultural landscapes will be exactly alike.

I have presented just a few (albeit important) examples of discourses and their physical manifestation in a particular American landscape. There obviously are other discourses intersecting in any given locale. Part of the task of landscape interpretation is to recognize the interpreter's (often unconscious) ordering of those discourses in terms of their centrality to any interpretation. Some would wish to see an *exact* ordering within a finite list of the most important discourses ordering any cultural landscape. Some will want to see this conceptual framework not as an aid to interpretation, but in a more empiricist or positivist sense, as a *model*, universally transportable and applicable. That is precisely what my proposed framework is not. I have not created a universal model of landscape analysis that would, positively and with certainty, identify *exactly* which discourses intersecting in the landscape ought to be first interrogated to find the exact origins, meaning, and intention of social (re)production inherent in the material environment. I propose not a master narrative but a framework with possibilities for contestatory readings and interpretations. To invoke an oft-cited quotation from Mann (1986): "Societies are not unitary . . . they are not totalities. . . . Societies are much *messier* than our theories of them" (see also Gregory 1989:84, Barnes and Duncan 1992a).

This framework and its tentative explication through the example of Ashland Park should be taken as a step beyond an objectivist, empiricist account of landscape description through model building. This essay might join a number of "signposts toward a poststructuralist geography" (Natter and Jones 1993), stemming, in part, from the so-called "crisis of representation" and the critique of positivism in the human sciences (Marcus and Fisher 1986). The notion of "signposts" in this case has a double meaning: first, where the presented framework stands as an intellectual argument informed by and contributing to recent geographic scholarship on postmodernism and poststructuralism specifically relating to the cultural landscape (e.g., Duncan and Ley 1993; Barnes and Duncan 1992b); and second, in a more material sense, where symbolic textual qualities of the landscape serve as tangible and visible "markers" of intertextuality. The landscape-as-text becomes one moment framed by and constitutive of larger discourses. In this sense, discourses are both enabling and constraining. As Barnes and Duncan (1992b:8) put it, "this is because discourses are not unified, but are subject to negotiation, challenge, and transformation. For power relations within a social formation are communicated, and sometimes resisted, precisely through the medium of particular discourses." And in this case, the cultural landscape quite literally stands as a "medium."

The cultural landscape serves to naturalize or concretize—to make *normal*—social relations as embodied in the various discourses and their combinations. This conceptual framework exposes the landscape, like Massey's (1993) "place," as the locus of articulated social relations. The landscape-as-mimesis is challenged in order to open the interpretive process. The act of landscape *interpretation* thus can become exactly that, situating interpreters within the cultural landscape they interpret, one part of a larger hermeneutic circle (e.g., Pickles 1992). A close reading of a landscape in a specific time and place will yield an interpretation of its position within a number of discourses. The key to understanding the place of landscape is *not* to present a totalizing, naturalized, and concretized (i.e., unchanging) image of a particular scene: the landscape as the sum total of history. The act of interpretation, and thus of understanding the place of the landscape, should be processual and reflexive, open to the challenge of new information and alternative interpretations. Ultimately the landscape-as-text is unstable and requires constant reinterpretation.

What makes the U.S. landscape a particularly difficult interpretive task is the seeming paradox between the ideals of liberal individualism and the cultural landscape's disciplinary capabilities. We want to simultaneously acknowledge our belief in free will and our rights to exercise social-political-economic choice; at the same time, we recognize the restrictions placed upon those choices within specific contexts. Every U.S. landscape involves countless individual, often self-interested decisions, each made from a range of choices often beyond our immediate control. The concept of the landscape as discourse materialized adopted and adapts the Foucauldian conception of power as both netlike and circulating. Understood as a material moment in a recurring flow of information/ideals/actions/power, the cultural landscape exists as a crucial point in and of power, as a place where action can contribute to, as well as be constricted by, the ideals that cohere the discursive network. Through the landscape, the human agent is both object and subject.

This attempt to capture through the landscape both the strategies and tactics of structural constraints and individual agency is at the same time an attempt to move beyond the structure-agency dualism. Like the question of the chicken and the egg, it is not possible to ascertain "which came first." Structure/agency and strategy/tactics are mutually constitutive, just as the landscape, as materialized discourse, embeds both disciplinary constraints and the possibilities for human agency.

The American cultural landscape *is* interesting for what it has to "say" about "us", just as its normalizing, normative capabilities simultaneously make the landscape central to the ongoing production and reproduction of place and identity (individual and collective). The heuristic utility of the "discourse" avoids a superorganic concept of either the landscape or the "culture" that produced it/is reproduced by it. It directs us instead to the actual people, events, and landscapes of particular places within wider geographical contexts.

Notes

1. Although this point is indisputable, it nevertheless may represent a general humanistic retreat into the realm of "the subjective" at a time when positivism and rational, scientific inquiry threatened to dominate the social sciences (see Pickles 1985). Furthermore, Meinig's seemingly autonomous observer requires theorizing, perhaps as a situated subject within geographies of knowledge, including systematized networks of knowledge which simultaneously inform and are informed by such thematic empirical observations (e.g., see Thrift 1985).
2. See Sauer (1941) on the importance of immersing oneself in a study area.
3. Primary materials used to support this section include: (1) Beveridge, Charles E. and Carolyn F. Hoffman, *The Master List of Design Projects of the Olmsted Firm, 1857–1950*, Boston: Massachusetts Association for Olmsted Parks, 1987. (2) Report to Professor M.A. Scovell, President Board of Park Commissioners, Lexington, Kentucky by Olmsted Brothers, 1904. Typescript copy in the records of the Ashland Neighborhood Association, Lexington, Kentucky. (3) Kerr, Bettie L., *National Register of Historic Places Inventory—Nomination Form*. 1985. From the archives of the Kentucky Heritage Council, Frankfort, Kentucky. (4) Correspondence of William C. McDowell. Xerox reproductions from the collections of the Manuscript Division, Library of Congress. In the records of the Ashland Neighborhood Association, Lexington, Kentucky, including: (a) Letters from McDowell to the Olmsted Brothers firm dated 9-4-1925, 11-23-1925, 6-5-1926 (telegram), 6-17-1926 (telegram), 11-9-1926; (b) letters from the Olmsted firm to McDowell, dated 9-10-1925, 11-18-1925, 12-4-1925, 6-5-1926 (telegram), 6-7-1926, 10-22-1926, 11-12-1926; (c) Report of a visit by F. B. Smith to William C. McDowell, 6-30-1926.
4. Primary materials used to support this section include: (1) *Lexington Planning Commission Minutes*, 1928–30. Microfilm copies held in the Special Collections department of M. I. King Library, University of Kentucky, Lexington. (2) *Lexington-Fayette Urban County Government (LFUCG) Zoning Ordinance* (1991), published by the LFUCG and available in the Government Documents department of M. I. King Library, University of Kentucky.
5. Primary materials used to support this section include: (1) Kerr 1985 (note 3). (2) An interview with Graham Pohl conducted on March 3, 1994. (3) The December 2, 1994 edition of the "Ashland News," a xeroxed newsletter distributed locally to Ashland Park residents. (4) A 1987 typescript manuscript of National Register Historic Districts prepared by and available at the LFUCG Planning Division.
6. Primary materials used to support this section include: (1) "Ashland News," issues from October 1991 through May 1995. (2) Letter and public documents from open meetings of the Fayette County Neighborhood Council (FCNC), dated February 15, 1994 (FCNC, 360 Hampton Court, Lexington, KY 40508). (3) *Neighborhood Association Handbook* published by and available at the LFUCG (1993).

References

Agnew, John. 1993. Representing Space: Space, Scale, and Culture in Social Science. In *place/culture/representation*, ed. James Duncan and David Ley, pp. 251–71. New York: Routledge.

Anderson, Kay, and Gale, Fay. 1992. *Inventing Places*. Melbourne: Longman Cheshire.

Babcock, Richard F. 1966. *The Zoning Game*. Madison: University of Wisconsin Press.

Barnes, Trevor J., and Duncan, James S., eds. 1992a. *Writing Worlds: Discourse, Text and Metaphor in the Representation of Landscape*. New York: Routledge.

——— and ———. 1992b. Introduction: *Writing Worlds*. In *Writing Worlds: Discourse, Text and Metaphor in the Representation of Landscape*, ed. Trevor J. Barnes and James S. Duncan, pp. 1–17. New York: Routledge.

Bondi, Liz. 1992. Gender Symbols and Urban Landscape. *Progress in Human Geography* 16:157–70.

Bourdieu, Pierre. 1977. *Outline of a Theory of Practice*. New York: Cambridge University Press.

Brown, Michael P. 1993. The Possibility of Local Autonomy. *Urban Geography* 13:257–79.

de Certeau, Michel. 1984. *The Practice of Everyday Life*. Berkeley: University of California Press.

Caro, Robert A. 1974. *The Power Broker*. New York: Knopf.

Casey, Russell. 1993. Letter to the Editor. *Lexington Herald-Leader*, January 30, p. A12.

Clark, Thomas D.; Magrish, Robert; and Green, Terry L. 1996. *Ashland, the Henry Clay Estate*. Greendale, IN: The Creative Company.

Conzen, Michael P., ed. 1990. *The Making of the American Landscape*. Boston: Unwin Hyman.

Cook, Robert S., Jr. 1980. *Zoning for Downtown Urban Design*. Lexington, MA: Lexington Books.

Cosgrove, Denis E. 1984. *Social Formation and Symbolic Landscape*. Totowa, NJ: Barnes and Noble.

———. 1985. Prospect, Perspective and the Evolution of the Landscape Idea. *Transactions of the Institute of British Geographers N.S.* 10:45–62.

———. 1993. Commentary on "The Reinvention of Cultural Geography" by Price and Lewis. *Annals of the Association of American Geographers* 83:515–17.

——— and Stephen Daniels, eds. 1988. *The Iconography of Landscape*. New York: Cambridge University Press.

Crary, Jonathan. 1990. *Techniques of the Observer.* Cambridge, MA: MIT Press.

Daniels, Stephen. 1989. Marxism, Culture, and the Duplicity of Landscape. In *New Models in Geography,* vol. 2, ed. R. Peet and N. Thrift, pp. 196–220. London: Unwin Hyman.

Davis, Darrell Haug. 1927. *The Geography of the Blue Grass Region of Kentucky.* Frankfort, KY: Kentucky Geological Survey.

Davis, Mike. 1990. *City of Quartz*. New York: Verso.

Duncan, James S. 1980. The Superorganic in American Cultural Geography. *Annals of the Association of American Geographers* 70:181–98.

———. 1990. *The City as Text: The Politics of Landscape Interpretation in the Kandyan Kingdom*. New York: Cambridge University Press.

———. 1993. Commentary on "The Reinvention of Cultural Geography" by Price and Lewis. *Annals of the Association of American Geographers* 83:517–19.

Duncan J., and Duncan, N. 1988. (Re)reading the Landscape. *Environment and Planning D: Society and Space* 6:117–26.

——— and Ley, David, eds. 1993. *place/culture/representation.* New York: Routledge.

Foucault, Michel. 1980. *Power/knowledge,* ed. Colin Gordon. New York: Pantheon Books.

Froehling, Oliver. 1993. Living in a Park: The Development of a Suburb in Lexington, Kentucky. Manuscript, Department of Geography, University of Kentucky, Lexington, Kentucky.

Giddens, Anthony. 1990. *The Consequences of Modernity*. Cambridge: Polity Press.

Gottdiener, Mark. 1985. *The Social Production of Urban Space*. Austin: University of Texas Press.

Gregory, Derek. 1989. Areal Differentiation and Post-Modern Human Geography. In *Horizons in Human Geography*, ed. Derek Gregory and Rex Wolford, pp. 67–96. London: Macmillan.

Hareven, Tamara. 1991. The Home and the Family in Historical Perspective. *Social Research* 58:253–85.

Harley, J. B. 1988. Maps, Knowledge, and Power. In *The Iconography of Landscape*, ed. Denis Cosgrove and Stephen Daniels, pp. 277–312. New York: Cambridge University Press.

Hart, John Fraser. 1975. *The Look of the Land*. Englewood Cliffs, NJ: Prentice Hall.

Jackson, John Brinckerhoff. 1984. *Discovering the Vernacular Landscape*. New Haven: Yale University Press.

———. 1972. *American Space*. New York: W.W. Norton.

Jackson, Kenneth. 1985. *Crabgrass Frontier*. New York: Oxford University Press.

Jackson, Peter. 1993. Berkeley and Beyond: Broadening the Horizons of Cultural Geography. *Annals of the Association of American Geographers* 83:519–20.

Jenkins, Virginia Scott. 1994. *The Lawn: A History of an American Obsession*. Washington: Smithsonian Institution Press.

Kenny, Judith. 1992. Portland's Comprehensive Plan as Text: The Fred Meyer Case and the Politics of Reading. In *Writing Worlds: Discourse, Text and Metaphor in the Representation of Landscape*, ed. Trevor J. Barnes and James S. Duncan, pp. 176–92. New York: Routledge.

Kniffen, Fred B. 1965. Folk Housing: Key to Diffusion. *Annals of the Association of American Geographers* 55:549–77.

Knox, Paul L., ed. 1993. *The Restless Urban Landscape*. Englewood Cliffs, NJ: Prentice Hall.

Lefebvre, Henri. 1991. *The Production of Space,* trans. Donald Nicholson-Smith. Cambridge, MA: Blackwell.

Lewis, Peirce. 1983. Learning from Looking: Geographic and Other Writing about the American Cultural Landscape. *American Quarterly* 35:242–61.

———. 1979. Axioms for Reading the Landscape. In *The Interpretation of Ordinary Landscapes,* ed. D.W. Meinig, pp. 11–32. New York: Oxford University Press.

Lipietz, Alain. 1993. The Local and the Global: Regional Individuality or Interregionalism? *Transactions of the Institute of British Geographers* 18:8–18.

Mann, Michael. 1986. *The Sources of Social Power*. Cambridge: Cambridge University Press.

Marcus, George E., and Fischer, Michael M. J. 1986. *Anthropology as Cultural Critique.* Chicago: University of Chicago Press.

Marsh, Margaret. 1990. *Suburban Lives.* New Brunswick, NJ: Rutgers University Press.

Massey, Doreen. 1984. Geography Matters. In *Geography Matters!: A Reader*, ed. Doreen Massey and John Allen, pp. 1–11. New York: Cambridge University Press.

———. 1993. Power-Geometry and a Progressive Sense of Place. In *Mapping the Futures: Local Cultures, Global Change,* ed. Jon Bird et al., pp. 59–69. New York: Routledge.

Meinig, D. W., ed. 1979a. *The Interpretation of Ordinary Landscapes*. New York: Oxford University Press.

———. 1979b. The Beholding Eye. In *The Interpretation of Ordinary Landscapes,* ed. D. W. Meinig, pp. 33–50. New York: Oxford University Press.

———. 1979c. Symbolic Landscapes. In *The Interpretation of Ordinary Landscapes*, ed. D. W. Meinig, pp. 164–94. New York: Oxford University Press.

Meyer, David R. 1990. The New Industrial Order. In *The Making of the American Landscape*, ed. Michael P. Conzen, pp. 249–68. Boston: Unwin Hyman.

Mills, C. 1988. 'Life on the Upslope': The Postmodern Landscape of Gentrification. *Environment and Planning D: Society and Space* 6:169–90.

Mitchell, Don. 1994. Landscape and Surplus Value: The Making of the Ordinary in Brentwood, California. *Environment and Planning D: Society and Space* 12:7–30.

Murray-Wooley, Carolyn, and Raitz, Karl. 1992. *Rock Fences of the Bluegrass*. Lexington: University Press of Kentucky.

Nast, Heidi, and Wilson, Mabel. 1994. Lawful Transgressions: This Is the House That Jackie Built . . . *Assemblage* 24:48–56.

Natter, Wolfgang, and Jones, John Paul III. 1993. Signposts toward a Poststructuralist Geography. In *Postmodern Contentions: Epochs, Politics, Space*, ed. John Paul Jones III, Wolfgang Natter, and Theodore Schatzki, pp. 165–203. New York: Guilford.

Noble, Allen G. 1984. *Wood, Brick, and Stone: The North American Settlement Landscape*. Amherst: University of Massachusetts Press.

Paddison, R. 1985. Local Democracy in the City. In Progress in Political Geography, ed. Michael Pacione, pp. 216–42. London: Croom Helm.

Pickles, John. 1985. *Phenomenology, Science, and Geography: Spatiality and the Human Sciences*. New York: Cambridge University Press.

———. 1992. Texts, Hermeneutics, and Propaganda Maps. In *Writing Worlds: Discourse, Text and Metaphor in the Representation of Landscape*, ed. Trevor J. Barnes and James S. Duncan, pp. 193–230. New York: Routledge.

Price, Marie, and Lewis, Martin. 1993a. The Reinvention of Cultural Geography. *Annals of the Association of American Geographers* 83:1–17.

——— and ———. 1993b. Reply: On Reading Cultural Geography. *Annals of the Association of American Geographers* 83:520–22.

Raitz, Karl B. 1980. *The Kentucky Bluegrass*. Chapel Hill, NC: University of North Carolina at Chapel Hill, Studies in Geography, No. 14.

Reisenweber, Julie, and Hudson, Karen. eds. 1990. *Kentucky's Bluegrass region*. Frankfort, KY: Kentucky Heritage Council.

Relph, Edward. 1987. *The Modern Urban Landscape*. Baltimore: Johns Hopkins University Press.

Reps, John W. 1965. *The Making of Urban America*. Princeton, NJ: Princeton University Press.

Ristow, Walter W. 1968. United States Fire Insurance and Underwriters Maps, 1852–1968. *Quarterly Journal of the Library of Congress* 25:194–218.

Roberts, Susan M. Forthcoming. Scalar Dynamics. In *Disciplining Boundaries*, ed. W. Natter. New York: Guilford Press.

——— and Schein, Richard. 1993. The Entrepreneurial City: Fabricating Urban Development in Syracuse, New York. *The Professional Geographer* 45:21–32.

Sack, Robert. 1992. *Place, Modernity, and the Consumer's World*. Baltimore: Johns Hopkins University Press.

Samuels, Marwyn S. 1979. The Biography of Landscape. In *The Interpretation of Ordinary Landscapes*, ed. D.W. Meinig, pp. 51–88. New York: Oxford University Press.

Sanborn Map Co. 1886. *Lexington, Kentucky*. New York: Sanborn Map and Publishing Co.

———. 1986. *Insurance Maps of Lexington, Kentucky*. New York: Sanborn-Perris Map Co.

———. 1907. *Insurance Maps of Lexington, Fayette County, Kentucky*. New York: Sanborn Map Co.

———. 1934. *Insurance Maps of Lexington including Greendale Kentucky*. New York: Sanborn Map Co.

———. 1936. *Surveyor's Manual for the Exclusive Use and Guidance of Employees of the Sanborn Map Company*. New York: Sanborn Map Co.

———. 1953. *Description and Utilization of the Sanborn Map*. New York: Sanborn Map Co.

———. 1958. *Insurance Maps of Lexington*. New York: Sanborn Map Co.

Sauer, Carl O. 1925. The Morphology of Landscape. Reprinted in *Land and Life: A Selection from the Writings of Carl Ortwin Sauer*, ed. John Leighly, pp. 315–50. Berkeley: University of California Press, 1963.

———. 1941. Foreword to Historical Geography. *Annals of the Association of American Geographers* 31:1–24.

Schein, R. H. 1993. Representing Urban America: 19th-Century Views of Landscape, Space, and Power. *Environment and Planning D: Society and Space* 11:7–21.

Schneider, Mark. 1980. *Suburban Growth*. Brunswick, OH: King's Court Communications, Inc.

Sies, Mary Corbin. 1987. The City Transformed: Nature, Technology, and the Suburban Ideal, 1877–1917. *Journal of Urban History* 14:81–111.

Soja, Edward. 1989. *Postmodern Geographies*. New York: Verso.

Sorkin, Michael, ed. 1992. *Variations on a Theme Park*. New York: Noonday Press.

Staddon, Caedmon. 1993. Gentrification and Historic Preservation: The View from Lexington, Kentucky. Paper presented at the Annual Meeting of the Association of American Geographers, Atlanta, Georgia.

Stilgoe, John R. 1982. *Common Landscape of America, 1580–1845*. New Haven: Yale University Press.

———. 1988. *Borderland: Origins of the American Suburb, 1820–1930*. New Haven: Yale University Press.

Stipe, Robert E., and Lee, Antoinette J. eds. 1987. *The American Mosaic: Preserving a Nation's Heritage*. Washington: U.S. Committee/International Council on Monuments and Sites.

Thrift, Nigel. 1985. Flies and Germs: A Geography of Knowledge. *In Social Relations and Spatial Struc-*

tures, ed. Derek Gregory and John Urry, pp. 366–403. London: Macmillan

Ulack, Richard; Raitz, Karl B.; and Hopper, Hillary Lambert. 1994. *Lexington and Kentucky's Inner Bluegrass Region*. Indiana, PA: National Council for Geographic Education, Pathways in Geography Series, No. 10.

Village of Euclid v. Ambler Realty. 1926. 272 U.S. 365.

Walker, Richard A. 1981. A Theory of Suburbanization: Capitalism and the Construction of Urban Space in the United States. In *Urbanization and Urban Planning in Capitalist Society*, ed. Michael Dear and Allen J. Scott, pp. 383–429. New York: Methuen.

Weaver, Clifford L., and Babcock, Richard F. 1979. *City Zoning: The Once and Future Frontier*. Washington: Planners Press.

Williams, Raymond. 1983. *Keywords*. London: Fontana Paperbacks.

Wright, John D., Jr. 1982. *Lexington: Heart of the Bluegrass*. Lexington, KY: Lexington-Fayette County Historic Commission.

Zelinsky, Wilbur. 1988. *Nation into State*. Chapel Hill: University of North Carolina Press.

Zube, Ervin H., and Zube, Margaret J. 1977. *Changing Rural Landscapes*. Amherst: University of Massachusetts Press.

Zukin, Sharon. 1991. *Landscapes of Power: From Detroit to Disney World.* Berkeley: University of California Press.

Correspondence: Department of Geography, University of Kentucky, Lexington, KY 40506-0027, email schein@pop.uky.edu.

[6]

Japanese gardens in Edwardian Britain: landscape and transculturation

Setsu Tachibana, Stephen Daniels and Charles Watkins*

School of Geography, University of Nottingham, University Park, Nottingham NG7 2RD, UK

Abstract

This article examines ideas of gardening, landscape and transculturation in Edwardian Britain through the fashion for Japanese gardens. Emphasis is placed on the writing and practice of two influential figures: Josiah Conder (1855–1920) and Reginald Farrer (1880–1920). Conder was one of the leading proponents of Japanese gardens and his book *Landscape Gardening in Japan* (1893) was a crucial source of ideas about Japanese gardens in the English speaking world. Farrer, who wrote extensively on rock gardens and developed his own plant nursery, was strongly influenced by his travels in East Asia and Japan. We focus on the creation of three different gardens in Britain: Newstead Abbey, Nottinghamshire; Cowden Castle, Clackmannanshire and Ingleborough Hall, Yorkshire. We show how Ethel Webb, the owner of Newstead Abbey, was strongly influenced by Conder's book and used it to produce a pattern book Japanese garden. At Cowden Castle, the employment of Japanese garden designers and gardeners helped in the creation and management of an 'authentic' Japanese garden. At Ingleborough Hall, Farrer created his own 'natural' rock gardens using exotic alpine plants. His gardening and estate management around Clapham village can be seen as an attempt to create a 'hybrid' landscape garden.

Introduction

> A garden in Japan is a representation of the scenery of the country, though it is essentially a Japanese representation. Favourite rural spots and famous views serve as models for its composition and arrangement. The laws of natural growth and distribution are closely studied and punctiliously applied in the management of even the smallest detail. The artificial hills, rocks, lakes, torrent beds, and cascades of gardens are copied from striking features in the varied landscape of the country. (Josiah Conderig 1964)[1]

* Corresponding author.
E-mail address: charles.watkins@nottingham.ac.uk (C. Watkins).

Not to plunge into the bottomless sea of their mysticism and symbolism in design, the sight of a Japanese garden is enough to bring tears of ecstasy to the eyes of any garden-lover. No distortions, no abortion, no discords are there, but some corner of landscape carefully copied to scale, with a sense of harmony and perspective so perfect that in a cottage yard four yards square you will seem to have half a mountainside. (Reginald Farrerig 1904)[2]

Japanese gardens were highly fashionable in Edwardian Britain, and in the decade before the outbreak of the First World War became something of a craze. Like the term 'English garden' which it helped redefine 'Japanese garden' denoted a variable national style, a shifting blend of designs and materials, of varying historical and geographical origin, packaged for consumption in a thriving horticultural market. While central to a sense of native, even insular, national identity, gardens have proved one of the main contact zones for cultural exchange between Britain and the world, both its imperial rivals like Holland and France and its subject territories. The process of transculturation in gardens has operated at the level of design, in both silent appropriation of styles and their conscious reproduction as 'Italianate', 'Dutch', 'American' and so on, and at the level of materials, notably plants, with physical adaptation and hybridisation. As ever, this process was enabled by broader transfers of expertise, patronage, technology and capital.[3]

In this paper, we consider how the exotic style of Japanese gardens was incorporated and hybridised in Britain. We use detailed archival and field evidence for three gardens, which display different stories of cultural exchange and assimilation. The makers of the three gardens all visited Japan and transferred different aspects of Japanese gardens to their estates. At Newstead Abbey and Cowden Castle, the gardens were strongly influenced by Josiah Conder's popular text *Landscape Gardening in Japan* of 1893. At Cowden, authenticity was further attempted through the employment of Japanese garden designers and a Japanese gardener. Reginald Farrer at Ingleborough Hall used a horticultural and philosophical understanding of Japanese rock gardens to modify his Yorkshire landscape and create hybrid gardens. Through an understanding of the making of these three gardens, the paper shows how British ideas of Japanese landscapes and nature were informed by a variety of transcultural flows of ideas, images and plants.

The fashion for Japanese gardens spread throughout Europe and the United States, stimulated by a series of related developments following the opening of Japan's ports in the mid-19th century which ended 250 years self-imposed isolation: the introduction and domestication of trees, shrubs and flowers; the vogue for design, arts and crafts known as 'Japonisme'; the publication of travel books and tourist guides; the flourishing export market in mature plants and bulbs and stone ornaments; and, perhaps most significantly, the display of gardens in international exhibitions, which received visitors in their millions and extensive coverage in the popular press, including the many illustrated garden magazines.[4] Western nations made competing claims on Japanese culture, as they did to influence its modern development. In the United States, where it was routinely claimed that Japan was effectively 'opened' by an American squadron led by Commodore Perry, Japanese artefacts were seen as a challenge to European cultural supremacy, and an incentive to American ambitions in the Pacific. In Britain, Japan was increasingly regarded as a global partner, with a comparable imperial history in which gardens featured strongly (Fig. 1)[5].

In the 1860s nurserymen John Gould Veitch and Robert Fortune travelled to Japan to gather plants new to the West from nurseries in the horticultural centre of Yedo, temple gardens on the city periphery and, from an expedition to the summit of the sacred mountain Fuji-yama. In reports, to the garden press and Royal Geographical Society, they marvelled at a garden culture among all ranks that appeared to

Fig. 1. Commemorative plate from the Japan British Exhibition of 1910, showing two Island Empires. Source: Private collection.

mirror that in their own country.[6] They made a fortune auctioning plants and stocking their nurseries, advertising them in gardeners' magazines and horticultural shows, fuelling the fashion for Japanese flowers, trees and shrubs such as lilies, magnolias, conifers, maples, bamboos and flowering cherries.[7] Indeed one of the earliest Japanese gardens was created out of the Veitch nurseries in Coombe Wood, Surrey, if, with its spreading shrubs, colourful underplanting and bright orange pavilions and bridges, it bore little resemblance to the consciously authentic, designed gardens promoted from the end of the century. While some Japanese plants were valued as specimens which demanded carefully controlled conditions, others were valued for 'hardiness', flourishing in uncultivated environments including smoky industrial cities, marshland and heathland, and showing throughout the year. Some plants, like *Aucuba japonica*, were thoroughly assimilated, others like the Japanese larch, physically hybridised.[8] Such so called 'wild gardening', with hardy exotics combined with native flora, was popularised in a revivified English landscape art.[9]

A Japanese garden displayed at the 1873 International Exhibition in Vienna was re-assembled in the grounds of London's Alexandra Palace, with the idea that it would be staffed by Japanese from whom visitors would be able to purchase Japanese goods and curiosities. The design itself recalled the English invented, stereotyped far-eastern landscape, familiar on porcelain plates hanging in many suburban villas. 'There is the little mound, the bridge, the stream with which are familiar on the Chinese and Japanese plates and dishes', noted *The Garden*, 'we seem only to want the Weeping Willow to make it complete'.[10] Connoisseurs of *japonisme* in gardens, as in other aspects of culture, sought to strip away the taint of *chinoiserie*. With its fall from imperial power, China was increasingly regarded as a decayed and degraded civilisation, its treatment of nature typically tortuous and perverse, in contrast to what was seen as a vital, natural culture of Japan.[11] Respect for Japan increased with its rise as the major imperial power in the East, after victorious wars with China (1895) and Russia (1905) and the establishment of colonies on the Asian mainland and Pacific islands. Unequal treaties were renegotiated; an Anglo-Japanese alliance of 1902 was renewed in 1905 and 1911.[12] Commentators adjusted their cultural evaluations, seeking similarities in such realms as racial physiognomy and physical geography, including climatic and topographical conditions regarded as favourable to the culture of gardening. By the turn of the century, exhibitions of Japanese culture in Europe were sponsored by the Japanese government as sites of cultural exchange.[13]

From the 1880s, the Meiji government instituted a policy of employing western experts, known as *oyatoi*, to help modernise administration and infrastructure, and many aspects of national culture, including architecture, banking, planning, education, law and transportation.[14] 'English' parks and gardens were laid out around the occidental buildings of the treaty port of Yokohama and the rebuilt and renamed capital Tokyo, sometimes on the sites of the *daimyo* gardens of the former feudal elite.[15] Europeans were employed to staff new art schools, introducing western conventions in the theory and practice of landscape art. The writings of John Ruskin had a profound influence on Japanese landscape tastes, notably for mountain scenery and its scientific appreciation. Japanese texts like *Nihon Fukeiron* (A Discourse on Landscape) by Shiga Shigetaka were keen to seek equivalencies or precedents in Japanese culture for western conventions.[16] Not surprisingly some of the resident *oyatoi*, as well as visiting western tourists and artists, were attracted to what was called the Old Japan, in contrast to the New Japan which made their views possible. Conflicts between *kaika*, enlightenment and *injun* traditions were a frequent subject of commentary in Japan, for example, in popular prints, and in reaction to official occidentalism where some affirmation and renewal of traditions, not least in garden design took place (Fig. 2).[17] The government encouraged an export-oriented horticultural industry. In Yokohama, large nurseries were established, such as L Boehmer's (originally owned by an *oyatoi* commissioned for his forestry expertise) and the Yokohama Nursery Company. British based nurseries stocked an increasing range of Japanese plants and ornaments, as did the leading arts and crafts design store Liberty's which advertised Japanese stone lanterns alongside its Celtic style bird baths.[18]

A series of English landscape painters and illustrators visited Japan, making reputations in both Tokyo and London with exhibitions and publications. Alfred Parsons visited Japan in 1892, exhibiting his watercolours of temple gardens, woodland groves and flower shows at the Tokyo Art School, which he later published in English and American magazine articles and books. Parsons promoted a deeper cultural understanding of gardens in Japan than was current in the West, but also helped recast gardens for western consumption through his picturesque compositions and focus on flowers (in Japan traditionally seen as of secondary importance to stonework). The pictures framed Parson's own increasingly involvement in garden culture in England, often in a consciously 'English' style, both as illustrator and designer.[19] In turn, some Japanese artists took up residence in London, notably Markino Yoshio, renowned in literary circles for his sage conversation and writings as well as paintings of the mist shrouded city.[20] The planning and building of the first 'Japanese gardens' in Britain was publicised in the horticultural press. In the *Journal of the Royal Horticultural Society*, James Hudson,

Fig. 2. Japanese popular print, Kaika injun kohai kagami, c 1873. Source: Kobe-shiritsu-Hakubutsukan (Eds) (Kobe City Museum eds) *Meiji Tetsudo Nishiki-e* (*Colored Woodblock Prints on Railway in the Meiji Era*) (Kobe 1994).

overseeing the Japanese Garden at Gunnersbury, Middlesex, for Leopold de Rothschild, revealed that the idea originated with some photographs Rothschild had taken of a Japanese garden by Lake Como, and paintings of gardens in Japan by Ella Du Cane exhibited in London. Hudson followed these rather than the 'lines... laid down by some authorities on Japanese gardens'. Much of the article was devoted to plants, and the photographs of the garden show, in five years, a profusion of foliage (Fig. 3).[21]

Held under the auspices of the Japanese government, the Japan–British exhibition at the White City, London in 1910 sought in the official declaration, to 'cement and make greater and more lasting friendship between two great 'Island Empires''.[22] There were displays of the New Japan, palaces of finance, industry, railways and education and a Palace of the Orient (including exhibits on Japanese colonies in Formosa, Manchuria, Kwantung and Korea), tableaux of imperial history, including celebrations of recent military victories over China and Russia and reconstructions of two 'native villages' of subject minorities, in Formosa and Hokkaido. Sited at the entrance were the two main gardens, the Garden of Peace and Garden of Floating Islands. Designed by Tokyo landscape architects Hannosuke Izawa and Keijiro Ozawa, one of the key figures in rescuing and restoring traditional gardens in Japan,[23] and built by Japanese workmen, the progress of their building was reported frequently in the garden press and upon completion they became the main icons of the exhibition. The Garden of Peace, named 'in harmony with the spirit in which Japan has participated in the... Exhibition' had shrine-topped hills and a lake, with a house, wisteria arbour and iris pond. The Garden of Floating Islands was more dramatic, with rocks and precipices and cascades, and a series of bridges joining the islands, and some tea-houses along the edge. Views of Shepherd's Bush were screened by canvas painted with scenic pictures of pine forests. Plants, trees and goldfish were shipped from Japan; rocks and boulders transported from Devon and Derbyshire.[24] For visitors inspired to create their own Japanese gardens, or just give established gardens Japanese highlights, the Yokohama Nursery Company exhibited over 2000 plants, including miniature trees, and scores of ornaments, and distributed their brilliantly illustrated mail order catalogues. Some of the resident elite Japanese community, and some Japanese officials, established a distance from the blend of commercial, cultural and political values on display. The Japanese Department of Education was careful to say:

Fig. 3. Postcard of the Japanese garden at the 1910 White City Exhibition. Source: Private collection.

'The gardens are not purely Japanese. They manifest the good feeling existing between the horticulturalists of England and Japan; equally they symbolise the alliance between our two countries, for Japan supplied the ideas and the plants while Britain contributed the site and materials'.[25]

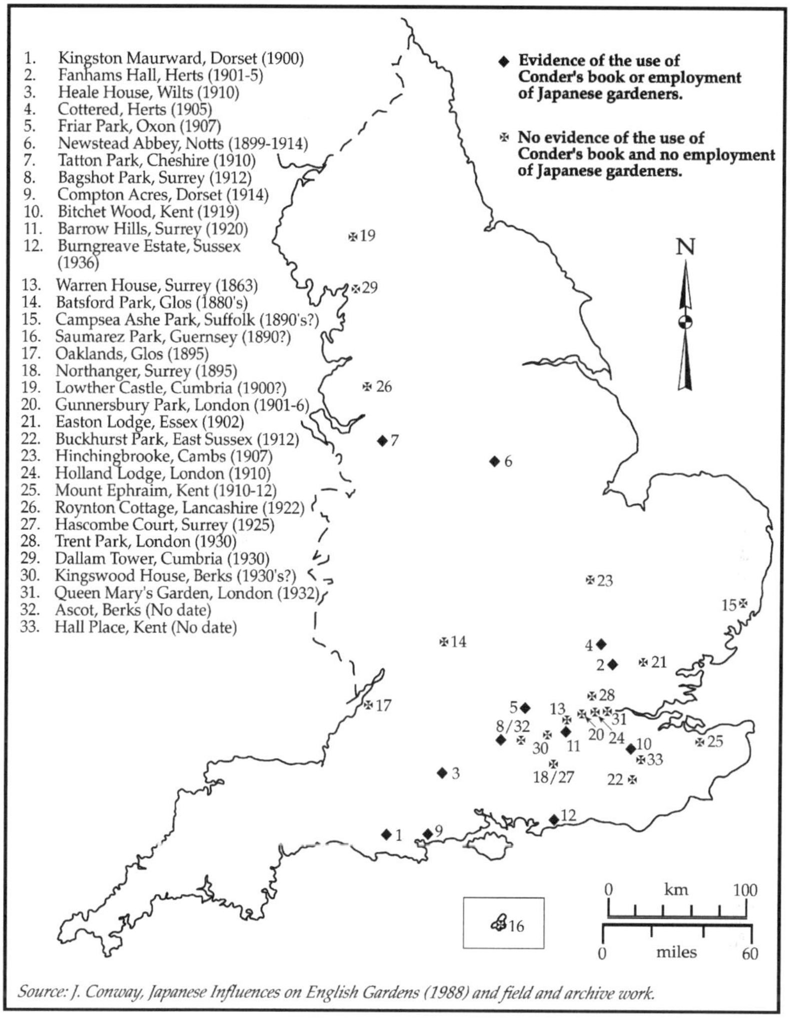

Fig. 4. Japanese Gardens in England (1860–1939).

With the powerful impression that these exhibition gardens made on the British public, and the increasing popularity of Japanese plants and ornaments, discussions of Japanese gardens focussed on issues of authenticity, and on degrees and kinds of Japanese influence (Fig. 4). The most spectacular, complete examples, some commissioned by businessmen, were extremely expensive to create and maintain, and attracted the insinuations of *nouveau riche* vulgarity which had been made a century earlier towards those deploying Chinese and Indian styled gardens. On the other hand, vulgarity could be the result of the very availability of a few Japanese ingredients within the price range of a small suburban gardener. The editor of *Country Life* announced in 1915 that 'the disposition of a few typical ornaments, of a bronze stork here and a stone lantern there, does not make a Japanese garden; it only makes an English garden speak with a Japanese voice'.[26] Purist advocates of the style objected to features arranged without the deeper meanings—of artistry, scholarship, etiquette and philosophical wisdom—attributed to traditional Japanese culture, or the more arcane aura which attracted devotees of mysticism. The leading authority on English garden design, Gertrude Jekyll, kept the style of gardens in Japan at a respectful distance, maintaining that it would be folly to attempt to reproduce it.[27] Some propagandists were keen to cite a Japanese precedent for many established features of gardens in Britain, from an iris parterre to the use of native plants. A series of articles by Hayward in *The Gardeners' Magazine* from 1909 to 1911 argued that the established importance of water, rock and wild gardening made garden culture in Britain peculiarly receptive to Japanese influence. The 'old English garden' would be reclaimed from two centuries of continental European influence, turned progressively 'into a living picture'. Japanese influence would 'widen the scope of the landscape gardener', bringing a 'vitality and transforming power': 'It is virile, throbbing with life and beauty. The possibilities stretch out beyond the horizon full and free'.[28]

The writings of Josiah Conder and Reginald Farrer, which are quoted at the head of this article, both focussed and reformed the fashion for Japanese gardens in Edwardian Britain. Each claimed in contrasting ways to offer an authentic account of the principles of gardening in Japan. While each emphasised the centrality of 'landscape' to gardens in Japan, they did so in different ways. In this paper, we will consider Conder's and Farrer's texts in relation to broader conventions of garden design in Edwardian Britain, and their influence in the making of Japanese gardens on landed estates in contrasting sites and regions, at Newstead Abbey, Nottinghamshire, Cowden Castle, Clackmannanshire and Ingleborough, Yorkshire.

Landscape gardening in Japan

Josiah Conder (1855–1920) arrived in Japan in 1877 at the invitation of the Meiji government as one of the *oyatoi*. Trained as an architect in London, Conder became the first Instructor of Architecture, and eventually Professor, in the Engineering Department of the new Imperial University, and served as an advisor to several government departments including those of Public Works, the Interior and Agriculture and Commerce (Fig. 5). He is credited with introducing western style architecture in Japan, designing many institutional buildings, including museums, embassies, clubs and churches, and a series of private mansions, in a variety of European styles—gothic, classical, Tudor, Italianate. Conder married a Japanese traditional dancing teacher in 1880, and remained in Tokyo for the rest of his life, becoming more absorbed with Japanese cultural forms, as he spread western forms through his practice (his own house was a mixture of English and Japanese styles). He learned traditional painting techniques, wrote on vernacular architecture and published his first book on Japanese flowers and floral arrangement. He wrote on Japanese gardens in 1886 in the *Transactions of the Asiatic Society of Japan*, the principal learned society for British residents. *Landscape Gardening in Japan* was first issued by the Anglo-Japanese publisher Kelly and Walsh which specialised in books for globe-trotting tourists in 1893 and reissued in a revised edition in 1912 (Fig. 6).[29]

Fig. 5. Photograph of Josiah Conder. Source: T. Fujimori, *Nihon no kindai kenchiku* (*Modern architecture in Japan*) (Tokyo 1993).

Landscape Gardening in Japan was the first serious and systematic exposition of the subject in English, and proved influential throughout the anglophone world. The word 'landscape gardening' had no equivalent in Japanese and Conder constructed the concept from a variety of garden styles with a scenic element, from a thousand year history of gardens with varying social and religious functions.

Fig. 6. Book cover of *Landscape Gardening in Japan* (1893).

Conder based his book on Japanese texts, mainly from the 19th century[30] and on observing historic gardens, especially the *daimyo* of the former feudal elite. Many *daimyo* provided the sites of new buildings in Tokyo, including those designed by Conder, but some had survived, renovated as grounds for new mansions.[31] Thus, while the book sampled from Japan's extensive history and geography of gardens, it focussed on a period and social stratum concurrent with the classic style of landscape gardening in the west and sought universal principles to revitalise it. Conder saw his book as a treatise,

Fig. 7. 'Plate XXXI Tea Garden' in Josiah Conder (1893) *Landscape Gardening in Japan*.

setting out 'the rules and theories of the Art of Landscape Gardening in Japan', one which showed the 'meaning to methods' which might otherwise 'possess merely the empty charm of novelty' but which included much that was not 'peculiar to Japanese design': 'beneath the quaint and unfamiliar aspect of these eastern compositions, there lie universally accepted Art truths'. 'Robbed of its local garb and mannerisms', the Japanese method deployed 'aesthetic principles applicable to the gardens of any country', including a 'modified form of Western gardening' (Fig. 7).[32]

Landscape Gardening in Japan catalogues the main elements of gardens—stones, lanterns, buildings, fences, bridges, water features, plants—illustrating their variety in detailed line drawings. It also shows how they are combined in scenic compositions, with illustrations taken from the texts consulted, modern lithographs and a supplement of specially commissioned photographs by the 'well known Japanese Photographic Artist K. Ogawa' of the 'best remaining gardens' together with 'natural views' to make clearer the 'faithfully representative character of these artificial landscapes'. While a few photographs display the compositional structures of traditional Japanese prints, with which western artists strove to modernise their art, most are conventionally western-style picturesque views, a style which *kaika*-conscious Japanese deployed to modernise their art too. Thus, *Landscape Gardening in Japan* was carefully primed as both a guide book for those touring gardens in Japan, and as a pattern book for those who wished to make their own Japanese Garden.

A garden modelled on those in *Landscape Gardening in Japan* was expensive to make and maintain. The most famous was that in the grounds of Friar Park, near Henley-on-Thames, constructed around 1906 for the eccentric millionaire Sir Frank Crisp. This was made according to the rules of 'Hill Garden—Finished Style' in Conder's book. It was the latest addition to a series of miniature gardens at Friar Park, fully described in the catalogue for the paying public. These included a rock garden for alpines with a scale model Matterhorn, and later a Dutch garden, Elizabethan garden and a series of gardens based on Medieval and Renaissance illustrations. Crisp's taste for scale model scenery extended to underground caves, peopled with terracotta gnomes, starting a fashion in garden ornaments with rivalled that of Japanese tin birds. It became conventional to situate Japanese gardens in such spectacular geographical and historical series.[33] The gardens described in *Landscape Gardening in Japan*, along with much of large scale Edwardian garden design, resonated with the exhibition culture of the time.

Newstead Abbey

The Japanese garden at Newstead Abbey was created between 1899 and 1914 under the direction of Ethel Webb (1862–1915) (Fig. 8). Webb was one of a number of wealthy Edwardian women who practised and promoted advanced garden design and horticultural knowledge; no less than global travel, it was a display of female independence.[34] Webb took over supervision of the Newstead Abbey estate upon the death of her father William Fredrick Webb in 1899 and ownership in 1910.

William Frederick Webb spent much of his wealth travelling in Africa and the Far East, exploring and big-game hunting, reporting on his exploits to the Royal Geographical Society and bringing home trophies of dead animals and living heroes to the house at Newstead Abbey. David Livingstone, who rescued an ailing Webb during one expedition, stayed for seven months at Newstead Abbey where he wrote his second book *The Zambesi and its Tributaries*.[35] Another noted visitor Sir Roderick Murchison made a timely observation that the estate was rich in coal measures and income was appreciably enhanced by leasing land for mines. Webb continued his travels with his daughters, falling fatally ill during a trip through Egypt. Webb had spent, or rather overspent, the rest of his fortune making improvements to a house and grounds, famous through its association with a former owner, the poet Lord Byron.[36] The year before his death, *Country Life* reported Webb was 'well versed in gardening' and had created a series of different gardens beyond those around the house, gardens devoted to bamboo, azaleas, alpines, carnations and a Spanish Garden.[37] After his death, his daughters instituted a regime of careful estate management designed to repair and maintain the grounds and, if possible to raise revenue, by the sale of lands and planting of trees. Central to this regime was a programme of hydrological improvement, taking in the many lakes, ponds and conduits. This helped drain lands for pasture and woodland in the lower part of the grounds and, between the two main lakes, provided a site for the creation of a Japanese garden set around ponds and cascades.[38]

The project for a Japanese garden may have been a family decision, for Japanese screens were brought back from one trip to the Far East and installed in a Japanese style room, but Ethel Webb, a committed horticulturalist, is credited with overseeing its creation, and giving it the greatest attention. It seems much of the work was carried out when she took over ownership of the estate in 1910, the year of the Japan–British Exhibition. Information on the design of the garden was scarce until detailed notes and sketches in Ethel Webb's hand on Newstead Abbey headed notepaper recently came to light. These show that the garden was largely, and often literally, based on Conder's *Landscape Gardening in Japan*. The sketches are close copies, sometimes transcriptions, of Conder's directions and illustrations for hill gardens, water gardening and a Tea Garden. The basic scheme comprises a conventional pairing of Master's Isle (*Shujin-to*) and Guest's Isle (*Kiakujin-to*) easily accessible by stepping stones and bridge to the banks.[39] The Master's Isle was in Japan appropriated to the owners of the garden, but there is no evidence this was part of the plan at Newstead (Fig. 9).

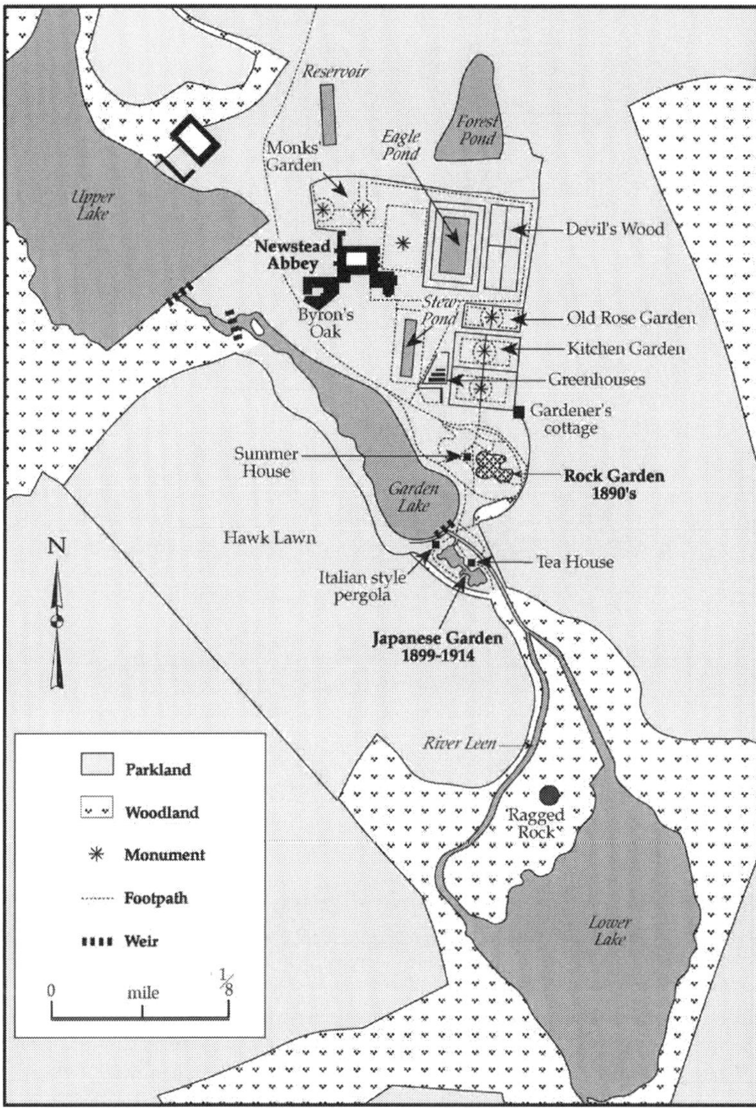

Fig. 8. Newstead Abbey showing the new rock gardens and Japanese garden, 1890s–1914.

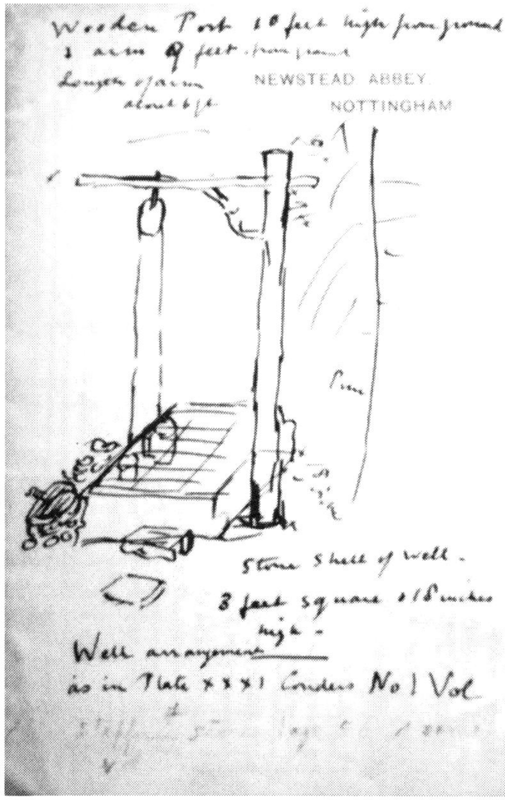

Fig. 9. The first page of Ethel Webb's manuscript notes, showing the rustic well. Source: Newstead Abbey Collection.

A thatched Tea House was built with accompanying 'rustic well', in an ensemble designed for the ritual of the tea ceremony, although it seems to have been used no differently from summerhouses elsewhere at Newstead, as a place to rest and view scenery. It housed a metal cormorant purchased from the Exhibition of 1910. The overall layout of the Japanese garden was credited in *Country Life* to 'a skilled Japanese horticulturalist brought over for the purpose'.[40] The entrance to the garden by the cascade from the lake gave a bird's eye view of its layout. Paths wound around the lake, past a succession of Japanese plants and trees, including pines, maples, lilies and irises, stone lanterns and occasional views out into the rest of the grounds.[41] As visitors exited from the garden, up thc stcps aside the cascade to the lake, and its Italianate pergola, they were given a spectacular view of the grounds rising to the gothic house. As in other pleasure grounds of the time, the Japanese garden was part of a succession of descending gardens, with axes and vistas punctuated by loggias and summerhouses.[42] The Guidebook to Newstead of 1916, issued a year after the death of Ethel Webb, reported that such was the popularity of the Japanese garden, that visits had to be curtailed.[43] In its 1917 feature on Newstead *Country Life* reported:

> The garden on which the greatest care and attention has been lavished in recent years—before the war, which has made the upkeep of all the gardens very difficult—is the Japanese... It drew from the Japanese Ambassador the magnificent compliment that it was the only thing he had seen since he left Japan which made him feel homesick.[44]

Cowden Castle

The Japanese garden at Cowden Castle, Clackmannanshire, on the slopes of the Ochils, was the creation of Ella Christie (1867–1944) from a wealthy mine-owning family (Fig. 10). Her father John Christie emphasised the educational virtues of travel taking Ella and her sister Alice to visit industrial works as well as places of historic interest in Europe, and insisting they learn foreign languages for travels abroad. Ella Christie did not marry and took the opportunity to travel throughout Asia. Dressed in her usual Scottish homespun (meeting her on Dollar Station, a friend thought she was off to Edinburgh for the day), accompanied by her maid Humphries, and armed with some travel advice from a friend in the War Office and a few handy phonetic phrases such as *Nicki e Dome Banzai* 'Long live Japanese

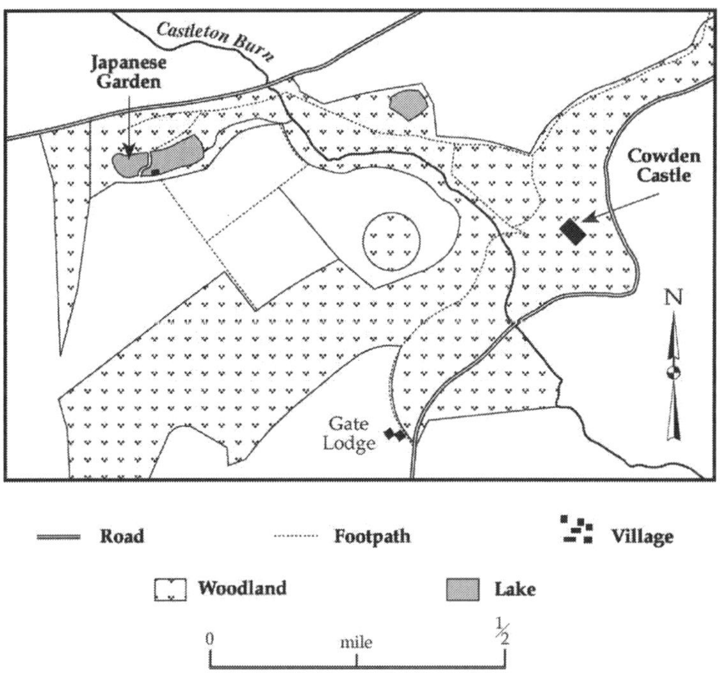

Fig. 10. Japanese garden at Cowden Castle, Dollar, Clackmannanshire.

and British alliance', Ella Christie set out for a year's sojourn in Japan in 1907, and set down her impressions in her journal and letters home to her sister.[45]

Christie stayed in the old imperial capital of Kyoto in April and the new one of Tokyo the following month, meeting the Du Cane sisters, Ella and Florence, there to compile material for *The Flowers and Gardens of Japan* (1908), who advised her on travel arrangements. Dismayed by 'the process of 'Europeanizing'', 'I daily long to see the *old* Japan', she declared, Christie eagerly visited gardens, both old ones of temples and *daimyo* and new ones in a traditional style. She was intrigued by the way new gardens faithfully copied old ones and the speed at which they gained the patina of age: 'the whole thing is laid out with all the marks of positive antiquity' she wrote to her sister, 'Not a plant looks as if it had ever been moved and the mossy stones go on mossing'. She caught the cherry blossom and in Tokyo 'saw the azaleas and wisteria in their glory' and regretted missing the irises by a week. With Ella Du Cane, 'one of the few women one can travel with', and a native boy guide, she took a nine hour walking tour around Mount Fuji. 'The path was a mere track through a tangle of shrubbery quite lovely in variety. We passed camphor trees, the largest thing in timber that I have ever seen, deutzias, camellias, hydrangeas, tiny roses, and bamboos'. From the outset, Ella Christie was on the look out for ideas to create a garden at Cowden.

> I went to see a temple on which is a show of wisteria, the flowers about eighteen inches long and trained on verandahs over the water. Miss Du Cane came to paint it. If feasible I shall bring a Jap home to lay out my pond. It could be a dream of beauty...[46]

While in Japan, Christie wrote to Josiah Conder for advice about designing a garden at home at Cowden Castle. He cautioned her against using Tokyo gardeners, for they 'lack both organization and acquaintance with matters of foreign export' and she would also be put into the 'hands of guides or unreliable interpreters'. Conder advised her to apply to the Yokohama Nursery Company; there she would find 'English speaking experts acquainted will all branches of gardening who will if asked, I think, supply you with suitable Japanese designs for gardens as well as advising you upon selections of trees, shrubs, stones, lanterns, etc.' (Fig. 11).[47]

The pond at Cowden Christie wanted to transform into a Japanese garden had been made out of a marshy field. It was over half a mile from the house, well beyond other gardens, but the very isolation and backdrop of woodland and misty hills gave it a romantic aura, which she wished to enhance. She employed Taki Honda, from the Royal School of Garden Design at Nagoya to create 'a real Japanese garden'.

> Realizing the possibilities of the somewhat uninteresting stretch of ground about the lake, and how, if the breath of life were breathed into it, it would become a living soul, I found a Japanese woman who would undertake the task and wave the magic wand. For six weeks she toiled and planned, while to the untrained eye apparently shapeless mounds arose, and stones were sought for and found and placed in the orthodox grouping.[48]

Part of the layout was planned according to the so called 'ancient rule' of the Imperial Palace Gardens, other parts copied from Conder's *Landscape Gardening in Japan* (Fig. 12). As at Newstead Abbey, a Master's Isle and Guest's Isle were created according to the conventions of water gardening. The approach was marked by an early 19th century granite lantern imported from a Kyoto antique dealer,

Fig. 11. Ella Christie standing in the Japanese garden at Cowden Castle, c.1930. Source: Private Collection of Robert Stewart.

who provided detailed advice about maintaining moss, if local insect hunting wrens ensured it did not survive. A red lacquered *torii*, the pergola-like Shinto shrine structure, was imported in sections. One embankment was modelled to denote the slopes of Fuji-yama. Local limestone was shaped into scores of symbolic stones, carefully positioned to articulate the space as (in Christie's phrase) a 'sacred domain'. Plants were selected with a view to their successful establishment, so native evergreens were mixed with exotics, including Korean pines raised from seed she collected on her travels, and substitutes sought

Fig. 12. Photograph of the Japanese garden at Cowden Castle, showing the bridge c.1930. Source: Private collection of Robert Stewart.

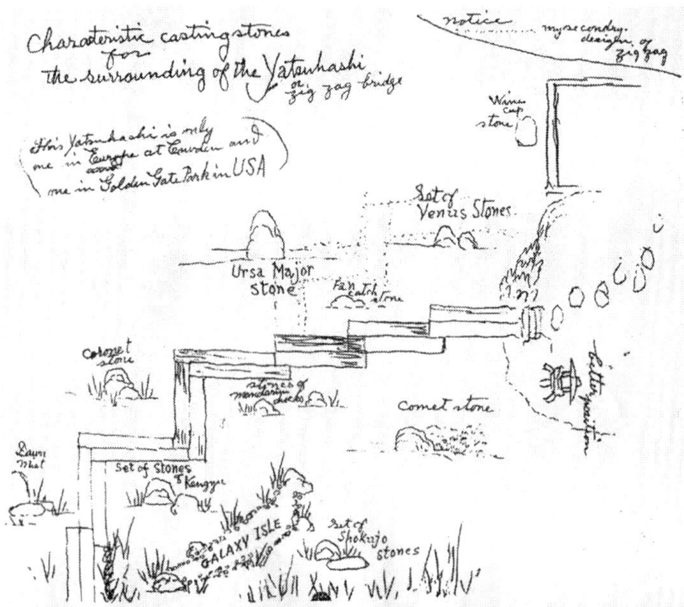

Fig. 13. Suzuki's letter to Ella Christie, showing design plan of the zig-zag bridge. Source: Stewart-Christie Papers, Acc 5058, National Library of Scotland.

for Japanese plants, for example, laburnum in the place of wisteria. Vistas were cut through beechwoods to reveal the surrounding hills and moorland. Its name *Sha-rak-uen*, 'a place of pleasure and delight', was inscribed at the entrance.[49]

The Japanese garden was modified over the years, for the rest of Christie's life, attracting a stream of visitors, local gentry keen to try the style, Edinburgh celebrities and fellow globe-trotters, including minor European aristocrats. She employed a Japanese designer Suzuki for occasional consultations. Suzuki was a Professor at the Zen Soami School of Imperial Design in Nagoya who also stayed for long periods in London to give advice on Japanese plants and gardens throughout Britain, increasingly so in the 1930s as the fashion for Japanese gardens in Britain waned, and financial conditions in Japan meant that 'gentlemen do not spend money on their gardens'.[50] He used Conder's book in his advice for Cowden, but, as with his recommendation of the zig-zag bridge (*yatsu-hashi*) modified it with supplementary historical and literary evidence (Fig. 13). He assured her that in his opinion, and that of knowledgeable visitors, it was one of the best in the western world. On Suzuki's recommendation, Christie employed a Japanese gardener Shinzaburo Matsuo, who had worked on other gardens in Europe. Matsuo never learned English but communicated largely by sign language; one visitor to Cowden recalled Christie and Matsuo 'standing together by a bed of irises, deep in a kind of hybrid conversation'.[51]

Rock garden truth

Reginald Farrer (1880–1920) travelled to Japan in 1903, where he was inspired to reform the tradition of English rock gardening (Fig. 14). He became interested in mountain plants as a boy roaming the limestone hills of his family's estate at Ingleborough, Yorkshire, to collect and catalogue specimens,

Fig. 14. Photograph of Reginald Farrer. Source: Illingworth J and Routh J (Eds). *Reginald Farrer: Dalesman, Planthunter, Gardener* (Lancaster 1991).

recording some for the first time. At St John's College, Oxford, he found in the Bursar, Henry Jardine Bidder, a fellow enthusiast for alpine flora and he helped Bidder make the extensive rock garden there. After graduating in 1902, Farrer set off on the first of his long journeys to the Far East in 1903, embarking on a consciously strenuous life of plant collecting, mainly in mountainous regions of Asia and Europe. The Craven Nursery he ran on the family estate, stocked with some of the plants he collected, was not a commercial success, and while he publicised the rock gardens he created there, his parents, concerned with the management of the estate as a whole, were reluctant to underwrite his horticultural enthusiasms. His expeditions were largely funded with proceeds from his writings, articles for magazines and books on travel and horticulture, and later sponsorship from a syndicate of garden enthusiasts, including the Royal Horticultural Society and several private patrons. Farrer converted to Buddhism on a trip to Sri Lanka in 1907. Later in his life, he became known as a painter of alpine plants in their habitats, exhibiting over 200 watercolours at the Fine Art Society in 1918. He died 2 years later during a plant hunting expedition in the Sino-Himalayas.[52]

Farrer was a singular figure in the horticultural world of Edwardian Britain, but in many ways a traditional one, following enterprising plant-huntsmen and nurserymen like Veitch and Fortune. Rock gardening, and a taste for mountainous plants, had been stimulated and reshaped in Japan by the reception of John Ruskin's writings, which are also (despite his disclaimers) a discernible influence on Farrer's sensibility, and a Society for the Study of Alpine Plants was established in Tokyo in 1902.[53] Farrer published his account of his first trip to the Far East as *The Garden of Asia: Impressions from Japan* (1904). 'My object has been to give true pictures rather than elaborate statistics. I have tried to render Japan as I saw it and felt it; not Japan as manuals and guide books try to present it'. Farrer sought out a range of gardens, including *daimyo* gardens and nurseries, but focussed on those hitherto unseen or overlooked by globe-trotting tourists or design books. Searching for a residence, he came across a typically modest garden in Tokyo, 'perhaps three yards square'.

> It was not ambitious. It did not aspire to rivulets and bridges and paraphernalia. But it was perfect. There were little bushes of azalea, and primeval-looking mossy stones that had all the effect of rocks, and a peeping fern, and mother-of-thousands, and tufts of grass, and a tree of the lovely *Camellia sasanqua*, with rose-pink blossoms like an enormous dog rose.[54]

More fascinating still was witnessing the way such gardens could be rapidly assembled. Farrer's residence proved to have only a few overgrown and sick-looking plants, so he ordered a new garden. Workmen brought mature trees and shrubs for background and built piles of rock work, where they carefully positioned smaller plants, 'till the whole had the air of a rocky clearing in a jungle'. While plants were carefully cultivated, rocks and stones were more coveted than flowers, with highly priced 'precious stones' from particular rivers that looked like 'common pebbles' to the 'untutored eye'. 'The Japanese garden is a paradise of stones rather than blossom. If a flower happened to come, well and good; but its bush was not put there to blossom so much as to set off the contrast between two lines of rock'. 'A good Japanese garden of the ordinary sort is one where the rocks are of perfect size, shape and disposition to replicate a landscape'; such rock gardens were 'very different from the careful cossetting-ground for ill-tempered little Alpine plants that we mean by the name'.[55]

The very stringent use of plants in Japan presented Farrer with an opportunity. He wrote home to his mother:

> I am literally dazed and staggered at the possibilities for border plants and especially Alpines. Japan is an absolutely virgin country! The few plants which are occasionally introduced from Japan by Veitch and so forth, with much blowing of trumpet, and at a preposterous price, are not found by them in the wilds of the mountains but in the cabbage gardens of Tokyo and Kyoto. All this I got from Alfred Unger [the head of Boehmer's Nursery Company] here, who does not know the importance rock gardening has assumed in England, and told me, that as Alpines have no commercial value, no one has ever approached the Alpine Flora of Japan. Now we know what a commercial value they have in England, so *what* a business opening for a person with discernment—and money![56]

As well as seeking out alpine plants in the wild, Farrer was excited to find sturdy specimens on the stalls of the night fairs of Tokyo.

> And there, before my eyes, in the streets of Tokio, shone the high alpine, unmistakable—stout rosettes of leathery, Galax-like leaves, and tall stems each carrying a dozen flowers or so—nodding great fringy bells, rose-pink, and marbled with crimson—far more dainty and more gorgeous than any Soldanella that ever pierced the snow. Of course I fell upon the treasure (so utterly different from the stunted little miseries that so far had represented *Schizocodon's* best efforts in England) and ordered hundreds. They arrived next day—stalwart, tall-growing, splendid clumps, costing two pence each or so. My heart was filled with joy, and I sent them off to Yokohama to be packed for home. And there in the summer heats of that unutterable place, they all sickened and died.[57]

Farrer reported this in his second and most popular book, *My Rock Garden*, first published in 1907, when he was able to add that he had now received at Ingleborough 'another lot of excellent plants', although he was 'beginning to look upon them anxiously' (Fig. 15). Farrer found a number of glorious plants 'always more or less at war with me and my soil' and resented the success of others like Mr Hindmarsh of Newcastle who managed to make the 'melancholy, shy rooted' *Shortia uniflora* grow 'like a groundsel' and showed off their success in the photographs in horticultural journals. Farrer had now received a 'superb importation' of *S. uniflora* from Japan which he had planted out and was hopeful of thriving, recalling the first time he saw the plant, pointed out by a hotelkeeper in Shoji, 'six miles or more across the country, in the dense pall of jungle that flows so royally down the western slope of Fuji-yama'. Farrer himself collected a Japanese variety of *Cypripedium* from the slopes of the mountain, 'as beautiful a thing as Fuji-yama ever held forth'. He secured a 'goodly batch' and a photograph in the book shows a Ruskin-style close-up view of the plant in the rock garden, 'three months after arrival from Japan' (Fig. 16). *My Rock Garden* advocates a combination of Japanese regard for imitative rock scenery and English regard for flowers. The ideal rock garden would be

> A Japanese garden stocked with European alpines; and when I say Japanese garden, I don't mean a silly jungle of bamboos, with Tori, and a sham tea-house, and Irises and a trellis—I mean a rocky glen, a pinnacled flank of mountain such as every other cottage in Kioto possesses, and had possessed for half a dozen generations.

Farrer would thus reform the 'shapeless, anarchic ideal' of the English rock garden 'held up for imitation by extremists of the Landscape school'. Moreover, Farrer criticised the entire tradition of

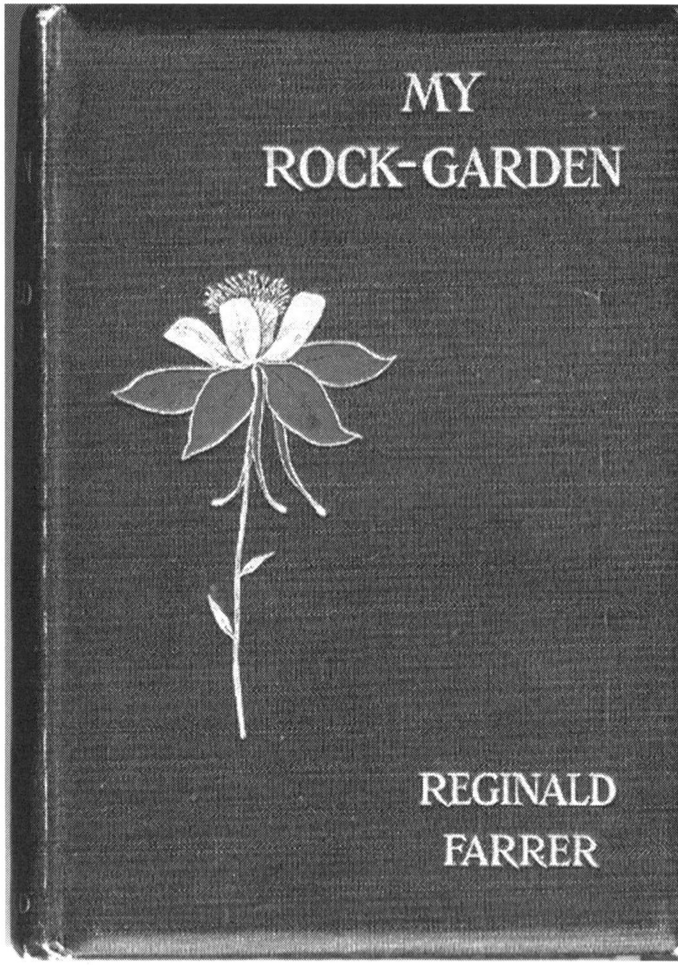

Fig. 15. Book cover of *My Rock Garden* (Lancaster 1893).

the 'Jardin Anglais—sham landscape, sham wildness, sham ruins, wobbling walks, pagodas, lakes, pools—you will find it all in *Headlong Hall* and *Sense and Sensibility*'.[57]

After his Buddhist pilgrimage to Ceylon in 1907, Farrer adopted a more metaphysical if more broadly orientalist view of Japanese gardens. Faced with the increasing craze for Japanese gardens in the wake of the 1910 exhibition, 'Japanolatory' he called it, Farrer set out his views of a more respectful use of the culture of gardens in Japan.[58] The Royal International Horticultural Society Exhibition of 1912,

CYPRIPEDIUM MACRANTHUM, VAR: VENTRICOSUM (THREE MONTHS AFTER ARRIVAL FROM JAPAN).

Fig. 16. Photograph of 'Cypripedium Macranthum' from *My Rock Garden* (London 1907).

the first since 1866, showed a series of Japanese gardens. In his catalogue essay on 'Rock Gardens and Garden Design', Farrer took the opportunity to pronounce on the fashion, and spell out a broader views of gardening in England.

Only in the last 30 years, as it finally freed itself from the dead hand of continental culture, had there been 'an original sense of garden design in England', Farrer claimed. The 'rooted, personal passion for flowers' was nowhere matched 'on this side of China', save for 'our kinsmen in Holland'. The Far East also offered better lessons of scenic representation than in conventional English landscape gardening. 'The East has sadly mocked our ignorance in this matter', he observed.

> The 'Japanese' gardens at the White City [Exhibition of 1910] were quite the most cruel and insolent practical jokes that I have ever seen leveled at our native innocence—vast puddings of unrelated pebbles, shapeless, ridiculous, peppered with toys and bronzes and haphazard shrubs. How their creators must have laughed in the building... 'Very pretty' said a Japanese Ambassador one day in England, wishing to say pleasant things about a famous 'Japanese' garden through which he was being conveyed by its complacent owner. 'Very pretty, yes. We have nothing like this in Japan!'.

'We have never yet in England seen any fair or reasonable reproduction of a Japanese garden' because of the 'wrongheadedness of 'copying' Japanese gardens', he continued.

> The West copies by sitting in front of a model and imitating it soullessly, outward line by line. When the copy is finished, nothing remains in it of the spirit that had inspired the original. The East...

'copies' by sitting in front of an original, absorbing it, studying it, grasping its principles and its essential greatness thoroughly; then by going on to translate these acquisitions into more or less new form, in terms of one's own taste and time... This is rather translation than 'copying' as we understand the word; and it is even more indeed than we mean by translation... Oriental copying is a perpetual renewal of life. Moving perpetually through varying manifestations and developments... And this is the only sense in which I could bear to talk of 'copying' a Japanese garden. Very likely we should arrive at something not resembling any one of them.

Such a garden might be incorporated in an English–Japanese collaboration:

Let Mr Lutyens or Detmar Blow endeavour to fit some Surrey villa of their building with a congruous rock garden... the result, if they are worth their names, would fund its triumph rooted in discovery or intuition of Oriental knowledge and balance. With the aid of Mr Yamanaka [a London antique dealer] and Mr Markino [the resident painter] (but without their tongues in cheek) the result would be yet more striking and illuminating.

Oriental harmony and modesty would help close the division between art and life, make us appreciate that expensive things were not more beautiful in themselves than 'a farthing paper fan from Kyoto'.

All this is true of our gardens. They are not part of ourselves; they are things we buy... Go round the International, in amazement at the wonderful paradises for sale. But they are not yet inspired, they are not yet fertile. How are they to become so? Why, if we care to make them works of art, our very caring will help make them so... It will all come from a recognition of rock garden truth, its dignity and high possibility, as well as from the engendering of a race of men more specially fitted to cope with it by long and profound contemplation of the principles and unities and rhythms that the art of the East (especially) holds most readily at the command of students in every department of its existence.[59]

Ingleborough

The Ingleborough fells above Clapham, Yorkshire, were already renowned for their limestone scenery and alpine plants when James Farrer, a lawyer in London, purchased the Ingleborough estate in the 18th century (Fig. 17).[60] In the early 19th century the grounds were landscaped in a picturesque style, with a lake and woodland walk to the show caves for increasing numbers of tourists. Reginald Farrer's father James Anson Farrer (1849–1923), a progressive landowner, undertook a campaign of improvements including hydraulic engineering of the lake and waterfalls and extensive timber-tree planting, which made the estate increasingly profitable. James Anson Farrer was also a noted Liberal and anti-imperialist, writing a series of comparative studies of culture, religion and ethnicity, which may well have influenced the views of his son. Reginald Farrer was committed to the estate he was due to inherit, and close to his parents, but his letters home from his frequent and prolonged plant hunting trips abroad reveal frictions about the resources his father was prepared to grant him, both in money and space, and the control they were prepared to allow him planting and gardening the estate[61].

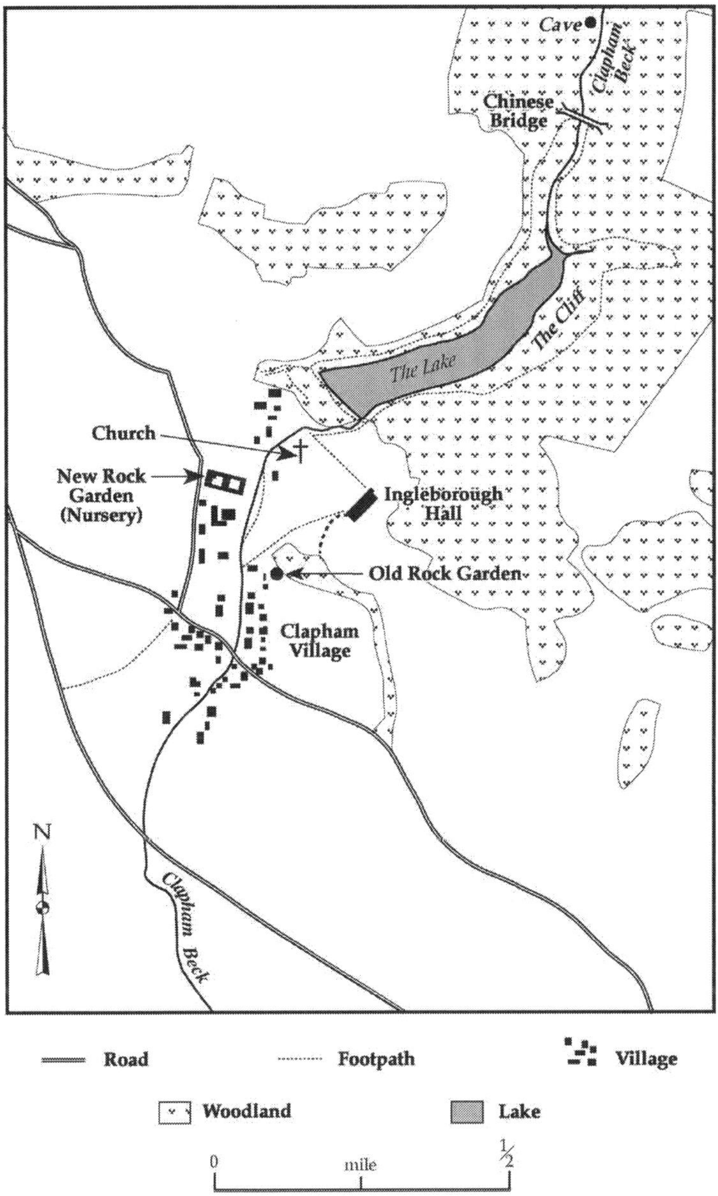

Fig. 17. Ingleborough Hall, Clapham, Yorkshire.

A GLEN IN THE OLD GARDEN.

Fig. 18. Photograph of 'Old Rock Garden' from *My Rock Garden* (London 1907).

Farrer's gardening at Ingleborough took two forms. In the lower part of the grounds, below the mansion, he created two enclosed rock gardens, with large limestone boulders brought down from the fells. The so-called Old Garden was on the site of a former quarry (Fig. 18). Boulders were positioned to create miniature glens for alpines and a pond constructed, although this was removed to create a boggy bed to raise irises. As he noted in *My Rock Garden* plants seldom thrived here on the thin soils, 'died, sulked, sat still'. By contrast, even delicate plants transported long distances thrived in the New Garden, built against the walls of a former kitchen garden and forming the lower part of his Craven Nursery. Plants grew precociously in the 'good fat soil' resulting from the site's former use. Here Farrer positioned four big blocks of stone to create another gorge landscape, filled with stone chips for sharp drainage and topped with good soil.[62] In the actual gorge above the house, Farrer modified the semi-natural woodland with exotics. The path followed the north side of the Craven Fault crossing a small outlier of acid soils. Here Farrer established the varieties of rhododendrons, bamboos and azaleas he collected. This garden was accented by a single stone bridge taking the path over the beck, a so called 'Chinese Bridge', confirming its generally oriental character. Farrer quarrelled with his parents about management when he has away. 'For the 10 or 12 parties who are annually shown my Cliff, you have four or five hundred who are allowed up to see the cave, so do leave off the attempt to make a grievance of that', he complained in 1910 'You are spoiling my work; making it hideous and ridiculous... leave me my valley, rocks and cliff; go and stick Polygonums elsewhere. You have fourteen thousand acres: leave me two!' In 1920, the year of his death plant-hunting in the Himalayas, he wrote that 'we must go on improving the collection' and in another 10 years, 'the place will become a show-garden... people who on account of my books expect to see a fine garden at Ingleborough simply laugh'.[63]

Conclusion

The gardens at Newstead Abbey, Cowden Castle and Ingleborough were expressions of the confidence in the culture of gardening in Japan to revision and revive that in Britain, and in this there are significant continuities with a longer cultural history of gardening in Britain. The Japanese gardens at Newstead and Cowden performed like the classically inspired gardens in 18th century landscape parks, with their statues, inscriptions, temples and meditative lakeside walks. Inspired by the Grand Tours of the day, such gardens were designed to evoke an aura of antique virtue, of a glorious imperial culture. Like Georgian grand tourists, the globe-trotters of the Edwardian era also attempted to reclaim a heritage which they saw as betrayed by contemporary developments, if this was not the degenerate Italy and Greece which dismayed 18th century gentlemen, but the modern, westernized Japan. The gardens at Newstead and Cowden were consciously rule governed, following designs in Conder's *Landscape Gardening in Japan*, using native born Japanese designers and consultants and commercially made ornaments and structures.

Reginald Farrer struck the traditional posture of landed gentlemen who prided themselves on manfully going it alone in their own garden designs, without recourse to pattern books, consultants, metropolitan fashion, or limitless funds. Indeed his disenchantment with fashionably theatrical 18th century gardening is reminiscent of those quarrelsome picturesque theorists Uvedale Price and Richard Payne Knight, who prided themselves on both superior, cosmopolitan landscape connoisseurship and a careful regard for the native plants and local ecology of their estates.[64] Farrer deployed his first hand observation of Japanese plants and Japanese culture to mount a critique of the vogue for 'Japanese gardens'. A more authentic appreciation of the garden culture of Japan, including its conventions of copying, would encourage the cultivation of scenic gardens which might not seem recognisably Japanese at all.

In enhancing the idea of imperial landscape, the craze for Japanese gardens in Edwardian Britain echoed that for Chinese gardens in the Georgian period. The principal 18th century advocate of 'oriental gardening', Sir William Chambers in the guise of 'Tan Chet-Qua Gent.', argued that if British landowners adopted the 'manner' of Chinese gardens, they could exceed their stature:

> the whole kingdom might soon become one magnificent vast Garden, bounded only by the sea...an empire transformed into a splendid Garden, with the imperial mansion towering on an eminence in the center, and the palaces of the nobles, scattered like pleasure pavilions amongst the plantations, infinitely surpasses any thing that even the Chinese ever attempted.[65]

Chinoiserie, although resisted by garden theorists and designers who cleaved to more local, native styles, reshaped the idea of English gardening, especially in Europe.[66] Not all exotic designs were successfully assimilated. In 1805, Humphry Repton, as ambitious as Chambers to turn the whole country into one landscape garden, confidently predicted that the popularity of the scenery and architecture of India would promise a 'great future change' in English landscape gardening. Repton was soon proved wrong as the Napoleonic Wars drained the resources of clients for landscape gardens of any kind.[67]

The craze for specialist Japanese gardens declined with the fashion for many kinds of expensive gardening during the First World War. After the war anti-Japanese sentiment in Britain increased, a reflection of hostility in the dominions, especially Australia and Canada, to Japanese expansion. The Anglo-Japanese alliance was formally terminated in 1923. The Second World War dealt a lasting blow to popular British regard for Japanese culture. Post war gardening in Britain was consciously vernacular, if the plant stock was full of naturalised exotics. Now, with a popular enthusiasm for 'traditional' oriental culture and a vogue for design-conscious gardens, Japanese gardens in Britain are undergoing a revival. Japanese gardens of the Edwardian period are being rediscovered and restored and new ones created. Moreover, English style gardens along with soccer, classic clothes and landscape imagery are now fashionable in Japan.[68]

Acknowledgements

We wish to thank John and Joan Farrer for showing us Reginald Farrer's gardens and grounds at Clapham and discussing the landscape history of the Ingleborough Estate, Clapham, Yorkshire; Robert Stewart for providing information and showing the Japanese garden at Cowden Castle, Dollar, Clackmannanshire; and Haidee Jackson (Keeper of the collection at Newstead Abbey, Nottinghamshire) and Helen Ward, (Lindley Library at Royal Horticultural Society) for their help with primary sources.

Notes

1. Josiah Conder, in: C. Lancaster (Ed.), *Landscape Gardening in Japan*, New York, 1964, 13–41, Supplement (np).
2. Reginald Farrer, *My Rock Garden*, London, 1904, 234.
3. On contact zones and transculturation, see M.L. Pratt, *Imperial Eyes: Travel Writing and Transculturation*, London, 1992, esp. 6-7; On the broader process of cultural exchange between Japan and Britain, see S. Tachibana, *Travel, plants and cross-cultural landscapes: British representations of Japan 1860–1914*, Unpublished PhD Thesis, University of Nottingham, 2000.

4. A. Herries, *Japanese Gardens in Britain*, Princes Risborough, 2001; B. Elliott, *Victorian Gardens*, London, 1986, 199–202; D. Ottewill, *The Edwardian Garden*, New Haven and London, 1989, 55–58; J. Conway, *Japanese influences on English gardens*, Unpublished Thesis submitted for the Conservation course at the Architectural Association, London, 1988.
5. A. Jackson, Imagining Japan: the Victorian perception of Japanese culture, *Journal of Design History* 5 (1996) 245–256; On the United States, including Japanese gardens, see C. Lancaster, *The Japanese Influence in America*, New York, 1983.
6. R. Fortune, *Yedo and Peking; A Narrative of A Journey to the Capitals of Japan and China*, London, 1863; J.G. Veitch, From the Mr J.G. Veitch's letters, *The Gardener's Chronicle*, 1860–1862; Appendix E. Notes on the agriculture, trees, and flora of Japan, by Veitch in R. Alcock, *The Capital of Tycoon*, Vol. II, London, 1863, 490–495.
7. On activities of plant collectors and nurseries, J.H. Veitch, *Hortus Veitchii*, London, 1906; A.M. Coats, *The Quest for Plants*, London, 1969; Y. Shirahata, *Puranto-hanta (Plant Hunters)*, Tokyo, 1994.
8. W.J. Bean, *Trees and Shrubs Hardy in the British Isles*, vol. I–III, London, 1947; H.J. Elwes and A. Henry, *The Trees of Great Britain and Ireland*, vol. II, Edinburgh, 1907; S. Tachibana, *Travel, plants and cross-cultural landscapes: British representations of Japan 1860–1914*, Unpublished PhD Thesis, University of Nottingham, 2000, 120–146; A. Herries, *Japanese Gardens in Britain*, Princes Risborough, 2001, 14–18.
9. A. Helmreich, Re-presenting nature: ideology, art and science in William Robinson's 'Wild Garden', in: J. Wolschke-Bulmahn (Ed.), *Nature and Ideology: Natural Garden Design in the Twentieth Century*, Washington, DC, 1997, 81–112; R. Preston, 'The scenery of the torrid zone': imagined travels and the culture of exotics in nineteenth century British gardens, in: F. Driver and D. Gilbert (Eds), *Imperial Cities: Landscape, Display and Identity*, Manchester, 1999, 194–215.
10. R. Carrington, *Alexandra Park and Palace: A History*, London, 1975, 34.
11. C. Clunas, Nature and ideology in western descriptions of the Chinese garden, in: J. Wolschke-Bulmahn (Ed.), *Nature and Ideology: Natural Garden Design in the Twentieth Century*, Washington, DC, 1997, 21–35.
12. P. Lowe, *Great Britain and Japan 1911–15: A Study in British Far Eastern Policy*, London, 1969.
13. P. Greenhalgh, *Ephemeral Vistas: the Expositions Universelles, Great Exhibitions and World's Fairs, 1851–1939*, Manchester, 1988, 74, 96, 148.
14. T. Shimada, et al. (Eds), *Za Yatoi: O-yatoi-gaikokujin no sogoteki kenkyuu (The Yatoi: A Comprehensive Study of Hired Foreigners)*, Kyoto, 1987; A.W. Burks (Ed.), *The Modernizers: Overseas Students, Foreign Employees and Meiji Japan*, London, 1985.
15. J. Iinuma, Y. Shirahata, *Nihon bunka to shite no koen (Public Park as Japanese cultures)*, Tokyo, 1993; H. Maruyama, *Kindai Nihon Koen-shi no Kenkyu (Study of the History of the Japanese Modern Public Park)*, Kyoto, 1994; H. Suzuki, *Tokyo no (chirei Genius loci) (Genius loci in Tokyo)*, Tokyo, 1990; Y. Shirahata, *Daimyo Teien (Gardens of Daimyo)*, Tokyo, 1997; Tachibana, *Travel, plants and cross-cultural landscapes*, 42–55; J. Hendry, Nature tamed: gardens as a microcosm of Japan's view of the world, in: P.J. Asquith and A. Kalland (Eds), *Japanese Images of Nature: Cultural Perspectives*, London, 1997, 83–106, quotations on 99–101.
16. M. Kimura, *Einichi Rasukin kanshin hikaku (Comparative interest in Ruskin between England and Japan) in Idem., John Ruskin and Victorian Art*, Tochigi, 1993; T. Watanabe (Ed.), *Ruskin in Japan 1890–1940: Nature of Art, Art for Life*, Sheffield, 1997.
17. H. Kato, A. Maeda, *Meiji Media-ko (Studies of Meiji Media)*, Tokyo, 1983; Tachibana, *Travel, plants and cross-cultural landscapes, 30–56;* J. Harada, *The Gardens of Japan*, London, 1928; B. Taylor (H Osgood), *Japanese Gardens*, London, 1912.
18. *The Gardeners' Chronicle*, 28 September 1907, 234; A. Herries, *Japanese Gardens in Britain*, Princes Risborough, 2001, 32–34; B. Elliot, Japanese nurseries and export trade, in: A. Farrer (Ed.), *A Garden Bequest—Plants from Japan*, London, 2001, 19–22.
19. A. Parsons, *Notes in Japan*, London, 1986; On Parsons' Japanese sojourn, see N. Milette, *Landscape painter as landscape gardener: the career of Alfred Parsons RA*, Unpublished PhD Thesis, University of York, 1997, 190–199.
20. M. Yoshio, *Japanese Artist in London*, London, 1910; T. Sato and T. Watanabe, *Japan and Britain: An Aesthetic Dialogue 1850–1930*, London, 1999.
21. J. Hudson, A Japanese garden in England, *Journal of the Royal Horticultural Society* 32 (1907) 1–10.
22. A. Hotta-Lister, *The Japan–British Exhibition of 1910: Gateway to the Island Empire of the East*, London, 1999.

23. Ozawa also designed a 'Japanese garden' at Hibiya public park in 1889, one of the first western style public parks planned by Tokyo City. This park also had an English landscape garden, a French formal style garden and carpet flower bedding.
24. Gardening at the Japan-British Exhibition, *The Gardeners' Chronicle*, 24 September 1910, 227–228; Japanese gardening, *The Gardeners' Magazine*, 9 July 1910, 530–532; *Official Report of the Japan–British Exhibition at the Great White City*, London, 1911.
25. A. Herries, *Japanese Gardens in Britain*, Princes Risborough, 2001, 21.
26. Herries *Japanese Gardens in Britain*, 37.
27. G. Jekyll, *Garden Ornament*, London, 1918, 449–452.
28. P.S. Hayward, The new theory in garden design, *The Gardeners' Magazine*, 28 January 1911, 68; in the same vein see also his Japanese art in English gardens, *The Gardeners' Magazine*, 6 March 1909, 187; The Japanese ideal, *The Gardeners' Magazine*, 14 July 1909, 580; The Japanese ideal in practice, *The Gardeners' Magazine*, 8 July 1911, 408; Ornamentation in the Japanese garden, *The Gardeners' Magazine*, 27 December 1913, 978. We owe these references to Elliott, 265-6, n. 73.
29. D. Finn, Josiah Conder (1852–1920) and Meiji architecture, in: H. Cortazzi, G. Daniels (Eds), *Britain and Japan 1859–1991*, 1991, London, 1852, 86–91; T. Watanabe, Vernacular expression or western style: Josiah Conder and the beginning of modern architectural design in Japan, in: N.C. Bowe (Ed.), *Art and the Natural Dream*, Duke, 1993, 43–52; J. Conder, in: C. Lancaster (Ed.), *Landscape Gardening in Japan*, New York, 1964, vi–vii.
30. K. Ono and W. Edwards, Landscape gardening in Japan in shirusareta 'Tsukiyama Teizoden' (Zenpen) shosai no yakuseki (Garden stone terms in *Tsukiyama-teizoden* described in *Landscape Gardening in Japan*), *Nihon teien gakkaishi (Journal of the Academic Society of Japanese Garden)* 8 (1999) 26–32.
31. T. Fujimori, *Meiji no Tokyo keikaku (Tokyo planning in Meiji)*, Tokyo, 1982; T. Fujimori, *Nihon no Kindai kenchiku (Modern architecture in Japan)*, Tokyo, 1993.
32. Josiah Conder, in: C. Lancaster (Ed.), *Landscape Gardening,* 15.
33. Herries, *Japanese Gardens in Britain,* 26–27 B. Elliott, *Victorian Gardens*, London, 1986, 228–229. D. Ottewill, *The Edwardian Garden*, New Haven and London, 989, 55.
34. D. Kellaway (Ed.), *The Illustrated Virago Book of Gardeners*, London, 1995, 8–15.
35. A.Z. Fraser, *Livingstone and Newstead*, London, 1913.
36. R. Coope, The Webb family of Newstead Abbey, *Transactions of the Thoroton Society* 105 (2001) 137–154.
37. Newstead Abbey, Nottinghamshire, *Country Life Illustrated*, 26 February 1898, 242–243.
38. Memorandum on the Administration of the Newstead and Cawton Estates, 1899–1915, 1915; Household Account Book. Newstead Abbey Acc 1122/12-13; W.F. Webb, Obituary, *Mansfield Reporter*, 3 March 1899.
39. Newstead Abbey collection, Nottinghamshire.
40. Newstead Abbey, Nottinghamshire, *Country Life*, 24 November 1917, 492–497.
41. A.J. Lloyd, *Guide to Newstead Abbey*, Nottingham, 1916; Memoirs of Miss Webb, Newstead Abbey Acc 3153 1133/6.
42. D. Ottewill, *The Edwardian Garden*, New Haven and London, 1989, 104–105.
43. A.J. Lloyd, *Guide to Newstead Abbey*, Nottingham, 1916, 44.
44. Newstead Abbey, Nottinghamshire, *Country Life*, 24 November 1917, 492–497 (497).
45. E. Christie and A. Stewart, *A Long Look at Life, by Two Victorians*, London, 1940, 17–56; A. Stewart, *'Alicella': A Memoir of Alice King Stewart and Ella Christie*, London, 1955, 3–20, 188–189.
46. A. Stewart, *'Alicella': A Memoir of Alice King Stewart and Ella Christie*, London, 1955, 203–210.
47. Josiah Conder to Ella Christie 9 May 1907, Stewart Christie Papers, Acc 5058, National Library of Scotland.
48. E. Christie and A. Stewart, *A Long Look at Life, by Two Victorians*, London, 1940, 234–235.
49. Box 10, XIII, Miscellaneous, Stewart Christie Papers; E. Christie and A. Stewart, *A Long Look at Life, by Two Victorians*, London, 1940, 234–239; A. Stewart, *'Alicella': A Memoir of Alice King Stewart and Ella Christie*, London, 1955, 210–216; E. Christie, Japanese gardens, *Dollar Magazine* (nd), 164–166; R. Stewart, *The Japanese Gardens at Cowden*, Dollar, 1955.
50. J. Suzuki to Ella Christie, 18 December 1932, Stewart Christie Papers.
51. Stewart, *'Alicella'*, 216.
52. J. Illingworth and J. Routh (Eds), *Reginald Farrer: Dalesman, Planthunter, Gardener*, 1991, Occasional Paper 19, Centre for North-west Regional Studies, Lancaster University.

53. T. Watanabe (Ed.), *Ruskin in Japan 1890–1940: Nature of Art, Art for Life*, Sheffield, 1997, 374.
54. R. Farrar, *The Garden of Asia: Impressions from Japan*, London, 1904, 15.
55. Farrer, *The Garden of Asia*, x, 15–16, 23.
56. Quoted in J.L. Illingworth, The correspondence of Reginald Farrer, in: J. Illingworth and J. Routh (Eds), *Reginald Farrer: Dalesman, Planthunter, Gardener*, 1991, Occasional Paper 19, Centre for North-west Regional Studies, Lancaster University, 72–80 (74).
57. R. Farrer, *My Rock Garden*, London, 1904, 233–234, 277–288,10, 4.
58. Reginald Farrer, On Japanese gardens, *The Gardeners' Chronicle*, 2 November 1912, 330–331.
59. R. Farrer, Rock gardens and garden design, in: R. Cory (Ed.), *The Horticultural Record*, London, 1914, 3–18 (14,15,18).
60. J. Roberts, Alpine plants in the Ingleborough area, in: J. Illingworth and J. Routh (Eds), *Reginald Farrer: Dalesman, Planthunter, Gardener*, 1991, Occasional Paper 19, Centre for North-west Regional Studies, Lancaster University, 89–93.
61. S. Mason, The Ingleborough estate: home of Reginald Farrer, in: J. Illingworth and J. Routh (Eds), *Reginald Farrer: Dalesman, Planthunter, Gardener*, 1991, Occasional Paper 19, Centre for North-west Regional Studies, Lancaster University, 81–89.
62. Farrer, *My Rock Garden*, 10–15.
63. Quoted J.L. Illingworth, The correspondence of Reginald Farrer, in: J. Illingworth and J. Routh (Eds), *Reginald Farrer: Dalesman, Planthunter, Gardener*, 1991, Occasional Paper 19, Centre for North-west Regional Studies, Lancaster University, 76, 78.
64. S. Daniels, C. Watkins, Picturesque landscaping and estate management: Uvedale Price at Foxley, 1770–1829, *Rural History* 2 (1991) 141–169; Idem, *The Picturesque Landscape: Visions of Georgian Herefordshire*, Nottingham, 1994.
65. W. Chambers, *A Dissertation on Oriental Gardening, to which is annexed An Explanatory Discourse by Tat Chet-Qua, of Quang-Chew-Fu, Gent London*, 1773, 133–134.
66. O. Siren, in: H. Honour (Ed.), *China and Gardens of Europe*, Washington, DC, 1990.
67. S. Daniels, *Humphry Repton: Landscape Gardening and the Geography of Georgian England*, New Haven, 1999, 181–205.
68. Herries, *Japanese Gardens in Britain*; R. Alexander, A natural affinity: the Japanese and British connection, in: A.Farrer (Ed.), *A Garden Bequest—Plants from Japan, London*, 2001, 23–26; D. Cosgrove and S. Daniels (Eds), *Fukei no Zuzogaku (The Iconography of Lanscape), translated by M. Senda and T. Uchida*, Kyoto, 2001; J. Potter, *Lost Gardens*, London, 2000, 55–75.

Part III
Making and Marking Identities

[7]

URBAN GRAFFITI AS TERRITORIAL MARKERS*

DAVID LEY AND ROMAN CYBRIWSKY

ABSTRACT. Wall graffiti can be indicators of attitudes, behavioral dispositions, and social processes in settings where direct measurement is difficult. The autographed inscriptions of inner city "graffiti kings" in Philadelphia are analyzed in terms of their style, motivation, and preferred setting. Graffiti written by teenage gangs delineate their turf or area of control; their content may indicate contested space and gang violence. Graffiti in an ethnic neighborhood identify tension zones related to social change. KEY WORDS: *Attitudes, Gangs, Graffiti, Indicators, Inner City, Neighborhoods, Philadelphia, Territoriality.*

IN an era of massive social change and unmitigated erasure of the past, one small feature of man's heritage is making a resurgence—the practice of marking graffiti in public places. Graffiti might be regarded, perhaps, as a rather whimsical element in the sum total of cultural baggage of interest to the social scientist, yet in inner city Philadelphia they provide accurate indicators of local attitudes and social process in areas where more direct measurement is difficult. The quality and location of graffiti display regularities. They manifest the distribution of various social attitudes and intimate subsequent behavior in space; as such, certain types of graffiti forecast both potential and actual behavior. To borrow a current graffito: "Today's graffiti are tomorrow's headlines."

The contemporary reappearance of graffiti dates from 1965, or perhaps a little earlier. The number of articles and commentaries on graffiti in popular magazines and the *New York Times* increased from only one over the fifteen-year period, 1950–64, to five in 1969, and then to forty in 1972. The rapid increase in 1972 accompanied an "urban epidemic" in East Coast cities.[1] Popular media often are more sensitive to the public mood than are the social sciences, and as yet there is limited indication of scholarly interest in graffiti. During the 1960s articles on graffiti had dealt with fairly traditional forms, the amorous, the erotic, the political, the historic, and the intellectual, but a new trend appeared in 1970. Graffiti writing had spread to the inner city. Almost all of the graffiti reports discovered for 1972 and half of those for 1971 were concerned with the newfound popularity of spray-painting among inner city youth.

Accepted for publication 8 February 1974.

Dr. Ley is Assistant Professor of Geography at the University of British Columbia in Vancouver, Canada V6T 1W5, and Dr. Cybriwsky is Assistant Professor of Geography at Temple University in Philadelphia, PA 19122.

* We wish to express our appreciation to Robert Sommer and Jack Eichenbaum for their comments on an earlier draft of this paper, and to the staff of Philadelphia's Youth Conservation Service for help in interpreting data.

[1] By the middle of the year the President of the New York City Council was suggesting a monthly Anti-Graffiti Day, much like Earth Day, under the auspices of the Environmental Protection Administration; *The New York Times*, May 21, 1972, p. 66.

This rediscovered form of expression will form the basis of our discussion.[2]

Graffiti have not traditionally been regarded as a societal indicator, but rather as a folk symbol in their own right. Consequently, their diagnostic significance has been neglected, and they have been primarily treated in anecdotal and jocular vein in popular magazines. Although it is scarcely surprising that the people's slogans should appear in the people's press, a broader treatment of graffiti is justified. The graffito represents part of "a twilight zone of communication," an outlet for often deeply felt but rarely articulated sentiments and attitudes. Wall inscriptions reveal "developments, trends, and attitudes in man's history . . . little insights, little peepholes into the minds of individuals who are spokesmen not only for themselves but for others like them."[3]

THE NEW URBAN EPIDEMIC[4]

In 1970 the *New York Times* first referred to the growing problem of graffiti defacement of subway stations and trains.[5] In 1970 vandalism cost the New York Transit Authority $2.6 million, of which $250,000 was spent on removing graffiti.[6] The following year graffiti removal accounted for over $300,000, and in 1972 this figure was estimated to have risen above $500,000.[7] In New York graffiti are regarded primarily as a subway problem, for rapid transit provides one of the few meeting points of the cultural worlds of the inner city and the suburb. As long as "defacement" was limited to low-income residential areas, there was apparently no problem; certainly the media maintained an unconcerned silence. Indignation appears only when the inner city impinges upon the suburbanite; thus an enraged reader of the *New York Times* responded to a cautious editorial suggesting that perhaps graffiti offered "a preservative of social history," by condemning the "cowardly vandalism" and the "defacing morons" responsible for it.[8]

Graffiti are even more prevalent in Philadelphia, which makes a rather abashed claim to be the graffiti capital of the world. Defacement from spray-painting on the subway system is costing the transit authority $1 million a year, and cleaning and removal elsewhere in the city accounts for three times as much again.[9] Since 1971 the city police department has maintained a twenty-five man "graffiti squad." Remedial measures have included the establishment of a Graffiti Alternatives Workshop, and a competition to devise a new design for the transit authority's bus fleet.[10] The pastime nevertheless seems to continue unchecked, and *Time* estimated that Philadelphia might have as many as 10,000 more or less habitual graffiti artists.[11] Stiffer penalties for culprits were an oft-mentioned element in Mayor Rizzo's platform in the 1971 mayoralty election, and early in 1972 legislation was introduced into the City Council to control the sale of spray-paint cans.[12]

THE GRAFFITI LONER

Perhaps the least geographical inner city inscriptions are the products of single individuals, or pairs of artists. In Philadelphia and New York these inscriptions include the work of graffiti celebrities, the self-proclaimed kings of the walls. The majority are black youngsters who range widely through the city, the downtown, the airport, the inner suburbs, and even beyond. They make little claim to the black residential sections away from the main thoroughfares. Their "territory" is overwhelmingly linear, following the main transport arteries, and their targets are city-wide public structures (Fig. 1). Their goal is to leave a mark on exotic space, to make a claim to the

[2] Urban graffiti are not new; Philadelphia walls, for example, were heavily inscribed in the 1850s; *Philadelphia Evening Bulletin*, January 12, 1973, p. G-3.

[3] R. Reisner, *Graffiti: Two Thousand Years of Wall Writing* (New York: Cowles Book Company, 1971), p. 1.

[4] The graffiti "problem" had developed to this point according to a newspaper editorial in mid-1972; *The New York Times*, May 26, 1972, p. 34.

[5] *The New York Times*, July 15, 1970, p. 38.

[6] *The New York Times*, March 30, 1971, p. 69.

[7] It is reported that every station porter in the New York subway system spends an hour a day removing graffiti, and the problem has become so acute that priorities for removal of different classes of graffiti have been established; *The New York Times*, February 11, 1972, p. 39.

[8] *The New York Times*, February 28, 1972, p. 30, and March 9, 1972, p. 40.

[9] *Time*, March 13, 1972, p. 44.

[10] *The New York Times*, February 11, 1972, p. 39.

[11] *Time*, March 13, 1972, p. 44.

[12] In New York City such legislation was enacted in 1972. The bill was introduced by Mayor Lindsay and toughened in passing by City Council. It prohibits the carriage of opened spray cans and has moved toward closer control of sales. Penalties, including imprisonment, will be enforced.

FIG. 1. Autographs of two of the most prominent graffiti loners on a bridge, part of a main city thoroughfare.

world outside the ghetto. They leave inner city residential blocks to the local street gang, which has its own claim to the walls of its turf.

The more brazen the spatial conquest, the greater the status, so that graffiti kings seek to emulate each other in the inaccessibility of locations they invade (Fig. 2). Tity Peace Sign sprayed his name in red on an elephant's back side at Philadelphia Zoo; Bobby Kidd sprayed a police car while a friend held the officers' attention; and at the airport Cornbread sprayed a TWA jet which took off to the south bearing his name upon its wings.[13] The signs at the airport read "Cornbread welcomes you to Philadelphia." So esteemed is his reputation as King of the Walls that when a city newspaper carried a false report of his death in a gang fight, a number of "Cornbreads" contacted the press to report they were still alive.[14]

The integral characteristics of the graffiti kings are their nicknames and their signatures. The names they select are revealing; in 1971 leading Philadelphia exhibitors included Dr. Cool No. 1, Cool Earl, Bobby Cool, Kool Kev,

FIG. 2. A cartoon by Tony Auth which appeared in the *Philadelphia Inquirer*, August 6, 1971. (Reproduced courtesy of the *Inquirer*.)

[13] S. Padwe, "The Aerosol Autographers: Why They Do It," *Today Magazine, Philadelphia Inquirer*, May 2, 1971, pp. 8 and 12; and *Philadelphia Daily News*, March 5, 1971, p. 27.

[14] *Philadelphia Tribune*, March 2, 1971, p. 1.

Sir Smooth. "Cool" denotes confidence, style, suavity, a mastery of the intricate signal language of street life.[15] A title is a common affectation; another is to paint a crown above the nickname. Nicknames often show considerable creativity and originality; North Philadelphia boasts a Sherlock Holmes and Dr. Watson, and even Baron Eric von Schwenk makes a graceful appearance in a solidly black neighborhood.

The signature is distinctive. It is usually spray-painted from an aerosol can and is highly accentuated, embellished with elegant curves and generous serifs. The letters, like the name itself, convey a message of "style," yet imply frustrated ambition, a bittersweet theme.

The conquest of territory, even in fantasy, is always an act performed for an audience. Locations have a meaning; to claim access to an inaccessible location is to make a claim of primacy for oneself. In Cool Earl's words:[16]

> I started writing . . . to prove to people where I was. You go somewhere and get your name up there and people know you were there, that you weren't afraid.

The same theme was repeated by Taki 183, a teenager living on 183rd Street in Manhattan, an artist with hundreds of imitators:[17]

> I don't feel like a celebrity normally . . . but the guys make me feel like one when they introduce me to someone. "This is him" they say. The guys knows who the first one was.

Inaccessibility rather than socioeconomic status determines difficult space, space worthy of entry. Thus Cornbread or an imitator made a nocturnal swing through Fairmount, a working class white neighborhood on the edge of black North Philadelphia, closely patrolled by its own white teenage groups, with a history of antiblack harassment. These very characteristics made the area worthy of invasion. Fairmount youngsters strongly resented Cornbread's impertinent foray, and a number of his inscriptions were amended with an obscenity directed against the intruder. An assertive and defamatory message against Cornbread appeared on the wall of a house near the interface with the black neighborhood, in essence providing the conqueror with his laurel.

The mastery and occupation of space might well be a behavioral primitive. Climbing mountains, descending to the ocean depths, landing astronauts on the moon who leave behind their own territorial marker, colonial adventures, riding the freeway, possessing a home on a large lot—middle-income Americans have ample opportunity to sublimate their territorial needs, but many of these options are closed to the inner city dweller. For him, spatial mastery is often possible only in the realms of the phantasmagorical. Blocked access to the mainstream culture encourages a projective fantasy.[18] Thus Cool Earl and Kidd have legitimate artistic ambitions, but inner city life provides no access to such a realization. "We'd like to get into commercial art . . . but no one has told us how to go about it."[19] Hence, like some other inner city residents, they follow the main highway of make-believe. An exotic nickname, a garnished signature, identify a new personality and a space conquered, a mark left. "A man has to win a battle somewhere."[20]

GRAFFITI OF THE TEENAGE STREET GANG

The complementarity of the relationship between space and society is a recurrent theme in accounts of inner city life.[21] The power to claim territory endues the claimant with both identity and status among his peer group. Aspirants will expend considerable effort and even endure discomfort and stress in order to capture new territory. Wall-marking permits inner city teenagers, who are denied access to

---

[15] Inner city signals are discussed in G. Suttles, *The Social Order of the Slum: Ethnicity and Territory in the Inner City* (Chicago: University of Chicago Press, 1968).

[16] Padwe, op. cit., footnote 13, p. 10; and R. Meltzer, "Graffiti: The Magnificent Obsession," *Philly Talk*, March 1971, p. 13.

[17] *The New York Times*, July 21, 1971, p. 37.

[18] This argument reappears constantly in black inner city literature; A. H. Fauset, *Black Gods of the Metropolis: Negro Religious Cults in the Urban North* (Philadelphia: University of Pennsylvania Press, 1971); and O. Williams, *Metropolitan Analysis: A Social Access Approach* (New York: The Free Press, 1971).

[19] Padwe, op. cit., footnote 13, p. 8.

[20] R. Sommer, "People's Art," *Natural History*, 80 (February 1971), pp. 40–45.

[21] D. Ley, *The Black Inner City as Frontier Outpost: Images and Behavior of a Philadelphia Neighborhood*, Monograph Series, No. 7 (Washington, D.C.: Association of American Geographers, forthcoming); F. Boal, "Territoriality on the Shankill-Falls Divide, Belfast," *Irish Geography*, Vol. 6 (1969), pp. 30–50; and Suttles, op. cit., footnote 15.

legitimate mastery over space, to claim control of a more ephemeral and chimeric nature. Spray-painting is one example of the adroitness of inner city culture in creating its own institutions in which desired status roles may be enacted.[22]

To some extent there is a complementarity between the graffiti king and the teenage street gangs. Both vie for the allegiance of adolescents; Cornbread turned to graffiti only after he had rejected the option of joining a gang:[23]

> There isn't much choice of what to do. . . . I did it because there was nothing else. I wasn't goin' to get involved with no gangs or shoot no dope, so I started writin' on buses. I just started with a magic marker an' worked up.

In each instance, escalation identifies the "adept" participant, whether the path leads from magic marker to aerosol can, from loudmouthing to cutting, or from glue to scag. Like the mainstream society from which it has long been excluded, the independent society of inner city adolescents offers recognition to the man who excels, but to many inner city youth only one attainable avenue is offered, to excel at being bad.

The complementarity of graffiti king and street gang is revealed also in their complementary claims for space. The signatures of the graffiti kings are commonly on main thoroughfares, on downtown buildings, and on exotic targets outside the inner city, but they rarely stray into the residential blocks. Should the signature of a black but nonlocal graffiti artist appear within the turf of a street gang in a black neighborhood, it would often receive the same obscene amendment as the Fairmount boys' response when Cornbread made a nocturnal circuit of their white working-class neighborhood.

In contrast to the graffiti kings, who are free-ranging and make temporary claims to space, the street gangs occupy a more fixed and permanent territory. Some gangs can trace a continuous identity over several decades; a few have shown their strength by reforming uptown after urban renewal cleared their original neighborhood.[24] Street gangs are strongly place specific; in Philadelphia most of them take their names from an intersection near the center of their territory. Though the relationship is not as exclusive as it once was, the gang remains a block and neighborhood-based organization.[25] Its sphere of movement, fixed rather than fluid, confined rather than expansive, is a close approximation to the movement patterns of many inner city residents.

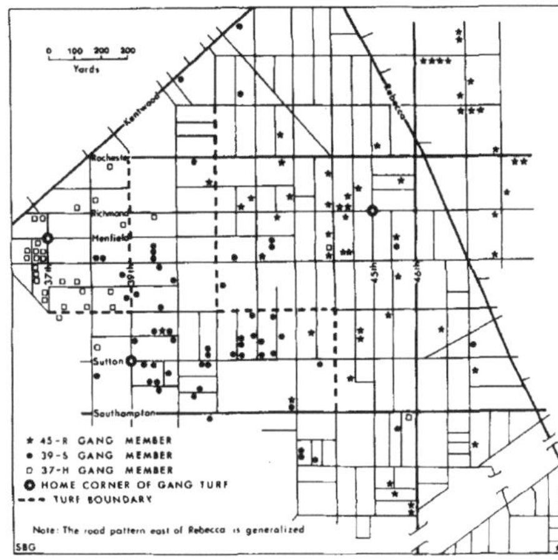

FIG. 3. Gang territories and residences of gang members in a North Philadelphia neighborhood.

A gang's claim to its turf is never in doubt in the minds of gang and nongang youth. Neighborhood teenagers have to take care in the paths they follow and the areas they use. A university junior living in a section of North

[22] This is a well-established causal theory for inner city juvenile delinquency; A. Cohen, *Delinquent Boys* (New York: Free Press, 1955); J. Short, ed., *Gang Delinquency and Delinquent Subcultures* (Chicago: University of Chicago Press, 1968); and C. Ward, ed., *Vandalism* (London: Architectural Press, 1973).

[23] Padwe, op. cit., footnote 13, p. 8.

[24] The Barbary Coast gang in West Philadelphia has a continuous history of over thirty years; the Uptown Norris gang in North Philadelphia was named after dislocation uptown by renewal and the expansion of Temple University, from their former turf on Norris Street.

[25] There is an abundant literature on teenage gangs; for Philadelphia in the mid-nineteenth century, see B. Laurie, "Fire Companies and Gangs in Southwark: The 1840's," in A. Davis and M. Haller, eds., *The Peoples of Philadelphia* (Philadelphia: Temple University Press, 1973), pp. 71–87; for Chicago in the 1920s, F. Thrasher, *The Gang* (New edition; Chicago: University of Chicago Press, 1963); for Boston in the late 1930s, W. F. Whyte, *Street Corner Society* (Chicago: University of Chicago Press, 1955); for New York City at the end of the 1950s, L. Yablonsky, *The Violent Gang* (New York: Macmillan, 1962); for Scotland today, J. Patrick, *A Glasgow Gang Observed* (London: Eyre-Methuen, 1970).

FIG. 4. A wall roster of members of the 12th and Poplar gang on an abandoned store.

Philadelphia with three fairly active street gangs (Fig. 3) expressed the dangers to free movement.

> It used to be terrible, man. It used not to be safe to walk beyond 41st Street without some guys getting at you. I live in 37th's territory but I was never with them. I know cats on Sutton and some on Richmond. Like I was between all three, and that's cool.

In Philadelphia, wall graffiti offer an accurate indicator of turf ownership. As a general rule, the incidence of gang graffiti becomes denser with increasing proximity to the core of a territory. Overwhelmingly graffiti consist of signatures, a nickname, often followed by the gang name (Fig. 4). Occasionally, boastful slogans are painted near the core of the turf. This is true, for example, of the large 39th and Sutton Street gang; despite its size, 39-S is vulnerable, for unlike 45-R and 37-H, which enlist outside groups into temporary alliances, 39-S has no allies.[26]

The evidence of the walls gives a good approximation of the extent of each territory. Boundaries compiled from the relative incidence of gang graffiti found a ready acceptance by neighborhood youth as an accurate portrayal of each gang's area of control, and the residences of gang members living in the neighborhood also show close agreement with them (Fig. 3). There is some overlap, with several gang members living in an alien turf, but this is both unusual and dangerous; a member of the 45-R gang who lived in 39-S territory fell an easy victim in the incident-ridden summer of 1970.

[26] The street and gang names are fictional; life in this neighborhood is discussed in Ley, op. cit., footnote 21.

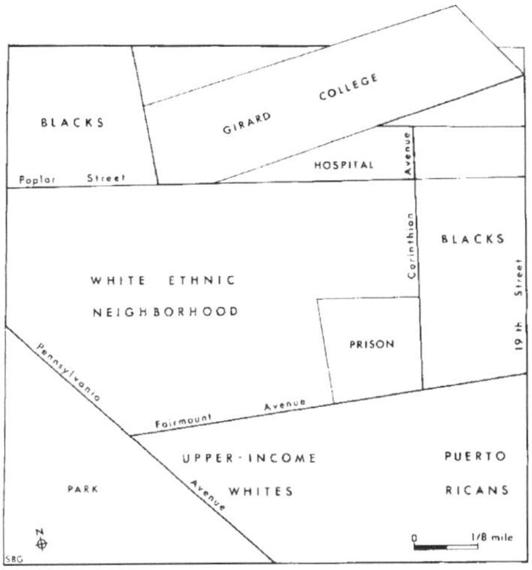

FIG. 5. Social groups in Fairmount and vicinity.

We have used graffiti to provide insights into the social and spatial order with equal success in a second Philadelphia neighborhood. Fairmount is a working-class white enclave on the edge of the North Philadelphia ghetto; black residential areas are north and east, a weakening Puerto Rican neighborhood is southeast, and a section of young white professionals, recent in-migrants, is to the south (Fig. 5).[27]

The territories of the Puerto Rican groups south of Fairmount Avenue, the Navaka Soul Gents and 20-G, are not well defined (Fig. 6). Their territorial claims show considerable overlap. Territorial indistinctness mirrors social weakness; both groups are weak, and neither appears in the files of the City's gang control units.[28] Both are on the western edge of the territory of the powerful 16-W gang; 16-W graffiti are scattered throughout their area, showing that neither 20-G nor the Gents have been able to assert their independence from their more powerful neighbor.

FIG. 6. Gang graffiti in the eastern portion of Fairmount and vicinity.

North of Fairmount Avenue, and east of the jail, there is a sharp break from a biracial population to a solidly black neighborhood. Despite a large number of vacant and cleared rubble-strewn lots, the intensity of graffiti increases, for this is part of the territory of the Moroccans, one of the larger and more violent North Philadelphia gangs.[29] At Corinthian Avenue their territory abuts abruptly against the white neighborhood of Fairmount, and there is very little graffiti transgression across this boundary (Fig. 6).

---

[27] R. Cybriwsky, "Social Relations and the Spatial Order in the Urban Environment: A Study of Life in a Neighborhood in Central Philadelphia," unpublished doctoral dissertation, Pennsylvania State University, 1972.

[28] 20-G claims to be the main drafting unit for local Puerto Rican youth, and sporadic minor coercion is directed against Puerto Rican teenagers who resist enlistment. The Navaka Soul Gents are unknown to city gang workers.

[29] The Moroccans and 16-Wallace have a long history of feuding, usually just east of the study area on 18th street.

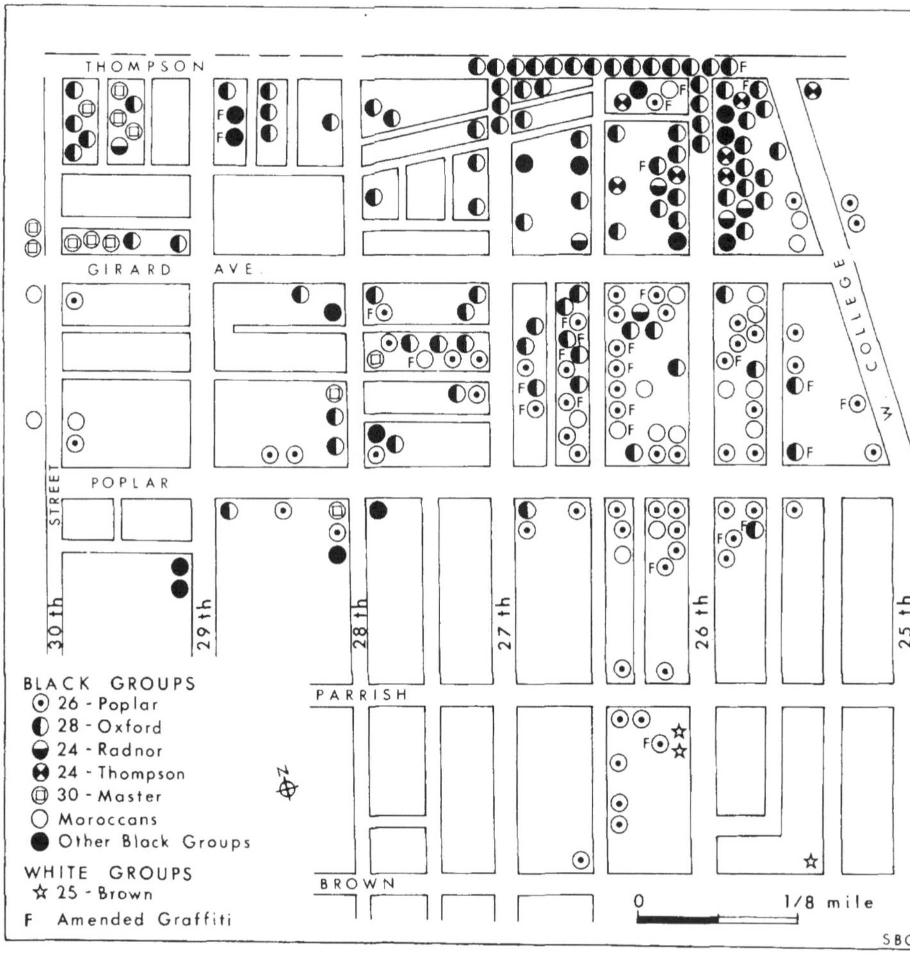

FIG. 7. Gang graffiti in the northwestern portion of Fairmount and vicinity.

Graffiti again become intense west of 25th and north of Poplar, in an area which has recently become black. Although a number of small or distant groups (including the Moroccans) make a weak claim to parts of this area, the strongest claimants are the powerful 28-Oxford gang, centered four blocks north, and a recent local group taking the name of the 26th and Poplar intersection (Fig. 7). The thrust of 26-P graffiti south of Poplar on 26th Street and a few blocks west indicates the beginning of the movement of the ghetto edge into Fairmount. The distribution of graffiti in Fairmount itself, more limited but extremely concentrated, has been omitted from the map, and will be analyzed separately. The middle class area south of Fairmount Avenue and west of 22nd has few graffiti, and these inscriptions include a predictably higher representation from the nongang graffiti kings.

GRAFFITI AND ZONES OF GANG CONTACT

Graffiti for more than twenty identifiable nonwhite groups were recorded for the neighborhood around Fairmount. Most of these groups had few territorial markers in the

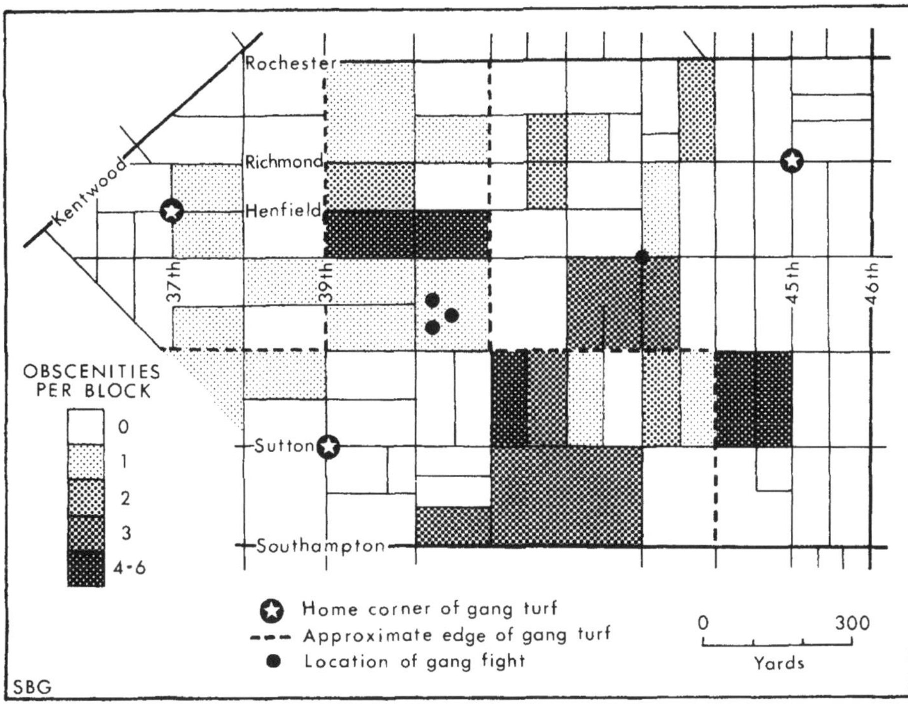

FIG. 8. Obscenities and gang fight locations in a section of Monroe.

neighborhood, either because their graffiti represented stray outliers from a distant turf, or because the smallness of the group precluded any real gang status. There were six nonwhite local gangs of some substance (Figs. 6 and 7). In the east the powerful Moroccan gang had uncontested supremacy; Fairmount Avenue marked a sharp southern boundary insulating the Moroccans from close contact with 16-W. The amorphous western boundary of 16-W was confused by the two relatively weak Puerto Rican gangs.

The most interesting boundaries were between the black gangs 26-Poplar and 28-Oxford (Fig. 7). The latter is a relatively well-established North Philadelphia unit with a moderate territory and more than twenty members. The smaller 26-P group is far more recent; in 1960 an overwhelmingly white population lived within a two-block radius of the 26th and Poplar intersection. The emergence of 26-P might be seen as spontaneous once the area contained a majority of nonwhite residents, but the dynamics are complicated; the 28th and Oxford intersection is only seven blocks away, and as one of the more aggressive gangs, 28-Ox might well have expected to expand south following the ghetto edge. Northward expansion was thwarted by powerful neighbors; between 1966 and 1970 eleven serious incidents involving 28-Ox and rival gangs to the north and east were known to city authorities.[30] The appearance of 26-P would clearly have been an obstacle to such southern expansion, and a certain amount of tension between the two groups might be expected.

The area which we might expect to have been contested consists of nine blocks or part-blocks which have tipped from predominantly white to predominantly nonwhite over the past decade. Both gangs have embossed on the walls, and thereby legitimized, their claims to this "virgin territory." Graffiti indicate the northern third has been effectively absorbed

[30] Youth Conservation Service, City of Philadelphia.

FIG. 9. A blunt warning at a turf boundary in North Philadelphia—"Do Not Enter."

by 28-Ox and the southern third by 26-P. The status of the central section (Girard-College-Poplar-28th) is more uncertain; the proportion of 26-P graffiti relative to 28-Ox ranges from twenty-five percent to seventy-five percent. The graffiti in these central blocks change from self-reinforcing autographs to aggressive epithets directed against the other gang (Fig. 7). These blocks have the tone of a contested marchland. What is the meaning of these aggressive graffiti? Interpretation could clearly lie at any point on the attitude-behavior continuum, including the sublimation of open hostility.

Evidence from the second inner city study area in North Philadelphia suggests that aggressive graffiti represent more than simply hostile attitudes; they indicate dispositions to overt behavior. There are three street gangs in this neighborhood, and turf boundaries again have the highest incidence of aggressive wall markings (Figs. 8 and 9). There is a consistent pattern to the victims of aggressive taunts; wall epithets are exchanged between 45-R and 39-S, and between 37-H and 39-S. If graffiti are indicators of behavior we would expect 45-R and 37-H to be common antagonists of 39-S. Behavioral data substantiate this expectation. 39-S is the largest gang with perhaps seventy-five members; 45-R has around sixty adherents and 37-H has thirty-five.[31] On a number of occasions 45-R and 37-H have formed an alliance against 39-S; in May, 1971, for example, after a week of gang fighting, fifty to sixty boys gathered on 37th Street and carried out a sweep along Sutton Street. This conflict has existed for a number of years. City files record ten serious incidents between 39-S and 45-R between 1966 and 1970, and a further ten between 39-S and 37-H, but no incidents between 45-R and 37-H. The articulation on the walls has indeed found its manifestation in the streets.

It would be intriguing if graffiti could also be used to predict the location of potential conflict. The zone of most probable conflict should coincide with the zone of aggressive graffiti, but we do not have enough data to

[31] Official estimates of gang size fluctuate between 85 and 200.

FIG. 10. Externally directed racial epithets in Fairmount.

make this claim. Incidents involving single boys or small groups are displaced either to a neutral place like school, or away from the boundary by the use of cars. Less spontaneous confrontations, however, involving larger groups of boys do show some concentration within the aggressive graffiti zone. A playground at a point of low aggregate travel for all three gangs is well known as the most common site for gang rumbles in this neighborhood. Three of four fights with known locations occurred in or near the playground (Fig. 8).

GRAFFITI AND THE DEFENDED NEIGHBORHOOD

A third type of inner city graffiti is wall inscriptions and emblems of the defended neighborhood, a neighborhood of some homogeneity, a recognized identity, and with authenticity as a distinct unit in its "foreign relations" with other parts of the city.[32] The wall markings of the defended neighborhood are internally supportive and externally aggressive. They are messages for two audiences. Such graffiti are commonly boundary markers; they delineate an interface, the edge of socially claimed space, a boundary which in the North American city is often ethnic, and commonly racial.[33] Fairmount is a defended neighborhood, a pocket of first and second generation immigrants experiencing encroachment from all quarters; to the north and east from the bursting North Philadelphia black community, and to the south from a burgeoning population of middle-class townhouse dwellers. The threat from the north and east is most keenly felt in Fairmount; the townhouse dwellers are reluctantly viewed as allies who at best will maintain property prices and the racial structure of the neighborhood.

The visible walls articulate invisible sentiments. In a defended neighborhood fettered by

[32] G. Suttles, *The Social Construction of Communities* (Chicago: University of Chicago Press, 1972).

[33] Graffiti in Belfast and Londonderry are solemn heralds of the spatial as well as the social structure of intergroup relations; curbs specially painted for the July 12th Orange celebrations in Protestant neighborhoods take on much of their meaning as discriminatory markers against an excluded Catholic minority; Boal, op. cit., footnote 21.

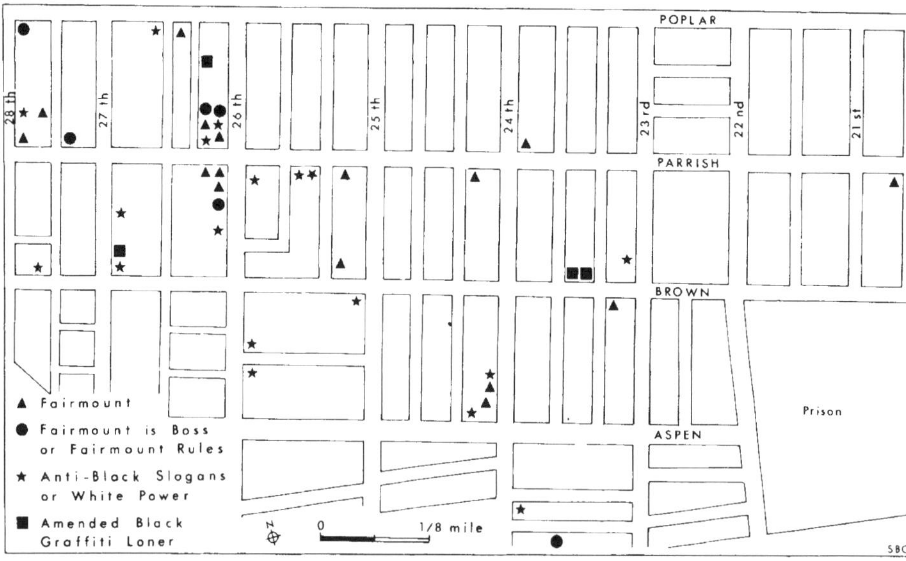

FIG. 11. The distribution of Fairmount graffiti.

more powerful neighbors, graffiti promote an introspective self-consciousness. "Fairmount Rules," "Fairmount is Boss," or simply "Fairmount," are the self-reinforcing slogans on the walls of a neighborhood with an uncertain identity. Externally directed racial epithets and obscenities are a neighborhood counterpart of aggressive gang graffiti (Fig. 10).

Fairmount graffiti are strongly localized (Fig. 11). Twenty-three of forty-four recorded occurrences of assertive or aggressive graffiti appear on the walls north of Brown Street and west of 26th. At first sight this distribution is surprising, for the ghetto front is more distinct at the eastern than at the northwestern edge of Fairmount. Blocks at the eastern edge of Fairmount are uniformly more than seventy percent nonwhite, and this proportion exceeds ninety percent on a number of blocks (Fig. 12); the nonwhite proportion on the northwestern boundary is much lower, and in 1970 most blocks had a substantial majority of white residents. Why should graffiti be concentrated where the "threat" in terms of absolute numbers is much less? We assumed that white fears of ethnic encroachment would show a more or less linear relationship with the size of the adjacent black community, and that consequently the zone adjacent to a homogeneous black population would display the highest density of aggressive graffiti. This was not the case; there was no relationship between graffiti density and the homogeneity of the adjacent nonwhite population. Either graffiti were poor indicators of white resistance, or the assumption that a monolithic black community would produce such resistance was erroneous. Investigation revealed that the assumption rather than the indicator was faulty; the northwest quadrant was indeed the section of highest white stress. The reason lies with the areal incidence of racial change. During the 1960s the eastern ghetto edge consolidated, and there was no fresh encroachment; every block which had been ninety-five percent white in 1960 remained ninety-five percent white in 1970 (Fig. 13).

Events were very different on the northwestern boundary, an area which eluded community territorial control in at least one important manner. Almost ninety percent of residential units in Fairmount are owned by neighborhood residents, a very high figure which permits the existence of a largely closed housing market and a firm control on real estate transfers.[34] Outside ownership has implied an open housing market; preliminary trends indicate that as much as eighty percent of the apartment units rented by blacks have outside owners. The fact that the northwest quadrant does contain housing units with outside owners certainly aided the initial percolation of nonwhites into this section, and the southward advance of the ghetto edge by at least two blocks during the 1960s.

In 1960 two anomalous outliers of nonwhite population already existed in the northwestern quadrant, two blocks south of Poplar between 26th and Bambrey. Both had one-third nonwhite households. Two developments in the 1960s gave added concern to neighboring white residents. The first was the steady transformation of the section north of Poplar from a white majority to a black majority, a racial turnover which averaged five percent a year and accelerated as the decade progressed (Fig. 13). The second development was the assembling of black youths who eventually claimed the gang status of 26-P on the northern end of these two blocks.[35]

White response has consisted of repeated attempts to "freeze out" black families, followed by sporadic harassment of black residents on and around the two blocks. In some senses this harassment attained its goals; nonwhite households decreased slightly, from thirty-four percent to thirty-one percent, between 1960 and 1970, but the number of black households on adjacent streets south of Poplar continued to rise. A zone of tension appeared which is located exactly by the evidence on the walls. Fourteen cases of assertive or aggressive graffiti, about one-third of the Fairmount total, are within one block of the 26th and Parrish intersection (Fig. 11). At this point a breach has occurred in the invisible walls defending white Fairmount; the high density of wall graffiti represent an aggressive territorial display by the present claimants to assert continued possession against the newcomers.

The wall display is not idle sublimation but a real threat which promises to have an overt consequence. Several new black residents have been physically harassed near 26th and Parrish. At least one case was reported in the black Philadelphia press:[36]

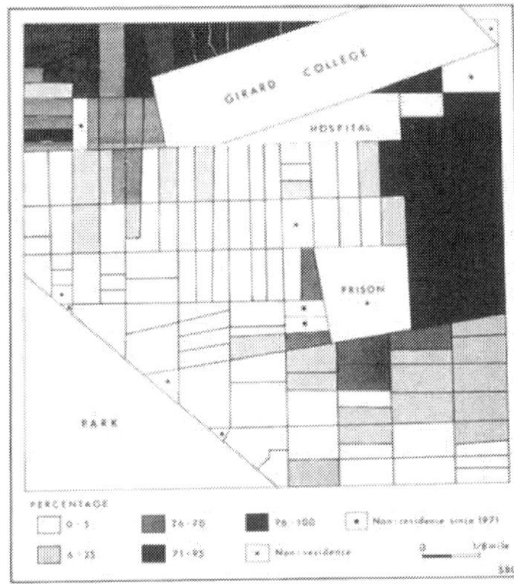

FIG. 12. Proportion of Negro occupants in Fairmount and vicinity, 1970.

> Declaring that she "might as well be living in Hell," Mrs. Emma Black said Wednesday that she will move from her predominantly white neighborhood in the Fairmount section of the city. Since she moved into the house at 814 N. Taney St., four months ago, Mrs. Black said that she has been a target for the abuse of surrounding whites.

In the late summer of 1970 continued harassment escalated into a racial confrontation involving perhaps as many as 400 white residents of Fairmount, and serious enough for a state of emergency to be declared. Two years later fresh racial graffiti indicated that the northwestern section remained a zone of racial tension.

At the level of the defended neighborhood, wall graffiti in Fairmount continue to be diagnostic indicators of an invisible environment of attitudes and social processes. They indicate far more than fears, threats, and prejudices; they are a prelude and a directive to open behavior,

[34] Detailed examples of the closed housing market are presented in Cybriwsky, op. cit., footnote 27, pp. 373–385.

[35] Liquor and drugs have been weakening 26-P; the state liquor store on Girard and West College has the graffito "26 P's Store."

[36] *Philadelphia Tribune,* October 31, 1970, p. 1: the victim's name has been changed; Mrs. Black's Negro neighbors had been forced from their home two weeks before: her own departure was precipitated by an attack in which her son was burnt by a firecracker. Other sources suggest prior provocation by Mrs. Black; *Philadelphia Evening Bulletin,* August 30, 1970, p. 1.

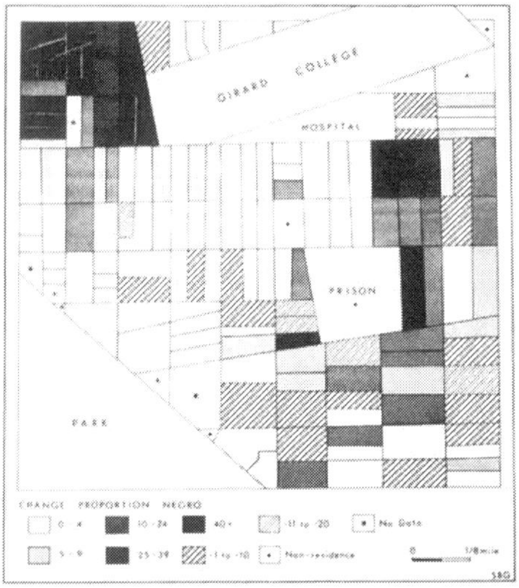

FIG. 13. Racial change in Fairmount and vicinity, 1960–1970.

and clues to the bounds and intensity of a community's control of its territory. In Fairmount, at least, they pinpoint the location where that behavior will occur.

CONCLUSION

The humble graffito is making a resurgence; our analysis of Philadelphia could profitably be extended to other cities in North America and Europe.[37] An understanding of this resurgence and of the distributions of inner city graffiti requires a delving into social process. In the same way that the biogeographer is not satisfied to record vegetation zones without an attempt at explanation, so the urban geographer turns to neighborhood social processes to explain his distribution maps. Process, however, is more equivocal than form, so that our explanation of the incidence of urban graffiti might best be regarded as an interpretation, although it is based on detailed knowledge of the neighborhoods concerned and a broad and supportive literature.[38]

In our discussion the mutual relationship between social process and the spatial order in the inner city has been a repetitive theme. The graffiti king, the street gang, and the defended neighborhood are all social groups which assert a territorial jurisdiction; each makes a public claim to space through an open declaration on the walls.

Territoriality is a process which has attracted far more speculation and hyperbole than detailed empirical treatment.[39] Earlier literature has described the elaborate territorial rituals and boundary markers of some primitive societies, but in the Western world empirical work has rarely escaped the built environment, commonly the schizophrenic ward or the simulated submarine.[40] Valuable as these and other room scale studies have been, "much remains to be done to identify the basic common features of human signalling behaviour related to territoriality as such."[41] Martin continues:[42]

> One field of research which could be extremely fertile is the analysis of artefacts where the signals have been transferred from the human body to an object which is used in a territorial context.

Sommer has shown how such objects are used to defend personal space at the room scale.[43] Territorial graffiti have an identical function at the neighborhood scale, that of ascribing a proprietary meaning to space.

The acceleration of urban crime, information overload, and a complex and often oppressive

---

[37] In Britain a number of these observations would apply to at least Glasgow and Birmingham; for example, David White, "Brum's Mobs," *New Society,* 21 October 1971, pp. 760–63.

[38] The works cited in footnote 22 are a sample of this literature, and are consistent in their findings with interpretations offered here; for the neighborhood studies in Philadelphia, see Cybriwsky, op. cit., footnote 27, and Ley, op. cit., footnote 21.

[39] A recent succinct review of part of the literature is A. Esser, "A Biosocial Perspective on Crowding", in J. Wohlwill and D. Carson, eds., *Environment and the Social Sciences: Perspectives and Applications* (Washington, D.C.: American Psychological Association, 1972), pp. 15–28; for more general classificatory discussions, see D. Stea, "Space, Territory and Human Movements", *Landscape,* Vol. 15 (1965), pp. 13–16; and S. Lyman and M. Scott, "Territoriality: A Neglected Sociological Dimension", *Social Problems,* Vol. 15 (1967), pp. 236–49.

[40] A. Esser et al., "Territoriality of Patients on a Research Ward," in J. Wortis, ed., *Recent Advances in Biological Psychiatry* (New York: Plenum Press, 1965).

[41] R. Martin, "Concepts of Human Territoriality," in P. Ucko et al., eds., *Man, Settlement and Urbanism* (London: G. Duckworth, 1972), pp. 427–45.

[42] Martin, op. cit., footnote 41, p. 443.

[43] R. Sommer, *Personal Space: The Behavioral Basis of Design* (Englewood Cliffs, N.J.: Prentice-Hall, 1969).

urban experience compound the uncertainty and confusion of inner city life. The restoration of some predictability to one's environment is a prescription for a territorial imperative. To define a small space of one's own permits a higher level of social control to be maintained, by surveillance, by restricting entry to recognized friends, and by limiting the range of acceptable behavior within the area.[44] Establishing the territory generates security; maintaining or embellishing it guarantees status.

The graffiti king conquers a place momentarily; maintenance is impossible, hence embellishment is maximized in the capture of an exotic, inaccessible place with a garnished signature. The gang turf and the defended neighborhood are more permanent territories controlling the flux of the confusing city, but they need visible and unequivocal cues to identify ownership and notify outsiders that they are entering a protected place and must respect the integrity of claimed property. He who is king of the walls claims also to be king of the streets and master of their use. The walls are more than an attitudinal tabloid; they are a behavioral manifesto.

Graffiti also have a more explicitly methodological interest to the human geographer. It is recognized that the emerging themes of man-environment interaction are conditioned not only by a tangible physical environment, but also by the invisible contours of a behavioral environment.[45] But how does one identify, let alone measure, this experiential topography? Graffiti are a visible manifestation of a group's social space. Moreover, assertive or aggressive graffiti represent more than attitudes. They are dispositions to behavior, and as such impress a bolder outline on the fuzzy transition between perception and action.[46]

The scholar must be sensitive to the nuances of such native guides as inner city graffiti.[47] If he is unable to interpret the visible, then the invisible meaning of place will be beyond his grasp. We must understand the behavioral environment, the complex of socially and culturally determined beliefs and perceptions, and learn to read its diagnostic indicators if we wish to develop social and behavioral theory.[48] To breathe life into the aggregate structural models of the 1960s will require such sensitivity, and perhaps a new microregional geography where the important contours are experiential, those of society and of the mind.[49] Important distributions might then be those showing surfaces of novelty, stress, status, security, opportunity, and social control in the city. Such existential qualities increasingly define the new gradients and the new boundaries to behavior in the contemporary American metropolis.

---

44 Compare the parallel conclusions in Oscar Newman, *Defensible Space* (New York: Macmillan, 1972).

45 W. Kirk, "Problems in Geography," *Geography*, Vol. 47 (1963), pp. 357–71; H. Brookfield, "On the Environment as Perceived," in C. Board et al., eds., *Progress in Geography*, Volume 1 (London: Edward Arnold, 1969), pp. 51–80; and A. Buttimer, "Social Space in Interdisciplinary Perspective," *Geographical Review*, Vol. 59 (1969), pp. 417–26.

46 M. Fishbein, "Attitude and the Prediction of Behavior," in M. Fishbein, ed., *Readings in Attitude Theory and Measurement* (New York: John Wiley, 1967), pp. 477–92.

47 For a more general discussion of indicators, see A. Patterson, "Unobtrusive Measures: Their Nature and Utility for Architects," in J. Lang et al., eds., *Behavioral Aspects of Architectural Design* (Stroudsburg, Pa.: Dowden, Hutchison, Ross, forthcoming).

48 For another example, see D. Ley and R. Cybriwsky, "The Spatial Ecology of Stripped Cars," *Environment and Behavior*, Vol. 6 (1974), pp. 53–67.

49 A classical example would be R. Morrill, "The Negro Ghetto: Problems and Alternatives," *Geographical Review*, Vol. 55 (1965), pp. 339–61.

[8]

MONUMENT AND MYTH*

DAVID HARVEY

ABSTRACT. The Basilica of Sacré-Coeur, which strategically and symbolically dominates the northern skyline of Paris, has a strange and tortured history. Conceived of in the course of the Franco-Prussian War of 1870–71, its construction was seen by many as an act of penitence for the moral decadence of the Second Empire of Napoleon III as well as for the supposed excesses of the Paris Commune of 1871. The cult of the Sacred Heart and the movement to build a Basilica for its glorification drew much of its support from ultraconservative Catholics as well as from those espousing the cause of monarchist restoration. Legislation pushed through at the height of the political reaction which followed the Commune turned the construction of the Basilica into a national project. This legislation permitted those in command of the project to select the heights of Montmartre as a site for its construction. The Basilica was thus built upon the very spot where some of the most significant opening and closing events of the Paris Commune occurred. In the years that followed the laying of the foundation stone in 1875, the predominantly republican population of Paris sought to stop what they perceived as a manifestation of reactionary monarchism from being erected in their midst. They all but succeeded, in 1882, in emasculating the whole project. The Basilica, finally completed and consecrated in 1919, was for many years seen as a provocation to civil war and to this day is still interpreted as a political symbol.

STRATEGICALLY placed atop a hill known as the Butte Montmartre, the Basilica of Sacré-Coeur occupies a commanding position over Paris. Its five white marble domes and the campanile that rises beside them can be seen from every quarter of the city. Occasional glimpses of it can be caught from within the dense and cavernous network of streets which makes up old Paris. It stands out spectacular and grand to the young mothers parading their children in the Jardins de Luxembourg, to the tourists who painfully plod to the top of Notre Dame or who painlessly float up the escalators of the Centre Beaubourg, to the commuters crossing the Seine by metro at Grenelle or pouring into the Gare du Nord, to the Algerian immigrants who on Sunday afternoons wander to the top of the rock in the Parc des Buttes Chaumont. It can be seen clearly by the old men playing "boule" in the Place du Colonel Fabien, on the edge of the traditional working class quarters of Belleville and La Villette—places that have an important role to play in our story.

On cold winter days when the wind whips the fallen leaves among the aging tombstones of the Père Lachaise cemetery, the Basilica can be seen from the steps of the tomb of Adolph Thiers, first President of the Third Republic of France. Though now almost hidden by the modern office complex of La Défense, it can be seen from more than twenty kilometers away in the Pavillion Henri IV in St Germain-en-Laye where Adolph Thiers died. But by a quirk of topography, it cannot be seen from the famous Mur des Fédérés in that same Père Lachaise cemetery where on May 27th, 1871, some of the last few remaining soldiers of the Commune were rounded up after a fierce fight among the tombstones and summarily shot. You cannot see Sacré-Coeur from that ivy-covered wall now shaded by an aging chestnut. That place of pilgrimage for socialists, workers, and their leaders is hidden from a place of pilgrimage for the

Dr. Harvey is Professor of Geography in the Department of Geography and Environmental Engineering at The Johns Hopkins University in Baltimore, MD 21218.

* I would like to acknowledge the support of the Guggenheim Foundation for a year's study leave in Paris, 1976–77.

FIG. 1. The Basilica of Sacré-Coeur.

Catholic faithful by the brow of the hill on which stands the grim tomb of Adolph Thiers.

Few would argue that the Basilica of Sacré-Coeur is beautiful or elegant (Fig. 1). But most would concede that it is striking and distinctive, that its direct Byzantine style achieves a kind of haughty grandeur which demands respect from the city spread out at its feet. On sunny days it glistens from afar and even on the gloomiest of days its domes seem to capture the smallest particles of light and radiate them outwards in a white marble glow. Floodlit by night it appears suspended in space, sepulchral and ethereal. Thus does Sacré-Coeur project an image of saintly grandeur, of perpetual remembrance. But remembrance of what?

The visitor drawn to the Basilica in search of an answer to that question must first ascend the steep hillside of Montmartre. Those who pause to catch their breath will see spread out before them a marvelous vista of rooftops, chimneys, domes, towers, monuments—a vista of old Paris that has not changed so much since that dull and foggy October morning in 1872, when the Archbishop of Paris climbed those steep slopes only to have the sun miraculously chase both fog and cloud away to reveal the splendid panorama of Paris spread out before him. The Archbishop marvelled for a moment before crying out loud: "It is here, it is here where the martyrs are, it is here that the Sacred Heart must reign so that it can beckon all to it!"[1] So who are the martyrs commemorated here in the grandeur of this Basilica?

The visitor who enters into that hallowed place will most probably first be struck by the immense painting of Jesus which covers the dome of the apse. Portrayed with arms stretched wide, the figure of Christ wears an image of the Sacred Heart upon his breast. Beneath, two words stand out directly from the Latin motto—GALLIA POENITENS. And beneath that stands a large gold casket, containing the image of the Sacred Heart of Jesus, burning with passion, suffused with blood and surrounded with thorns. Illuminated day and night, it is here that pilgrims come to pray.

Opposite a life-size statue of Saint Marguerite-Marie Alacoque, words from a letter written by that saintly person—date, 1689, place, Paray-le-Monial—tell us more about the cult of the Sacred Heart:

> THE ETERNAL FATHER WISHING REPARATION FOR THE BITTERNESS AND ANGUISH THAT THE ADORABLE HEART OF HIS DIVINE SON HAD EXPERIENCED AMONGST THE HUMILIATIONS AND OUTRAGES OF HIS PASSION DESIRES AN EDIFICE WHERE THE IMAGE OF THIS DIVINE HEART CAN RECEIVE VENERATION AND HOMMAGE

Prayer to the Sacred Heart of Jesus which, according to the scriptures, had been exposed when a centurion thrust a lance through Jesus's side during his suffering upon the cross, was not unknown before the seventeenth century. But Marguerite-Marie, beset by visions, transformed the worship of the Sacred Heart into a distinctive cult within the Catholic Church. Although her life was full of trials and suffering, her manner severe and rigorous, the predominant image of Christ which the cult projected was warm and loving, full of repentance and suffused with a gentle kind of mysticism.[2]

Marguerite-Marie and her disciples set about propagating the cult with great zeal. She wrote to Louis XIV, for example, claiming to bring a message from Christ, in which the King was asked to repent, to save France by dedicating himself to the Sacred Heart, to place its image upon his standard and to build a chapel to its glorification. It is from that letter of 1689 that the words now etched in stone within the Basilica are taken.

The cult diffused slowly. It was not exactly in tune with eighteenth-century French rationalism which strongly influenced modes of belief among Catholics and stood in direct opposition to the hard, rigorous, and self-disciplined image of Jesus projected by the Jansenists. But by the end of the eighteenth century it had some important and potentially influential adherents. Louis XVI privately took devotion to the Sacred Heart for himself and his family. Imprisoned during the French revolution, he vowed that within three months of his deliverance he would publicly dedicate himself to the Sacred Heart and thereby save France (from what, exactly, he did not, nor did he need to say). And he vowed to build a chapel to the worship of the Sacred Heart. The manner of Louis XVI's deliverance did not permit him to fulfill that vow. Marie-Antoinette did no better. The Queen delivered up her last prayers to the Sacred Heart before keeping her appointment with the guillotine.

These incidents are of interest because they presage an association, important for our story,

[1] R. P. Jonquet, *Montmartre, Autrefois et Aujord'hui* (Paris: Dumoulin, no date).

[2] Jonquet, op. cit., footnote 1, provides considerable background on the cult. See also Adrien Dansette, *Histoire Religieuse de la France Contemporaine* (Paris: Flammarion, 1965).

between the cult of the Sacred Heart and the reactionary monarchism of the *Ancien Régime*. This put adherents to the cult in firm opposition to the principles of the French revolution. Believers in the principles of liberty, equality, and fraternity, who were in any case prone to awesome anticlerical sentiments and practices, were, in return, scarcely enamored of such a cult. Revolutionary France was no safe place to attempt to propagate it. Even the bones and other relics of Marguerite-Marie, now displayed in Paray-le-Monial, had to be carefully hidden during these years.

The restoration of the monarchy in 1815 changed all that. The Bourbon monarchs sought, under the watchful eye of the European powers, to restore whatever they could of the old social order. The theme of repentence for the excesses of the revolutionary era ran strong. Louis XVIII did not fulfill his dead brother's vow to the Sacred Heart, but he did build, with his own moneys, a Chapel of Expiation on the spot where his brother and his family had been so unceremoniously interred—GALLIA POENITENS. A society for the propagation of the cult of the Sacred Heart was founded, however, and proceedings for the glorification of Marguerite-Marie were transmitted to Rome in 1819. The link between conservative monarchism and the cult of the Sacred Heart was further consolidated.

The cult spread among conservative Catholics but was viewed with some suspicion by the liberal progressive wing of French Catholicism. But now another enemy was ravaging the land, disturbing the social order. France was undergoing the stress and tensions of capitalist industrialization. In fits and starts under the July Monarchy (installed in 1830 and just as summarily dispensed with in the revolution of 1848) and then in a great surge in the early years of the Second Empire of Napoleon III, France saw a radical transformation in certain sectors of its economy, in its institutional structures, and in its social order.[3] This transformation threatened much that was sacred in French life, since it brought within its train a crass and heartless materialism, an ostentatious and morally decadent bourgeois culture, and a sharpening of class tensions. The cult of the Sacred Heart now assembled under its banner not only those devotees drawn by temperament or circumstance to the image of a gentle and forgiving Christ, not only those who dreamed of a restoration of the political order of yesteryear, but also all those who felt threatened by the materialist values of the new social order.

To these general conditions, French Catholics could also add some more specific complaints in the 1860s. Napoleon III had finally come down on the side of Italian unification and committed himself politically and militarily to the liberation of the central Italian states from the temporal power of the Pope. The latter did not take kindly to such politics and under military pressure retired to the Vatican, refusing to come out until such time as his temporal power was restored. From that vantage point, the Pope delivered searing condemnations of French policy and the moral decadence which, he felt, was sweeping over France. In this manner he hoped to rally French Catholics in the active pursuit of his cause. The moment was propitious. Marguerite-Marie was beatified by Pius IX in 1864. The era of grand pilgrimages to Paray-le-Monial began. The pilgrims came to express repentence for both public and private transgressions. They repented for the materialism and decadent opulence of France. They repented for the restrictions placed upon the temporal power of the Pope. They repented for the passing of the traditional values embodied in an old and venerable social order. GALLIA POENITENS.

Just inside the main door of the Basilica of Sacré-Coeur in Paris, the visitor can read the following inscription:

THE YEAR OF OUR LORD 1875
THE 16TH JUNE
IN THE REIGN OF HIS HOLINESS POPE PIUS IX IN ACCOMPLISHMENT OF A VOW FORMULATED DURING THE WAR OF 1870–71 BY ALEXANDERLEGENTIL AND HUBERT ROHAULT DE FLEURY RATIFIED BY HIS GRACE MGR. GUIBERT ARCHBISHOP OF PARIS; IN EXECUTION OF THE VOTE OF THE NATIONAL ASSEMBLY OF THE 23RD JULY 1873 ACCORDING TO THE DESIGN OF THE ARCHITECT ABADIE; THE FIRST STONE OF THIS BASIL-

[3] Roger Price, *The Economic Modernization of France* (London: Croom Helm, 1975) and Fernand Braudel and Ernest Labrousse, eds., *Histoire Economique et Sociale de la France, Tome III* (Paris: Presse Universitaire de France, 1976).

ICA ERECTED TO THE SACRED HEART OF JESUS WAS SOLEMNLY PUT IN PLACE BY HIS EMINENCE CARDINAL GUIBERT . . .

Let us flesh out that capsule history and find out what lies behind it.

As Bismarck's battalions rolled to victory after victory over the French in the summer of 1870, an impending sense of doom swept over France. Many interpreted the defeats as righteous vengeance inflicted by divine will upon an errant and morally decadent France. It was in this spirit that the Empress Eugene was urged to walk with her family and court all dressed in mourning, from the Palace of the Tuilleries to Notre Dame, to publicly dedicate themselves to the Sacred Heart. Though the Empress received the suggestion favorably, it was, once more, too late. On September 2nd, Napoleon III was defeated and captured at Sedan, on September 4th the Republic was proclaimed on the steps of the Hôtel-de-Ville and a Government of National Defense was formed. On that day also the Empress Eugene took flight from Paris having prudently, and at the Emperor's urging, already packed her bags and sent her more valuable possessions on to England.

The defeat at Sedan ended the Empire but not the war. The Prussian armies rolled on and by September 20th they had encircled Paris and put that city under a seige that was to last until January 28th of the following year. Like many other respectable bourgeois citizens, Alexander Legentil fled Paris at the approach of the Prussian armies and took refuge in the provinces. Languishing in Poitiers and agonizing over the fate of Paris, he vowed in early December that "if God saved Paris and France and delivered the sovereign pontiff, he would contribute according to his means to the construction in Paris of a sanctuary dedicated to the Sacred Heart." He sought other adherents to this vow and soon had the ardent support of Hubert Rohault de Fleury.[4]

The terms of Legentil's vow did not, however, guarantee it a very warm reception, for as he soon discovered, the provinces "were then possessed of hateful sentiments towards Paris." Such a state of affairs was not unusual and we can usefully divert for a moment to consider its basis.

Under the *Ancien Régime*, the French state apparatus had acquired a strongly centralized character which was consolidated under the French Revolution and Empire. This centralization thereafter became the basis of French political organization and gave Paris a peculiarly important role in relation to the rest of France. The administrative, economic, and cultural predominance of Paris was assured. But the events of 1789 also showed that Parisians had the power to make and break governments. They proved adept at using that power and were not loath, as a result, to regard themselves as privileged beings with a right and duty to foist all that they deemed "progressive" upon a supposedly backward, conservative, and predominantly rural France. The Parisian bourgeois despised the narrowness of provincial life and found the peasant disgusting and incomprehensible.[5]

From the other end of the telescope, Paris was generally seen as a center of power, domination, and opportunity. It was both envied and hated. To the antagonism generated by the excessive centralization of power and authority in Paris were added all of the vaguer small town and rural antagonisms towards any large city as a center of privilege, material success, moral decadence, vice and, social unrest. What was special in France was the way in which the tensions emanating from the "urban-rural contradiction" were so intensely focused upon the relation between Paris and the rest of France.

Under the Second Empire these tensions sharpened considerably. Paris experienced a vast economic boom as the railways made it the hub of a process of national spatial integration. At the same time, falling transport costs and the free trade policies signalled by the Anglo-French Treaties of Commerce in 1860 brought the city into a new relationship with an emerging global economy. Its share of an expanding French export trade increased dramatically, and its population grew rapidly, largely through a massive inmigration of rural laborers.[6] Concentration of wealth and power proceeded apace as Paris became the center of financial, speculative, and commercial operations. The contrasts

[4] Hubert Rohault de Fleury, *Historique de la Basilique du Sacré-Coeur*, 3 Vols, printed but not published (Paris: Bibliothèque Nationale, 1903, 1905, and 1907).

[5] Theodore Zeldin, *France, 1848–1945*, 2 Vols. (London: Oxford University Press, 1973 and 1977).

[6] Jeanne Gaillard, *Paris, La Ville, 1852–70* (Paris: Honoré Champion, 1977).

between affluence and poverty became ever more startling and were increasingly expressed in terms of a geographical segregation between the "bourgeois" quarters of the west and the working class quarters of the north, east, and south. Belleville became a foreign territory into which the bourgeois citizens of the west rarely dared to venture. The population of that place, which more than doubled between 1853 and 1870, was pictured in the bourgeois press as "the dregs of the people" caught in "the deepest depths of poverty and hatred" where "ferments of envy, sloth and anger bubble without cease."[7] The signs of social breakdown were everywhere. As the economic boom ran out of steam in the 1860s and as the authority of Empire began to fail, Paris became a cauldron of social unrest, a prey ripe for agitators of any stripe.

And to top it all, Haussman, at the Emperor's urging, had set out to "embellish Paris" with spacious boulevards, parks, and gardens, monumental architecture of all sorts. The intent was to make Paris a truly imperial city, worthy not only of France but of western civilization. Haussman had done this at immense cost and by the slipperiest of financial means, a feat which scarcely recommended itself to the frugal provincial mind. The image of public opulence which Haussman projected was only matched by the conspicuous consumption of a bourgeoisie, many of whom had grown rich speculating on the benefits of his improvements.[8]

Small wonder, then, that provincial and rural Catholics were in no frame of mind to dig into their pockets to embellish Paris with yet another monument, no matter how pious its purpose.

But there were even more specific objections which emerged in response to Legentil's proposal. The Parisians had with their customary presumptuousness proclaimed a republic when provincial and rural sentiment was heavily infused with monarchism. Furthermore, those who had remained behind to face the rigors of the seige, were showing themselves remarkably intransigent and bellicose, declaring they would favor a fight to the bitter end, when provincial sentiment showed a strong disposition to end the conflict with Prussia.

And then the rumors and hints of a new materialist politics amongst the working class in Paris, spiced with a variety of manifestations of revolutionary fervor, gave the impression that the city had, in the absence of its more respectable bourgeois citizenry, fallen prey to radical and even socialist philosophy. Since the only means of communication between a besieged Paris and the nonoccupied territories was pigeon or balloon, abundant opportunities arose for misunderstanding, which the rural foes of republicanism and the urban foes of monarchism were not beyond exploiting.

Legentil therefore found it politic to drop any specific mention of Paris in his vow. Even so, the vow gained adherents slowly. But towards the end of February, the Pope endorsed it and from then on the movement gathered some strength. And so on March 19th, a pamphlet appeared which set out the arguments for the vow at some length.[9] The spirit of the work had to be national, the authors urged, because the French people had to make national amend for what were national crimes. They confirmed their intention to build the monument in Paris. To the objection that the city should not be further "embellished" they replied: "were Paris reduced to cinders, we would still want to avow our national faults and to proclaim the justice of God on its ruins."

The timing and phrasing of the pamphlet proved fortuitously prophetic. On March 18th, Parisians had taken their first irrevocable steps towards establishing self-government under the Commune. The real or imagined sins of the communards were subsequently to shock and outrage bourgeois opinion. And as much of Paris was indeed reduced to cinders in the course of a civil war of incredible ferocity, the notion of building a basilica of expiation upon these ashes became more and more appealing. As Rohault de Fleury noted, with evident satisfaction, "in the months to come, the image of Paris reduced to cinders struck home many times."[10] Let us rehearse a little of that history.

The origins of the Paris Commune lie in a whole series of events which ran into each other in complex ways. Precisely because of its political importance within the country, Paris had

[7] C. Lepidis and E. Jacomin, *Belleville* (Paris: Veynier, 1975).

[8] David Pinkney, *Napoleon III and the Rebuilding of Paris* (Princeton: Princeton University Press, 1958).

[9] Rohault de Fleury, op. cit., footnote 4, Volume 1, pp. 10–13.

[10] Rohault de Fleury, op. cit., footnote 4, Volume 1, pp. 10–13.

long been denied any representative form of municipal government and had been directly administered by the national government. For much of the nineteenth century, a predominantly republican Paris was chafing under the rule of monarchists (either Bourbon "legitimists" or "Orleanists") or authoritarian Bonapartists. The demand for a democratic form of municipal government was long-standing and commanded widespread support within the city.

The Government of National Defense set up on September 4, 1870, was neither radical nor revolutionary.[11] But it was republican. It also turned out to be timid and inept. It labored under certain difficulties, of course, but these were hardly sufficient to excuse its weak performance. It did not, for example, command the respect of the monarchists and lived in perpetual fear of the reactionaries of the right. When the Army of the East, under General Bazaine, capitulated to the Prussians at Metz on October 27th, the general left the impression that he did so because, being monarchist, he could not bring himself to fight for a republican government. Some of his officers who resisted the capitulation saw Bazaine putting his political preferences above the honor of France. This was a matter which was to dog French politics for several years. Rossel, who was later to command the armed forces of the Commune for a while, was one of the officers shocked to the core by Bazaine's evident lack of patriotism.[12]

But the tensions between the different factions of the ruling class were nothing compared to the real or imagined antagonisms between a traditional and remarkably obdurate bourgeoisie and a working class that was beginning to find its feet and assert itself. Rightly or wrongly, the bourgeoisie was greatly alarmed during the 1860s by the emergence of working class organizations and political clubs, by the activities of the Paris branch of the International Working Men's Association, by the effervescence of thought within the working class and the spread of anarchist and socialist philosophies. And the working class—although by no means as well organized or as unified as their opponents feared—were certainly displaying abundant signs of an emergent class consciousness.

The government of National Defense could not stem the tide of Prussian victories or break the siege of Paris without widespread working class support. And the leaders of the left were only too willing to give it in spite of their initial opposition to the Emperor's war. Blanqui promised the government "energetic and absolute support" and even the International's leaders, having dutifully appealed to the German workers not to participate in a fratricidal struggle, plunged into organizing for the defense of Paris. Belleville, the center of working class agitation, rallied spectacularly to the national cause, all in the name of the republic.[13]

The bourgeoisie sensed a trap. They saw themselves, wrote a contemporary commentator drawn from their ranks, caught between the Prussians and those whom they called "the Reds." "I do not know," he went on, "which of these two evils terrified them most; they hated the foreigner but they feared the Bellevillois much more."[14] No matter how much they wanted to defeat the foreigner, they could not bring themselves to do so with the batallions of the working class in the vanguard. For what was not to be the last time in French history, the bourgeoisie chose to capitulate to the Germans, leaving the left as the dominant force within a patriotic front. In 1871, fear of the "enemy within" was to prevail over national pride.

The failure of the French to break the siege of Paris was first interpreted as the product of Prussian superiority and French military ineptitude. But as sortie after sortie promised victory only to be turned into disaster, honest patriots began to wonder if the powers that be were not playing tricks which bordered on betrayal and treason. The government was increasingly viewed as a "Government of National Defection."[15]

The government was equally reluctant to respond to the Parisian demand for municipal

---

[11] Henri Guillemin, *Cette Curieuse Guerre de 1870* (Paris: Gallimard, 1956).

[12] Edith Thomas, *Rossel (1844–1871)* (Paris: Gallimard, 1967).

[13] Prosper-Olivier Lissagaray, *Histoire de la Commune de 1871* (Paris: Maspero, 1976 edition).

[14] Quoted in Jean Bruhat, Jean Dautry, and Emile Tersen, *La Commune de 1871* (Paris: Editions Sociales, 1971), p. 75.

[15] Marx uses this phrase to telling effect in his passionate defense of the Commune; Karl Marx, *The Civil War in France* (New York: International Publishers, 1968 edition). The idea was widespread throughout Paris at that time; see Marcel Cerf, *Edouard Moreau* (Paris: Denoel, 1971).

democracy. Since many of the respectable bourgeois had fled, it looked as if elections would deliver municipal power into the hands of the left. Given the suspicions of the monarchists of the right, the Government of National Defense felt it could not afford to concede what had long been demanded. And so it prevaricated endlessly.

As early as October 31st, these various threads came together to generate an insurrectionary movement in Paris. Shortly after Bazaine's ignominious surrender, word got out that the government was negotiating the terms of an armistice with the Prussians. The population of Paris took to the streets and as the feared Bellevillois descended en masse, took several members of the government prisoner, agreeing to release them only on the verbal assurance that there would be municipal elections and no capitulation. This incident was guaranteed to raise the hackles of the right. It was the immediate cause of the "hateful sentiments towards Paris" which Legentil encountered in December. The government lived to fight another day. But, as events turned out, they were to fight much more effectively against the Bellevillois than they ever fought against the Prussians.

So the siege of Paris dragged on. Worsening conditions in the city now added their uncertain effects to a socially unstable situation. The government proved inept and insensitive to the needs of the population and thereby added fuel to the smoldering fires of discontent.[16] The people lived off cats or dogs, while the more privileged partook of pieces of Pollux, the young elephant from the zoo (forty francs a pound for trunk). The price of rats—the "taste is a cross between pork and partridge"—rose from sixty centimes to four francs a piece. The government failed to take the elementary precaution of rationing bread until January, when it was much too late. Supplies dwindled and the adulteration of bread with bone meal became a chronic problem which was made even less palatable by the fact that it was human bones from the catacombs which were being dredged up for the occasion. While the common people were thus consuming their ancestors without knowing it, the luxuries of café life were kept going, supplied by hoarding merchants at exorbitant prices. The rich that stayed behind continued to indulge their pleasures according to their custom, although they paid much more dearly for it. The government did nothing to curb profiteering or the continuation of conspicuous consumption by the rich in callous disregard for the feelings of the less privileged.

By the end of December, radical opposition to the Government of National Defense was growing. It led to the publication of the celebrated *Affiche Rouge* of January 7th. Signed by the central committee of the twenty Parisian arrondissements, it accused the government of leading the country to the edge of an abyss by its indecision, inertia, and foot-dragging, suggested that the government knew not how to administer or to fight, and insisted that the perpetuation of such a regime could end only in capitulation to the Prussians. It proclaimed a program for a general requisition of resources, rationing, and mass attack. It closed with the celebrated appeal: "Make way for the people! Make way for the Commune!"[17]

Placarded all over Paris, the appeal had its effect. The military responded decisively and organized one last mass sortie which was spectacular for its military ineptitude and the carnage left behind. "Everyone understood," wrote Lissagaray, "that they had been sent out to be sacrificed."[18] The evidence of treason and betrayal was by now overwhelming for those close to the action. It pushed many an "honest patriot" from the bourgeoisie, who put love of country above class interest, into an alliance with the dissident radicals and the working class.

Parisians accepted the inevitable armistice at the end of January with sullen passivity. It provided for national elections to a constituent assembly which would negotiate and ratify a peace agreement. It specified that the French army lay down its arms but permitted the National Guard of Paris, which could not easily be disarmed, to remain a fighting force. Supplies came into a starving city under the watchful eye of the Prussian troops.

---

[16] Louis Lazare, *La France et Paris* (Paris: Etudes Historiques et Municipales, 1872) and George Becker, ed., *Paris under Siege, 1870–71; From the Goncourt Journal* (Ithaca: Cornell University Press, 1969).

[17] There is a voluminous literature on the Commune and an extensive bibliography is provided in Bruhat, Dautry, and Tersen, op. cit., footnote 14. A recent work in English is Stewart Edwards, *The Paris Commune* (Chicago: Quadrangle Books, 1971).

[18] Lissagaray, op. cit., footnote 13, p. 75.

In the February elections, the city returned its quota of radical republicans. But rural and provincial France voted solidly for peace. Since the left was antagonistic to capitulation, the republicans from the Government of National Defense seriously compromised by their management of the war, and the Bonapartists discredited, the peace vote went to the monarchists. Republican Paris was appalled to find itself faced with a monarchist majority in the National Assembly. Thiers, by then seventy-three years old, was elected President in part because of his long experience in politics and in part because the monarchists did not want to be responsible for signing what was bound to be an ignoble peace agreement.

Thiers ceded Alsace and Lorraine to Germany and agreed to a huge war indemnity. He was enough of a patriot to resist Bismarck's suggestion that Prussian bankers float the loan required. Thiers reserved that privilege for the French and turned this year of troubles into one of the most profitable ones ever for the gentlemen of French high finance.[19] The latter informed Thiers that if he were to raise the money, he must first deal with "those rascals in Paris." This he was uniquely equipped to do. As Minister of the Interior under Louis Philippe, he had, in 1834, been responsible for the savage repression of one of the first genuine working class movements in French history. Ever contemptuous of "the vile multitude," he had long had a plan for dealing with them—a plan which he had proposed to Louis Philippe in 1848 and which he was now finally in a position to put into effect.[20] The plan was simple. He would use the conservativism of the country to smash the radicalism of the city.

On the morning of March 18th, the population of Paris awoke to find that the remnants of the French army had been sent to Paris to relieve that city of its cannons in what was obviously a first step toward the disarmament of a populace which had, since September 4th, joined the National Guard in massive numbers (Figs. 2 and 3). The populace of working class Paris set out spontaneously to reclaim the cannon as their own. On the hill on Montmartre, weary French soldiers stood guard over the powerful battery of cannons assembled there, facing an increasingly restive and angry crowd. General Lecomte ordered his troops to fire. He ordered once, twice, thrice. The soldiers had not the heart to do it, raised their rifle butts in the air, and fraternized joyfully with the crowd. An infuriated mob took General Lecomte prisoner. They stumbled across General Thomas, remembered and hated for his role in the savage killings of the June days of 1848. The two generals were taken to the garden of No. 6, rue des Rosiers and, amid considerable confusion and angry argument, put up against a wall and shot.

This incident is of crucial importance to our story. The conservatives now had their martyrs. Thiers could brand the insubordinate population of Paris as murderers and assassins. But the hilltop of Montmartre had been a place of martyrdom for Christian saints long before. To these could now be added the names of Lecomte and Clement Thomas. In the months and years to come, as the struggle to build the Basilica of Sacré-Coeur unfolded, frequent appeal was to be made to the need to commemorate these "martyrs of yesterday who died in order to defend and save Christian society."[21] On that 16th day of June in 1875 when the foundation stone was laid, Rohault de Fleury rejoiced that the Basilica was to be built on a site which, "after having been such a saintly place had become, it would seem, the place chosen by Satan and where was accomplished the first act of that horrible saturnalia which caused so much ruination and which gave the church two such glorious martyrs." "Yes," he continued, "it is here where Sacré-Coeur will be raised up that the Commune began, here where generals Clement Thomas and Lecomte were assassinated." He rejoiced in the "multitude of good Christians who now stood adoring a God who knows only too well how to confound the evil-minded, cast down their designs and to place a cradle where they thought to dig a grave." He contrasted this multitude of the faithful with a "hillside, lined

[19] Henri Guillemin, *L'Avènement de M. Thiers et Réflexions sur la Commune* (Paris: Gallimard, 1971), pp. 308–11; Bruhat, Dautry, and Tersen, op. cit., footnote 14, pp. 104–05; and R. Dreyfus, *Monsieur Thiers contre l'Empire, La Guerre, La Commune* (Paris: Grasset, 1928), p. 266.

[20] John Allison, *Monsieur Thiers* (New York: Norton, 1932), chapter 9; and Guellemin, op. cit., footnote 19.

[21] This phrase was actually used by the Committee of the National Assembly appointed to report on the proposed law which would make the Basilica a work of public utility. See Rohault de Fleury, op. cit., footnote 4, Volume 1, p. 88.

FIG. 2. The hillside of Montmartre on the eve of March 18, 1871.

FIG. 3. Members of the Parisian National Guard with one of the famous cannon of Montmartre.

with intoxicated demons, inhabited by a population apparently hostile to all religious ideas and animated, above all, by a hatred of the Church."[22] GALLIA POENITENS.

Thiers' response to the events of March 18th was to order a complete withdrawal of military and government personnel from Paris. From the safe distance of Versaille, he prepared methodically for the invasion and reduction of Paris. Bismarck proved not at all reluctant to allow the reconstitution of a French army sufficient to the task of putting down the radicals in Paris and released prisoners and material for that purpose.

Left to their own devices, and somewhat surprised by the turn of events, the Parisians, under the leadership of the Central Committee of the National Guard, arranged for elections on March 26th. The Commune was declared a political fact on March 28th. It was a day of joyous celebration for the common people of Paris and a day of consternation for the bourgeoisie.

The politics of the Commune were hardly coherent. While a substantial number of workers took their place as elected representatives of the people for the first time in French history, the Commune was still dominated by radical elements from the bourgeoisie. Comprised as it was of diverse political currents shading from middle of the road republican, through the Jacobins, the Proudhonists, the socialists of the International, and the Blanquist revolutionaries, there was a good deal of factionalism and plenty of contentious argumentation as to what radical or socialist path to take. Much of this proved moot, however, since Thiers attacked in early April and the second siege of Paris began. Rural France was being put to work to destroy working class Paris.

What followed was disastrous for the Commune. When the Versaille forces finally broke through the outer defenses of Paris—which Thiers had had constructed in the 1840s—they swept quickly through the bourgeois sections of western Paris and cut slowly and ruthlessly down the grand boulevards that Haussman had constructed into the working class quarters of the city. So began one of the most vicious bloodlettings in an often bloody French history. The Versaille forces gave no quarter. To the deaths in the street fighting, which were not, by most accounts, too extensive, were added an incredible number of arbitrary executions without judgement. The Luxemburg Gardens, the Barracks at Lobau, the celebrated and still venerated wall in the cemetery of Père Lachaise, echoed ceaselessly to the sound of gunfire as the executioners went to work. Between twenty and thirty thousand communards died thus.[23] GALLIA POENITENS—with vengence (Figs. 4 and 5).

Out of this sad history there is one incident which commands our attention. On the morning of May 28th, an exhausted Eugene Varlin—bookbinder, union and food cooperative organizer under the Second Empire, member of the national guard, intelligent, respected and scrupulously honest, committed socialist, and brave soldier—was recognized and arrested. He was taken to that same house on rue des Rosiers where Lecomte and Clement Thomas died. Varlin's fate was worse. Paraded around the hillside of Montmartre, some say for ten minutes and others for hours, abused, beaten and humiliated by a fickle mob, he was finally propped up against a wall and shot. He was just thirty-two years old. They had to shoot twice to kill him. In between fusillades he cried, evidently unrepentent, "Vive la Commune!" His biographer called it "the Calvary of Eugene Varlin." The left can have its martyrs too. And it is on that spot that Sacré-Coeur is built.[24]

The "bloody week," as it was called, also involved an enormous destruction of property. Paris burned. To the buildings set afire in the course of the bombardment were added those deliberately fired for strategic reasons by the retreating communards. From whence arose the myth of the "incendiaries" of the Commune who recklessly took revenge, it was said, by burning everything they could. The communards, to be sure, were not enamoured of the privileges of private property and were not averse to destroying hated symbols. The Vendôme Column—which Napoleon III had doted upon—was, after all, toppled in a grand ceremony to symbolize the end of authoritarian rule. The painter Courbet was later held responsible for this act and condemned to pay for the re-

[22] Rohault de Fleury, op. cit., footnote 4, Volume 1, p. 264.

[23] See Bruhat, Dautry, and Tersen, op. cit., footnote 14, and the works they cite for full accounts of the Commune.

[24] M. Foulon, *Eugene Varlin* (Clermont Ferrand: Editions Mont-Louis, 1934).

FIG. 4. Gouache by Alfred Darjon of the executions at the Mur des Fédérés in Père Lachaise cemetery (Musée Carnavelet).

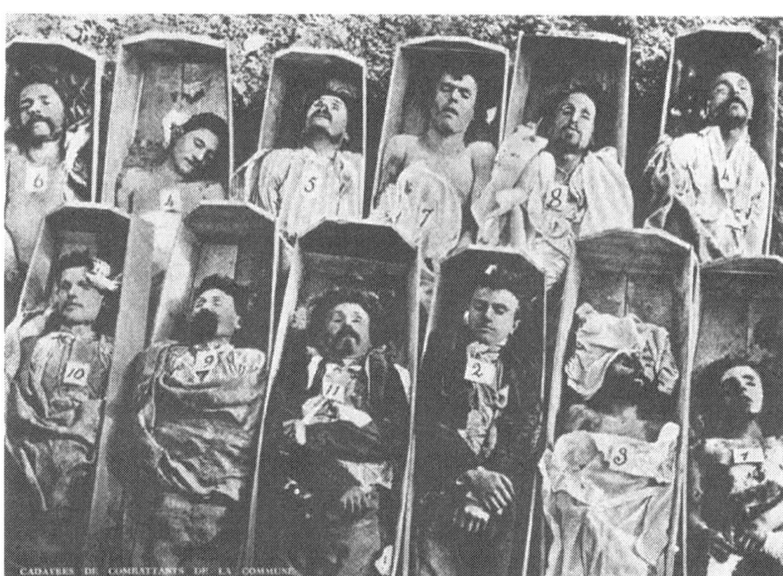

FIG. 5. Bodies of communards shot by the Versailles troops, May, 1871 (Musée Carnavelet).

construction of the monument out of his own pocket. The communards also decreed, but never carried out, the destruction of the Chapel of Expiation by which Louis XVIII had sought to impress upon Parisians their guilt in executing his brother. And when Thiers had shown his true colors, the communards took a certain delight in dismantling his Paris residence, stone by stone, in a symbolic gesture which de Goncourt felt had an "excellent bad effect."[25] But the wholesale burning of Paris was another matter entirely (Figs. 6 and 7).

No matter what the truth of the matter, the myth of the incendiaries was strong. Within a year, the Pope himself was describing the communards as "devils risen up from hell bringing the fires of the inferno to the streets of Paris."[26]

The ashes of the city became a symbol of the Commune's crimes against the Church and were to fertilize the soil from which the energy to build Sacré-Coeur was to spring. No wonder that Rohault de Fleury congratulated himself upon that felicitous choice of words—"were Paris to be reduced to cinders . . ." That phrase could strike home with redoubled force, he noted, "as the incendiaries of the Commune came to terrorize the world."[27]

The aftermath of the Commune was anything but pleasant. The bloodletting began to turn the stomachs of the bourgeoisie until all but the most sadistic of them had to cry "stop!" The celebrated diarist, Edmond de Goncourt, tried to convince himself of the justice of it all when he wrote:[28]

> It is good that there was neither conciliation nor bargain. The solution was brutal. It was by pure force. The solution has held people back from cowardly compromises . . . the bloodletting was a bleeding white; such a purge, by killing off the combative part of the population defers the next revolution by a whole generation. The old society has twenty years of quiet ahead of it, if the powers that be dare all that they may dare at this time.

These sentiments were exactly those of Thiers. But when de Goncourt passed through Belleville and saw the "faces of ugly silence," he could not help but feel that here was a "vanquished but unsubjugated district." Was there no other way to purge the threat of revolution?

The experiences of 1870–71, when taken together with the confrontation between Napoleon III and the Pope and the decadent "festive materialism" of the Second Empire, plunged Catholics into a phase of widespread soul-searching. The majority of them accepted the notion that France had sinned and this gave rise to manifestations of expiation and a movement of piety that was both mystical and spectacular.[29] The intransigent and ultramontane Catholics unquestionably favored a return to "law and order" and a political solution founded on respect for authority. And it was the monarchists, generally themselves intransigent Catholics, who held out the promise for that law and order. Liberal Catholics found all of this disturbing and distasteful, but they were in no position to mobilize their forces—even the Pope described them as the "veritable scourge" of France. There was little to stop the consolidation of the bond between monarchism and intransigent Catholicism. And it was such a powerful alliance that was to guarantee the building of Sacré-Coeur.

The immediate problem for the progenitors of the vow was, however, to operationalize a pious wish. This required official action. Legentil and Rohault de Fleury sought the support of the newly appointed Archbishop of Paris.

Mgr. Guibert, a compatriot of Thiers from Tours, had required some persuading to take the position in Paris. The three previous archbishops had suffered violent deaths. The first during the insurrection of 1848, the second by the hand of an assassin in 1863, and the third during the Commune. The communards had early decided to take hostages in response to the butchery promised by Versaille. The Archbishop was held as a prime hostage for whom the communards sought the exchange of Blanqui. Thiers refused that negotiation apparently having decided that a dead and martyred archbishop (who was a liberal catholic in any case) was more valuable to him than a live one exchanged against a dynamic and aggressive Blanqui. During the "bloody week," the communards took whatever vengeance they could. On May 24th, the archbishop was shot. In that final week, seventy-four hostages were shot, of which twenty-four were priests. That awesome anticlericalism was as alive under the Commune as it had been in 1789. But with the massive purge

---

25 Becker, op. cit., footnote 16, p. 288.

26 Dansette, op. cit., footnote 2, p. 343.

27 Rohault de Fleury, op. cit., footnote 4, Volume 1, p. 13.

28 Becker, op. cit., footnote 16, p. 312.

29 Dansette, op. cit., footnote 2, pp. 340–45.

FIG. 6. View of Paris burning from above the Père Lachaise cemetery.

FIG. 7. The toppling of the Vendôme Column by the communards.

which left more than twenty thousand communards dead, nearly forty thousand were held prisoner and countless others fled, Thiers could write reassuringly on June 14th to Mgr. Guibert —"the 'reds,' totally vanquished, will not recommence their activities tomorrow; one does not engage twice in fifty years in such an immense fight as they have just lost."[30] Reassured, Mgr. Guibert came to Paris.

The new archbishop was much impressed with the movement to build a monument to the Sacred Heart. On January 18, 1872, he formally accepted responsibility for the undertaking. He wrote to Legentil and Rohault de Fleury thus:[31]

> You have considered from their true perspective the ills of our country. . . . The conspiracy against God and Christ has prevailed in a multitude of hearts and in punishment for an almost universal apostasy, society has been subjected to all the horrors of war with a victorious foreigner and an even more horrible war amongst the children of the same country. Having become, by our prevarication, rebels against heaven, we have fallen during our troubles into the abyss of anarchy. The land of France presents the terrifying image of a place where no order prevails, while the future offers still more terrors to come. . . . This temple, erected as a public act of contrition and reparation . . . will stand amongst us as a protest against other monuments and works of art erected for the glorification of vice and impiety.

By July 1872, an ultraconservative Pope Pius IX, still awaiting his deliverance from captivity in the Vatican, formally endorsed the vow. An immense propaganda campaign unfolded and the movement gathered momentum. By the end of the year, more than a million francs were promised and all that remained was to translate the vow into its material, physical representation.

The first step was to choose a site. Legentil wanted to use the foundations of the still-to-be-completed Opera which he considered "a scandalous monument of extravagance, indecency and bad taste."[32] Rohault de Fleury's initial design of that building had, in 1860, been dropped at the insistence of Count Walewski ("who had the dubious distinction of being the illegitimate son of Napoleon I and the husband of Napoleon III's current favorite").[33] The design that replaced it (which exists today) most definitely qualified in the eyes of Legentil as a "monument to vice and impiety" and nothing could be more appropriate than to efface the memory of Empire by constructing the Basilica on that spot. It probably escaped Legentil's attention that the communards had, in the same spirit, toppled the Vendôme column.

By late October, 1872, however, the Archbishop had taken matters into his own hands and selected the heights of Montmartre because it was only from there that the symbolic domination of Paris could be assured. Since the land on that site was in part public property, the consent or active support of the government was necessary if it was to be acquired. The government was considering the construction of a military fortress on that spot. The Archbishop pointed out, however, that a military fortress could well be very unpopular while a fortification of the sort he was proposing might be less offensive and more sure. Thiers and his ministers, apparently persuaded that ideological protection might be preferable to military, encouraged the Archbishop to pursue the matter formally. This the latter did in a letter of March 5th, 1873.[34] He requested that the government pass a special law declaring the construction of the Basilica a work of public utility. This would permit the laws of expropriation to be used to procure the site.

Such a law ran counter to a long-standing sentiment in favor of the separation of church and state. Yet conservative catholic sentiment for the project was very strong. Thiers prevaricated. But his indecision was shortly rendered moot. The monarchists had decided that their time had come. On May 24th, they drove Thiers from power and replaced him with the archconservative royalist, Marshall Mac-Mahon who, just two years before, had led the armed forces of Versaille in the bloody repression of the Commune. France was plunged, once more, into political ferment—a monarchist restoration seemed imminent.

The Mac-Mahon government quickly reported out the law which then became part of its program to establish the rule of moral order in which those of wealth and privilege—who therefore had an active stake in the preservation of society—would, under the leadership of the

30 Guillemin, op. cit., footnote 19, pp. 295–96; and Rohault de Fleury, op. cit., footnote 4, Volume 2, p. 365.

31 Rohault de Fleury, op. cit., footnote 4, Volume 1, p. 27.

32 Jonquet, op. cit., footnote 1, pp. 85–87.

33 Pinkney, op. cit., footnote 8, pp. 85–87.

34 Rohault de Fleury, op. cit., footnote 4, Volume 1, p. 75.

king and in alliance with the authority of the church, have both the right and duty to protect France from the social perils to which it had recently been exposed and thereby prevent the country falling into the abyss of anarchy. Large-scale demonstrations were mobilized by the Church as part of a campaign to reestablish some sense of moral order. The largest of these demonstrations took place on June 29, 1873 at Paray-le-Monial. Thirty thousand pilgrims, including fifty members of the National Assembly, journeyed there to dedicate themselves publicly to the Sacred Heart.[35]

It was in this atmosphere that the Committee formed to report on the law presented its findings on July 11th to the National Assembly, a quarter of whose members were adherents to the vow. The Committee found that the proposal to build a basilica of expiation was unquestionably a work of public utility. It was right and proper to build such a monument on the heights of Montmartre for all to see, because it was there that the blood of martyrs—including those of yesterday—had flowed. It was necessary "to efface by this work of expiation, the crimes which have crowned our sorrows" and France, "which has suffered so much" must "call upon the protection and grace of Him who gives according to His will, defeat or victory."[36]

The debate which followed on July 22nd and 23rd in part revolved around technical-legal questions and the implications of the legislation for State-Church relations. The intransigent Catholics recklessly proposed to go much further. They wanted the Assembly to commit itself formally to a national undertaking which "was not solely a protestation against the taking up of arms by the Commune, but a sign of appeasement and concord." That amendment was rejected. But the law passed with a handsome majority of 244 votes.

A loan dissenting voice in the debate came from a radical republican deputy from Paris:[37]

> When you think to establish on the commanding heights of Paris—the fount of free thought and revolution—a catholic monument, what is in your thoughts? To make of it the triumph of the Church over revolution. Yes, that is what you want to extinguish—what you call the pestilence of revolution. What you want to revive is the Catholic faith, for you are at war with the spirit of modern times. . . . Well, I who know the sentiments of the population of Paris, I who am tainted by the revolutionary pestilence like them, I tell you that the population will be more scandalized than edified by the ostentation of your faith. . . . Far from edifying us, you push us towards free thought, towards revolution. When people see these manifestations of the partisans of monarchy, of the enemies of the Revolution, they will say to themselves that Catholicism and monarchy are unified, and in rejecting one they will reject the other.

Armed with a law which yielded powers of expropriation, the Committee formed to push the project through to fruition acquired the site atop the Butte Montmartre. They collected up the moneys promised and set about soliciting more so that the building could be as grand as the thought that lay behind it. A competition for the design of the Basilica was set and judged. The building had to be imposing, consistent with Christian tradition, yet quite distinct from the "monuments to vice and impiety" built in the course of the Second Empire. Out of the seventy-eight designs submitted and exhibited to the public, that of the architect Abadie was selected. The grandeur of its domes, the purity of the white marble, and the unadorned simplicity of its detail impressed the Committee—what, after all, could be more different from the flamboyance of that awful Opera House?

By the Spring of 1875, all was ready for putting the first stone in place. But radical and republican Paris was not, apparently, repentent enough even yet. The Archbishop complained that the building of Sacré-Coeur was being treated as a provocative act, as an attempt to inter the principles of 1789. And while, he said, he would not pray to revive those principles if they happened to get dead and buried, this view of things was giving rise to a deplorable polemic in which the Archbishop found himself forced to participate. He issued a circular in which he expressed his astonishment at the hostility expressed towards the project on the part of "the enemies of religion." He found it intolerable that people dared to put a political interpretation upon thoughts derived only out of faith and piety. Politics, he assured his readers, "had been far, far from our inspirations; the work had been inspired, on the contrary, by a profound conviction that politics was powerless to deal with the ills of the country. The causes of these ills are moral and religious and the remedies

[35] Dansette, op. cit., footnote 2, pp. 340–45.

[36] Rohault de Fleury, op. cit., footnote 4, Volume 1, p. 88.

[37] Rohault de Fleury, op. cit., footnote 4, Volume 1, p. 88.

must be of the same order." Besides, he went on, the work could not be construed as political because the aim of politics is to divide "while our work has for its goal the union of all. . . . Social pacification is the end point of the work we are seeking to realize."[38]

The government, now clearly on the defensive, grew extremely nervous at the prospect of a grand opening ceremony which could be the occasion for an ugly confrontation. It counselled caution. The Committee had to find a way to lay the first stone without being too provocative. The Pope came to their aid and declared a day of dedication to the Sacred Heart for all Catholics everywhere. Behind that shelter, a much scaled-down ceremony to lay the first stone passed off without incident. The construction was now under way. GALLIA POENITENS was taking shape in material symbolic form.

The forty years between the laying of the foundation stone and the final consecration of the Basilica in 1919 were often troubled ones. Technical difficulties arose in the course of putting such a large structure on a hilltop rendered unstable by years of mining for gypsum. The cost of the structure increased dramatically and as enthusiasm for the cult of the Sacred Heart diminished somewhat, financial difficulties ensued. And the political controversy continued.

The Committee in charge of the project had early decided upon a variety of strategems to encourage the flow of contributions. Individuals and families could purchase a stone and the visitor to Sacré-Coeur will see the names of many such inscribed upon the stones there. Different regions and organizations were encouraged to subscribe towards the construction of particular chapels. Members of the National Assembly, the army, the clergy, and the like, all pooled their efforts in this way. Each particular chapel has its own significance.

Among the chapels in the crypt, for example, the visitor will find that of Jesus-Enseignant, which recalls, as Rohault de Fleury put it, "that one of the chief sins of France was the foolish invention of schooling without God."[39] Those who were on the losing side of the fierce battle to preserve the power of the Church over education after 1871 put their money here. And next to that chapel, at the far end of the crypt, close to the line where the rue des Rosiers used to run, stands the Chapel to Jesus-Ovrier.

That Catholic workers sought to contribute to the building of their own chapel was a matter for great rejoicing. It showed, wrote Legentil, the desire of workers "to protest against the fearsome impiety into which a large part of the working class is falling" as well as their determination to resist "the impious and truly infernal association which, in nearly all of Europe, makes of it its slave and victim."[40] The reference to the International Working Men's Association is unmistakable and understandable since it was customary in bourgeois circles at that time to attribute the Commune, quite erroneously, to the nefarious influence of that "infernal" association. Yet, by a strange quirk of fate, which so often gives an ironic twist to history, the chapel to Jesus-Ouvrier stands almost exactly at the spot where ran the course of the "calvary of Eugene Varlin." Thus it is that the Basilica erected on high in part to commemorate the blood of two recent martyrs of the right, commemorates unwittingly in its subterranean depths a martyr of the left.

Legentil's interpretation of all of this was in fact somewhat awry. In the closing stages of the Commune, a young Catholic named Alfred de Munn, watched in dismay as the communards were led away to slaughter. Shocked, he fell to wondering what "legally constituted society had done for these people" and concluded that their ills had in large measure been visited upon them through the indifference of the affluent classes. In the spring of 1872, he went into the heart of hated Belleville and set up the first of his *Cercles-Ouvriers*.[41] This signalled the beginnings of a new kind of Catholicism in France —one which sought through social action to attend to the material as well as the spiritual needs of the workers. It was through organizations such as this, a far cry from the intransigent ultramontane Catholicism that ruled at the center of the movement for the Sacred Heart, that a small trickle of worker contributions began to flow towards the construction of a Basilica on the hilltop of Montmartre.

---

[38] Rohault de Fleury, op. cit., footnote 4, Volume 1, p. 244.

[39] Rohault de Fleury, op. cit., footnote 4, Volume 1, p. 269.

[40] Rohault de Fleury, op. cit., footnote 4, Volume 1, p. 165.

[41] Dansette, op. cit., footnote 2, pp. 356–58; and Lepidis and Jacomin, op. cit., footnote 7, pp. 271–72.

The political difficulties mounted, however. France, finally armed with a republican constitution (largely because of the intransigence of the monarchists) was now in the grip of a modernization process fostered by easier communications, mass education, and industrial development. The country moved to accept the moderate form of republicanism and became bitterly disillusioned with the backward looking monarchism that had dominated the National Assembly elected in 1871. In Paris the "unsubjugated" Bellevillois, and their neighbors in Montmartre and La Villette, began to reassert themselves rather more rapidly than Thiers had anticipated. As the demand for amnesty for the exiled communards became stronger in these quarters, so did the hatred of the Basilica rising in their midst. The agitation against the project mounted.

On August 3, 1880, the matter came before the city council in the form of a proposal—a "colossal statue of *Liberty* will be placed on the summit of Montmartre, in front of the church of Sacré-Coeur, on land belonging to the city of Paris." The French republicans at that time had adopted the United States as a model society which functioned perfectly well without monarchism and other feudal trappings. As part of a campaign to drive home the point of this example, as well as to symbolize their own deep attachment to the principles of liberty, republicanism, and democracy, they were then raising funds to donate the Statue of Liberty that now stands in New York harbor (Fig. 8). Why not, said the authors of this proposition, efface the sight of the hated Sacré-Coeur by a monument of similar order?[42]

No matter what the claims to the contrary, they said, the Basilica symbolized the intolerance and fanaticism of the right—it was an insult to civilization, antagonistic to the principles of modern times, an evocation of the past, and a stigma upon France as a whole. Parisians, seemingly bent on demonstrating their unrepentent attachment to the principles of 1789, were determined to efface what they felt was an expression of "Catholic fanaticism" by building exactly that kind of monument which the Archbishop had previously characterized as a "glorification of vice and impiety."

FIG. 8. The Statue of Liberty stands in its Paris workshop preparatory to its shipment to the United States. It was just such a statue that the Paris Council wished to erect in front of the Basilica of Sacré-Coeur.

By October 7th the city council had changed its tactics. Calling the Basilica "an incessant provocation to civil war," the members decided by a majority of sixty-one to three to request the government to "rescind the law of public utility of 1873" and to use the land, which would revert to public ownership, for the construction of a work of truly national significance. Neatly sidestepping the problem of how those who had contributed to the construction of the Basilica—which had hardly yet risen above its foundations—were to be indemnified, it passed along its proposal to the government. By the summer of 1882, the request was taken up in the Chamber of Deputies.

Archbishop Guibert had, once more, to take to the public defense of the work. He challenged what by now were familiar arguments against the Basilica with familiar responses. He insisted that the work was not inspired by politics but by Christian and patriotic sentiments. To those who objected to the expiatory character of the work he simply replied that no one can ever afford to regard their country as infallible. As to the appropriateness of the

[42] Ville de Paris, Conseil Municipal, Procés Verbal, 3rd August, 7th October, and 2nd December, 1880 (Paris: Hôtel de Ville).

cult of the Sacred Heart, he felt only those within the Church had the right to judge. To those who portrayed the Basilica as a provocation to civil war he replied: "Are civil wars and riots ever the product of our Christian temples? Are those who frequent our churches ever prone to excitations and revolts against the law? Do we find such people in the midst of disorders and violence which, from time to time, trouble the streets of our cities?" He went on to point out that while Napoleon I had sought to build a temple of peace at Montmartre, "it is we who are building, at last, the true temple of peace."[43]

He then considered the negative effects of stopping the construction. Such an action would profoundly wound Christian sentiment and prove divisive. It would surely be a bad precedent, he said (blithely ignoring the precedent set by the law of 1873 itself), if religious undertakings of this sort were to be subject to the political whims of the government of the day. And then there was the complex problem of compensation not only for the contributors but for the work already done. Finally, he appealed to the fact that the work was giving employment to six hundred families—to deprive "that part of Paris of such a major source of employment would be inhuman indeed."

The Parisian representatives in the Chamber of Deputies which, by 1882, was dominated by reformist republicans such as Gambetta (from Belleville) and Clemenceau (from Montmartre) were not impressed by these arguments. The debate was heated and passionate. The Government for its part declared itself unalterably opposed to the law of 1873, but was equally opposed to rescinding the law since this would entail paying out more than twelve million francs in indemnities to the Church. In an effort to defuse the evident anger from the left, the Minister went on to remark that by rescinding the law, the Archbishop would be relieved of the obligation to complete what was proving to be a most arduous undertaking at the same time as it would provide the Church with millions of francs to pursue works of propaganda which might be "infinitely more efficacious than that to which the sponsors of the present motion are objecting."

The radical republicans were not about to regard Sacré-Coeur in the shape of a white elephant, however. Nor were they inclined to pay compensation. They were determined to do away with what they felt was an "odious" manifestation of pious clericalism and to put in its place a monument to liberty of thought. They put the blame for the civil war squarely on the shoulders of the monarchists and their intransigent Catholic allies.

Clemenceau rose to state the radical case. He declared the law of 1873 an insult, an act of a National Assembly which had sought to impose the cult of the Sacred Heart on France because "we fought and still continue to fight for human rights, for having made the French Revolution." The law was the product of clerical reaction, an attempt to stigmatize revolutionary France, "to condemn us to ask pardon of the Church for our ceaseless struggle to prevail over it in order to establish the principles of liberty, equality and fraternity." We must, he declared, respond to a political act by a political act. Not to do so would be to leave France under the intolerable invocation of the Sacred Heart.[44]

With impassioned oratory such as this, Clemenceau fanned the flames of anticlerical sentiment. The Chamber voted to rescind the law of 1873 by a majority of 261 votes to 199. It appeared that the Basilica, the walls of which were as yet hardly risen above their foundations, was to come tumbling down.

The Basilica was saved by a technicality. The law was passed too late in the session to meet all the formal requirements for promulgation. The Government, genuinely fearful of the costs and liabilities involved, quietly worked to prevent the reintroduction of the motion into a Chamber which, in the next session, moved on to consider matters of much greater weight and moment. The Parisian republicans had gained a symbolic but Pyrrhic parliamentary victory. A relieved Archbishop pressed on with the work.

Yet somehow the matter would not die. In February, 1897, the motion was reintroduced.[45] Anticlerical republicanism had by then made great progress as had the working class movement in the form of a vigorous and growing socialist party. But the construction atop the hill had likewise progressed. The interior of the Basilica had been inaugurated and opened for

[43] Rohault de Fleury, op. cit., footnote 4, Volume 2, pp. 71–73.

[44] Rohault de Fleury, op. cit., footnote 4, Volume 2, pp. 71 et seq., gives the text of the debate.

[45] Paul Lesourd, *Montmartre* (Paris: Editions France Empire, 1973), pp. 224–25.

worship in 1891 and the great dome was well on the way to completion (the cross which surmounts it was formally blessed in 1899). Although the church was still viewed as a "provocation to civil war," the prospect for dismantling such a vast work was by now quite daunting. And this time it was none other than Albert de Mun who defended the Basilica in the name of a Catholicism that had, by then, seen the virtue of separating its fate from that of a fading monarchist cause. The Church was beginning to learn a lesson and the cult of the Sacred Heart began to acquire a new meaning in response to a changing social situation. By 1899, a more reform-minded Pope dedicated the cult to the ideal of harmony among the races, social justice, and conciliation.

But the socialist deputies were not impressed by what they saw as maneuvers of cooptation. They pressed home their case which would bring down the hated symbol, even though almost complete, and even though such an act would entail indemnifying eight million subscribers to the tune of thirty million francs. But the majority in the Chamber blanched at such a prospect. The motion was rejected by 322 to 196.

This was to be the last time the building was threatened by official action. With the dome completed in 1899, attention switched to the building of the campanile, which was finally finished in 1912. By the spring of 1914, all was ready and the official consecration set for October 17th. But war with Germany intervened. Only at the end of that bloody conflict was the Basilica finally consecrated. A victorious France —led by the fiery oratory of Clemenceau—joyfully celebrated the consecration of a monument conceived of in the course of a losing war with Germany a generation before. GALLIA POENITENS at last brought its rewards.

Muted echoes of this tortured history can still be heard. In February, 1971, for example, demonstrators pursued by police took refuge in the Basilica. Firmly entrenched there, they called upon their radical comrades to join them in occupying a church "built upon the bodies of communards in order to efface the memory of that red flag that had for too long floated over Paris." The myth of the incendiaries immediately broke loose from its ancient moorings and an evidently panicked rector summoned the police into the Basilica to prevent the conflagration. The "reds" were chased from the church amid scenes of great brutality. Thus was the centennial of the Paris Commune celebrated on that spot.[46]

And as a coda to that incident, a bomb exploded in the Basilica in 1976, causing quite extensive damage to one of the domes. On that day, it was said, the visitor to the cemetery of Père Lachaise would have seen a single red rose on August Blanqui's grave.

Rohault de Fleury had desperately wanted to "place a cradle where [others] had thought to dig a grave." But the visitor who looks at that mausoleum-like structure that is Sacré-Coeur might well wonder what it is that is interred there. The spirit of 1789? The sins of France? The alliance between intransigent catholicism and reactionary monarchism? The blood of martyrs like Lecomte and Clement Thomas? Or that of Eugene Varlin and the twenty thousand or so communards mercilessly slaughtered along with him?

The building hides its secrets in sepulchral silence. Only the living, cognizant of this history, who understand the principles of those who struggled for and against the "embellishment" of that spot, can truly disinter the mysteries that lie entombed there and thereby rescue that rich experience from the deathly silence of the tomb and transform it into the noisy beginnings of the cradle.

All history is, after all, the history of class struggle.

---

[46] Lesourd's account, op. cit., footnote 45, pp. 239–45, is rather one-sided as evidenced by his erroneous insistence that no communards died on Montmartre.

[9]

Back to the land: historiography, rurality and the nation in interwar Wales

Pyrs Gruffudd

This paper presents an explanation of the appeal of the rural in interwar Wales and the emergence of 'back to the land' tendencies. The context is the fluid and contested understanding of the nation and of national identity. The 'moral topography' of academics at Aberystwyth, most notably the geographer H J Fleure, is outlined. A view of rural society as essentially stable and spiritually virtuous emerges from this work. The development of such ideas within Welsh nationalist politics is outlined and 'back to the land' proposals, from academics and politicians, are discussed. A consideration of the relationship between the traditional and the modern and the role of the rural within this relationship forms the conclusion.

key words national identity Wales geography ruralism back to the land modernity

Lecturer in Geography, Department of Geography, University College of Swansea, Singleton Park, Swansea, West Glamorgan SA2 8PP

revised manuscript received 20 August 1993

Introduction

Several recent texts have argued that the nation and national identity are fluid, contextual and contested. They must, it is argued, be read in the context of discourses as diverse as textual narrative (Bhabha 1990), sexuality (Parker *et al.* 1992) and patriotism (Samuel 1989). Nations are now as much imagined as they are material entities; they are 'imagined communities' (Anderson 1983). It is this process of cultural and geographical imagining which is the theme of this paper. According to Anthony Smith (1986), 'legends and landscapes' are integral features of the national imaginings of both 'ethnic' and 'civic' nations. Territory is nationalized through ethnic and historical associations and achieves significance in a symbolic sense: 'a land of dreams is far more significant than any actual terrain' (1986, 28). Within cultural geography, the landscape has been seen as an arena for national symbolism: 'as exemplars of moral order and aesthetic harmony, particular landscapes achieve the status of national icons'. (Daniels 1993, 5). But whilst Daniels stresses the role of the arts in the articulation and negotiation of national identity, he notes that scholars and professionals – geographers included – have been enlisted too. National identity is, therefore, a complex zone of convergence of a number of discourses – political, artistic, academic – which are not merely reflections of some social reality but serve to constitute that reality.

My concern in this paper is not with aesthetic representations of the nation but with the imagined grounding of a nation in a particular environment and the presumed moral attributes of that environment. The paper is concerned with 'back to the land' tendencies in interwar Wales and with some of their political resonances. A yearning for the 'spiritual wholeness' of the countryside is a common theme amongst nationalists and intellectuals and it motivated the 'invention of tradition' in Wales during the Romantic period (Morgan P 1983). But I would argue that the rural imagination presented here is not an uncomplicated romanticism. It is, rather, a dynamic engagement with a 'place on the margin', to borrow from Rob Shields (1991). The rural becomes almost a liminal zone which is seen as occupying a ground between tradition and modernity and the societies they represent. The polarity between the two cultural outlooks is blurred and the possibility of a new social formation emerges.

This paper outlines the construction of a particular Welsh cultural identity in the interwar period. I begin by outlining the 'moral topography' (Livingstone 1992, 285) of academics at Aberystwyth and their notions of rural 'virtue'. I then consider how this work was echoed in the political programme of Plaid Cymru, the Welsh nationalist party founded in 1925, including its adoption – in 'back to the land' campaigns – of policies on rural revitalization. Finally, I argue that these various 'back to the land' philosophies, far from being anachronistic responses to 'modernization', actually represented utopian fusions of tradition and modernity which challenged the polarized notion of rural stagnation and urban modernization.

Essences of locality

Wales and her people[1]

According to Campbell and Livingstone (1983), it is neo-Lamarckism which underpins most post-Darwinian geography: a belief in the direct modification of organisms due to changes in the natural environment and in the heritability of those modifications. Whereas in the United States neo-Lamarckism had its origins in the biological sciences, British neo-Lamarckism drew on the human and social sciences. This characteristic, Campbell and Livingstone argue, is clear in the case of the geographer H J Fleure (1877–1969) (Fig. 1). Hugely influential in educational circles and a major proponent of geographical humanism (Freeman 1987), Fleure became President of the Royal Anthropological Society in 1945 and was the first vice president of the International Geographical Union. He is one of Stoddart's (1986) makers of modern academic geography.

Figure 1. H J Fleure (courtesy of the Geographical Association)

Fleure went to Aberystwyth in 1897 to study the natural sciences. Born in Guernsey, his origins made him sensitive, he claimed, to the Welsh cultural tradition (Fleure 1955; see also Freeman 1987). After time spent studying zoology in Zurich Fleure returned to Aberystwyth to lecture on zoology and geology but, by 1908, was lecturing in geography (Campbell 1972). In 1917 he was appointed to the Gregynog Chair of Geography and Anthropology[2]. Fleure insisted that the Chair should bear the title 'Anthropology' as he had been influenced by the evolutionary theories of Darwin and Huxley and sought to apply them to human society. He modified A J Herbertson's 'natural regions' to 'human regions', recognizing the importance of human will and 'the continuous interaction and interpenetration of man and environment and the cumulative alteration of man and earth with the unfolding of history' (Fleure 1919a, 94). To Fleure, 'the idea of fitting the human story into a physical frame seemed open to criticism' (Fleure 1965, 348) and through his stress on human consciousness and social evolution he countered the excesses of the environmental determinists and those imperialists for whom the political map was the basis of all explanation. His sense of the interplay between environment and society was influenced by Vidal de la Blache's treatment of the nature-nurture debate and by his poetic regional descriptions of France but a more profound influence came from Patrick Geddes, the biologist and sociologist[3]. It was probably Geddes's interest in the social implications of the Darwinian thesis that attracted Fleure; as Campbell and Livingstone (1983, 283) suggest, Fleure's friendship with Geddes and his circle 'served to increase his propensity for this version of evolutionary theory, as various aspects of his geography clearly reveal'. The two collaborated often; Fleure helped Geddes, whom he called 'a catalyst of the mind, a pilgrim-apostle' (Fleure 1953a, 13) organize a 'Civics' exhibition in Dublin in 1914 and Geddes lectured at Fleure's Aberystwyth Summer Schools for geography teachers. Fleure's first book (1918a) – *Human geography in Western Europe: a study in appreciation* – was published in Geddes and Victor Branford's 'The making of the future' series. In it, Fleure expressed the pervasive influence of evolutionary theory in the early twentieth century (Kern 1983), arguing that if the geographer will

> take time into consideration, if he will strive towards an evolutionist standpoint, he may look out upon the world as the field of expression, and especially of cumulative expression, of the human spirit. (Fleure 1918a, 1)

This theoretical position explains Fleure's support for Geddes's Regional Survey movement based on the 'Place-Work-Folk' triad of the French sociologist, Le Play (see Fleure 1915; 1918b; 1919b; see also Matless 1992). The region, like human life, was conceived out of a continuous stream of influences and experiences. It was not enough, Fleure argued, to study places according to the dictates of logic; an awareness of the human soul was also a necessary foundation. Furthermore, the aims of regional survey should be the enrichment of human life and its spiritual expression.

Fleure was elected to a Fellowship of the Royal Society in 1936 largely, so Campbell and Livingstone (1983) claim, due to his work in physical anthropology and the study of British physical types[4]. However, this research supported what has been called the social-Lamarckism of Patrick Geddes, an extended understanding of human types, including mental, spiritual and social characteristics. It was in this way that Fleure contributed to debates on identity in Wales, an identity historically seen as being rooted in the rural and in the traditional. Between 1905 and 1916 2500 individuals were surveyed in Wales by Fleure and his colleague T C James (Fleure and James 1916), with name, sex, location, family history and a total of 19 physical characteristics – including head shape and skin pigmentation – being recorded. Data relating to individuals, however, were mapped only if all four grandparents came from within a 12–15 mile radius; in this way, people were read as 'concentrated essences of that locality' (Fleure and James 1916, 42). Mapping demonstrated that Wales was characterized by marked regional differentiation (Fig. 2), understood as the result of interplay between heredity and environment. The latter had protected local distinction, Wales having experienced only limited effects of modern population movements. What Fleure called the simple folk of Wales represented types of humankind whose distinctions dated from a remote past. These types were easily distinguishable; Welshmen could instinctively 'tell' someone was from a particular district. This subterranean geography of Welshness was further strengthened by Fleure's humanist insistence that local types be studied in relation to natural regions rather than administrative units. In this way, Welshness was constructed as an organic unity between humans and environment. The basic Welsh physical inheritance, Fleure claimed, was from the Palaeolithic era when north-west Europe after the Ice Age was in the hands of a remnant population – strongly built people with dark colouring, long heads and deep-set eyes (Fig. 3). Fleure argued that 'such people survive on the Plynlymon [sic] moorlands and in the Black Mountain country of Carmarthenshire', (Fleure 1926a, 7). The scattered farms on Pumlumon had in fact 'yielded more than seven adult male cases of unusually complete survival of physical characteristics we generally associate with the earliest type of modern man' (Fleure 1926b, 15). Upland Wales – its geography as rugged and inhospitable as in prehistory – was seen, therefore, as a refuge from what Fleure called 'the new world-life outside' (Fleure 1926a, 7). The fundamental type – the 'little dark people' – were a predominant element in both rural and industrial Wales. But on the coastal lands could be found a type which revealed something of the historical and cultural geography of Wales (Fig. 3). These men of 'stalwart build' and dark colouring were found 'in nearly all the fishing harbours' of Cornwall (Fleure 1923a, 102) and also in the Hebrides, Ireland, Brittany, Spain and southern Italy, proving Welsh links with seafaring European nations and the process of cultural diffusion along the western seaways[5].

It should be stressed that Fleure challenged notions of racial purity and that his analysis of Welsh physical types was based in large part on the effects of culture contact and mixing. He consistently attacked the idea of national types, arguing in 1922 that all humans were

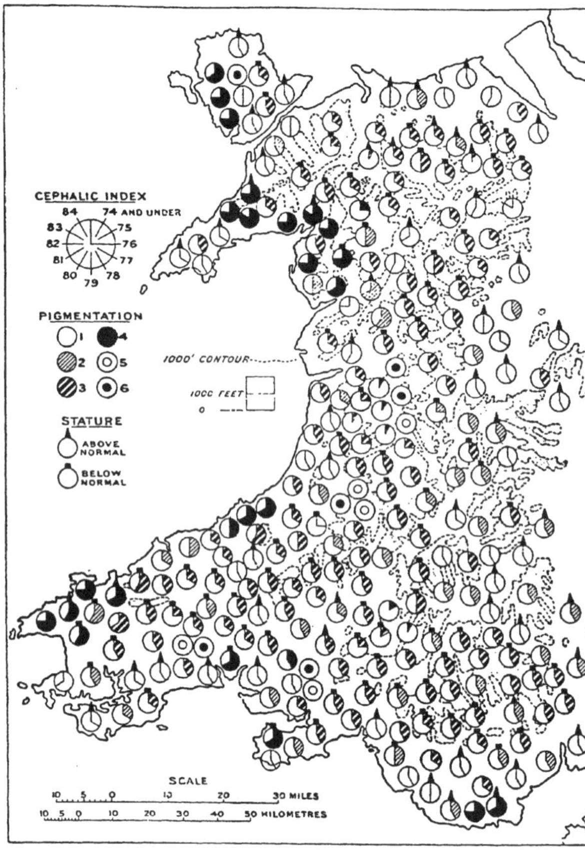

Figure 2. The localized distribution of physical types in Wales. No. 4, for example, relates to people with very dark hair and eyes.
Source: Ogilvie 1953, p. 253

> mosaics of inheritances and that a 'race-type' exists mainly in our own minds and should not be used without great reserve in scientific discussion. (Fleure 1922, 94)

He was active in anti-racist campaigns and was particularly critical of racial theories, like the Nazis' Nordic Myth, where science was used to veil sinister political propaganda (Fleure 1930–1; 1935–6). Whilst there is an ongoing debate about the historiography of eugenics (Jones G 1986), we can, nonetheless, locate Fleure within a progressive, reforming strand of a movement which was not, according to Searle (1976, 2), peripheral or crankish but 'an important challenge to politicians and academic theorists alike'[6]. Whilst Fleure rejected the notion of a 'Celtic race', he still felt that 'Celtic types' were physically and socially disadvantaged within industrial society (Fleure 1922a) and his 'moral geography', discussed below, was a feature of 'positive' eugenics concerned with social and environmental reform.

The spirit of the land

Unlike Fleure, the agricultural scientist George Stapledon (1882–1960) did entertain notions of genetic purity. Stapledon joined the Agricultural Department at Aberystwyth in 1912 and became Director of the Welsh Plant Breeding Station in

Figure 3. Welsh physical types. The upper is the remnant type from Pumlumon and the lower is the coastal, Mediterranean type.
Source: Fleure and James 1916

1919 from where, according to Anna Bramwell (1989), he revolutionized farming and ecological practices all over the world. He was a renowned geneticist but did not see science in a linear or causal fashion, believing rather in the play of a diversity of extraneous influences. Bramwell (1989) discerns beliefs shared with Patrick Geddes and Lewis Mumford and, according to Waller (1962), Stapledon would – had he read Jung – have seen synchronicity as a more appropriate way of understanding the relationships between natural phenomena. Cosgrove (1990) has suggested that Jung's synchronicity forms part of a strand of holistic environmental thought that stretches from the Renaissance to post-modernism. Despite denials – later deleted (Stapledon 1943) – elements of this holism find clear expression in Stapledon's work:

> I am not a theosophist, I am not a spiritualist, not a Rudolf Steinerite – I am of this earth, most extraordinarily earthy. But when I ride my fat pony over the Welsh hills I realize how much of sympathy there is between the pony and myself . . . Animals in some strange way are in harmony with the universe, they are happy and enjoy themselves, modern man is not in harmony with his man-made environment, and is not happy, and does not enjoy himself . . .. (Stapledon 1937, 107) (see Fig. 4)

Stapledon had read Henri Bergson

> and was deeply imbued with the feeling that an evolving Spirit of Life was immanent in Nature: that it was only by reference to this all-embracing and unifying Spirit that the details of Nature could ultimately be explained. (Waller 1962, 74)

This vitalism and stress on the non-material needs of peoples had, Stapledon argued, been effectively eradicated by urban existence but lived on in the rural population.

The mystical qualities which Stapledon attributed to the rural population are discussed below but he certainly attributed genetic qualities to the rural population as a whole. The 'country stock' was sounder than the urban because it was more inbred and was, therefore, a pure foundation stock, be it Welsh, Scottish or English. It was a honeycomb of genetic purities and

> Every breeder knows that a foundation of pure stocks constitutes an essential reservoir upon which to draw in the improvement and development of a race. (Stapledon 1944, 231)

In this sense, Stapledon also reflects some of the concerns of the eugenics movement. Within eugenics there was a struggle between biological and environmental explanations of 'racial' degeneration. Stapledon argued that phenotypical attributes like personality were affected by environment but that genotypical characteristics were not and remained potentially pure. The countryside of Britain, he argued, 'carries in its population the genes, unsullied and uncontaminated, that maintain and perpetuate our national vigour and national characteristics' (1944, 231). Stapledon's notion of a honeycomb approaches Fleure's 'mosaic' but it is the stress on *purity* that distinguishes the former's ideology and, whilst he does not erect any notion of a national type such as the Nazis' Nordic Myth, he nonetheless echoes that myth's romanticism in establishing the notion of rural genetic purity. Although not explicitly stated, there are clear links here to the programme of 'negative' eugenics – that is the intervention in the breeding patterns of those urban

Figure 4. George Stapledon's contemplation.
Source: Stapledon (1944) *The land now and tomorrow*

peoples deemed 'unfit' – alongside the promotion of the 'fit' rural population.

A moral topography

Whilst Fleure rejected the notion of rural genetic purity he agreed with Stapledon on the moral and sociological strength of rural Britain. Wales in particular was 'a refuge of old ways and old types' (Fleure 1923b, 233) where continuity and persistence were revealed by archaeology and anthropology. Geography caused this continuity for

> The physical features of the country, the framework of mountain-moorland that separates the Wye and Severn region from the valleys that radiate out to the sea, have broken the face of many waves of change ere they have reached the quiet western cwms. (Fleure 1926a, 1)

In the remote western areas could be found racial remnants and a persisting folk way of life, a localism distilled by undisturbed centuries. In *The personality of Britain*, the archaeologist Cyril Fox (1932), Director of the National Museum of Wales, drew on Fleure's work to argue that Highland and Lowland Britain were influenced by different culture streams. The Lowlands were susceptible to rapid cultural change but in the Highland region new cultures were absorbed thus generating cultural continuity. For Fox

> There is a wide range of evidence of this such as the survival of the Celtic language, the persistence in the west of very old racial stocks, and the persistence of tribal custom, the importance of kinship and clan. (1932, 32)

More broadly, Fleure saw the Celtic fringe of Europe as

> the ultimate refuge in the far west, wherein persist, among valleys that look towards the sunset, old thoughts and visions that had else been lost to the world. (1926a, 1)

In this broad and admittedly fragmentary sense the work of Fleure and his colleagues echoed a broader European conceptual tradition of 'habitat, economy and society' (Langton 1986). Elsewhere, the countryside, and in particular the hill country, was seen as keeping alive inheritances from the past. The German countryside, for instance, was seen 'as a reservoir of traditionalism, and . . . the peasantry as an arsenal of pre-modern characteristics' (Farr 1986, 9). This was later corrupted into the Nazis' 'Blood and Soil' philosophy (Bramwell 1985). In France, rural sociology and ethnology flourished after the Great War (Levi-Strauss and Mendras 1974) and in Scandinavia the study of agrarian history and the roots of folk life was perceived to be of great contemporary importance. Fleure praised Denmark's role as 'a laboratory of experiments in the modernisation of peasant life without setting the peasants adrift' (1943, 61).

The stream of inspiration

But this European concern with the rural was not necessarily a nostalgic response to modernization. In many cases it represented an attempt to theorize the perceived spiritual importance of the remote rural areas and their peoples, seen as wellsprings of civilization. Fleure studied what Campbell (1972) calls 'bio-social psychology': biological in that he tentatively attempted to link psychological with physical types; sociological in the sense that Fleure asserted the primacy of the social group. The Little Dark People, the basic Welsh type, contributed large numbers of church ministers, for the moorland people's idealism 'usually expresses itself in music, poetry, literature and religion rather than in architecture, painting and plastic arts generally' (Fleure and James 1916, 119). Fairer, Nordic types and the darker, coastal types were more prominent in commerce and the former were also astute politicians. So the geography that allowed for recovery of 'survivals' on Pumlumon also allowed for the recovery of a storehouse of values protected by social continuity. In this sense, the regionalizing ritual (Livingstone 1992) was also a moral one. Fleure was convinced that the peasantry cherished universal and abiding values, and that peasant life retained a vital diversity. To urbanites and suburbanites it might, Fleure admitted, seem mere fond sentimentalism to admit to an interest in old countryside traditions. Times, they would stress, had changed but Fleure noted that

> some of us doubt whether change is always progress and whether change, as it affects us, is not often a specialisation in certain directions that cuts off possibilities in others. (Fleure 1940, 883)

Fleure was not anti-urban; he saw the potential of the city as the social expression of the better elements of the human soul (1918b) but, under the conditions of industrial capitalism, the effects on both people and the environment were devastating. Hence Fleure's support for various town planning movements, including Patrick Geddes's 'Civics'. According to Fleure, modern society threatened to neglect the spirit in a pursuit of materialism: 'It is a case of cheap goods and cheap food for cheap people in cheap houses and cheap towns' (Fleure 1916). Modernization was seen as having a detrimental effect on personality in contrast to the rich diversity of peasant life. Whilst the latter's life-long sequences perhaps limited initiatives, it also protected people from becoming the flotsam and jetsam of slum and suburb. In the rural west, according to Fleure, the personality was fully developed and even simple working folk would eagerly discuss philosophy or religion.

Stapledon was less measured than Fleure in his criticism of the urbanism that had caused 'every sort of personal and social neurosis, and every conceivable evil' (1943, 115). His anti-urbanism was founded on an essentialist reading of both rural genetics and psychology. Stapledon too recalled encountering an educated peasantry but he claimed that townsfolk regarded the country personality – its apparent slowness and lack of enterprise – with an arrogance that ignored intuitive qualities of shrewdness and caution. Rural wisdom, Stapledon argued, arose from experience and contemplation rather than from the classroom. It was a sturdy individualism 'based primarily on the solid rock of deeply felt personal experiences' (1944, 233). Stapledon argued that the innate wisdom of the rural population was an aspect of the 'pre-mental' side of human character nourished by contact with and contemplation of wild nature. The rural population, unlike the urban, was in harmony with the yearning of the collective unconscious for 'the sounds of nature, the sea on the shore, the babbling brook, the wind in the willows' as opposed to 'the intellect-made din and screeching of modern streets and factories' (1937, 111). Thus, the rural intellect was 'sustained by emotions whose roots draw their nourishment from sources that are natural, pure and utterly devoid of perversion' (1937, 110)[7].

In Fleure's opinion, the peasantry was of importance in combating the materialism of *laissez-faire*. In 1921 he argued that civilization's one hope of avoiding collapse was to have a stream of supply from the rural areas where the treasures of ancient inspiration survived (Fleure 1921). Wales was seen as

> a fount whence may well up streams of inspiration refreshing to the jaded and overstrained business life of our perplexed modern England. (Fleure 1922b, xi)

Fleure argued that British life had been enriched by migrants from the Celtic west seeking employment in arenas like the army, the churches or politics where the social characteristics attributed to their physical types might best be utilized. The Celtic west had

> been that spring of an ancient cultural tradition with its vision and its dreams that has given its men a quality we need to keep us fresh. The miners of South Wales have been preserved from some of the worst evils of industrialism by these contacts and few who have known them well would dispute the statement that they are a specially valuable element in our British population. (Fleure 1940, 884)

Stapledon similarly foresaw calamity following a decline in what he called Britain's 'coefficient of ruralicity':

> Let rural Britain die completely and the whole superstructure will totter to ruin. It is just and only just not too late to stop the rot, but only heroic endeavours will suffice. (Stapledon 1943, 94)

In contrast to Fleure, Stapledon treasured the notion of the countryside as sustaining free trade. In the interests of the nation and of the Empire, Stapledon saw the most urgent need for a vigorous and explicitly male rural population as a prop to the great cities and as a recruiting ground for the trades and professions. Countrymen, he claimed, were in demand for posts demanding special reliability and, furthermore, 'the country-born do not in the main contribute to the town unemployed' (1944, 232). Most importantly, the rural population was a means of sustaining the nation's vigour and sanity. Fleure and Stapledon clearly disagreed politically: the former was a socialist and an enthusiast for the League of Nations, the latter a conservative British imperialist. The two disagreed on the ethical role of the rural in British society. Fleure saw the rural as an essential challenge to materialism and to industrial values. Stapledon, whilst stressing the importance of what he called 'the non-material' in his nature mysticism, nonetheless argued for notions of vigour embracing both economics and culture. However, the two *did* share a view of a British society sustained by rural values and both, as we shall see, argued for the rebuilding of modern, British society on a rural foundation. It is to the complex relationship between Britishness, Welshness and rurality that I now turn.

Nationalism and ruralism

In both Fleure and Stapledon's work, Wales is constructed as the remote hinterland of Britain, its cultural 'other' which exported poets, visionaries and teachers. Politically, this 'otherness' was complex. Both Fleure and Stapledon were arguably part of the 'rebirth of the nation' (Morgan KO 1982) – the emergence of an intellectual and social vitality in Wales and the establishment of national institutions in places like Aberystwyth. Fleure was opposed to the way nation states like Britain suppressed the linguistic and cultural diversities which reflected the variety of biological units and ancient inheritances (Fleure 1921). He believed that the encouragement of small nations like Wales was in the highest interest of civilization, given that they could have no military ambitions and would, therefore, concentrate on spiritual expression (1918a). Like Harold Peake, with whom he wrote *The corridors of time* series on historical anthropology, Fleure advocated the re-organization of Europe along devolved lines. Peake supported the devolution of powers to Wales, Brittany and Tuscany – provinces which drew on common memories and experiences in appealing to their peoples' 'historic sense' (Peake 1919, 104).

But Wales's cultural inheritance was increasingly threatened. According to Fleure, the geography of Wales was a cultural barrier 'likely to count for less and less in these days of powerful charabancs, listening-in and universal education' (1926a, 2). For many, the fate of the Welsh language symbolized a broader concern about these socio-cultural changes allied to continued rural depopulation. The Aberystwyth geographer, DT Lewis, argued that the 1921 Census showed that north-west Wales 'remains the fortress of the old language. Outside of this region the language is in decline' (1926, 417–8). DT Williams (1937) likewise noted the declining efficacy of topography in the maintenance of Welsh cultural distinctiveness and focused attention (e.g. Williams 1936) on the 'linguistic divide' drawn through the zone of conflict between Welsh and English.

In the work of Iorwerth Peate (1901-82), this changing cultural and political context is interpreted through an intellectual heritage derived, as Peate frequently acknowledged, from Fleure. Through Peate the ideas of Fleure and Stapledon were incorporated into broader political discourses on Wales and Welshness. Peate was one of Fleure's first students of geography and anthropology at Aberystwyth, and he later pursued doctoral research under his supervision on the links between physical type and Welsh language dialect (Peate 1926a)[8]. A carpenter's son from rural Wales, Peate shared Fleure's concern about cultural insensitivity and particularly the tendency of modernization to eradicate local differences. He was a passionate defender of the *gwerin* – the common folk – and shared with Fleure and Stapledon a belief in their essential wisdom and spirituality. According to Peate, to discuss religion, literature or politics in Wales, one naturally went to the carpenter, shoemaker, truly Welsh miner or blacksmith (Peate 1932). Peate saw rural society as emerging from a living tradition and, ultimately, a living language. Thus folk life assumed immense significance in the context of modern challenges to cultural continuity.

For Peate, Wales was a refuge from the waves of new cultures advancing from the east. In the west could be found

> folk songs, superstitions, crafts, the gentle bearing of the poor, and a host of other things which are like the fragments of a dream lost in the uproar of industry's juggernaut. (Peate 1931a, 2)

Peate's particular interest was the craft industries of

rural Wales. When he edited Fleure's *Festschrift* in 1930 (Peate 1930a), he contributed a discussion of Welsh wood turners and their trade in which he outlined the geography of production and marketing and traced European influences on style (again tying Wales into a broader European network). But new influences were increasingly evident:

> the introduction of German-made spoons into Pembrokeshire, and of Woolworth spoons into the large towns is slowly destroying the remnants of the turner's trade. (Peate 1930b, 183)[9]

As the English observer of rural life, George Sturt (1943), had observed, aesthetic changes reflected cultural change. Peate was convinced that the spiritual basis of Welsh rural life – and indeed civilization as a whole[10] – depended on the preservation of rural industrial organization, based on a combination of agriculture, industry and the crafts. The village had always been, he argued, a self-sufficient community

> where work and leisure, individual enterprise and co-operation were combined to produce a rural polity which seems to be far nearer perfection than the unhealthy striving of those communities where poverty is extreme and wealth out of all proportion to the needs of those who enjoy it. (Peate 1928, 104)

The community was an organic, self-sufficient and cooperative system which encouraged courtesy, artistry and kindness. However,

> the shoddy furniture of the cities and the short-lived manufactures of the mass-production firms have found their way into the countryside, and the result is not only a deterioration of the common necessities of life, but a disintegration also of rural society. (Peate 1928, 103)

Stylistic changes, therefore, mirrored social and cultural changes which would lead to the disintegration of rural society and its characteristic moral geography.

'A call to nationhood'

Peate was amongst the first members of Plaid Cymru, the Welsh nationalist party, founded in 1925 and helped edit *Y ddraig goch*, its monthly paper. According to D H Davies (1983), during its first twenty years Plaid Cymru was not a political party at all but a cultural and educational movement seeking to elicit a sense of common ethnic identity and to 'resist and reverse all those trends that were assimilating Wales into England' (vii). Whilst this sense of identity was overwhelmingly focused on the language, the concept of cultural continuity was understood in geographical terms which echoed the work of Fleure and others. Key components in this reconstruction of identity were the appeal of the rural and of the *gwerin*. *Y ddraig goch* was one method through which this geographical and cultural message was diffused[11]. In 1926, Peate argued in its pages that Wales was an 'immortal nucleus' (1926b, 4) containing the core of the western World's traditions and that it was incumbent on the Welsh to ensure its perpetuation by resisting English cultural encroachment. Saunders Lewis, the party's President, emphasized this view, arguing that Wales was a European nation rather than a British region. He claimed that in Britain the European Latin tradition was only represented by the Welsh and, like Fleure, chose to regard the Welsh 'not as a people driven headlong to the West and the mountains before a swift and irreversible Anglo-Saxon onslaught' but as Britons that had identified themselves with Roman cultural and spiritual ideals including Christianity (Jones D G 1973, 33)[12]. Modern Celtic cultural continuity in western Britain was, therefore, a rebuttal of Anglo-Saxon culture and politics. Given a choice between the Empire and the League of Nations, the Welsh – claimed Lewis – would opt for the League and for a quasi-federal Europe of small nations. He urged that the claims of Welsh nationhood be considered in the context of European history, with the Middle Ages before the rise of the modern state, as the ideal (Davies D H 1983).

Thus Saunders Lewis and Iorwerth Peate tied Wales firmly into the historical and contemporary currents of European civilization. Differences between these two should not be overlooked, most notably Lewis's Catholic, High Art notion of European culture as contrasted with Peate's Nonconformist, Folk Art notion (Lord 1992). Lewis was critical of Labour and Liberal politics and saw the Conservatives as the English nationalist equivalent to Plaid Cymru, whereas Iorwerth Peate was a socialist who yearned for the formation of a Welsh Labour Party (Davies D H 1983). Nonetheless, the 'call to nationhood' embraced diverse positions centring on issues of national identity. As part of this identity a politically-resonant European and rural historiography of Wales and of Welshness emerged which distanced Wales from Anglicized metropolitan values and which was seemingly supported by Fleure's and Peate's elaborations of the historical geography of Wales (see Gruffudd 1989).

This view of Wales in a Europe of the Nations was but part of a broader attachment to the continent shared by Lewis and other party members. *Y ddraig goch* contained European and world political analysis and, whilst some looked to Ireland, many drew inspiration from European culture and politics. Iorwerth Peate, working from the National Museum of Wales and later as Curator of the Welsh Folk Museum, had many links with scholars of folk culture, particularly in Scandinavia. He was the only British delegate to attend the opening of the German National Museum of Folk Culture in 1935, by which time the 'Blood and Soil' philosophy was well on its way to the heart of state ideology (Peate 1935). Saunders Lewis was influenced by Maurice Barrès, a French conservative politician, described by Tombs (1991) as organicist and racist, who argued for steadfast commitment to province and *patrie* (Burns 1991). Lewis's conversion to Catholicism in 1932 prompted hostility from party members rooted in liberal nonconformity and, in the context of a broader attraction to Europe, was seen as evidence of Fascist leanings. Plaid Cymru officially rejected the deification of state power represented by both Fascism and Communism but some party members did flirt with quasi-Fascist groups. Ambrose Bebb was influenced by Charles Maurras, the extreme-Right mentor of *Action Française* who 'propagated nationalism as a substitute religion culminating in a mystical cult of the earth and the dead' (Alter 1989, 38)[13]. Bebb had also succumbed to the cult of strong leadership:

> Strength and power are the only things that count in the myriad troubles and battles of this iron age. The age of Lenin! The age of Mussolini! Indeed it is a Mussolini that Wales needs. Just as he had to strike fearlessly to save Italy, so will Wales's Mussolini when he comes. And coming he is. (Quoted in Davies D H 1983, 97)

Back to the land

Amongst the currents of political and social thought strong in interwar Europe was, as we have seen, an idealization of the rural population and of the rural areas as sustaining 'national' characteristics. This idealization underpinned various movements aimed at shifting the orientation of society, both ideologically and physically, 'back to the land'. Many of these movements were on the Right, the extreme example being the Nazi 'Blood and Soil' ideologues (Bramwell 1985). British Fascist and Conservative groups also promoted similar ideas (Bramwell 1988; Cullen 1987). But this rural idealism was also central in many socialist and distributist movements. The Danish cooperative movements, for instance, influenced the work of many Plaid Cymru members, and in Britain the Arts and Crafts and Garden City movements were founded on the romantic socialist opposition between industrial oppression of the proletariat and rural liberation[14]. In the South Wales coalfield, the Quakers sought to alleviate deprivation by introducing crafts workshops and farm units. These fragmentary influences contributed to an anti-industrial and anti-urban sentiment within Plaid Cymru which argued that these influences were anathema to Welshness. Ambrose Bebb – echoing George Stapledon – blamed the education system for causing rural depopulation by failing to inculcate rural values:

> How sad it is ... to see arising generation after generation of boys and girls, who swarm together to the schools of the plains, there to drink from a poison which visibly weakens them and makes them unable to perceive the majesty of the high pastures and the shepherd's life, the romance of farming the land, and of passionately smelling the fertile soil; but who rather set off in cowardly fashion, with neither valour nor heroism, for the lazy, inert abodes of the towns and cities. (Bebb 1927, 4)

In Bebb's opinion

> One of Wales's greatest needs today is not only to keep her sons on the land, but to bring back from the city to the land the masses who flowed there during recent years. (Quoted in Davies D H 1983, 91)

The party's chief agricultural adviser, Moses Gruffudd, argued that

> Placing the people back on the land is not only appropriate, but is essential if the Welsh nation is to live. The Welsh nation is a nation with its roots in the country and the soil. (Quoted in Davies D H 1983, 92)

This back to the land sentiment clearly echoed the academic interpretation of rural virtue. But academics like Fleure and Stapledon also proposed plans for rural rejuvenation which also came to influence broader political discourses, in part through the agency of Iorwerth Peate. Fleure was committed to the active role of education in the life of the community, thus maintaining a tradition established at Aberystwyth when the college was founded in 1872 with over 100 000 donations of under half a crown from the *gwerin* (Morgan K O 1982). He believed that Britain's well-being depended on the return of vitality to provincial life and that the universities might be instrumental in this renewal as, he argued, had been the case at Aberystwyth. His work in adult education (see Bowen 1957) and with the Regional Survey movement and his belief in the university's role as provider of specialist advice, were part of this reciprocal relationship between institution and community. Survey should, he argued, foster social renewal by inculcating citizenship and a spiritual awareness of place:

> My first plea then is that in our work we should cultivate the master [*sic*] light of memories and traditions, the deep intuitions of life, and that we can do this very forcefully by encouraging direct observation and study of the surroundings in which we live. (Fleure 1918b, 32)

This awareness and local patriotism would then facilitate a re-birth founded on tradition. In a report for an influential planning campaign group, Fleure (1917) addressed the social and economic future of Cardigan Bay and thus applied his philosophical understanding of rural society. He blamed *laissez faire* for the decay of the traditional industries and noted some of the consequences, most notably the increasing role of tourism in coastal villages. A population which came to depend entirely on the city tourist was socially degenerate and Fleure urged action for reasons as much moral and civic as financial. He advocated scientific support from the university and a centralized and cooperative authority to control local industries. Themes in Geddes's sociology, such as 'Civic sympathy' and 'social reintegration', were crucial elements in Fleure's vision of social evolution. Technology also played a part. Afforestation and hydro-electric power might evolve alongside fisheries – a pattern then characteristic of the Alpine countries and Scandinavia – and indeed hydroelectric power (HEP) was seen as crucial in transforming what Fleure called 'regions of difficulty'. This was part of a broader trend of evangelizing on behalf of the new technology which Bill Luckin (1990) has termed 'techno-arcadianism' whereby the old, moral order is re-established on modern, technological foundations[15].

George Stapledon also harboured visionary plans for the countryside. He began, as always, with the philosophical construction of attitudes to nature, and argued that modern education had lost touch with the country and its heritage of wisdom and sturdy individualism (Stapledon 1944). But Stapledon argued most vigorously for material acts like land reclamation. He admired Mussolini's reclamation schemes, seeing them as creative acts having

> a stimulating influence not only on all who are directly concerned, but also upon the nation as a whole. To turn land incapable of maintaining man or beast into productive and smiling acres is a creative act that stands to the credit of the nation with the heart to initiate such work not merely for decades but for hundreds, perhaps thousands, of years. (1944, 181)

The visionary monumentalism of Fascism stood in stark contrast to the pragmatism of British planning and, in his writings, Stapledon reflected the contemporary lure of the charismatic, masculine leader though he warned against dictatorship. Like Ambrose Bebb, Stapledon admired Mussolini's 'visionary' leadership, and his blend of modernism and rural conservatism. He talked of the emergence of a great architect to give visual expression to the coming of the new age, an architect who will be

> a great countryman, a man imbued with the urge blindly striving for self-expression that is the driving force behind the great pilgrimage. (1944, 276)

Above all, there is the figure of the engineer, 'a man of action; he does things and he creates things . . . he is not for ever bothering with ultimates' (1937, 105). The engineer's fundamentalism, Stapledon claimed, was akin to the pre-mental side of human character that was attuned to nature[16]. In visionary vein, Stapledon argued for comprehensive schemes of reclamation and revitalization in upland Britain. He called for the regrouping of isolated farmhouses into hamlets serviced by communications, welfare and entertainment. Roads and motor vehicles might then invigorate the uplands for

> The bus, the motor car and the lorry are something absolutely tremendous – and incidentally these have made hilly villages, cottages and hamlets accessible . . . But we have not done with the internal combustion engine yet, for we have also the tractor, the engine which must inevitably revolutionize the pastoral no less than the agricultural practices of . . . every corner of the world. (Stapledon 1933, 18)

The rebirth of the nation

The key to Stapledon's vision for the uplands was accessibility, thus conflicting with Fleure and Peate, both of whom argued that it was the very inaccessibility of upland Wales which had preserved its cultural characteristics. However, Peate's vision was far from being a nostalgic retreat to the past. He again provides a link between academic and political discourses on the nature of Welsh rural society. Peate's proposed action to reinvigorate rural Wales was based on his academic studies of crafts and social organization. In a critique of the 1942 Scott Report on *Land utilization in rural areas,* Peate (1943) attacked the division between urban and rural, claiming it was sentimental and indeed immoral, having no application in Wales. Rural Wales, Peate argued, had always been characterized by the dual foundation of agriculture and industry. No-one, given care in planning and design, needs fear the destruction of beauty by industrialization. Industry could indeed beautify:

> We must face these facts rather than live in a sentimental mist and be content with the persistent feebleness of the countryside. There are dynamic foundations to true beauty. (1943, 14)

Peate, like Fleure, called for the development of HEP and forestry and argued for the introduction into rural Wales of new 'mobile' industries like plastics, located in well-planned additions to existing settlements. But this was not merely a countryside planning argument; it was calculated to re-establish the moral geography of the organic community. Small factories could breathe new life into declining districts, stemming population flow and re-establishing the old social organization and its moral basis of cooperation on a new, technological foundation. As Luckin (1990, 83) puts it, triumphalist enthusiasts for electricity

> came to stress 'natural' connections between farming, the revival of the 'organic' village community, and the new form of energy as a stimulant to rural crafts and industries.

This back to the land ideology was adopted by Plaid Cymru who derived widespread inspiration for their notion of a national plan. Nationalist economic analysts (Davies and Davies 1939), drawing on the work of George Stapledon and on the experiences of other small European nations, argued in 1939 that Wales could afford self-government. Another inspiration was the integrated, democratic

Figure 5. 'The TVA Way'
Source: Welsh nationalist party 1945

development scheme of the Tennessee Valley Authority (Welsh Nationalist Party 1945) (Fig. 5). Saunders Lewis argued for a Welsh National Development Council to guide the de-industrialization under way in the depression years for the benefit of Wales. Former industrial workers should, he argued, be settled in farming colonies with the policy operating alongside slum clearance in the urban areas (Davies D H 1983). But the policy was not solely a response to industrial decline. It had both economic and ideological coherence as illustrated by Lewis's *Ten points of policy,* published in 1938. He argued that agriculture should be the primary industry of Wales 'and the foundation of its civilization' and that South Wales must be de-industrialized 'for the moral health of Wales' and of the region's population (Lewis S 1985, 16). A cartoon (Fig. 6) from *Y ddraig goch* gives visual representation to this economic and cultural stance and to the political appeal of the *gwerin*. It juxtaposes two ways of life. The peasant farmer has an elemental beauty in contrast with the awkward ugliness of the capitalist. The farmer ploughs quiet fields with his horse whilst the capitalist stands in front of his domain of polluting industry. The caption indicates that the capitalist clearly foresees the diminution of his influence as the peasant becomes self-sufficient. Not only does this image make a point about the nobility of a pastoral life, it argues that this is the route to Welsh self-sufficiency, giving a base for political sovereignty. To leave the currents of industrial capitalism was to leave English influence.

Conclusion

Throughout the debates on the primacy of rural values, Peate argued that the traditions of the past

Cyfalafydd: "Er — mwyn — masnach rydd, — peidiwch cynhyrchu eich anghenion eich hun."

Figure 6. Back to the Land. The caption reads 'Capitalist: "For the sake of the free market – don't become self-sufficient" '
Source: *Y ddraig goch* 1936, 10 7

had to be acknowledged in the solution of present problems, one of the themes of the Regional Survey movement. In all replanning, the Welsh nation should, he argued, look to its own traditions. In *Y ddraig goch* he advocated the role of agriculture and market gardening as complements to mining and industry in the revival of South Wales. HEP and large-scale forestry could also form the basis of a new, Welsh rural culture:

> We cry for old methods in vain: we attempt to revive the dead in vain, but on the grave of the old methods, we can build new factories and keep alive, in the sound of the machines of this age, the spirit of the rich culture we have inherited from the old craftsmen of Wales. (Peate 1929b, 8)

In Peate's techno-arcadianism, the old order was revived on the foundation of new, modern industries. Such ideas, expressed also by Fleure and Stapledon, begin to make 'the rural' more complex in the interwar period. In essence, we cannot see the relationship between country and town as a simple polarization of tradition and modernity, or stagnation and progress. In his study of Weimar and Nazi Germany, Jeffrey Herf (1948b) identifies what he calls 'reactionary modernism' – an assimilation of technological advance into anti-capitalist romanticism. The Right were particularly adept at straddling the tradition-modernity divide, harnessing historical idealism in conjunction with the promises of an ordered, scientific twentieth century, producing 'a strange contrast . . . between modern and anti-modern themes' (Cullen 1987, 124).

But the key idea here is, perhaps, the tension between materialism and idealism. Herf claims that reactionary modernists did not see materialism and technology as identical. Technology and idealism could, therefore, be reconciled and, whilst the Right reconciled them around National Socialism, elsewhere in interwar Europe – including Wales – they were reconciled around the notion of a 'moral geography'. Modernist notions of progress, utopianism and democracy represented for some by technology were allied to a rural idealism focused on morality and cultural continuity. The move back to the land was not necessarily a regressive or reactionary step, but one which challenged dominant ideas of 'progress': ideas based on industrial capitalism and urban life. It asserted that certain values of community and artistry, apparently denied by urban civilization, could be re-captured in the rural areas. In the case of Welsh nationalism, a move back to the land could also reunite a culture with its European influences. Whilst modernity is generally cast in opposition to notions of romanticism or nostalgia, this move back to the land advocated its own version of progress, founded on an utopian fusion of past and future. Iorwerth Peate (1959) summarized this new relationship between the traditional and the modern by drawing on Lewis Mumford's *Technics and civilization*. As Mumford himself put it (1946, 427)

> [W]ith a change in ideals from material conquest, wealth, and power, to life, culture, and expression, the machine like the menial with a new and more confident master, will fall back into its proper place: our servant, not our tyrant.

Welshness

Historiographical writing and academic research in the interwar period were both part of the process of remaking Welsh national identity according to cultural inheritance and geographical rootedness. Geographers located continuity and virtue in the countryside and theorized the social contribution of the *gwerin* and the organic communities it had formed. These themes were reflected in the work of a political party expressly established to challenge the cultural assimilation of Wales. But, as Anthony Smith has argued (1991, 14)

> A national identity is fundamentally multi-dimensional; it can never be reduced to a single element, even by particular factions of nationalists, nor can it be easily or swiftly induced in a population by artificial means.

There were other powerful identities in existence that challenged the rural. Particularly strong was that identity centred on the industrial valleys of South Wales. According to Gwyn A Williams (1985), there are two archetypal myths of the Welsh – the *gwerin* and the working class. The former relies essentially on notions of the organic community, the latter on notions of class solidarity. Fleure in no sense rejected the industrial working class as un-Welsh but, in both biological and spiritual terms, he was concerned about the degenerative effects of life in the coalfield. Like Fleure, Peate was a socialist but he too saw both cultural and political salvation in a return to earlier, rural and more equitable forms of social organization. The attempt to replace class with community identity found its clearest expression in the early work of Plaid Cymru and, in particular, of Saunders Lewis who identified industrial capitalism not only with materialism but with Englishness. The relationship between these two enduring discourses of *gwerin* and working class is perhaps summed up by Fleure's closing lines (1953b, 263) in his contribution to Ogilvie's *Great Britain: essays in regional geography*, first published in 1928 but still in print as late as 1953:

> The establishment of an Adult School on residential lines, not unlike the Danish Folk High Schools, was one of the noteworthy events in Welsh life in 1927; its attention is specially focused on the problems of the miners, but it is characteristically situated at Harlech, N. Wales, in the beautiful country between the mountains and the sea.

Acknowledgements

I would like to thank Denis Cosgrove for comments on early versions of this material and three anonymous referees and Roger Lee for their suggestions on this paper. Versions of it were also presented at the IBG annual conference in Swansea and at the ICHG conference in Vancouver, and I am grateful to all those who commented on those occasions. Thanks also to Alan Cutliffe for his photographic work.

Notes

1. This is the title of a book published by Fleure in 1926 and the title, in Welsh, of a book published in 1931 by Iorwerth Peate. The gendering of Wales will be the subject of a later paper. See note 8 below.
2. The Chair was endowed by Margaret and Gwendoline Davies who also donated Gregynog to the University of Wales. Their brother, David Davies, was a coal-owner and Liberal MP and one of David Lloyd George's 'mafia' of Welsh advisers known as the 'garden suburb'. This was particulary appropriate in Davies's case as he funded a number of organizations committed to improving Welsh housing conditions.
3. On Geddes see Boardman (1978), Fleure (1953a) and Robson (1981). In Fleure, Geddes found, according to Stevenson (1978, 59), 'a mind that equalled his own in breadth of interests and synthetic power'.
4. Livingstone (1992, especially Chapter 8) argues that anthropology and anthropometry were of central importance in early twentieth-century Geography. As well as Fleure, anthropometry was also practised by other Aberystwyth geographers including Estyn Evans who attempted to identify Irish 'racial' types; E G Bowen, who studied the relationship between physical type and susceptibility to disease; and Iorwerth Peate, who examined the relationship between physical type and Welsh dialect. Fleure later worked on the geographical distribution of blood types.
5. Fleure's anthropological work at Aberystwyth was complemented by a number of other studies. These included E G Bowen's research on the influence of the western seaways on the historical geography of Wales. In his study of artefactual remains, Bowen linked in to a morphological tradition extending from Ratzel and Vidal to Sauer. His comprehensive studies eventually fostered the idea of a Welsh *pays* (see Bowen 1959; Thomas 1985). Mention must also be made of Estyn Evans's application of similar ideas to Ireland.
6. On eugenics see also, for example, Searle (1979) and Stepan (1982).

7. This idealization of the noble savage is, of course, a recurrent theme in Western romantic thought. Stapledon's thinking parallels the Neo-Romantic movement in British art (see Mellor 1987) which found inspiration in Celtic Britain. Graham Sutherland found Pembrokeshire peasant life to be 'almost biblical in its sober dignity' (*Sutherland yng Nghymru* 1976, 14) and the St Ives school in Cornwall lauded the primitive paintings of retired fisherman Alfred Wallis.
8. Although better known for his work in the museum world, Peate was a vigorous advocate of Geography. His book *Cymru a'i phobl* (Wales and her people) (1931a) was a conscious attempt to write the kind of 'new' human geography developed by Fleure and, in a paper published the same year (1931b), he made a vigorous defence of the synthesizing powers of his subject. Elsewhere (1929a) he drew implicitly on the work of Vaughan Cornish in elaborating on the role of the frontier in the development of Welsh nonconformism.
9. In *The Welsh house: a study in folk culture* (1944), Peate argued that the humble peasant dwelling was richly expressive of a tradition of localism. However, English influences could also be traced architecturally, be they the black and white houses of the Severn valley, the 'English' quaintness of the Vale of Glamorgan, or the 'bungaloid growth' of the seaside.
10. 'It is my belief that if our civilization is to be saved, the study of the "little clan", its social organization, its skills, its spiritual beliefs and visions, will play a vital part in that salvation. As students of folk life, it is by that faith that we should work'. (Peate 1959, 109)
11. Other methods included the Party's summer schools. Other political groupings also promoted the idea of the rural. The Workers' Educational Association's magazine *Cambria* frequently discussed rural life, craftsmanship and the role of rural culture in shaping Welsh national consciousness. So did the non-partisan *Welsh outlook*, founded in 1914 by ex-Deputy Cabinet Secretary Thomas Jones, a friend of Fleure's who also founded Coleg Harlech, the adult education centre, in an attempt to address the cultural deprivation of the South Wales coalfield (Morgan K O 1982).
12. Fleure was in agreement with this revised historiography:

    For lack of interpretative research there grew up the idea that post-Roman super-men founded a new civilization in Britain, and drove out the decadent Britons to the 'celtic fringe', i.e. that the English-speaking population was inherently superior. The harm done by this view is writ very large in the political history of Britain. (Fleure 1931, 52)
13. Although Benedict Anderson (1983), of course, argues that all nationalisms are essentially cults of the dead.
14. On socialism and environmentalism, see Gould (1988). Less reverently, George Orwell (quoted in Cunningham 1988, 207) suggested that

    It would help enormously if the smell of crankishness which still clings to the Socialist movement could be dispelled. If only the sandals and the pistachio-coloured shirts could be put in a pile and burnt, and every vegetarian, teetotaller and creeping Jesus sent home to Welwyn Garden City to do his yoga exercises quietly!
15. On the political and cultural complexities surrounding electricity see Gruffudd 1990.
16. For a comparison with the idealization of the engineer by German 'reactionary modernists' see Herf (1984a, 1984b).

References

Alter P 1989 *Nationalism* Edward Arnold, London

Anderson B 1983 *Imagined communities: reflections on the origins and spread of nationalism* Verso, London

Bebb A 1927 Bywyd gwledig Cymru heddiw *Y ddraig goch* 1 (12) 4

Bhabha H K ed. 1990 *Nation and narration* Routledge, London

Boardman P 1978 *The worlds of Patrick Geddes: biologist, town planner, re-educator, peace-warrior* Routledge and Kegan Paul, London

Bowen E G 1957 For Herbert John Fleure, FRS *Geography* 42 137–40

Bowen E G 1959 Le pays de Galles *Inst Br Geogr Trans Pap* 26 1–26

Bramwell A 1985 *Blood and soil: Richard Walther Darre and Hitler's 'green party'* Kensal Press, Bourne End

Bramwell A 1989 *Ecology in the 20th century: a history* Yale University Press, New Haven

Burns M 1991 Families and fatherlands: the lost provinces and the case of Captain Dreyfus in **Tombs R** ed. *Nationhood and nationalism in France: from Boulangism to the Great War, 1889-1918* Harper Collins Academic, London 50–62

Campbell J A 1972 Some sources of the humanism of H J Fleure University of Oxford, School of Geography, Research Papers No 2

Campbell J A and Livingstone D N 1983 Neo-Lamarckism and the development of geography in the United States and Great Britain *Trans Inst Br Geogr* NS 8 267–94

Cosgrove D 1990 Environmental thought and action: pre-modern and post-modern *Trans Inst Br Geogr* NS 15 344–58

Cullen S 1987 The development of the ideas and policy of the British Union of Fascists, 1932–40 *J Contemp Hist* 22 115–36

Cunningham V 1988 *British writers of the thirties* Oxford University Press, Oxford

Daniels S 1993 *Fields of vision: landscape imagery and national identity in England and the United States* Polity Press, Cambridge

Davies D H 1983 *The Welsh Nationalist Party 1925–1945: a call to nationhood* University of Wales Press, Cardiff

Davies D J and N 1939 *Can Wales afford self-government?* Swyddfa'r Blaid, Caernarfon

Farr I 1986 'Tradition' and the peasantry: on the modern historiography of rural Germany in **Evans R J and Lee W R** eds *The German peasantry: conflict and community in rural society from the eighteenth to the twentieth centuries* Croom Helm, London 1–36

Fleure H J 1915 *Regional surveys in relation to geography* Holywell Press, Orford

Fleure H J 1916 Geography and citizenship Lecture by H J Fleure to NUT Caernarfonshire County Association 24.6.16 H J Fleure, Offprints of Articles, National Library of Wales

Fleure H J 1917 Inshore fisheries and their development *Welsh housing and development yearbook 1917* WHDA Cardiff 60–5

Fleure H J 1918a *Human geography in western Europe: a study in appreciation* Williams & Norgate, London

Fleure H J 1918b Regional survey preparatory to town planning *J Tn Plan Inst* 4 31–43

Fleure H J 1919a Human regions *Scott Geogr Mag* 35 94–105

Fleure H J 1919b Regional surveys I & II *Sociol Rev* 11 28–34

Fleure H J 1921 Countries as personalities *Nature* 108 573–5

Fleure H J 1922a Some aspects of race study *Eugen Rev* 14 93–102

Fleure H J 1922b The land of Wales in **Muirhead F** ed. *Wales: the blue guides* Macmillan, London xi-xxvi

Fleure H J 1923a *The races of England and Wales: a survey of recent research* Benn Brothers, London

Fleure H J 1923b Problems of Welsh archaeology *Archaeologia Cambrensis* 3 225–42

Fleure H J 1926a *Wales and her people* Hughes and Son, Wrexham

Fleure H J 1926b The people of Cardiganshire *Cardiganshire Antiquarian Soc Trans* 4 15–21

Fleure H J 1930–1 The Nordic myth: a critique of current racial theories *Eugen Rev* 22 117–21

Fleure H J 1931 The place of folk-lore in a regional survey *Folklore* 42 51–4

Fleure H J 1935–6 Race and politics *Eugen Rev* 27 319–20

Fleure H J 1940 The Celtic west *J R Soc Arts* 88 882–4

Fleure H J 1943 Peasants in Europe *Geography* 28 55–61

Fleure H J 1953a Patrick Geddes (1854–1932) *Soc Rev* NS1 5–13

Fleure H J 1953b, originally 1928 Wales in **Ogilvie A G** ed. *Great Britain: essays in regional geography* Cambridge University Press, Cambridge 237–63

Fleure H J 1955 U C W Aberystwyth: an old student looks at the college *Trans Hon Soc Cymmrodorion* 1955 Session 57–63

Fleure H J 1965 Recollections of A J Herbertson *Geography* 50 348–9

Fleure H J and James T C 1916 Geographical distribution of anthropological types in Wales *J R Anthrop Inst* 46 35–153

Fox C 1932 *The personality of Britain: its influence on inhabitant and invader in prehistoric and early historic times* National Museum of Wales, Cardiff

Freeman T W 1987 Herbert John Fleure, 1877–1969 *Geogr Biobib Stud* 11 35–51

Gould P C 1988 *Early green politics: back to nature, back to the land, and socialism in Britain 1880–1900* Harvester Press, Sussex

Gruffudd P 1989 Landscape and nationhood: tradition and modernity in rural Wales, 1900–1950 Unpublished PhD thesis, University of Loughborough

Gruffudd P 1990 'Uncivil engineering': nature, nationalism and hydro-electrics in north Wales in **Cosgrove D and Petts G** eds *Water, engineering and landscape: water control and landscape transformation in the modern period* Belhaven Press, London 159–73

Herf J 1984a The engineer as ideologue: reactionary modernists in Weimar and Nazi Germany *J Contemp Hist* 19 631–48

Herf J 1984b *Reactionary modernism: technology, culture and politics in Weimar and the Third Reich* Cambridge University Press, Cambridge

Jones D G 1973 His politics in **Jones A R and Thomas G** eds *Presenting Saunders Lewis* University of Wales Press, Cardiff 23–78

Jones G 1986 *Social hygiene in twentieth century Britain* Croom Helm, London

Kern S 1983 *The culture of time and space 1880-1918* Harvard University Press, Cambridge Mass

Langton J 1986 Habitat, economy and society revisited: peasant ecotypes and economic development in Sweden *Cambria* 13 5–24

Levi-Strauss L and Mendras H 1974 Rural studies in France *J Peasant Stud* 1 363–78

Lewis D T 1926 Sur la distribution du parler Gallois dans le Pays de Galles d'après le Recensement de 1921 *Annales de Géographie* 35 413–8

Lewis S 1985, originally 1938 *Canlyn Arthur: ysgrifau gwleidyddol* Gwasg Gomer, Llandysul

Livingstone D 1992 *The geographical tradition: episodes in the history of a contested enterprise* Blackwell, Oxford

Lord P 1992 *The aesthetics of relevance* Gomer Press, Llandysul

Luckin B 1990 *Questions of power: electricity and environment in inter-war Britain* Manchester University Press, Manchester

Matless D 1992 Regional surveys and local knowledges: the geographical imagination in Britain, 1918–39 *Trans Inst Br Geogr* NS 17 464–80

Mellor D ed. 1987 *A paradise lost: the Neo-Romantic imagination in Britain, 1935–55* Lund Humphries/ Barbican, London

Morgan K O 1982 *Rebirth of a nation: Wales 1880–1980* Oxford University Press, Oxford

Morgan P 1983 From a death to a view: the hunt for the Welsh past in the Romantic period in **Hobsbawm E and Ranger T** eds *The invention of tradition* Cambridge University Press, Cambridge 43–100

Mumford L 1946 *Technics and civilization* George Routledge & Sons, London

Ogilvie A G ed. 1953 originally 1928 *Great Britain: essays in regional geography* Cambridge University Press, Cambridge

Parker A Russo M Sommer D and Yaeger P eds 1992 *Nationalisms and sexualities* Routledge, London

Peake H J 1919 Devolution: a regional movement *Soc Rev* 11 97–113

Peate I C 1926a The Dyfi basin: a study in physical anthropology and dialect distribution *J R Anthrop Inst* 55 58–72

Peate I C 1926b Y genedl ac awdurdod *Y ddraig goch* 1 (2) 4

Peate I C 1928 The social organization of Welsh rural industries *Welsh housing and development yearbook 1928* WHDA, Cardiff 103–5

Peate I C 1929a Lle'r ffiniau yn natblygiad annibyniaeth yng Nghymru *Y cofiadur* 7 3–21

Peate I C 1929b Y crefftwr yng Nghymru *Y ddraig goch* 4 (1) 4 & 8

Peate I C ed. 1930a *Studies in regional consciousness and environment (essays presented to H J Fleure* Oxford University Press, Oxford

Peate I C 1930b Some Welsh wood turners and their trade in **Peate I C** ed. 1930a 174–88

Peate I C 1931a *Cymru a'i phobl* Gwasg Prifysgol Cymru, Caerdydd

Peate I C 1931b Daearyddiaeth – a'r haneswyr *Y llenor* 10 161–8

Peate I C 1932 Welsh folk culture *The Welsh outlook* 19 294–7

Peate I C 1935 The German national museum of folk culture *Museums J* 35 329–32

Peate I C 1943 Yr ardaloedd gwledig a'u dyfodol *Y llenor* 22 10–18

Peate I C 1944 *The Welsh house: a study in folk culture* Hugh Evans and Son/The Brython Press, Liverpool

Peate I C 1959 The study of folk life: and its part in the defence of civilization *Gwerin* 2 97–109

Robson B T 1981 Geography and social science: the role of Patrick Geddes in **Stoddart D R** ed. *Geography, ideology and social concern* Basil Blackwell, Oxford 186–207

Samuel R ed 1989 *Patriotism: the making and unmaking of British national identity* Routledge, London

Searle G R 1976 *Eugenics and politics in Britain 1900–1914* Noordhoff International, Leyden

Searle G R 1979 Eugenics and politics in Britain in the 1930s *Ann Sci* 36 159–69

Shields R 1991 *Places on the margin: alternative geographies of modernity* Routledge, London

Smith A D 1986 *The ethnic origins of nations* Basil Blackwell, Oxford

Smith A D 1991 *National identity* Penguin, Harmondsworth

Stapledon R G 1933 Climate and the improvement of hill-land *Geography* 18 17-25

Stapledon R G 1937 *The hill lands of Britain: development or decay?* Faber and Faber, London

Stapledon R G 1943 *The way of the land* Faber and Faber, London

Stapledon R G 1944 originally 1935 *The land now and tomorrow* Faber and Faber, London

Stepan N 1982 *The idea of race in science: Great Britain 1800–1960* Macmillan, London

Stevenson W I 1978 Patrick Geddes, 1854–1932 *Geogr. Biobib. Stud.* 2 53–65

Stoddart D R 1986 *On geography and its history* Basil Blackwell, Oxford

Sturt G 1943 originally 1923 *The wheelwright's shop* Cambridge University Press, Cambridge

Sutherland yng Nghymru/Sutherland in Wales 1976 Alistair McAlpine, London

Thomas C 1985 Landscape with figures: in the steps of E G Bowen *Cambria* 12 15–31

Tombs R ed. Introduction in **Tombs R** ed. *Nationhood and nationalism in France: from Boulangism to the Great War, 1889–1918* Harper Collins Academic, London 3–7

Waller R 1962 *Prophet of the new age: the life and thought of Sir George Stapledon FRS* Faber and Faber, London

Welsh Nationalist Party 1945 *TVA for Wales* Welsh Party Offices, Caernarfon

Williams D T 1936 Linguistic divides in North Wales: a study in historical geography *Archaeologia Cambrensis* 91 194–209

Williams D T 1937 A linguistic map of Wales according to the 1931 Census, with some observations on its historical and geographical setting *Geogr J* 89 146–51

Williams G A 1985 *When was Wales? A history of the Welsh* Penguin, Harmondsworth

[10]

Once-upon-a-Time in the Commodity World: An Unofficial Guide to Mall of America

Jon Goss

Department of Geography, University of Hawai'i

This paper provides an account of the contemporary operation of the commodity aesthetic through a critical reading of Mall of America, the largest themed retail and entertainment complex in the U.S. Inspired by Walter Benjamin's analysis of nineteenth century arcades, I argue that the modern megamall is a dreamhouse of the collectivity, where fantasies of authentic life are displaced onto commodities that are fetishized in the spatial, anthropological, and psychological senses. Vital to these processes is the construction of temporal-spatial contexts, or chronotopes of consumption, which include the spatial archetypes of Public Space, Marketplace, and Festival Setting, and temporal archetypes of original Nature, Primitiveness, Childhood, and Heritage. Within these contexts, fetishism operates through themes of transport, both bodily in terms of motion and travel, and imaginatively in the form of memory and magic. Following a critique of a failure of dialectic thinking in existing literature on commodity consumption, I provide a dialectical "reading" of Mall of America, and outline its implications for a progressive political engagement with the contemporary retail built environment. *Key Words: consumption, commodity aesthetic, fetishism, shopping malls, allegory.*

> Every society possesses heterotopias, or "counter-sites," a kind of effectively enacted utopia in which the real sites, all the other real sites that can be found in the culture, are simultaneously represented, contested, and inverted. Places of this kind are outside of all places, even though it may be possible to indicate their location in reality (Foucault 1986:24).

The contemporary shopping mall is an example of what Foucault (1986) calls "heterotopias of compensation," real and discrete "counter-sites" where multiple images of ideal times and places combine to create an illusion of a world standing outside of everyday life. Like the nineteenth-century arcade, which it incorporates and expands on an enormous scale, the late twentieth-century megamall expresses a spatial unity and temporal stability that contrasts with the dynamic and discontinuous, fragmented and segregated reality beyond its bounds (Benjamin 1985; Geist 1983). The modern retail built environment narrates our collective loss of a natural world of innocence and immanence, and promises restoration in a utopian community of consumption. In this paper, I examine the visual and verbal rhetoric deployed to this effect and consider the possibility of a critical-sympathetic engagement using the example of Mall of America, which was opened in 1992 in Bloomington, Minnesota (Figure 1).

While it is true that we should not loosely talk of *the* mall (Morris 1993:394), since shopping centers are differentiated to appeal to diverse consumer cultures and markets, I argue that Mall of America represents nothing less than the imagined totality of the geography and history of the consumer world, and that, by its name, scale, and ambition, it is the apotheosis of the modern mall.[1] It is very hard not to say *the* Mall of America (hereafter, The Mall), as its logo is an armorial ensign of red, white, and blue ribbons emblazoned with stars, and its marketing rhetoric evokes a quintessential American experience. It is "the largest fully enclosed retail and family entertainment complex in America" (MoA 1997a:1), and it receives more than forty million visitors per year—more than the Grand Canyon, Disneyland, and Graceland combined. In its enormity, it seems to contain the entire world, both of reality and fantasy. Its "Funtastic Facts" boast "unbelievable" and "incredible" superlatives and comparatives, such as how many St. Peter's Basilicas or

Figure 1. Mall of America, "the largest fully enclosed retail and family entertainment complex in the U.S." (Credit: Mall of America)

Yankee Stadiums it could accommodate, and how many miles of phone lines, gallons of paint, and live plants it contains, establishing not only its uniqueness as a destination, but also its status as a "marvelous real" (Buci-Glucksmann 1994:140). Its seventy-eight acres and four levels contain 4.2 million square feet of retail space in more than 520 stores, but it is much "more than just a fantastic place to shop" (MoA 1996a), since it also boasts a theme park, an aquarium, an 18-hole miniature golf course, an entertainment district, and even its own "transit station." It is described as "a city within a city," but even that is not enough, for here also "you can ride a roller coaster, get married, visit a rainforest, dive with over 10,000 fish and now [even] get a college degree" (MoA 1997b:n.p.). It seems rather that "Mall of America has it all" (MoA 1997a:8), hence it claims to "bring the world to Minnesota" (MoA 1997c:2), capturing, in an otherwise unremarkable suburb of Minneapolis, the entire realm of commodified possibility in miniature. Simply put, Mall of America is the "mother of them all."[2]

For readers unfamiliar with the American-English vernacular expression of exemplariness, I apologize, but in this tour of the megamall, I will frequently resort to wordplay, not merely for fun or academic fashion, but also to evoke the constant slippage and play of meaning characteristic of the retail built environment. This occurs in both verbal and visual texts in The Mall, i.e., in retail concepts ranging from the frankly inane, such as Successories ("executive novelties") and Aerosoles (footwear), to the more subtle and clever, such as The Gap and The Limited (both clothing stores), and in retail and restaurant mise-en-scènes, such as the obvious use of treasure chests in gift-store displays and used luggage in souvenir stores. In wordplay, as in the commodity aesthetic, surface forms conceal deeper meaning that resonates and multiplies through a chain of correspondences towards, but never quite reaching, its imaginary origin in the Real.

With a serious eye on the play of meaning, and with Walter Benjamin as my guide, I read The Mall as a "dreamhouse of the collectivity," a repository of cultural images of utopia. Under the regime of capitalist production, however, the collective dream of authentic life is not expressed in the political process but is distorted by ideology and harnessed to commodity consumption. First, under the rule of ideology, popular desire manifests as nostalgia, as anxiety felt about the direction and pace of social and technological change is sublimated in a poignant regret and melancholy longing for a lost past (Casey 1987:363; Lowenthal 1989:21). Unlike memory, which is arguably organic and intrinsic, nostalgia depends upon a narrative of loss (Stewart 1984). Shopping centers are examples of "hypernarrated" spaces (Boyer 1994:190), providing texts and contexts for commemoration through which are narrated the loss and restoration of authentic meaning (Nora 1989). The Mall, for example, systematically insinuates itself into the natural and cultural history of Minnesota, recreates an originary urbanness in the form of past and distant places, and evokes personal memory and collective mythology with the overall effect of evoking natural relations with self, other, and object-world in a mythical realm of immanence—a unity of materiality and meaning. Vital to the narrative, and to the imagined experience of this realm, is the theme of transport, both physical and fantastic, and, hence, The Mall is suffused with the aesthetics of motion and imagination.

Second, under the rule of the commodity, the dreaming collective is seduced by the phantasmagoria of capitalism, such that the "ineradicable drive toward collectivity" (Jameson 1979:148) is experienced in the distractions of spectacular entertainment, while shared hopes and dreams circulate in the form of commodities. Images of the good life thus live in goods. In The Mall, spatiotemporal themes provide spectacular mise-en-scènes in which the commodity is thoroughly fetishized: severed from its relations of production and distribution, it is invested with an aura of innocence and authenticity, directly through interpretive texts and indirectly through contiguity with antiques, art, and magic objects (Baudrillard 1990).

The brilliant combined effect of the narrative of nostalgia, spatial and temporal displacement through transport, and fetishization of commodities, is to contrive the experience of The Mall literally and figuratively as a shopping "trip," a bodily and imaginary journey to distant places and past times that is undertaken in states of reverie and distraction, and for which the purchase stands as souvenir (Benjamin 1985). In The Mall, commodities are the vehicles of narratives that unite individual biography with natural and cultural history (Stewart 1984:135), but the collective dream is negatively sublimated in individual material wants.

Following a brief note on my method and review of some problems in theorizing consumption that I characterize as "failure in dialectical thinking," I discuss how spatial and temporal archetypes operate to construct idealized contexts of consumption. I discuss first how they combine in the Minnesota "sense of place," and then consider how spatial themes of mobility and travel effect a fetishism of the commodity in the Marxist sense of masking its material origins. I do not, however, discuss the spatial archetypes of Public Space, Marketplace, and Festival Setting here. Although The Mall realizes them in new ways, and while my own thinking on the potential for genuine urbanity in the retail built environment has changed, I have discussed these themes elsewhere (Goss 1993, 1996). Instead, I focus on the operation of the temporal themes beginning with an analysis of memory and magic, and the fetishism of the commodity in an anthropological sense, that is, as the animation of the object world. I then consider the four temporal archetypes—Nature, Primitiveness, Childhood, and Heritage. In the last section, I consider the implications of representing the commodity as souvenir, first in terms of fetishism in the psychoanalytical sense of displacing and condensing repressed desire, and second in terms of Benjamin's aesthetics of allegory. In the conclusion, I speculate on the political implications of conceiving The Mall as a dream world. On the way, I will note some moments of potential awakening, although ultimately I am not as much concerned with dispelling dreams as living them consciously (Buck-Morss 1989:120).

A Note on Method

My analysis is based upon ten days of semiotic reading and participant observation in The Mall, during which I visited every tenant, reading displays and promotional materials, shooting almost one thousand frames of slide film, and, if opportune, informally interviewing retail assistants and waiting staff. I met management and custodial staff, and chatted with shoppers, moviegoers, and flaneurs. And, of course, I shopped, returning late at night with my purchases and other weary shoppers, by shuttle to our nearby hotel. I also spent many hours in The Mall's virtual incarnation, browsing its World Wide Web site and other sources. I feel it necessary, however, to defend my "privileged" semiotic reading against those enamored of the so-called "ethnographic moment" in the geography of consumption (Jackson and Thrift 1995:229).

First, the purpose of my critical reading of the retail built environment is not to generalize consumers' experiences of The Mall, any more than the task of literary criticism is to study meanings readers draw from books. An exemplary reading considers how the "text" works, how it draws upon antecedent texts and draws in the intended reader. Second, I do not see how studies of consumer experience are any less "privileged" interpretations than semiotic readings, since the ethnographer must herself make a reading of texts produced by consumers—from interviews to focus-group transcripts—the main difference being that nonreflexive researchers tend to efface the authority and discursive strategies upon which their selections and interpretations hinge (Morris 1990). If there is pleasure in the power of my reading, I do not intend in any way to "give voice" to other interpretation and use of these texts—my aims are normative rather than positive. Let me admit at the outset that my reading is privileged, primarily by my education and social status, but certainly also by race and gender, and that my aim is to use this privilege responsibly, just as I hope that ethnographers will responsibly wield the political and economic power that enables their funded research. Third, the anxiety of academics to show that consumers actively make meaning, rather than passively consuming dominant messages (Strathern 1992:x), is in part a necessary reaction to the fatalism of cultural conservatism and left asceticism, yet it shares dominant culture's faith in the transparency of popular culture, its amenability to apprehension and representation, and ultimately manipulation—hence it is no accident that qualitative studies of consumers, so new to academics, have long been a part of marketing research, which has also long appropriated the findings of ethnographic and behavioral studies of contemporary society. Finally, the ethnographic approach risks banality by reproducing the obvious finding that consumers make their own meanings, without engaging in positive and negative critique of the politics of meaning (Morris 1990:14). We do need studies of consumption that resist the "seduction of negation" (Miklitsch 1996:25), but only if we also avoid patronizing consumers with a simple inversion of fatal politics, and the trap of binary thinking that dominates consumer research.

A Failure in Dialectical Thinking

> There are two ways to look at it. . . . You can denounce it all completely, drone on about the lack of civic space, bemoan eroding of genuine cultural amenities in favor of numbing commerciality . . . or you can denounce the fuddy-duddy puritans of the old left and the new left and become the Next Left (Moore 1991: 213).

A persistent cliché of modern thought is that since the advent of monopoly capitalism and the bureaucratic state, culture has been progressively severed from nature, rationality from the imagination, and the present from the past. Prospects for genuine sociality, commerce, and celebration have been undermined as the price of material progress and expanded individual freedom. Modernity is the contradictory experience of a nostalgia born of alienation from authentic nature, Self, and other, and anticipation of the development of technological and social means to realize "immanent totality" (Lukacs 1971). Either way, of course, the present is marked by a felt lack. Even quite mundane acts of consumption are invested with this dialectic, as the very latest product promises timeless value. The nostalgic pole of the dialectic dominates contemporary consumption, however, in part as a response to the accumulation of material evidence in the form of obsolete objects and ruins around us (Buck-Morss 1989:95), but also, no doubt, due to complex historical and contingent determinations. The nostalgic mode of consumption—the consumption of commodity forms and values preserved or reproduced from the past—only exposes the consumer to a second homologous contradiction, however: that between the material and mental. Central to western metaphysics is the inherent corruption of the material world based on a "depth ontology," which generally elevates mind over matter, soul over body, structure over ornament, and use value over exchange value (Miller 1995:25). Dominant discourse, therefore, betrays a contempt for the materialism of consumption and willful surrender of the masses to the seductions of the consciousness industry. Modern critics have mercilessly attacked the "heritage industry" that substitutes commodified history for the real thing (Hewison 1987; Walsh 1992), and "illusions of publicity" for genuine civility (Boyer 1994).

The class implications of the high cultural critique are obvious, for relative authenticity is deployed as a mark of distinction (Bourdieu

1984). It is also at least implicitly gendered, since consumption, particularly shopping, remains dominated by women while the imagined prelapsarian rational public sphere is the domain of propertied men (Wilson 1991; Nava 1987: 162). More to the point, however, by arrogating to themselves a privileged understanding of the Real, critics establish and police the boundary between reality and fantasy, maintaining the very possibility of an "original authenticity" that the ever-expanding popular sphere seeks to attain. To compound the problem, in condemning consumer culture, critical discourse evinces its own nostalgia for an originary Real beyond representation, for authentic unmediated relations with nature and the past in the form of natural objects and real history. The critic only reproduces the structural contradiction of authenticity-inauthenticity around which revolve the homologous oppositions central to the desires of consumption (Frow 1991:136).

There are two main alternatives to this critique. First, in his "hyper-critique" of commodity consumption, Jean Baudrillard argues that passivity is an ironic response, a refusal of subjectivity tantamount to a "radical anti-metaphysics." The surrender of "all rationality of choice and all exercise of will, of knowledge, and of liberty" (Baudrillard 1985: 585) to the seductions of marketing is a fatal tactic of last resort which effectively subverts, by taking the logic of consumption to its extreme ("we will consume all and everything"). Baudrillard's notion of "passive agency" is not born out by empirical research. Second, "cultural populists" (McGuigan 1992) celebrate consumption as "acts of inversion," perhaps most obvious in the case of marginalized subcultures (Hebdige 1988; Fiske 1989), but even "ordinary" consumers can be shown to actively appropriate and transform meanings of commodities. Marketing discourse has responded by acknowledging socially responsible and ironic modes of consumption, but there still seems to be a commitment to rescue "rationalist" and, perhaps, masculinist notions of agency (Warren 1993:178), as, for example, in Miller's (1995) polemic that redirects attention to the "old-fashioned virtue of thrift," "practical perspicacity and morality," and "sacrifice" of heroic housewives shopping for survival. While a useful corrective to the "consumer as dupe" scenarios of dominant discourse, such arguments hardly succeed in grappling with awkward issues of desire, seduction, and pleasure, which surely inform even everyday purchases.

Consumption treads the dialectical knife-edge between actual realization and rational manipulation of desire (Slater 1993: 207). It is profoundly contradictory since, as in tourism, with which it shares a drive to authenticity, the realization of desire in consumption is progressively deferred by the very structures and relations that offer limited satisfactions (MacCannell 1976). Critics, both culturally conservative and populist, evince a failure in dialectical thinking, and fail to recognize how they, like consumers, are caught up in modern narratives and cultural desire. Nor should we want to escape, even were that possible, for while material and nostalgic desires are exploited for instrumental ends in consumption, they contain latent progressive tendencies that are vital to our humanity (Bloch 1995:162-63; Enzenberger 1974:110). The critic's task is not to rudely wake up the consumer to the reality outside of consumption, but to ourselves awaken to the potential of the dream inside of which we shop, and so to reveal the traces of ideals of collectively meaningful life that are so vulnerable to forgetting.

Combining method and theory, I approach the shopping mall, like the city, as an "unconscious," its spatial forms objectifying collective values and dreams of fulfillment. I attempt a "psychoanalysis of space" (Lefebvre 1991:36), recognizing the very real living presence that repressed desires obtain. I interpret them, not to dispel "false consciousness," but to recognize their latent truth content. There is, then, a third way of looking at The Mall, neither disdainful nor celebratory, but critical-participatory. It is not to ask what The Mall really is, to seek "the real" beneath the "plague of fantasies," nor is it to accept simulation as all there is and fatally surrender to "hyperreality" (Žižek 1997). Rather it is to be able to take pleasure [*jouisance*] in the play of reality and fantasy, while critically examining how things actually seem and how we actually believe in this distinction. I will return to this subject in the conclusion when I address the question, What is the redemptive significance of this brilliant illusion? I have personally learned to love The Mall, as from it, I have learned a great deal about myself: about my humanity, the values and beliefs of "my" culture, and my intimate desires.

The Archetypes of Authenticity

Within the development of culture and exchange economy, the search for authentic experience and

> correlatively, the search for the authentic object become critical. As experience is increasingly mediated and abstracted, the lived relation of the body to the phenomenological world is replaced by a nostalgic myth of contact and presence. "Authentic" experience becomes both elusive and allusive as it is placed beyond the horizon of present lived experience, the beyond in which the antique, the pastoral, the exotic, and other fictive domains are articulated (Stewart 1984:133).

The shopping mall brings together the archetypes of the "good world" with the world of goods, presenting the world of commodities apparently innocent of the commodification of the world (Agnew 1989:133). It constructs idealized contexts of consumption, a complex of what Mikhail Bakhtin calls "chronotopes," that is, generic spatiotemporal structures where stories take place (Stam 1989:11). The narrative of lost authenticity and its redemption takes place in spatial settings I have previously identified as Public Space, Marketplace, and Festival (Goss 1993, 1996), and in temporal settings I introduce here as Nature, Primitivism, Childhood, and Heritage. These spatiotemporal archetypes are combined in a number of intermediate forms—for example, the aquarium and museum incorporate themes of public space with nature and heritage, respectively; the bazaar joins primitiveness with the marketplace; and the fairground the realm of childhood with festival—and they are multiplied in retail themes carefully selected for their complementarity. They come together most effectively, however, in the Minnesota theme, through which, despite its artificiality and placelessness, The Mall is born in local natural history and cultural geography.

The official history tells how The Mall is built upon the site of Metropolitan Stadium, previous home of local sports teams, the Minnesota Twins (baseball) and Minnesota Vikings (American football), potent symbols of local identity, and a natural place of congregation. Note, for example, the remarkable continuity suggested in the themes and semantic choices in the first paragraph of The Mall's souvenir book:

> Long ago, the land beneath Mall of America was *home* to golden *fields* of grain and lush pastures. It then became *home field* to *thousands* of cheering fans as *the site* of a major sports complex, Metropolitan Stadium. Since Mall of America opened its doors, *the site* welcomes over 40 *million* visitors each year (MoA 1996a:3; italics mine).

There is even the original home plate and the stadium chair that marks the landing spot for Harmon Killebrew's record-breaking home run (MoA 1997c:1; Getting Around 1997:1). The developers are never identified, but the story tells of their "natural" response to the opportunity presented in the form of cheap land, excellent communications, local demographics, and an "understored" retail market, and of The Mall's natural evolution from nearby Southdale, the "first fully-enclosed, climate-controlled shopping mall in the United States" (MoA 1997c).[3]

An essentialist Minnesota "sense of place" pervades The Mall, from the main attractions to retail and restaurant themes (Table 1). Camp Snoopy, for example (from the Peanuts cartoon created by Charles M. Schultz, a famous native son of St. Paul), is "the nation's largest indoor theme park" (MoA 1997a:14) and The Mall's major attraction. It evokes "a rustic clearing, deep in Minnesota's northwoods" (MoA 1997f:n.p.), and convincingly reproduces a stretch of a northwoods stream (Figure 2). A Northwoods feeling is evoked by 30,000 plants and trees—the largest indoor planting in the world" (MoA 1997f:n.p.)—an artificial seventy-foot waterfall, plaster cliffs cast from the originals along the scenic St. Croix River, browsing animatronic moose, and bird noises. Camp Snoopy is a combined lumber and mining camp: with log buildings, abandoned mining and logging paraphernalia, themed fairground rides such as Paul Bunyan's Log Shute, Pepsi's Ripsaw Rollercoaster, and the Mystery Mine Ride, and the dining experience of Tall Timber Grille, which "combines logging lore with the sights and sounds of a Minnesota Lumber Camp" (MoA 1997d). UnderWater World, the second major attraction, is an aquarium that offers the prospect of "diving" into Minnesota Dawn, a cool glacial lake with morning mist, beaver dams, and local fish. Even where the regional theme is not directly evoked in the names of stores and restaurants, northern landscapes and ways of life are employed in the mise-en-scène, and retailers sell authentic souvenirs of The Mall's Minnesota experience.

The visitor is transported to "Minnesota" and other idealized time-spaces through bodily movement, as in motion and travel, and imagination, as in personal memory and magic. Again these systems overlap, for example, in the restored historic vehicles and fairground rides, or in Chrysler

Table 1. Minnesota and the "Northwoods" Theme

Theme	The Mall	Attractions	Tenants
Minnesota	Metropolitan Stadium Twins and Vikings themes	Underwater World Minnesota Dawn Camp Snoopy Peanuts Local sports	Concepts Explore Minnesota Peanuts Minnesota Mittens Destination Minnesota Love from Minnesota Minnesota Sweatshirt Co. City Shirts Minnesewta Minnesot-ah
Northwoods		Camp Snoopy Timber/mining camp Forest/riverscape Hunter's Paradise Golf Mountain Indigenous plantings	Concepts Lake Wobegon USA Northern Reflections Northern Getaway Caribou Coffee Cabin Fever Winona

Figure 2. The "northwoods" stream in Knott's Camp Snoopy.

Corporation's display of a concept car described as "your very own rollercoaster," conceits that seem to move us simultaneously to experience past modes of travel and old-fashioned fun. The chronotopes of exteriority and anteriority and themes of transport combine to effect fetishization of commodities on display, and while the mechanisms overlap and combine in the complex form of the souvenir, I will consider fetishism in the materialist sense as an effect of spatialization, in the anthropological sense as the effect of a temporal disjuncture, and in the Freudian sense as an effect of fantastic transport. Each will be discussed accordingly.

Spatial Transport: Motion and Travel

> It is signs and images . . . that tend to fill the interstices in question [between facilities of political power]. Signs of happiness, of satisfaction. Signs and images of nature, of Eros. Images and signs of history, of authenticity, of style. Signs of the world: of the other world and of another—a different—world. Neo-this and neo-that, consumed as novelties, and signs of the old, the venerated, the admirable. Images and signs of the future. Signs and images of the urban, of "urbanness" (Lefebvre 1991:389).

The Mall evokes authentic urbanity by reproducing its ideal spatial forms: Public Space is premised on social solidarity and the community of strangers in an authentic public realm; Marketplace promises exchange unmediated by rational imperatives and institutional structures; and Festival Setting offers the possibility of social liminality in street theater and "acting out" of individual and cultural identity. Again, I have discussed these elsewhere, so suffice it to say that modernity narrates the loss of genuine public interaction, transaction, and festivity under the progressive privatization, commercialization, and rationalization of urban life, and that The Mall attempts to recover their possibilities in associated, ideal urban forms. Consideration of spatial transport and fetishism is, nevertheless, necessary to my argument regarding the positioning of the commodity as souvenir, so I will revisit the role of motion and mobility, adding to these the themes of travel and tourism, which seem particularly significant in this case (see Table 2).

Like the arcade, only at greater speed and intensity, The Mall is a kinesthetic space, and contemporary flaneurs are literally moved to aggregate, shop, and celebrate: they are drawn across thresholds and along paths by the use of contrast in color and light, focal attractions, and linear design elements; carried away by escalators and elevators; directed by spatialized narratives in the form of waymarkers and sequential interpretive texts; whirled around and whipped along on fairground rides; and reflected endlessly in mirrors and glass surfaces. The aesthetics of motion are repeated in the decor, in colorful flags, banners, mobiles, and water features, and in modes of transport in The Mall's innumerable attractions and retail concepts (see Table 2). Examples include: fairground rides in Camp Snoopy, echoed in mobiles in The Mall; reproduction antique toys and ornaments of nineteenth-century streetcars, trains, and bicycles, in stores such as The Great Train Store and Simpler Times; the mise-en-scènes of Runkel Brothers American Garage, which offers "American industry wear and nostalgia" in an old gas station; and retail concepts such as Crazy About Cars, Pro Motor Sports, and NASCAR Silicon Motor Speedway. The latest experience of virtual motion—the contemporary equivalent to the panorama and diorama in the nineteenth-century arcade—is available in Chrysler Corporation's store ("Great Cars. Great Trucks") and Camp Snoopy's Secrets of the Lost Temple. There is also transport to the future, as in Starbase Omega's interactive intergalactic warfare, LEGO Imagination Center's space-station exhibit, and video games, as well as in the space-shuttle simulator planned for phase two (MoA 1997b). Time travel provides souvenirs in the form of cuddly alien visitors such as "E.T.," but it is mostly dystopian, and in The Mall, the possibilities of the future receive short shrift compared to the certainties of the past.

The speculative activity of shopping has been effectively combined with mobilities of tourism since at least the nineteenth century, when the retail built environment became a tourist destination, and tourist resorts, shopping centers (Freidberg 1993:110; Harris 1991:66).[4] The Mall, however, is the acme of this process. The number-one tourist attraction in the U.S. (Nelson 1998:xv), it operates its own Tourism Department and contains six Mall of America Gift Stores. The Mall is divided into districts based on tourist-retail destinations (Figure 3): West Market evokes a bustling European marketplace under a train station-style roof, with street furniture, vendors, carts, and traditional stores (Figure 4); North Garden reproduces a European landscaped garden with terraces, gazebos, and fine shops; South Avenue is based upon luxury hotels with carpeted walks and sophisticated stores; and East Broadway is modeled on an upbeat American city with its chrome, neon lights, and contemporary fashions. The transit station seems like an airport, and the parking areas are named after the fourteen leading tourism states in the U.S. Out-of-state visitors, including more than four hundred Japanese tour groups per year, account for eighty percent of The Mall's economic impact on the

Table 2. Themes of Mobility and Travel

Theme	The Mall	Attractions	Tenants
Motion	Elevators, escalators Decor Banners, flags, mobiles Water features	Camp Snoopy Winding paths Flags and Banners Water features Rides UnderWater World People mover Water features Fish Lego Imagination Center Mobiles Clockwork/mechanical objects	Concepts Air Traffic Kites and Games Alamo Flags Thingamagigs that Fly
Transport	Transit station Decor, mobiles	Camp Snoopy Rides Balloon Race, Camp Bus Bloomington Express Red Baron, Truckin' Lego Imagnation Center Biplanes Space vehicles Great Cars. Great Trucks Chysler automobiles	Concepts NASCAR Silicon Motor Speedway Pro Motor Sports Runkel Bros. American Garage Crazy About Cars Great Train Store Starbase Omega Biplanes Dapy Objects Toy vehicles Reproduction antique vehicles
Travel/tourism	Tourism Department Gift stores Car parks Travel promotions Meet and Greet services Orientation Tours Mall themes West Market East Broadway North Garden South Avenue	Camp Snoopy Souvenir/gift stores UnderWater World Diving holidays Gift shops Great Cars. Great Trucks Outdoor adventure vehicles Chrysler souvenirs Rainforest Cafe Safari adventure Gift shop	Destinations Alamo Grill California Cafe Arizona Accents Colorado Mandarin Garden Little Tokyo Rainforest Cafe Tropical Treasures Tucci Benuch European Gift Shop Irish Indeed Babuska Kiwi Beach Promotions Minnesota Travel Information Cruise Holidays Grand Casino Florida Vacation Store Accessories The Map Store Trappings Bentley's Luggage and Gifts URBAN Traveller Adventure clothing/shoes Britches Great Outdoors Timbuktu Journeys Overland Trading Co. The Walking Co Track 'n' Trail

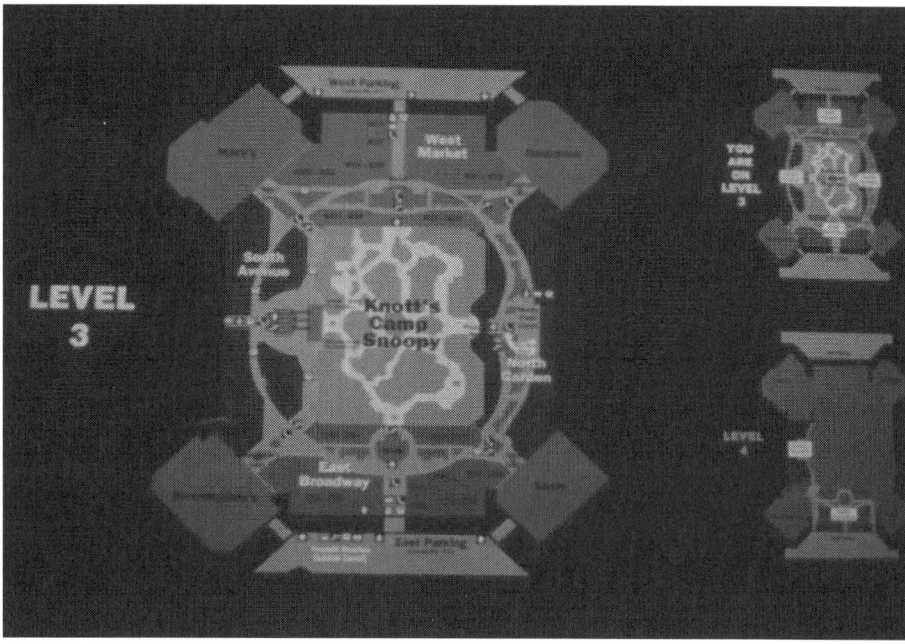

Figure 3. Mall of America plan.

state (MoA 1996b:sleeve). The Mall also actively promotes tourism: its "international extravaganzas" highlight other exotic locations, and vacation packages are available in its numerous travel agencies, including the public-funded Explore Minnesota Store.

Retailers in The Mall exploit travel themes to sell accessories, adventure clothing, and shoes. Perhaps the best example of this is Timbuktu Station, a women's clothing store whose products are displayed in antique suitcases and shipping cases decorated with whimsical literary quotations about travel. Concept restaurants and fast-food outlets are positioned as imaginary destinations in the consumer-tourist world. Rainforest Cafe, for example, is described as an "enchanted place for fun far away, that's just beyond your doorstep." Diners are issued "passports" for a meal that is a "safari" and are exhorted to complete their "trip" by buying a souvenir stuffed-animal toy in the adjacent "Retail Village."[5] Attractions, such as Camp Snoopy, UnderWater World, Chrysler Corporation, and Planet Hollywood also have gift shops selling souvenirs of their commodified experience.

Clichéd souvenirs from tourist destinations are sold in innumerable stores and kiosks, from European Gift Shop to Tropical Treasures, and from Irish Indeed! to Living on Island Time. The putative place of origin seems to play an important part in establishing the value of the product in stores from Colorado Pens to RCC Western Clothing, and from Bostonian Shoes to Pacific Sunwear of California. In order to appear an authentic souvenir of other times and places, however, narrative must effect a material and magical transport of the commodity itself from its material origins to the context of display. The process of fetishism describes the means by which the labor that has gone into its extraction, manufacture, and trade, and the "backstage" details of its presentation, are systematically obscured (Jhally 1987:49; Sack 1992:118).

Fieldwork in The Mall revealed two excellent examples of the combination of spatial structure, aesthetics of display, and marketing discourse to work this effect. In one case, there was a stuffed bear for sale in Love From Minnesota for $155: a tag on its right ear said "Minnesotans who live deep in the northwoods among the loons, wolves

Figure 4. The railway station aesthetic of West Market.

and scented pine trees, listen to the gentle lapping of the waves of the shoreline while they handcraft unique mementos of our homeland, like this one, to share with you"; a tag on the other ear said, "Bear made by Mary Meyer Corp., Townshend, Vermont. . . . Made in Indonesia." In a second case, when I picked up a coat made in Sri Lanka and imported by Columbia Sportswear of Oregon, and asked the assistant about the connection with the store's name, I was told, "It's just a name. Most people associate it with the outdoors. I mean, if we called it Iowa or something. . . . who'd wanna come?" The store is headquartered in New York and has no outlet in Colorado itself, but the slogan of the store explains everything: "Remember COLORADO isn't only a store: it's a state . . . of mind."

It should be noted, however, that shoppers are increasingly sophisticated, and the origins of some commodities are explicitly traced through authentic relations of production and distribution. The Body Shop, for example, makes beauty products from "natural" ingredients purchased from traditional communities whose cultural integrity it explicitly seeks to preserve (Kaplan 1995). The process of fetishization itself may be an attraction, and "kinky" consumers can, for a price, see "behind the scenes," that is, how products are delivered and stored, and how material waste is disposed.[6] On stage, however, The Mall's show is good, clean family fun as the crowd really does perform civility, exchange, and celebration in mise-en-scènes of Public Space, Marketplace, and Festival Settings. But in this paper, the concern is more with themes that are, in the words of one retail concept, "All About Time," that is, memory and magic, or historic and mythical imagination.

Temporal Transport: Memory and Magic

> In the course of social development, the old never sets itself off sharply from the new; rather, the latter, striving to set itself apart from the recently outmoded, renews archaic, ur-temporal elements. The utopian images that accompany the emergence of the new always concurrently reach back to the ur-past. In the dream in which every epoch sees in images before its eyes the one that follows it, the images appear wedded to elements of ur-history—that is, to classless society (Benjamin 1982:1225, trans. Buck-Morss 1989:116).

Under conditions of modernity, as a result of scientific faith in the objectivity of the material world and in deist religions that repress animism and idolatry, apparently "spirits are no longer seriously, truly thought to dwell in material items of human craft; they no longer animate objects, make them function, make them go" (Martin 1992:29). Enlightenment rationality and humanist religiosity has severed us from our original natures, when we shared an innocent faith in the enchantment of the world, but The Mall reconnects us to this magical realm via collective myth and personal memory. The Mall recalls times before the repression of spirit under the reality principles of modernity and adulthood, in the form of originary Nature, Primitivism, Childhood, and Heritage. I conceive of these forms as temporal since they operate through an imagined return to a putative innocence and whole(some)ness. There is some obvious overlap with spatial themes in The Mall: for example, nature is often place-specific and evokes a sense of place, primitiveness operates in the form of exotic Fourth

World peoples, and heritage is based on the preindustrial city. Also, the temporal forms overlap: for example, nature and primitiveness in the human-environment relations of native peoples, and childhood and heritage in the form of Victorian nursery toys. Still, these forms are distinct in terms of transport effected through memory and magic via the vehicles of commodities that effectively *re*-present lost powers in their material traces of natural nature and authentic labor.

Memory is jogged in The Mall in interpretive texts and retail themes such as Reminisce, Things Remembered, and Simpler Times, which ask us to recall the imagined innocence and plenitude of the past. Collective memory reaches back to the Eden of human origins in nature and primitiveness, to a time before nature and humans were separated through the process of civilization, and to the "natural economy" before exchange came to be structured by abstract equivalencies and motivations of profit. Cultural memory is evoked in themes of the nineteenth century, a time in U.S. history associated with a buoyant economy and cultural optimism (the teenage 1950s are also evoked as a minor nostalgic theme), but also with nonalienated labor in the form of hand-crafted products. Finally, personal memory connects us to childhood, a time when we all shared an innocent faith in the animation of the world. Everywhere there are discursive, visual, and material reminders of these pasts, signifying the presence of aura and authenticity in the objects on display (Table 3).

It has been commonly observed that the commodity aesthetic, and particularly advertising, corresponds remarkably with the magic systems of so-called primitive societies (Adorno and Horkheimer 1979; Williams 1980). That is, contemporary consumption depends upon a faith in power of objects such that one who rightly desires and rightfully takes possession is in turn possessed by desirable sexuality, physical prowess, and social status. The commodity is thus fetishized in the anthropological sense: it embodies aura, that is, an imagined ability to return our gaze and engage in relations with us (Wolin 1982:237), and is invested with "soul" (Dichter 1960), that is, a meaning and capacity for agency is possessed by the object itself rather than the social relations that enable its legitimate possession.

Magic is everywhere in The Mall. A form of "magical realism," the combination of technological modernization and traditional enchantment of the world, is evoked in the "magical waters" of UnderWater World and the "magical wonderland" of Rainforest Cafe, and in the two hologram stores and five science-education stores. The paraphernalia of the magic of illusion are for sale in Max Magic, the Fun Shop, and other novelty and toy stores. The magic of the supernatural is evoked in power-objects that seem to be everywhere: stars upon which to make wishes and masks to transform identity and decorate walls; angels and elves, fairies and pixies, and unicorns and leprechauns are for sale in miniaturized collectible forms or decorate displays of otherwise mundane objects (Figure 5); and a whole range of commodities—potions, jewels, crystals, clothes, shoes, and cars—promise to transcend the natural limits of individual physical and sexual capacity. Transformations are material and fantastic: the spa, beauty salons and health services create a new, younger, and healthier you, while Chrysler promises spatial and imaginative transport: "You are what you drive. Fantasize." The commodity makes all the difference, you just have to *buy it:* that is, both make the purchase and have faith in its power.

Aura circulates in the Mall in the form of art and personalized products, and magic in the form of power-objects, movies, and makeovers, although both forms frequently overlap in the enchantment of objects through animation and anthropomorphization. These tropes resonate, multiply, and combine to infect more mundane commodities on display in complex, though generally complementary, ways. Fine art, for example, expresses the aura of its creator and attests to the taste and social status of its owner. Original art is widely deployed in the public spaces, both in permanent displays and regular exhibitions, and can be purchased at some expense in several art galleries, while authentic reproductions of priceless originals are available in the Museum Store. Crafts on consignment from local artists and exotic natives can be purchased in Allie's Market and Pueblo Spirit. Paintings, sculptures, and jewelry pieces of natural landscapes and animals—wolves in particular—dominate the art of display, combining the aura of art and the living world.

Consumers are invited to invest their own aura in commodities: stores like Mrs. Anderson's Personalized Treasures and Things Remembered offer personalized gifts, one can buy personalized storybooks in Smokey Joe's Boutique, and San Francisco Music Box offers seventy-one "signature" tunes. The commodity becomes a vehicle of personal memory and meaning, and is

Table 3. Themes of Temporal Transport

Theme	The Mall	Attractions	Retail/Restaurant Mise-en-scène
Memory	Objects and decor Antiques, toys Reproduction street furniture	Camp Snoopy Victoriana Childhood past	Concepts Simpler Times Best of Times Moments and Memories Afterthoughts All About Time Reminisce Things Remembered
Aura/trace of authenticity	Decor Carousel horses Pushcarts and kiosks Storefronts "Cobblestones"	Camp Snoopy Antique tools/artifacts	Arts and crafts Thomas Kincade Artal Craft Art-a-Craft Grassland Gallery Pueblo Spirit Vaase Allie's Market Museum Store Modern Art Cafe
Essence of self		Camp Snoopy Personalized souvenirs Caricatures	Concepts Designs for You Monogramming State Your Name Lifetouch Lee's Gallery Personally Yours
Magic	Decor Masks, stars	Camp Snoopy Superstition Mystery Mine Ride Rainforest Cafe "Magical wonderland" Magic Mushroom Bar UnderWater World "Magical waters"	Concepts Max Magic shows Hologram Fantastic Hologram Land Elegant Illusions Objects Tracy the Talking Tree Magic Sets Various power objects
Media/makeover	Automated video recording Photo opportunities Movie location TV/radio recording Star appearances	Camp Snoopy Digital photos of rides Face painting	Concepts Old Time Photography Foto Fantasy Amazing Pictures Glamor Shots Glamor Kids Planet Hollywood Saturday Matinee Suncoast Motion Picture Co.

marked—literally, engraved—as a potential object of intergenerational memory. Nominal identity is also exploited by Names Unlimited, which offers certificates of the origins of first names, and the Historical Research Society, which provides certified genealogies for family names. These commodities trace the cultural memory of individual identity captured in names, and promise to reveal the essential meaning of the sign of oneself, perhaps even to realize the ultimate immanence, that is, unification of signifier and signified, recalling that original moment of seminal and semiotic origin, when the action and word of the Father conceives bodily and linguistic identity. The essence of self is also revealed and captured for posterity in Lifetouch portrait studio, or for amusement in the form of caricature by various "street" artists such as Lee's Gallery.

Enchantment through anthropomorphization occurs in masquerade widely deployed in carnival characters, employed to promote products and entertain passers-by in the public spaces of The Mall. Enchantment through animation includes: the full-size mechanical moose in Camp Snoopy, "fabulous animation" such as Tracy the Talking Tree in Rainforest Cafe, the robots and clocks of LEGO Imagination Center, and numerous battery-powered toys found in science and toy stores such as Cuddly Critters. It continues in cartoon theme stores such as the Disney Store and Warner Brothers, in giant Peanuts characters that decorate The Mall, in cartoon characters for sale as toys and on clothing design, and in the puppetry in the Rainforest Cafe and The Amazing Space. Stuffed toys and dolls are sold in a remarkable number of stores—especially old-fashioned and handcrafted forms, mass-produced cuddly incarnations of Africa megafauna, and cartoon characters—and they decorate commodity displays, signifying combinations of innocence, aura, and enchantment of the object world. Bears, particularly, seem to be everywhere—and this is not to exaggerate. They stand at doors welcoming visitors into stores and sit cutely on commodity displays (Figure 6). These objects bear powerful multiple meanings: they fit with the regional theme (most bears are manifestly black, or less appropriately, grizzly), they are signs of wild and fugitive nature, they are the archetypal cuddly toy of childhood, and their manufacture is a craft. More than a dozen separate stores sell handcrafted bears, making much of the aura of craft labor. A poster for one manufacturer, for example, announces affordable "antique reproduction," hand-cast and hand-painted ornamental bears, which "combine aesthetics, craftsmanship, and whimsey [sic]," and advises us to "always look for the pawprint [sic], the symbol of authenticity" (Boyds Bears and Friends 1995).

The enchantment of the object world is also brilliantly evoked in features such as Americana Carousel, a "hand-painted contemporary version of the classic American merry-go-round," in Camp Snoopy. Carousel horse ornaments and toys are dominant in window displays in stores such as American Pewter and San Francisco Music Box, and they also commonly decorate candy stores and gift shops (for example, there is a full-size

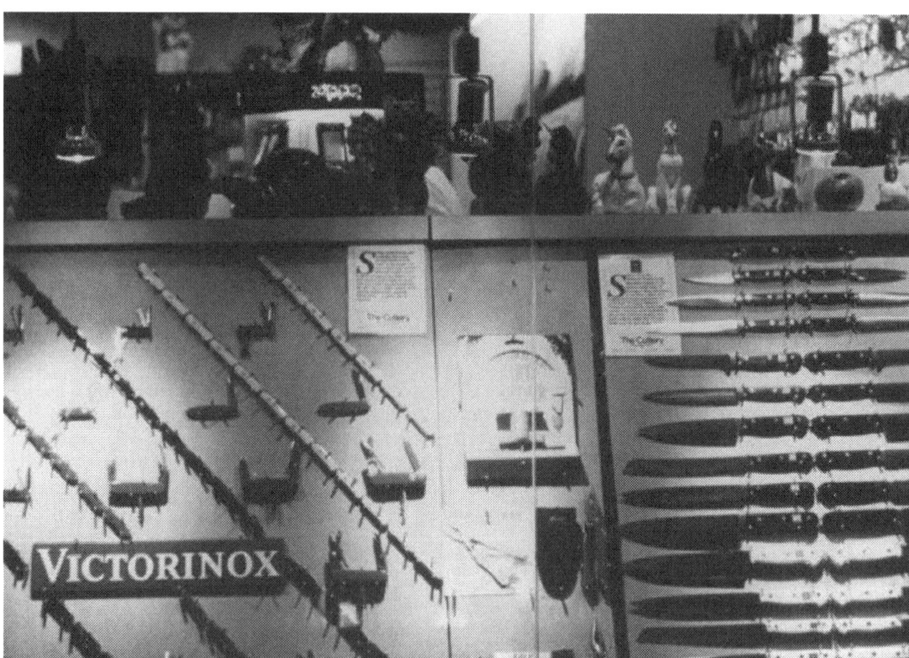

Figure 5. The contiguity of magic objects and mundane comodities on display.

Figure 6. Bears as signs of authenticity in commodity displays.

image of a carousel horse made from jellybeans in Peppermint Patty's Sweetshop). Such objects quote childhood stories in which innocent faith brings toys to life (Goss 1993), a trope remarkably explicit in the exhortation made by Moments and Memories, a gift shop in Camp Snoopy, to, "Remember when you were a child. It's all about the fantasy-filled feelings so often forgotten. Remind yourself and become a child again."

The magic of the movies is evoked to effect fantasized transport in The Mall. Promotional materials make much of the fact that movies such as *Jingle All the Way* and *Mighty Ducks* have been made in The Mall itself, and there are regular live broadcasts of radio and television shows. Brochures promise scheduled "celebrity appearances" and even an authentic personal contact with the aura of fame as "big name stars drop in just to see The Mall" (MoA 1997b). Stores and restaurants incorporate movie imagery, music, costumes, and action into their settings, sometimes explicitly applying a "movie treatment," hence cardboard cut-outs of movie characters in store windows, video monitors playing movies, licensed movie products, and movie star endorsements of products and retail concepts. Shop assistants act like actors: in The Train Store, for example, they wear caps and scarves reminiscent of locomotive engineers in the style of Casey Jones, while waiting staff in Tall Timber Grill are appropriately attired and introduce themselves to guests by announcing, "My name is John, and I'll be your lumberjack this evening." Several stores specialize in movie memorabilia (Suncoast Motion Picture Company, Studio Store, and Planet Hollywood). Makeover stores and "trick" photographers, from Glamour Shots to Foto Fantasy, and Glamour Kids to Face Painting, flatter adults and children with fantasies of their own discovery, transportation to stardom, and prospects of immortality through the filmic image, and there are several automated video and sound booths that record your own dancing and karaoke efforts. For a modest fee, one can become a "Guest Star" at The Mall with special discount privileges. Finally, given that the retailing of apparel and footwear is the bread and butter of The Mall, it is not surprising to find them explicitly connected to magical transformations of identity in the theme of dressing up, from The Amazing Space to Old Time Photography.

The photograph is the perfect souvenir because it re-*presents* an incontrovertible material trace of past and distant times and places. To photograph an object-moment is automatically to invest it with the pathos of loss (Sontag 1977). The photograph is literally a "still life," a dying moment photochemically immortalized. The photograph is also an important commodity, of course, and here, contrasting with some other shopping centers of my experience, visitors are "encouraged to photograph and videotape . . . scenes for their personal enjoyment" (MoA 1997e). The Mall hosts numerous stores offering rapid developing and finishing of colorful reminders of your trip, and automated kiosks vend instant images of fun. Camp Snoopy offers computer snapshots of your ride on the Ripsaw Rollercoaster. Knott's Old Time Photography ingeniously doubles the pathos of photography, the store's mise-en-scène taking one back to a time when photography itself was a craft, and its sepia-tinted images of the customers in "Wild West" or "Victorian Era" clothing and settings provide proof of your imaginative presence in the ideal past. While female customers dress in crinolines and carry parasols, the males wear military uniforms and sport weapons (in the style of the U.S. Cavalry or The Civil War), recreating times when femininity and manhood were more "real" and "natural," perhaps where women were ladies and men of war were heroes rather than victims.

Finally, we are asked to imaginatively transport ourselves: brochures, for example, exhort us to close our eyes and hear the roar of the crowd in a famous 1967 ballgame at Met Stadium, or to feel the romance in the Rotunda, where an average of ninety-two couples get married each year during a live radio show (MoA 1996a:5). Rainforest Cafe is described as a "magical wonderland" (Rainforest Cafe 1997), Camp Snoopy is an "enchanted wonderland," and The Mall "can fill days with adventure and imaginative escapes" (MoA 1997f). Not all transport is to the realm of innocence, however, and darker fantasies are satisfied in a section of Camp Snoopy's Mine Ride called Superstition, a simulation of fire, jungle, and a bat-ridden haunted cemetery ruled over by Elvira, Mistress of the Dark. Even Santa Claus has darker desires, according to the Victoria's Secret window, where he is surrounded by "Christmas bunnies" and accused of "secret fantasies," an image that neatly plays on the dialectics of innocence-knowledge and adult-child upon which imaginative transport depends.

If erotic potential is channeled into safe, heterosexist fun, The Mall, like the arcades, contains potential for the eruption of surplus libidinal energy. The "illusion" of public space and festivity has indeed been taken by sex professionals as an invitation to bring their trade into malls across America, and The Mall's older sibling in West Edmonton, Manitoba, has larger than life-sized statues of police arresting prostitutes on its imaginary Bourbon Street in New Orleans, enacting the repression of commodified sexuality as it acknowledges its presence. Teenagers and young adults form unpredictable crowds and, particularly under the influence of alcohol, threaten real carnivality. In The Mall, the fourteen-screen cinema and the Upper East Side entertainment district, consisting of bars and comedy clubs, is located on the fourth floor and intensely monitored to isolate and control disruptive sexual and social behaviors, but rowdy carousers still spill over into the food courts and main malls. As we shall see in the following discussion of the realms of Nature, Primitiveness, Childhood, and Heritage, as The Mall evokes imaginative transport through memory and magic for positive effect, it must also engage in constant vigilance in case the negative moments of alterity erupt.

Nature. As a trope for an originary innocence and contemporary scarcity (MacNaughten and Urry 1995: 211-12), nature plays a central role in commodity aesthetics, and the gap between nature and culture is vital to the discourse of nostalgia (Stewart 1984: 23). The Mall thus re-*stores* nature, and enacts its reconciliation with culture under the sign of the commodity. It evokes originary nature in the form of a bountiful pastoral that contrasts with the degraded environment of the modern urban world, reproducing an imagined paradise with subtropical temperatures and "organic pods," as the designers refer to islands of evergreen plants, colorful flowers, gently flowing waters, and natural light. The Mall seems to contain the whole natural world in its temperate forests and lakes, tropical forests, and oceans: UnderWater World, for example, invites you to "totally submerge yourself in a whole different world" (MoA 1997g), individual headsets instructing visitors on the productive ecology of riverine and marine ecosystems as they literally walk through (via an acrylic tube)"Minnesota to the Mississippi" and "The Gulf to the Caribbean"; in Rainforest Cafe, "you'll be immersed in a tropical wonderland with . . . whimsical butterflies, crocodiles, snakes and frogs, trumpeting elephants and other spectacular wildlife; all moving within the surroundings of larger than life banyan trees, with sounds and aromas of a tropical forest" (Rainforest Cafe 1997); and Camp Snoopy is a "natural wonderland . . . of babbling brooks, quiet ponds, [and] shady forest paths" (MoA 1997f:n.p.).

There are real animals for petting in Nature's Wonders, Wilderness Station, and Wilderness Theater ("featuring live animals from around the world"), and many stores specialize in animal-themed gifts and collectibles: the long list includes retail concepts from Butterflies to Wags and Whiskers, and from Dogs and Cats to Wild Bird (Table 4). There is literally an "A to Z" of nature stores, from Animal Lover to The Zoo Store. Stores evoke pristine and mysterious nature with names such as Forever Green, Natural Wonders, and Rhythms of the Earth, and sell, in particular, ornaments and paintings that incorporate images of the mammals and birds of North America, including buffalo, caribou, loons, eagles, and, of course, the ubiquitous wolves and bears. Interpretive texts highlight the fragility of nature, particularly evident in themes such as Precious Earth and The Endangered Species Store, the latter with a brilliant product line called "Extinction is Forever." This slogan appears on beautifully printed T-shirts with images of fugitive species, perfectly deconstructing the

Table 4. Themes of Nature

Theme	The Mall	Attractions	Retail/Restaurant Concepts
Natural Environment	Organic pods EcoFoundation	Camp Snoopy Northwoods UnderWater World Four ecosytems Rainforest Cafe Reef and rainforest Menu items	Concepts Into the Woods Zoo Store Forever Green Rhythms of the Earth Evergreen Mint Nature's Wonders Natural Wonders Nature Co. Earth Wear
Animals		Camp Snoopy Mooses, bird noises Wilderness Station WildernessTheater UnderWater World Marine/riverine life Rainforest Cafe Resident parrots Reef fish Stuffed animals UnderWater World Marine/riverine life	Concepts Wags and Whiskers All About Horses Wooden Bird Butterflies Nature's Wonders Big Dog Sportswear Dogs and Cats Wild Bird Objects Cuddly toys and gifts Welcoming bears
Fugitive species		Camp Snoopy Exotic petting zoo UnderWater World Albino specimems, turtles Rainforest Cafe Various and numerous	Concepts Wild Wings Endangered Species Objects and images Wolves, bears and loons "Vanishing Wild"
Stewardship	Recycling Environmental awareness Contributions	Camp Snoopy Environmental awareness Contributions UnderWater World Environmental awareness Conservation Rainforest Cafe Environmental awareness Conservation and captive breeding Contributions	Concepts RecyKIDables Junk Yard Ragstock Objects Body Shop's products Endangered Species products

nature-commodity aesthetic: buy one before it goes out of fashion! Similarly, UnderWater World boasts rare albino specimens and a Kemp's Ridley, "the most endangered species of sea turtle on the planet" (MoA 1997b).

In the nature stores, consumers purchase not only commodified nature, but an educational and spiritual experience, and a natural way of being. The plaque on the Nature Co. store, for example, explains that "Our goal is to provide you with products, insights and experiences which kindle your own sense of wonder and which help you feel good about yourself and the world in which you live." A piece of nature comes with the peace of mind that nature is somehow not exploited. We are told of the contributions both the individual stores and The Mall, through its EcoFountain coin collections, for example, make to environmental causes, and there are recycling bins throughout, so that the consumption of nature itself seems to contribute to its preservation. The trope of primitiveness echoes here in the form of natural relations with nature, or stewardship, and also travel, for ecoshopping, like ecotourism, is underpinned by a perverse "dialectic of love and money" (Haraway 1989:265). And fetishism is obviously at work: we purchase souvenirs made from commodified nature that deny their nature as commodities.

The Mall, like the dioramas of the nineteenth-century arcade, attempts to render a perfect imitation of nature through technical artifice (Benjamin, cited in Frisby 1985:242). Interpretive texts tell us of the scale of the endeavor, the immense investment of capital and technology involved in the authentic reproduction of nature in the main attractions: "Fun Facts and Quotables" tells us that in Camp Snoopy, for example, there are 1.2 miles of skylight, 55 semitruck loads of dirt, 30,000 live plants, and a "realistic" nine-foot graduation to the landscape (MoA 1997h); Rainforest Cafe's (1997) "Passport" describes "dazzling special effects: cool mists that permeate through cascading waterfalls; gentle tropical rainstorms; thunder and lightning . . . ;" and Under-Water World's souvenir guidebook describes its construction—four million pounds of dirt, six million pounds of concrete, 400,000 pounds of sand and gravel, several hundred tons of thick Japanese-made acrylic costing $750,000 (Under-Water World 1995:4). If Walter Benjamin saw nineteenth-century arcades as "human aquariums," suggesting that arcades are window displays of the dream world of commodities, while aquariums are display windows onto dreams of a non-commodified world, The Mall brilliantly combines both forms and processes to preserve life and meaning in nature and the commodity world from the forces of extinction and obsolescence. Crocodiles and sharks feature prominently in attractions such as Rainforest Cafe and Under-Water World (which promotes gift certificates using a remarkable image of a gift-wrapped shark), for they symbolize not only the danger and excitement associated with large predators, but the long-term survival of form, function, and meaning. But the epitome of both excitement and longevity is the dinosaur, for certain the most prevalent beast in The Mall.

Popular culture is obsessed with the dinosaur, and rightly so, not just because they are perfect monsters, nor due to the contingency of a strange marketing hype (Gould 1991). Part of the fascination is undoubtedly the function of their fate, as a hopeful allegory for human politics suggesting the inevitable decline of large-scale maladapted institutions (Montgomery 1991:23). More important, however, they allegorize the fate of the commodity, for the dinosaur both naturalizes the obsolescence of form built into the social product and yet maintains a dream of resurrection through faith in mythical nature, the rationality of science, and the aesthetics of the commodity. The dinosaur is thus the perfect sign of the tragic extinction of life and meaning in nature, and of its potential resurrection through sophisticated cultural technologies—the painstaking efforts of archaeology, the glamour of animatronics, or the fantasy of cloning fossilized DNA sequences—operating under the ubiquitous sign of commerce. The dinosaur is the natural equivalent to the antiques and collectibles on display in The Mall, and large-scale animated dinosaurs feature in LEGO Imagination Center (Figure 7), small-scale toys and accessories dominate toy stores, and there is even a retail theme called TEE REX. That life lives in inert nature is also suggested in skeleton models in various toy and educational stores, themes such as "Living Stones" (small ornaments of domestic animals sold in Camp Snoopy General Store), in the Nature Co.'s "Spirit Rocks," and presumably also in the new store called Fossil, coming to The Mall at the time of writing.

These themes echo remarkably with Walter Benjamin's characterization of the love-object's fate with the loss of aura under modernity, that is, its "petrifaction" in the "dream of stone." This describes a dual movement in which the concealment of the progressive work of death beneath superficial and artificial beauty produces a morbid fascination with the hidden object—hence the "erotics of decadence" in the skeleton and fossil (and also in treasure, antiquities, and antiques)—such that desire becomes impotent and perverse, desiring as much its own death as its realization (Buci-Glucksmann 1994:100-01). Put differently, and more simply, in order to waken the "spirit" of the past that is the object of nostalgic desire, the narrative must first have it die, and it is through the continuous play of its death that desire becomes eternalized (Stewart 1984:143). Extinction constantly threatens and is enacted in these "lost objects," but what is dead comes alive fleetingly in display and the imaginative act of acquisition: in the narrative of commodity aesthetics, life is forever being literally re-*stored*.

Figure 7. Petting monsters at the LEGO dinosaur exhibit.

Benjamin argues that the dream of awakening from petrifaction, from the temporary sleep of death, is the manifestation of a progressive desire for social awakening from the dream sleep of capitalism, a theme I will take up in the conclusion. Nature and death can never be tamed, of course, and their surplus "realities" escape The Mall's control. Dinosaurs are frightening monsters who threaten revenge for their premature extinction of life and meaning, and while some children pet the animated monsters, others cry at them. At Rainforest Cafe, children constantly must be told not to touch the parrots for fear of falling in the water or being bitten, and surely, at some point, a persistent child will be pecked by a beak that we are warned can break broom handles. Artificial nature is maintained at tremendous expense, but eruptions of real nature are impossible to contain—real wild birds that no expense can eliminate have made their homes in the trees in Camp Snoopy, for example, and a scandal was caused when hundreds of fish were accidentally killed by the aquarium. Animal activists have also highlighted the contradiction, protesting the sale of furs in Macy's.

Primitiveness. In the dominant modern imagination, preindustrial peoples are modernity's children, closer to nature and their natural selves, and freely exhibiting values and behaviors that civilization has repressed. Although it has powerful negative associations in the form of heathenism and barbarism, the positive image of primitiveness in colonial discourse carries considerable weight in the commodity aesthetic, where it is associated with stewardship relations with nature, a "natural" economy, and innocent faith in the animate powers of the object world. Like nature, primitiveness is endangered, which of course enhances its value, and like the tourist, consumers are belated travelers motivated by nostalgia for the natural world of the exotic other, seeking the signs of its authentic presence precisely as it ceases to exist (Behdad 1994). The exotic commodity is a souvenir of the disappearing exotic, "a bit of the other" (Table 5).

Like that between nature and culture, the gap between the primitive and modern can be bridged by science and technology, as may the contradiction between the economies of colonialism and exotic desire be potentially resolved in the

Table 5. Themes of Primitivenes

Theme	The Mall	Attractions	Retail/Restaurant Concepts
Native Americans		UnderWater World Native use of resources Chrysler Jeep Cherokee	Concepts Pueblo Spirit Wild Wings Objects Native American artifacts Sculpture and paintings
Native peoples		Rainforest Cafe Menu items	Concepts Alpaca Petes Objects: Body Shop's products Native handicrafts
Exoticism	Tropical plants/settings	Rainforest Cafe Menu items Tropical setting	Concepts Banana Republic Limbo Lounge Tropical Treasures Timbuktu Calido Chili Peppers Cacique Raising Cane

appropriate form of commodity aesthetic. This occurs remarkably in The Body Shop and Rainforest Cafe, where purchases of goods containing natural ingredients collected by indigenous peoples contribute to preservation of primitive traditions, and so appeal to consumers feeling guilty for their part in degrading the nature and ways of life of native peoples. The Body Shop (n.d), for example, claims to "dedicate [its] business to the pursuit of social and environmental change," and its (1995) Trade Not Aid program aims to "explore trade-based approaches to supporting sustainable development by sourcing materials from socially and economically marginal communities."

The existence of the gap is repressed entirely in the "imperialist nostalgia" of retail themes such as Banana Republic and Timbuktu Station, which seem to deny entirely the presence of the other except as a cipher for a narcissistic desire (Rosaldo 1989; Thomas 1994). Exhibiting a culture of amnesia, they reduce colonialism to a fashion show on tropical location, incorporating safaris and "Indiana Jones" archaeology, inevitably hunting big game and long lost power-objects from pre-Christian times. They also cleverly double alterity in the form of the past exotic, evoking a time of authentic travel that contrasts with contemporary tourism, such that the clothes and accessories sold are souvenirs of an authentic mode of the souvenir. The past and exotic are combined in themes such as Pueblo Spirit, which sells Native American artifacts, mostly Navajo, while the name evokes the Anasazi, the ghosts of North America's aboriginal people, and so promises souvenirs of prehistoric primitiveness.

The Native American is a powerful trope of primitivism and seems to stand simultaneously for the death of both nature and culture. Native Americans are "wild" like the buffalo, eagle, and wolf, and fugitive to the point of extinction: Wild Wings, for example, sells images of Native Americans alongside endangered animals, while in gift and art stores, there are paintings from which they stare in pathos like the wolves, and statuettes in which they raise their heads and hands in lament. In store displays, the Indian stands, like the bear, for welcome and authenticity: statues of very wooden "Indians" stand at the door of the Old Wood Factory, where, like everything else, they come with a price. Elsewhere there are native American artifacts, such as jewelry, authentic hand-crafted of course, and reproductions of weapons such as bows and arrows and tomahawks. There are at least two interesting parallels here: one with the authentic war depicted in Old Time Photography, and the other with the axes in the timber camp. Native American violence is primitive, and the hunter-warrior (there are no females at all to be seen) takes biological life, but leaves the spirit essence. He kills as he needs, not needlessly kills: the Native American stands for an authentic way of being in the world. None of the stores that sell "Indianness" are owned by

Native Americans, however, and I came across only one Native American retail assistant in The Mall.

Finally, primitiveness is exploited in two other ways. First, in the exotic names of menu items in Rainforest Cafe (such as The Congo Mogambo and Rasta Pasta), and although not, strictly speaking, primitive, the ethnic foods sold in food courts and stores, such as Calido Chili Traders, promise cannibalistic delights in eating a bit of the other (hooks 1992). Similarly, the use of foreign languages in retail themes, from Bellas Cosa to Zhivago Rugs, and including the tongue-in-cheek Maison du Popcorn, give pleasure of distinction to those who can pronounce them and know their meaning, while further satisfying an "oral fixation" with alterity. Second, primitive sexuality is evoked in stores such as Limbo Lounge, which uses a "natural" eroticism to sell clothes.

Despite the exploitation of primitiveness, what dominant culture takes as the negative manifestations of primitiveness are rarely acknowledged in The Mall. Remarkably, even though the Rainforest Cafe boasts larger than life "Magic Mushrooms," no mention is made of the widespread use of psychoactive plants in the Body Shop or other stores that source medicinal materials in the rainforest. If "bad" primitivism is repressed, the "worst" primitive seems to be the one who is not different enough, those who would cut down the rainforest to get their own bit of the other, that is, access to the commodity world. There is a sustained attack in the discourse of the nature stores on those who commit the ultimate sin of betraying their own primitive nature by destroying the forest, and so undermining the nature of the social relations that structure the contemporary commodity world (Ramos 1994:80; Thomas 1994: 30). The consumer who purchases plastic reproductions of "The Vanishing Wild" is strangely innocent of any part in the problem.

Childhood. In a corporate state, a place must be made for innocence, and its many uses. In developing an official version of innocence, the culture of childhood has proven invaluable. Games, fairy-tales, legends from history, all the paraphernalia of make-believe can be adapted and even embodied in a physical place (Pynchon 1974:488).

A persistent cliché in the critique of consumerism is the intellectual immaturity of the masses, whether in childish surrender or childlike insubordination to the exhortations of the consciousness industries (Marcuse 1964; Baudrillard 1985:588), and although I agree with the politics of neither assessment, the child does indeed provide an appropriate model for the consumer in The Mall. The child, of course, is not yet modern, still inhabiting the time of myth, and with primitiveness, is the ideal "subject supposed to believe," the other whose capacity for enchantment opens up a space for the rational adult, through the mechanism of transference, to experience faith in the animation of the world (Žižek 1997:107). While The Mall has something for everyone, from baby showers to health programs for the elderly, it is foremost a place of family fun organized around the experience of children—affectionately referred to as "Mall Rats" in promotional materials—and their capacity for enchantment (Table 6).

Like the arcades erected during the childhood of modernity, The Mall reproduces fairy tale (Buck-Morss 1989:317). Transformations of scale occur in both gigantism and miniaturism: The Mall boasts of its vastness and superlative scale, and there are giant Snoopy characters (Figure 8) and bears, a 19-foot-tall Paul Bunyan and a 22-foot-long Babe the Blue Ox (MoA 1997h). Attractions and retail themes contain models of the earth floating in space or reconstruct entire ecosystems, reflecting as I noted at the outset, The Mall's drive to totalization of historical and geographical experience. A brilliant example of the latter is provided by a commodity prominently displayed in toy shops and education stores: the Atlasphere made by Odyssey is a computerized globe with which you can "travel the world without leaving home," study "politics without border disputes," and go "around the world in an evening" (Odyssey 1997).[7] Convolutions of space are effected in labyrinths and mazes: The Mall contains miles of passages, even organizing "team building" scavenger hunts for those who want to explore them (MoA 1997a), and there is the "Where's Woodstock?" Adventure Maze in Camp Snoopy and the tunnel maze in The Amazing Space. Stores sell collectible figurines of characters from myth and magic, lending their aura of magic to commodities, while fairy-tale dramas are enacted in The Mall's rotunda. There are kindly tree spirits in Rainforest Cafe and FAO Schwarz, and in Camp Snoopy, there is Paul Bunyan, the lumberjack, or North American version of the woodcutter, who is the archetypal "rational" figure of the fairy tale: whether poor honest father or handsome hero, he invariably rescues

Table 6. Themes of Childhood

Theme	The Mall	Attractions	Retail/Restaurant Concepts
Family fun	Events	Camp Snoopy	Concepts
	Family fun	"World of family fun"	Albert's Family Restaurant
		Picnics and birthday parties	Filene's Family Fun Days
	Services	Family Fun Night	
	Lost children		
	Diaper changing		
Child's play	Decor	Camp Snoopy	Concepts
	Carousel horses, toys and kites	Camp SnoopyToys	Simon Sez
	Stuffed Snoopy characters	Cartoon characters	Amazing Space
	Cartoon characters	Masquerade	Mega Play Zone
		Peppermint Patty's	Air Traffic Control
			Great Train Stores
			Al's Farm Toys
		Rainforest Cafe	Funcoland
	Events	Entertainment for kids	Giggles Etc.
	Easter Bunny		Kids 'R' Us
	Santa Claus		Children's Place
	Halloween	LEGO Imagination Center	Cuddly Duds
	Story telling	Play areas	
	Pantomime	UnderWater World	Objects
		"Sleep with the sharks"	Toys, games
Education	Student Scavenger Hunt	Camp Snoopy	Concepts
	Student Education Tours	School Spectacular	Brainstorms
		Park and Recreation	Store of Knowledge
	Events:		Amazing Space
	VISA Tell Me a Story	LEGO Imagination Center	Scientific Revolution
		LEGO Workshops	Learningsmith
		Rainforest Cafe	Objects
		Educational products	Educational toys and books
		School outreach	

innocence from the forces of darkness, his "axe of reason" cleaving the primeval forest and admitting the light of reason (Buci-Glucksmann 1994:109). Giant axes decorate Camp Snoopy, while larger than life and somewhat sinister "creatures of the northwoods" stalk innocent woodsmen and visitors (Figure 9). Further evidence of the dark side is provided by gargoyle ornaments, and the "evil" characters of Disney cartoons and Star Wars. For example, Camp Snoopy's water show is choreographed to the theme music of Star Wars, and cardboard cut-outs of Luke Skywalker and Darth Vader, among others, stand in store windows of a magic shop and science store, as well as the more obvious video stores. Shoppers are presumably meant to feel "the Force."

Observing families shopping, it is obvious that to a considerable extent, children's desires lead parents in The Mall. The Mall is family friendly. It seems to work in part because adults are nostalgic for childhood trips to public places and because they want to share themes of their childhood with their children. The Mall explicitly encourages adults to regress and relive their own wonder at the object world. A brochure says: "If the child in you is struggling to be free, just wait until you stroll into the largest—and most remarkable—indoor theme park in the country" (MoA 1997f). The memory of childhood lost is obviously also evoked in the theme of the Victorian nursery, innocent play captured, for example, in the mechanical toys of Wind Me Up and Camp Snoopy Toys, which is designed as an old-fashioned toy factory. Adults endearingly display some rather childlike behavior in The Mall, from the men left "playing LEGO" with their children while their wives/mothers go shopping, to those screaming and making faces for the computer cameras on the Camp Snoopy rides. More subtly, the promotional rhetoric of The Mall and the retail concepts of its tenants suggest that, though easily pooh-poohed by modern critics, and repressed by adult consciousness, the fairy tales of consumption exert a powerful hold over the collective imagination. Gorgeous gowns and the perfect slippers still have the capacity to transform

Figure 8. Gigantism in Camp Snoopy.

Figure 9. The Hudag, one of the mythical beasts of the northwoods.

Cinderella into a princess, and hence the jingle on the promotional video that goes, "Who told you you can't have it all. . . . At Mall of America you'll have a ball" (MoA 1996b).

The rationality of adulthood is acknowledged and appeased through articulating fun and commodity consumption to educational purpose. I am struck by the heavy pedagogical messages in the nature and science stores, and the extraordinary technical details The Mall provides on its construction and operation, which I am told appeal particularly to men. The Mall offers college credits in an extension of the St. Paul campus of National American University, high school credits through Metropolitan Learning Alliance (Urban Educator 1995), and educational daycare in The Amazing Space. The numerous science stores, such as Learningsmith and Store of Knowledge, demonstrate instructional materials for children, and The Mall promotes educational goals in regular special events. Many of the toys and texts have undoubted pedagogical value, but they also teach valuable lessons on commodity aesthetics: Brainstorms and Scientific Revolution, for example, show that natural science produces cool toys such as scary skeletons and glow-in-the-dark night skies; in Rainforest Cafe, we learn that nature is endangered, but that twenty-five cents can save 1/100,000th of an Amazonian jaguar; and Snoopy tells us the centrality of consumption when he says that Christmas "is all about getting gifts." The prize, however, must go to VISA Read Me a Story, which uses traditional and modern fairy-tale characters to tell its story, brilliantly combining adult rationality and childlike wonder in an entertaining spectacle. In this modern pantomime, Peter Pan, "the boy who never grew up," sadly cannot read and so cannot enjoy fairy tales. Fortunately, he meets friends who sing lots of songs while taking him to Mother Goose's schoolhouse where a story is read. It is a multicultural show, and the manifest moral lesson for adults in the audience is the importance of reading to children, while for the children, it is the importance of reading. Another lesson, however, is about corporate philanthropy and socially conscious consumption, and the reconciliation of enlightened rationality and innocent enchantment through commodity consumption, since the fairy-tale spectacle and books given to children are donated by Citicorp in proportion to credit-card purchases. The VISA logo is prominently displayed over the set: it is the alchemy that transforms plastic into gold, and the spell that opens doors into the fairy tale of the commodity aesthetic. In this enchanted world, consumption literally creates a world of fairy tale, and we will all live happily ever after.

Fairy tales in The Mall are stripped of contradiction to become entertainment, and their characters, freed from inner conflicts, become caricatures. For example, one lesson of Cinderella regarding the obsolescence built into magic objects seems to have been forgotten. Still, it is not all innocent fun, and alternative moral lessons could be drawn from the text made available, perhaps in alternative performances by storytellers and artists. I saw no evidence of this, however. On the other hand, children are not always willing to be the adult's image of innocence, to be seduced by the fantasies of fairy tales or taught by educational texts and toys, and I have witnessed the highly publicized problems of unchaperoned teenagers, disrupting tenants and shoppers and being disrespectful to security guards (McCune 1997:3). The Mall introduced a "Youth Liaison

Program," in which security guards wore sweatsuits and offered outreach to public schools, and then the "Mighty Mom Program," in which local mothers of teenagers patrolled with security guards, later supplementing this with "Dedicated Dads." These programs attempted to replicate parental discipline but to little avail. The repressed other of childhood innocence persistently returned, and ultimately The Mall introduced its "Parental Escort Policy," which denies entry to those under 15 unescorted by an adult on Friday and Saturday after 6 pm.[8]

Heritage. Although the "heritage industry" evokes various nostalgic chronotopes, the retail built environment is dominated by that of the Victorian era, a time associated with the simple virtues, stable identity, and sense of public purpose of America's halcyon days.[9] Ironically, as the contemporary city finally "shakes off the last traces of its nineteenth-century self" (Hubbard 1994:26), it is recreated in The Mall in images of commercial capitalism, the birth of consumer society, and the expansion of travel and commerce, but more particularly through institutions of urban exhibitionary complexes: museums, aquariums and zoos, arcades and department stores, and expositions and world fairs (Bennett 1995).

The Mall incorporates forms associated with the economic and aesthetic revolution of early modernism, most obviously the arcade, but also forms such as the Victorian railway-station theme of West Market, and the Victorian-era mining and lumber camp of Camp Snoopy (Table 7). Antique waymarkers, "Ye Olde English" spelling and fonts, and names of stores from Victorian Sisters to Victoria's Secret can be read literally as signs of the past. Biplanes and steam trains in Camp Snoopy rides, and "authentic reproductions" of antique vehicles such as streetcars, velocipedes, and butterfly carriages in tenants such as A Simpler Time and the Museum Store, evoke transport in and to the past; carriage clocks and grandfather clocks in gift stores such as Reminisce and Best of Times tell today's time and tell of past times; clocks incorporating northwoods scenes and animals in stores such as Wooden Bird tell (of) natural time (Figure 10); and the many handicrafts stores from Allies' Market to Tiffany Connection that sell quilts, stained glass, pressed flowers, and hand-made paper, for example, evoke the pastimes of past times. A flavor of the past is found in stores selling old-fashioned candy, cookies, and ice cream, from Kemp's Cookies and Cream to Ruby Tuesday, which offers "good times you can taste." Camp Snoopy General Store and the Old Farmers Almanac General Store trade on the past, evoking trade in the past.

As illustrated by the presence of the Museum Store, a "museal sensibility" (Huyssen 1995: 14) has penetrated The Mall, and the commodity aesthetic has incorporated the museum. And as in the museum, the collection of objects in The Mall forms a narrative that *presents* the meaning of the past. The antique tools and products that

Table 7. Themes of Heritage

Theme	The Mall	Attractions	Retail/Restaurant Concepts
Historical retail	West Market Pushcarts, kiosks	Camp Snoopy General Store	Concepts Old Farmers Almanac Allies market
Victoriana	Architecture Street furniture Wrought ironwork	Camp Snoopy Mining/lumber tools Bric-a-brac "Old-fashioned rides"	Concepts Rockport Shoppe Victorian Sisters Victoria's Secret Objects Authentic reproductions
Traditional values	Events Thai handicraft exhibition	Camp Snoopy Moments and Memories "Old-fashioned amusement" Chrysler "Timeless qualities"	Concepts Basic Brown Bear Factory Pottery Barn, Candleman Museum Company Field of Dreams Old Navy Objects Handicrafts, antiques

Figure 10. Natural time, past time, and fossilized time, in Cabin Fever.

litter Camp Snoopy and casually decorate retail and restaurant settings suggest the currency of old-fashioned values of labor and the persistence of historic labor value, while retail concepts such as Field of Dreams and One-on-One, which sell autographed sporting memorabilia, attest to the contemporary economic value of traditional sporting values. The potential convertibility of sentimental and monetary value is made explicit in Al's Farm Toys, whose window shows us that children's toys from the past are serious investments for big boys today, and the immortality of meaning and value is suggested by Tomorrow's Treasures, which sells stuffed toys and cute gifts. This conceit extends even to the automobile, arguably the product that best illustrates the working of obsolescence under contemporary capitalism, as Chrysler claims to preserve past values by bringing "timeless qualities to modern interpretation" in its homage to basic American forms (MoA 1997i). As in the museum, interpretive texts tell of the rescue of the aura of craft labor from the ignominy of obsolescence in the restoration of material objects and moral-economic values. The ghost of long-dead labor haunts The Mall, its surplus aura possessing the commodities on display.

The Mall organizes demonstrations of craft labor, such as banana-leaf boat making, silk spinning, fruit carving, and umbrella painting during its Thai "international extravaganza," and consumers are even invited to invest their own labor in making teddy bears at the Basic Brown Bear Factory or caps at ZAP. If this is unlikely to motivate them to take up crafts as a living, or to make their own toys and headgear at home, it could at least provide an opportunity to imagine the real conditions under which craft persons labor. For example, handicraft workers even in such exhibitions are unsentimental about their labor, talking openly about it. And the fact that Basic Brown Bear Factory is promoted as "one of the few remaining manufacturers of stuffed animals in the United States" (BBBF n.d.), even while almost all of its accessories are foreign-made, may potentially raise questions about the international division of labor. These retail concepts surely also reproduce ideals of traditional labor even as they exploit them, and potentially sensitize consumers to critical political issues. It may well be that consumers will still consciously choose cheapness of purchase over social responsibility, and so "vote" for inequality and exploitation (Miller 1995:8), but the evidence from relatively successful protests of sweatshop relations of production in garment and shoe industries does suggest a sensitivity to the "real" qualities of the product that fantasies of fetishism would obscure.

The Souvenir and Allegory

Tourist destinations and gift shops, aquariums and nature stores, fairgrounds and toy stores, art museums and art galleries, and history museums and antique stores all share the same relationship: they represent the articulation of collective dreams of authenticity with the means of its private appropriation. The Mall incorporates distant places and times associated with the imagined possibility of authenticity, but its visual and verbal rhetoric displaces desire from its "natural object"—unmediated relations with the world—onto commodities, which then stand as souvenirs of the dreaming experience. While critics find the consumption experience wanting for lack of fealty to the "real" authenticity of other times and places, they miss the point: *the commodity is not a souvenir of the real experience of a past or distant reality, but a memento of the retail experience in which its possibility is imagined.* The measure of consumption is not authenticity per se, but the capacity to evoke and represent our dream of it.

On the other hand, it is trite to observe the uniqueness of individual experience in consumption, precisely because even mass-produced commodity-souvenirs effect a condensation of the dreaming experience, linking personal biography and geography with the collective narratives of nostalgia. The Mall makes it particularly obvious, given its emphasis on tourism, souvenirs, and gifts, but the commodity is always personal. My Mall of America "Shop till you drop" sweatshirt and Endangered Species "Extinction is Forever" T-shirt, are *not* souvenirs of Minnesota, the northwoods, or an experience of nature, but of my investment in the narratives evoked at the scene of their acquisition. They are testaments of my consumer faith—which is why the slogans so often seem to proselytize—and the means by which cultural history moves into private time and natural history into personal possession (Stewart 1984:138–40). It is partly for this reason that marketers stress the specificity of the consumer-product relationship, and hence that Chrysler Corporation proclaims, "You are an individual," and that Levi Strauss claims that for their jeans, "Every rip tells a story."

The form of the commodity is not entirely arbitrary, however, for cultural meanings condense only upon an imagined material trace of authenticity, an essential "fragment of absolute reality" (Baudrillard 1990:39) that remains despite the perceived loss of authentic meaning. This trace operates as a fetish in the psychoanalytical sense: it substitutes metonymically for the missing whole, the impossible object of desire that is the reconciliation of the oppositions Culture and Nature, Past and Present, Self and Other, Word and World, and Signifier and Signified (Barthes, cited in Buci-Glucksmann 1994:141). The Mall's promotional materials even reproduce the discourse of sexual fetishism, describing, for example, "loving attention" to "adorable detail," this perfection of parts particularly important in the always "authentic" recreation of nature and reproduction of heritage (MoA 1997h). Again, we can see the allegorical function of objects such as the dinosaur, reconstructed in its monstrous entirety from bone fragments or DNA sequences, and antique tools, restored from ruin to themselves, imaginatively construct lost objects and worlds of value.

The displacement of desire is never complete, however, and like the sexual fetish, the commodity simultaneously denies and affirms the lack of the real, reproducing the tension between imagined immanence and felt alienation that structures the field of modern desire (Benjamin 1985:55; Stewart 1984:140). Like sexual fetishists, consumers know very well the inadequacies of the commodity, but act as if they did not. This paradox is the origin of distraction and reverie, or the fact that, again like tourists, consumers typically effect a fantasized dissociation from the material enactment of consumption even as they take pleasures in its spiritual enchantments. As rational actors, consumers know very well that the object's effect is not real, and yet to imaginative beings, this is exactly how it seems. The child is vital in The Mall whether as the "subject supposed to believe" or the "inner child," the object upon whom adults project their contradictory desires.

Also like fetishists, however, shoppers attain no sustained satisfaction. Desire and pleasure depend upon the gap between the object and its referent so the commodity aesthetic breaks down when the product is taken out of context. Without the narratives that sustain nostalgia—when it is displayed at home as opposed to the shop window—it loses its luster, and we all surely recognize the lie that "wearing a Snoopy T-shirt is guaranteed to keep the child in you from disappearing long after you've left the park" (MoA 1997f). The T-shirt will remind you, however, of your desire to be a child again, to be of an age or live in an age that innocently accords power to material objects.

According to Benjamin (1985:54), the souvenir is the key figure in allegory, and like this defunct literary form, The Mall similarly offers an antidote to the perpetual ruination of nature and the meaninglessness of the social world (Wolin 1982:72). Youth and childhood are celebrated in The Mall, and the threat of old age and death is displaced through memory and bodily rejuvenation purchased in commodity consumption. A crude and cruel example is the much-publicized Golden Stars program for seniors, which incorporates promotional campaigns with health services. Senior citizens, if not at the peak of their physical powers are, of course, at the peak of their purchasing power (hence the name is metaphor for both age and value), and the program seems to prey on their fear of mortality. Sponsored by Blue Cross & Blue Shield, which operates a clinic in The Mall, the program includes mall walking and regular medical check-ups with promotions such as "Picture of Health," a demonstration of makeovers by Glamour Shots "that will bring out a fresher, younger-looking you" (Glamour Shots n.d.), and Shopping for Better Health, which promotes lifestyle products for seniors (MoA 1997j).

The Mall's very creditable recycling efforts (currently fifty-three percent of solid waste is sorted, processed, and recycled) also achieve prominent status in promotional materials, as the contradictory drives of Thanatos and Eros are reconciled under the sign of the commodity. The ever-present fear of absence of meaning in commodities, as in life itself, however, perhaps explains why, if otherwise complete in virtually every other feature of commodified public life, The Mall does not yet offer funeral services (Larsen 1997:2). It also begins to explain the appeal of the store whose name and retail concept seems pointedly at odds with the spatiotemporal themes I have identified, that is, the clothes store Fearless Factory, whose motto is "No Fear" and whose setting is an urban industrial garage. This is a clothing store for teenagers who seemingly evince an "in your face" attitude to death, even while they are inextricably bound up with its aesthetics, and, of course, that of the commodity.

Like allegory, the commodity aesthetic turns objects and persons into abstractions and bestows upon them its own arbitrary signification (sign value). But allegory is dialectical, demonstrating simultaneously a poignant appreciation of the transience of material objects and meanings, and a profound desire to redeem them for eternity (Wolin 1982:71): it exerts a will to make meaning out of the opaque materiality of the world, and confers upon objects "a power that makes them appear no longer commensurable with profane things, which raises them onto a higher plane, and which can, indeed, sanctify them" (Benjamin 1977:175). Anthropologists have observed that all cultures construct allegorical narratives or erect physical structures that both symbolize and attempt to contradict the irreversibility of time and the inevitability of death, and it is tempting to see The Mall as a modern temple where the past returns in symbolic structures and souvenirs, and where contemporary culture enacts ritual celebrations of the cycle of life, death, and rebirth. Like the nineteenth-century arcade, The Mall quotes elements of ecclesiastic architecture in its circles, arches, and cruciform constructions, and it incorporates the aquarium and the museum—mausoleums of natural and cultural history. The original meaning of consumption is progressive wasting and death from pulmonary tuberculosis, but in the contemporary commodity form, it simultaneously incorporates obsolescence and destruction with the dream of restorating of life and meaning.

The key to understanding the fetishism of commodities is not that we misperceive objects as reifications of social relations, or that relations between things displace relations between people, although this is true, but rather that complex relations between people and things, or better, people-things, are constitutive parts of the symbolic order (Žižek 1997:105). The fundamental principle is neither passivity nor activity, the manipulation or tactical resistance of the consumer, but interpassivity: we accede to enjoyment of The Mall and shopping-entertainment precisely because we can suspend our disbelief by projecting it upon others, and ultimately onto objects. There is no prospect of dispossessing us of objects and our collective faith in them, for this is our fate—the commodity-object stands for our subjectivity, and as inert material, for our own objectivity (Žižek 1997:115). In The Mall, "Objects 'R' Us."

Conclusion

> For a nostalgic to reach his or her goal of closing the gap between resemblance and identity, *lived* experience would have to take place, an erasure of the gap between sign and signified, an experience which would cancel out the desire that is nostalgia's reason for existence (Stewart 1984:145).

If The Mall fulfills vital religious and psychological functions in contemporary culture by bringing together collective myth and individual memory and fantasy, it also performs a modern secular function, for collective dreams are economic as well as spiritual and libidinal in content. That is, The Mall responds to the felt inadequacies of the social product and alienation in the social order of production, expressing deep fears we hold about the relationship between the alienated form of the object and the destruction of nature, culture, the past and the city (Benjamin 1985: 44–45). A critical-sympathetic reading recognizes that even as it systematically manipulates what Benjamin calls "the empathy of the soul with the commodity" (Benjamin 1982b:55 n). The Mall contains within it a critique of commodity aesthetics. The Mall mourns obsolescence of form and meaning, and laments the loss of innocence in consumption as it ritually enacts restoration and preservation. To merely denounce The Mall for its instrumentalism is to evince a failure of dialectical thinking and political will.

This is not to deny the commercialization of aesthetics and the privatization of public space in the retail built environment, but to reject the fatalism that ignores the political opportunities opened in its rhetorical commitment to universality, one that promises to contain everything and everyone in its whole world. It seems to me, in fact, that The Mall effectively makes public a private space and aestheticizes the commodity, and that like the museum, it presupposes and performs a public: it must appear inclusive and democratic in order to effectively symbolize community, and it must promote the spontaneity of crowds in order to evoke natural commerce of the marketplace (Duncan 1995:57; Slater 1993). Collective identity forged in consumption is kitsch, a self-conscious sharing of faith in material signs of a "real" community that is felt in its absence, but it also at least provides glimpses of imagined authenticity and keeps alive dreams of utopian possibility, a vital political function (Mannheim 1991; Bloch 1995). If, as Walter Benjamin argues, the dream sleep is the distracted state of the masses under commodity capitalism, upon awakening to a nostalgia conscious of itself, myths and symbols of what is lost will be vital to remembrance, and to the potential realization of real material needs and psychological desires of everyday life (Buck-Morss 1989:120).

The commodity aesthetic depends on fetishism, the condensation and fantastic reconciliation of contradictory desires in material objects, and the chronotopes of The Mall provide a context that contains desire within dominant narratives, locating the potential for authentic life in the imagined past-present and distant-here. The Mall, like allegory, represents the "dialectic at a standstill": it is a petrified landscape of frozen images and fixed meanings, of static oppositions and oxymoronic forms (Benjamin 1982b:171). If The Mall tends to stasis and totalization, however, the dynamic of the dialectic cannot be repressed, and I have shown how some contradictory meanings escape the dominant narratives. I believe that it is in such moments that consciousness potentially awakens, and so I advocate a progressive pedagogy based on a therapeutic approach. Montage and parody, in particular, can be vital in bringing together image and reality, past and present, and near and far in unintended and destabilizing fashion (Savage 1995:209).[10] I can attest from my own leafleting experience in local shopping malls that retailers are extremely vulnerable to simple tactics that highlight contradiction, for example, between the qualities and values of products in handicrafts stores and clothing outlets, or nature stores and fashion boutiques. The technical and semiotic perfection of illusion makes The Mall vulnerable to contamination and destabilization by "real" traces of the Real, such as specks of dirt and evidence of exploitative relations of production. On field trips to local shopping malls, my students learn for themselves how the commodity aesthetic depends upon a knowing willingness not to know, and they see for themselves behind the scenes of the collective fantasy. Such approaches are much more likely to reveal the monstrous nature of objects that the commodity aesthetics represses, that is, for example, disease and violence of the natural world, the cruelty of children, the barbarity of primitivism, and the phantasm of exploited labor in the artifact. The task is to recognize how the age-old fears of obsolescence—nature dies, children grow up, primitives are "civilized," and our heritage is lost—and dreams of immanence, live in the commodity aesthetic, not to eliminate them as so much "false consciousness," but to liberate them and live them more fully in really meaningful consumption.

Notes

1. It has been suggested by a representative of Jon Jerde Partnership, the architects of Mall of America, that this is perhaps the last gasp of the North American megamall—ongoing changes in development and retail presaging its ultimate demise (Magill 1997).
2. The gendered nature of this expression is only too appropriate since The Mall celebrates *mutterrecht* [the "maternal principle"] in its embrace of nature, its nurture of the child, its erotic community, and its womb-like architecture. The design even includes something the architect remarkably refers to as a "mother line," a serpentine pathway that links the mall's rectangular footprint and

lower floors with the third floor's circular form (see Figure 3).

3. And even if the Mall excludes the weather, it is a natural response to Minnesota's climatic extremes: according to associate general manager, Maureen Hooley Bausch, "When plans for Mall of America were announced, the overwhelming question was 'Why Minnesota?' but with the weather we have here, the answer seemed obvious" (McCune 1997:2). Thus, Phase II development will offer year-round ice-free fishing.
4. By the 1890s, for example, mass tourism was a significant source of retail profits in New York's department stores and arcades, and was an important motivation to travel to the city: "Commercial tourism was, in part, the tourism *of* commerce" (Harris 1991:66).
5. There is inevitably a long wait for tables at meal times, and parties of diners are given names with "nature" themes, so there are bizarre announcements like "Galapagos, your safari party of four is now leaving, please approach the elephant."
6. The Tourism Department offers a "VIP Marketing, Technical and Behind the Scenes Tour." For those without $50 to spare, there is also a virtual "Behind the Scenes" which describes the security monitoring, mail delivery, waste management and recycling, storage facilities, and details of aquarium maintenance (see "Behind the Scenes" 1997b).
7. The Atlasphere is also, alas, described as "the finest geography teacher you will ever meet" (Odyssey 1997).
8. At the time of writing, in response to representation from parents and teenagers, The Mall is considering establishing a youth center, a service that public institutions have failed to provide (McFarland 1997).
9. A minor theme in The Mall, for example, is that of the 1950s, a period of political-economic confidence and cultural optimism, typically articulated with teenage themes in stores such as Runkel Brothers, Johnny Rockets, Ruby Tuesday, Flashbaxx, and the sporting memorabilia stores.This era represents the last of America's youthful innocence before a cultural "coming of age" of the 1960s.
10. This might include what the Situationists, for example, referred to as *detournement* and *derive*, respectively the lifting of textual elements from their original contexts to make radically new meanings, and a mode of behavior involving vague wandering through and wondering at meanings in urban space (see Sussmann 1989).

References

Adorno, T., and Horkheimer, M. 1979. *Dialectic of Enlightenment*. London: Verso.

Agnew, J-C. 1989. A House of Fiction: Domestic Interiors and the Commodity Aesthetic. In *Consuming Visions: Accumulation and Display of Goods in America, 1880-1920*, ed. S. Bronner, pp.133–55. Winterthur, DE: Henry Francis du Pont Winterthur Museum.

BBBF. n.d. *Visit the Basic Brown Bear Factory*. Brochure. San Francisco: Basic Brown Bear Factory.

Baudrillard, J. 1985. The Masses: the Implosion of the Social in the Media. *New Literary History* 16: 577–89.

———. 1990. *Revenge of the Crystal. Selected Writings on the Modern Object and its Destiny, 1968-1983*. Trans. P. Foss and J. Pefanis. London: Pluto Press.

Behdad, A. 1994. *Belated Travellers: Orientalism in the Age of Colonial Dissolution*. Durham, NC: Duke University Press.

Behind the Scenes. http://twincities.sidewalk.com/link/11354. 1997. November 6, 1997.

Benjamin, W. 1977. *The Origin of German Tragic Drama*. Trans. J. Osborne. London: New Left Books.

———. 1982a. *Das Passagen-Werk*, ed. R. Tiedemann. Frankfurt am Main: Suhrkamp Verlag.

———. 1982b. Paris—the Capital of the Nineteenth Century. In *Charles Baudelaire: A Lyric Poet in the Era of Capitalism*, trans. H. Zohn. London: New Left Books.

———. 1985. Central Park. *New German Critique* 34:32–58.

Bennett, T. 1995. *The Birth of the Museum: History, Theory, Politics*. New York: Routledge.

Bloch, E. 1995. The Principle of Hope, vol. 1. Cambridge, MA: MIT Press.

The Body Shop. 1995. *Trading with Communities in Need*. Brochure. Wake Forest, NC: The Body Shop.

———. n.d. *We're Taking Styling to New Heights*. Brochure. Wake Forest, NC: The Body Shop.

Bourdieu, P. 1984. *Distinction*. New York: Routledge.

Boyds Bears and Friends. 1995. *Certificate of Authenticity*. Gettysburg, PA: Boyds Collection Ltd.

Boyer, C. 1994. *The City of Collective Memory: Its Historical Imagery and Architectural Entertainments*. Cambridge, MA: MIT Press.

Buci-Glucksmann, C. 1994. *Baroque Reason: The Aesthetics of Modernity*. Thousand Oaks, CA: Sage.

Buck-Morss, S. 1989. *The Dialectics of Seeing: Walter Benjamin and the Arcades Project*. Cambridge, MA: MIT Press.

Casey, E.S. 1987. The World of Nostalgia. *Man and World* 20: 361–84.

Dichter, E. 1960. *The Strategy of Desire*. New York: Boardman.

Duncan, C. 1995. *Civilizing Rituals: Inside Public Art Museums*. New York: Routledge.
Enzenberger, H-M. 1974. *The Consciousness Industry*. New York: Seabury.
Fiske, J. 1989. *Understanding Popular Culture*. Boston: Unwin Hyman.
Foucault, M. 1986. Of Other Spaces. *Diacritics* 16: 22–27.
Freidberg, A. 1993. *Window Shopping*. Berkeley: University of California Press.
Frisby, D. 1985. *Fragments of Modernity: Theories of Modernity in the Work of Simmel, Kracauer and Benjamin*. Cambridge: Polity Press.
Frow, J. 1991. Tourism and the Semiotics of Nostalgia. *October* 57:123–51.
Geist, J. 1983. *Arcades: History of a Building Type*. Cambridge, MA: MIT Press.
Getting Around. *http://twincities.sidewalk.com/link/11347*. 1997. October 6: 1–2.
Glamour Shots. n.d. *"Can I Look and Feel My Best?" Yes You Can!* Brochure.
Goss, J. 1993. The Magic of the Mall: Form and Function in the Retail Built Environment. *Annals of the Association of American Geographers* 83:18–47.
———. 1996. Disquiet on the Waterfront: Nostalgia and Utopia in the Festival Marketplace. *Urban Geography* 17:221–47.
Gould, S.J. 1991. The Dinosaur Rip-Off. In *Bully for Brontosaurus*, pp. 94-106. London: Penguin.
Haraway, D. 1989. *Primate Visions: Gender, Race, and Nature in the World of Science*. New York: Routledge.
Harris, N. 1991. Urban Tourism and the Commercial City. In *Inventing Times Square: Commerce and Culture at the Crossroads of the World*, ed. W.R. Taylor, pp. 66–82. New York: Russell Sage Foundation.
Hebdige, D. 1988. *Hiding in the Light*. London: Routledge.
Hewison, R. 1987. *The Heritage Industry: Britain in a Climate of Decline*. London: Methuen.
hooks, b. 1992. *Eating the Other: Desire and Resistance*. Boston: South End Press.
Hubbard, P. 1994. Re-imaging the City: The Transformation of Birmingham's Urban Landscape. *Geography* 81:26–36.
Huyssen, A. 1995. *Twilight Memories: Marking Time in a Culture of Amnesia*. New York: Routledge.
Jackson, P., and Thrift, N. 1995. Geographies of Consumption. In *Acknowledging Consumption: A Review of New Studies*, ed. D. Miller, pp. 204–37. New York: Routledge.
Jameson, F. 1979. Reification and Utopia in Mass Culture. *Social Text* 1(1):130–48.
Jhally, S. 1987. *The Codes of Advertising*. London: Frances Pinter.
Kaplan, C. 1995. "A World Without Boundaries": The Body Shop's Trans/national Geographics. *Social Text* 43:45–66.
Larsen, E. 1997. *Mall of America. http://twincities.sidewalk.com/mallofamerica.*
Lefebvre, H. 1991. *The Production of Space*. Trans. D. Nicholson-Smith. Cambridge, MA: Blackwell.
Lowenthal, D. 1989. Nostalgia Tells It Like It Wasn't. In *The Imagined Past: History and Nostalgia*, ed. C. Shaw and M. Chase, pp.18–32. Manchester: Manchester University Press.
Lukacs, G. 1971. *The Theory of the Novel*. Trans. A. Bostock. Cambridge, MA: MIT Press.
MacCannell, D. 1976. *The Tourist: A New Theory of the Leisure Class*. New York: Schocken.
MacNaughten, P., and Urry, J. 1995. Towards a Sociology of Nature. *Sociology* 29:203–20.
Magill, T. 1997. Designing the Mall of America. Paper presented at the conference, Learning from the Mall of America, November 20–23. Minneapolis: Weisman Art Museum.
Mall of America (MoA). 1996a. *Mall of America Souvenir Book*. Bloomington, MN: MoA Management Office.
———. 1996b. *Official Souvenir Video*. Glenwood Group.
———. 1997a. *Mall of America: Destination Planning Guide 1997–1998*. Bloomington, MN: MoA Management Office.
———. 1997b. *Celebrating 5 Years of Bringing the World to Minnesota!* News release. Bloomington, MN: Mall of America Management Office.
———. 1997c. *The History of the Mall of America. http://www.com/about/history.htm*, November 5:1–3.
———. 1997d. Dining at Knott's Camp Snoopy. *http:\www.campsnoopy-moa.com/dining.htm*, November 12:1–2.
———. 1997e. *Knott's Camp Snoopy general information. http://proto.webconn.com/mall/camp.snoopy/info.html*. October 6:1.
———. 1997f. *Mall of America: The Place for Fun in Your Life*. Brochure. Bloomington, MN: MoA Management Office.
———. 1997g. *Mall of America Attractions—UnderWater World. http://www.mallofamerica.com/attract/uwworld.htm*. November 12:1–2.
———. 1997h. *Fun Facts about Camp Snoopy. http://www.campsnoopy-moa.com/funfacts.htm*. November 12:1–2.
———. 1997i. *Mall of America Attractions—Chrysler Center. http://www.mallofamerica.com/attract/chrysler.htm*. November 12:1–2.
———. 1997j. *Golden Stars*. Brochure. Bloomington, MN: MoA Management Office.
Mannheim, K. 1991. *Ideology and Utopia: An Introduction to the Sociology of Knowledge*. London: Routledge.
Marcuse, H. 1964. *One-Dimensional Man*. Boston: Beacon Press.
Martin, C.L. 1992. *In the Spirit of the Earth: Rethinking History and Time*. Baltimore: Johns Hopkins University Press.

McCune, M.K. 1997. *Center Profile: Celebrating a Milestone at Mall of America. http://internetreview.com/pubs/scw/97jun/sew9706g.htm.*

McFarland, Teresa. Mall of America Public Relations. Interview, Nov. 19, 1997.

McGuigan, J. 1992. *Cultural Populism*. London: Routledge.

Miklitsch, R. 1996. Commodity-Body-Sign: Toward a General Economy of "Commodity Fetishism." *Cultural Critique* 33:5–40.

Miller, D. 1995. Consumption as the Vanguard of History: A Polemic by Way of Introduction. In *Acknowledging Consumption: A Review of New Studies*, ed. D. Miller, pp.1–57. New York: Routledge.

Montgomery, S. 1991. Science as Kitsch: The Dinosaur and Other Icons. *Science as Culture* 10: 7–58.

Moore, S. 1991. *Looking for Trouble: On Shopping, Gender and the Cinema*. London: Serpent's Tail.

Morris, M. 1990. Banality in Cultural Studies. In *Logics of Television*, ed. P. Mellencamp, pp. 14–43. London: British Film Institute.

———. 1993. Metamorphoses at Sydney Tower. In *Space and Place: Theories of Identity and Location*, ed. E. Carter, J. Donald, and J. Squires, pp.379–96. London: Lawrence and Wishart.

Nava, M. 1987. Consumerism and Its Contradictions. *Cultural Studies* 1:157–73.

Nelson, E. 1998. *Mall of America: Reflections of a Virtual Community*. Lakeville, MN: Galde Press.

Nora, P. 1989. Between Memory and History: *les Lieux des memoire. Representations* 26:7–25.

Odyssey, 1997. *Don't Believe Your Eyes*. Brochure. Santa Clara, CA: Explore Technologies, Inc.

Pynchon, T. 1974. *Gravity's Rainbow*. New York: Bantam.

Rainforest Cafe. 1997. *Your Passport to Adventure*. Hopkins, MN: Rainforest Cafe, Inc.

Ramos, A. 1994. From Eden to Limbo: The Construction of Indigenism in Brazil. In *Social Construction of the Past: Representations as Power*, ed. G.C. Bond and A. Gilliam, pp.74–88. New York: Routledge.

Rosaldo, R. 1989. Imperialist Nostalgia. *Representations* 26: 107–22.

Sack, R. 1992. *Place, Modernity and the Consumer's World: A Relational Framework for Geographical Analysis*. Baltimore: Johns Hopkins University Press.

Savage, M. 1995. Walter Benjamin's Urban Thought: A Critical Analysis. *Environment and Planning D: Society and Space* 13:201–16.

Slater, D. 1993. Going Shopping: Market, Crowds, and Consumption. In *Cultural Reproduction*, ed. C. Jencks, pp.188–209. New York: Routledge.

Sontag, S. 1977. *On Photogaphy*. New York: Farrar, Straus and Giroux.

Stam, R. 1989. *Subversive Pleasures: Bahktin, Cultural Criticism and Film*. Baltimore: Johns Hopkins University Press.

Stewart, S. 1984. *On Longing: Narratives of the Miniature, the Gigantic, the Souvenir, the Collection*. Durham, NC: Duke University Press.

Strathern, M. 1992. Forward: The Mirror of Technology. In *Consuming Technologies: Media and Information in Domestic Spaces*, ed. R. Silverstone and E. Hirsch, pp. vii-xiii. New York: Routledge.

Sussmann, E. 1989. *On the Passage of a Few People Through a Rather Brief Moment in Time: The Situationist International 1957–1972*. Cambridge, MA: MIT Press.

Thomas, N. 1994. *Colonialism's Culture: Anthropology, Travel and Government*. Princeton, NJ: Princeton University Press.

UnderWater World. 1995. *UnderWater World. Souvenir Guidebook*. Minnesota UnderWater World Aquarium Stores, Inc.

Urban Educator. 1995. *Bells Ring for Students at Mall of America. http://www.cgcs.org/newslett/educator/1995/dec/a2.htm.*

Walsh, K. 1992. *The Representation of the Past: Museums and Heritage in the Post-modern World*. New York: Routledge.

Warren, S. 1993. "This Heaven Gives Me Migraines": The Problems and Promise of Landscapes of Leisure. In *Place/Culture/Representation*, ed. J. Duncan and D. Ley, pp. 173–86. New York: Routledge.

Williams, R. 1980. Advertising: The Magic System. In *Problems of Materialism and Culture: Selected Essays*, p. 170–95. London: Verso.

Wilson, E. 1991. *The Sphinx in the City: Urban Life, the Control of Disorder, and Women*. London: Virago.

Wolin, P. 1982. *Walter Benjamin: An Aesthetic of Redemption*. New York: Columbia University Press.

Žižek, S. 1997. *The Plague of Fantasies*. New York: Verso.

Correspondence: Department of Geography, University of Hawai'i, 2424 Maile Way, Honolulu, HI 96822, e-mail jgoss@hawaii.edu.

[11]

Gendered Politics and Nationalised Homes: women and the anti-colonial struggle in Delhi, 1930–47

STEPHEN LEGG, *Homerton College, University of Cambridge, UK*

ABSTRACT *The aim of this article is to highlight the ways in which women achieved agency in a nationalist movement that, while encouraging female participation, attempted to spatially delimit this activity to the home. While nationalist historiographies have acknowledged those women who left the home and joined nationalist protests in public, there were many more women supporting those processions and meetings in private. While the home cannot be considered outside of the community, local or national scale activities that framed it, an examination of the varieties of political participation within the home should help revise Partha Chatterjee's suggestion that the Indian middle-class home was a site of silencing and resubjection. Combining interview material with archival evidence, this research investigates the ways in which women in Delhi between 1930 and 1947 responded to Gandhi's call to be politically active from the home. Using the work of bell hooks on 'homeplaces', the home is portrayed as a potentially political space of anti-colonial resistance, rather than an apolitical site of withdrawal, oppression or safety.*

Introduction

Colonial Indian society in the twentieth century witnessed fights for both political independence and improvements in women's standards of living; one superficially focusing on race and the nation, the other addressing sexuality and gender. Seen as separate movements, they can appear contradictory. The former criticised the British for being oppressive and uncivilised towards Indians, the latter criticised Indian men for acting likewise towards women. However, the two endeavours worked in tandem, although by no means always in harmony. Through uplifting India's women, Orientalist claims that Indian was savage and backward would be further undermined. Women could also serve not only to produce an independent India through reproducing and rearing nationalist children but could also, if educated, contribute directly to the fight for freedom. This was a fight that promised respect and protection for women in an independent era.

The aim of this article is to highlight the ways in which women achieved agency in a nationalist movement that, while encouraging female participation, attempted to spatially delimit this activity to the home. While nationalist historiographies have emphasised that many women left the home and joined nationalist protests in public, many more were supporting those processions and meetings in private. While the home cannot be considered outside of the community, local or national scale activity that framed it, an examination of the varieties of political participation within the home

Correspondence: Stephen Legg, Homerton College, Cambridge CB3 0DG, UK: e-mail: sil21@hermes.cam.ac.uk should help illustrate that the Indian middle-class home was not always a site of silencing and resubjection.

The investigation begins with a summary of the works on gender and nationalism, which shows that while women have been mobilised as representations of the nation, they have often been excluded from actively participating in nationalist movements. The 'gender irony' of supposedly egalitarian nationalisms attempting to silence their female populations is shown to apply to India in the late nineteenth and early twentieth centuries. However, it is argued that authors like Partha Chatterjee have conflated the *attempts* to instil a new patriarchy with the *achievement* of resubjection.

Counterpoised to Chatterjee's domestic site of tradition and silence is Gandhi's vision of home-based female agency. A summary of debates on the nature of the home suggests that the work of bell hooks on 'homeplaces' can help elucidate how women could be politically active from the home. This concept is used to present the empirical case studies of female domestic-political activity in Delhi between 1930 and 1947. Just as the home cannot be considered without external politics, so the activities within the home are shown to overlap with those outside of it. In conclusion, it is stressed that many women either rejected Gandhi's prescriptions or befell the fate outlined by Chatterjee of resubjection within a 'new patriarchy'.

Methodology

This research is based on an oral history project. Perks and Thomson (1998, p. ix) have stressed that while such projects can bring to light the evidence of those 'hidden from history', the reliability of memory and the nature of interviewing relationships can be highly problematic. After discussing the methodology behind this research, the engagement between historical narratives and 'subaltern fragments' will be briefly summarised.

The methodology adopted is introduced using Baxter and Eyles's (1996, p. 518) checklist of questions that can be deployed to evaluate qualitative interview research. Firstly, the *natural history* of this research is that of a broader investigation of the nationalist movement in Delhi, which was made the capital of India in 1911. The research concerns the interconnections between the conflictual, occasional and public nature of the nationalist movement, and its subjective, day-to-day, local and often private nature. During a research trip to Delhi between January and July 2001, I used various public archives to gather *data*, the second checklist question, regarding the development of colonial and nationalist politics between 1911 and independence in 1947. This information is used in footnotes to cross-check information received orally. I also conducted 42 interviews with people who could remember pre-1947 Delhi and with individuals who shared family memories and recollections. Interviews were recorded and then transcribed, with additional notes being made during the interviewing process.

Baxter and Eyles's third check, the *sampling* procedure, was complicated by the vast urban upheaval following partition in 1947 and Delhi's mass expansion in the next 50 years. Many of the original Delhian occupations had left the city walls, the focus area for most of my investigations. Much of the Muslim population had migrated to Pakistan while urban redevelopment has destroyed many of the communities that were so vital for the cultural geography of the nationalist movement. As such, interviews were based on recommendation, forming various networks that exposed the experiences of different areas and social groupings within the city. Family members helped with translation in most cases where it was necessary, although most interviewees could speak English (the class and caste relevance of which will be discussed later). The results *presented* here dwell on those recollections that dealt specifically with the domestic realm, although briefer references are made to other interviews.

The *credibility* of this research is impossible to consider outside of its specifically historical nature. Many interviewees were over 70 years old and found recollection a very draining experience, thus curtailing the opportunities for 'prolonged engagement'—few of the interviews lasted more than an hour and a half—or 'persistent observation' (Baxter & Eyles, 1996, p. 514). However, a historical study presents innumerable opportunities for 'triangulation' in which multiple sources, methods and theories can be deployed to examine statements and suggestions. Newspapers, records from colonial or nationalist archives, bibliographical guides and memoirs were all deployed to add credibility to these memories and recollections.

Similarly, while the 50 years since independence will have led to ideological reformulations of accounts in people's minds, a knowledge of these trends allows questions to be formulated that it is hoped will provoke novel forms of thought. Questions on the relatively under-investigated domestic realm tended to be unexpected and appeared thought-provoking. Such an approach should increase the chances of what Lummis referred to as the 'recall' of dormant events, rather than the performance of polished and familiar 'memories' (Lummis, 1998, p. 274). A historical perspective also presents greater opportunities for *transferability*, in which the relevance of data can be anticipated within the wider field, while the literature on a study area can help situate and frame the qualitative data.

Within the Indian context, the nature of historical recollection has been specifically addressed by the theories and investigations to emerge from the Subaltern Studies publications. The Subaltern Studies collective rejected colonialist, nationalist and Marxist teleological narratives that saw all actions and thoughts as somehow contributing to, respectively, colonial subjection, national independence or class consciousness (Prakash, 1990, p. 399). Instead, the autonomy of lower-class insurgency, those subalterns previously denied voice in the archive, was stressed (Ludden, 2001, p. 9).

The later studies expanded the scope of investigation from subaltern politics to subjectivity, examining different 'fragments' of the nation that did not have narratives of their own. Chakrabarty (1998, p. 475) has stressed that subaltern pasts belonged to marginal or inferior groups and that, as such, elite and dominated groups both have their subaltern elements. Chatterjee (1992, p. 42) has studied the 'subalternity of an elite' through stressing the 'middleness' of the Calcutta middle class. This class viewed itself as being dominated by the colonial authorities yet also as leaders of the new public sphere, the Indian National Congress [1] and the indigenous population. As such, the middle classes and higher castes can be legitimately studied for their internal divisions and opposition to the British, despite positions of influence. My research concerns this sector of the Delhi population that opposed the Raj, and attempted to direct the urban masses while simultaneously framing its own subaltern women.

Fragments of the subaltern survive not only in court transcripts, newspapers and memoirs, but also in shouts, rumours, names, temples and nomenclatures (Pandey, 1995, p. 225). Subaltern studies attempt to liberate minority languages and memories through interrogating archival sources and exploring new indigenous histories that could be oral, symbolic, vernacular or dramatic. This simultaneously involves challenging Eurocentric historical notions of what evidence is 'verifiable' (Chakrabarty, 1991, p. 105), while acknowledging that fragments of perspective are not 'true' or 'transparent' but are as power-laden and subjective as the official archive. As Amin (1995, p. 118) stated, 'The subaltern make their own memories, but they do not make them just as they please'.

Butalia and Amin have both used recollections to study events in the past, acknowledging that such memories must be 'tainted' and that the emphasis will be on 'remembrance' as much as 'fact' (Butalia, 2000, p. 8). Amin (1995, p. 194) stressed that his aim was to produce an 'enmeshed, intertwined and imbricated web of narratives' rather than a linear history. To do this he used the recollections of both direct witnesses and their families to access information about the past. To analyse the nature of the nationalist home, I have also used interviews from the sons and daughters of families involved in the nationalist movement in an attempt to preserve accounts that 'will soon be dead to the historian' (Amin, 1995, p. 194). The most common response to my request for further interviewees while in Delhi was 'if only you had been ten years earlier' or 'so many of them have died now'. Through triangulation and contextualisation, this research aims to suspend personal and subjective recollections in such a way as to complement and augment archival resources and theoretical concepts. While there are risks of memory distortion and hagiography, these must be faced if genealogies of women's contributions to the nationalist movement are to be constructed that successfully combine the 'erudite knowledge' of local Congress struggle with the 'disqualified knowledge' of the domestic realm (Foucault, 1980, p. 82). It is the path from the *transferable* and general musings of theory to the (in)*credible* and stubbornly local nature of Delhi's oral history that will now be traced.

Nationalism and Gender

Following Kandiyoti (2000, p. 492), it is possible to divide works investigating the relationships between women and nationalism into those that focus on women as *representations* of the nation and those that stress female *participation* in nationalist projects; that is, as the subjects or agents of nationalist movements. As subjects of nationalist discourses, women are represented as the biological reproducers of ethnic collectivities, as symbolic reproducers of the boundaries of these groups, and as signifiers of particular communities. In addition to these demarcations of the female subject position, Yuval-Davis and Anthias (1989, p. 7) acknowledge that women have also acted in the ideological reproduction of collectivities and as participants in national, economic, political and military struggles.

This agonistic positioning of women between agency and subjection is at the centre of what Mayer (2000, p. 2) has termed the 'gender ironies of nationalism'. Despite its egalitarian rhetoric, Mayer states that the nation is the property of men. The 'nation' and the 'woman' are social constructs that must be continually re-created and defended, yet the nation itself is also a gendered concept (Sharp, 1996; Radcliffe, 1999; Ranchod-Nilsson & Tétreault, 2000). Nations have 'gender regimes' (Connel, 1990, p. 523) that are suspended in divisions of labour, structures of power and emotional attachments. When aspiring nations attempt to articulate their gender regime, the 'ironies' at their heart become exposed. This exposition often takes the form of a contradiction between nationalist movements' claim to innovate in the name of the 'modern' and preserve in the name of 'tradition'. Especially in anti-colonial contexts, women were paraded as symbols of a new, modern nation that they not only represented but also re-created, while also embodying tradition and the past. As Kandiyoti has shown, 'Wherever women continue to serve as boundary markers between different national, ethnic and religious collectivities, their emergence as full-fledged citizens will be jeopardised' (Kandiyoti, 1991, p. 435). Before looking at how this tension was played out in Delhi, the wider Indian context will be summarised.

Antoinette Burton has recently suggested that, within colonial conditions, gender and sexuality maintained 'their capacity, as contingent and highly unstable systems of power, to interrupt, if not thwart, modernizing regimes' (Burton, 1999, p. 1). In nineteenth-century India, the British were not the only group cultivating a 'modernizing regime' and thus not the only group to struggle with concepts of gender and sexuality. Tanika Sarkar (1998) has documented how the Hindu middle class emerged from the traditional upper and literate classes, allowing caste and gender profiles to continue without major interruption in the mid-nineteenth century. These profiles were implicitly based on Brahmanical prescriptions of women's conduct that debarred widow remarriage, demanded extreme female chastity and often embraced the tradition of *purdah* (literally 'veil' but usually taken to mean female restriction to the home), which was widespread amongst the upper and upper middle classes across northern India.

From the 1870s onwards British and Indian reformers came to threaten these gender relations through challenging certain traditions, most famously that of *sati* (widow-burning), but also the nature of domestic arrangements. The suggestion that 'companionate marriage' should replace subservience in the home accompanied wide-ranging debate about the nature of the Hindu woman. Dipesh Chakrabarty (1994) has shown that this debate was very much about delineating the personal and domestic from the communal and public. Importantly, this division was a distinctly colonial one, born of administrative attempts by the Government to create a civil society, a public sphere in which reform would take place, separate to the private sphere in which traditional India would be preserved. Yet, in attempting to preserve this area of traditional India, the domestic tradition was itself reinvented, if not created. Grewal has suggested that accompanying the importation of the concept of the public sphere came that of the private, the home: 'For the Indians, what colonial discourse termed the *harem*, a space of opacity, became the *home*, a reconstituted Victorian space that was transparent in its clear manifestation of moral virtues as symbolised by Indian middle-class women' (Grewal, 1996, p. 25).

This home and the woman it contained were created through discussions largely between men. Women were discussed as objects for reform, the biological and symbolic reproducers of boundaries to which Yuval-Davis and Anthias (1989) referred. Community space and female identity were negotiated through discussions about gender and sexuality, whether focusing on female honour or cultural and biological reproduction. One of the core debates circulated around the following question: how to create a modern national culture that was not Western? In relation to women, Partha Chatterjee has suggested that the emancipatory possibilities of this question were foreclosed by the discussions about private and public spheres, the home and the world.

Chatterjee (1989, p. 238) suggested that the commonly rehearsed response involved drawing a distinction between the ultimately unimportant external and material realm against that of the internal and spiritual, which was 'genuinely essential'. The external could be modernised, industrialised and secularised while the inner should remain traditional, pure and Indian. He then suggested that when one applies 'the inner/outer distinction to the matter of concrete day-to-day living you get a distinction of the social space into ghar and bahir, the home and the world. The world is external, the domain of the material; the home represents our spiritual self, our true identity' (Chatterjee, 1989, p. 238). Within the gender divisions of social space, this meant that women and the home were to be preserved while men and the world would progress. Although not considered by Chatterjee, these divisions existed outside of the Hindu world. As Lateef (1990, p. 80) commented, 'the dual pressures on Muslim women were immediately apparent, for while the message to educate and participate in public affairs had been delivered, the appeal to tradition was being made to maintain community cohesion'.

The 'preservation' of women and traditions did not foreclose change; indeed, a 'new woman' was discursively fashioned to complement the new patriarchy within the new middle class (Chatterjee, 1993, p. 127). However, elements of modernity were selected that were compatible with the new female. She would be educated in domestic sciences, literate enough to read nationalist and religious texts and able to run the household in accord with outside conditions. She could even travel outside the home, as long as her spiritualism and discipline were marked in her dress, habits and eating (Chatterjee, 1993, p. 130).

Chatterjee *did* attempt to investigate how women struggled against subjection, analysing five autobiographies to show how women coped with changes and contributed from within the home (Chatterjee, 1993, pp. 135–147). However, he eventually concluded that the promise of nationalist emancipation ended, tragically, in 'betrayal' (Chatterjee, 1993, p. 154). Exclusion followed the formation of the cultural nation, while citizenship for some was followed by representation for the silenced others. The home as such was portrayed as the site of betrayal, resubjection and silencing for most Indian women.

Visweswaran (1996, p. 88) has made the absolutely crucial point that Chatterjee, Chakrabarty and the subalternists fail to break free from the discourse they analyse and are thus complicit in identifying and reproducing the domestication of women. While the emergence of a new patriarchy is in evidence, in assuming its success female agency is immediately curtailed, as is the project of conceiving an adequately gendered subaltern subject. Reproducing what Mayer termed the 'gender irony of nationalism', Chatterjee stresses the representation of the oppressed female subject, but does not fashion a space in which her agency can be demonstrated, or seriously investigated. As such, '"subaltern" agency for resistance among women, or their counter-discourses, are invisible in *Subaltern Studies*' (Bannerji, 2000, p. 917).

Even within the middle classes to which he referred, Chatterjee's work requires substantial revision to be applicable to the wider nationalist movement. This is, in part, due to the much-altered conditions of the twentieth century, to which he does not directly refer, although his work is stated to generally refer to the 'colonial and post-colonial' worlds. From 1919 onwards, the nationalist movement affected a wider audience and witnessed the meteoric rise of Mahatma (Great Soul) Gandhi. In his earlier work, Chatterjee (1986, pp. 93 and 113) stressed that Gandhi's opinions and actions were not within the conceptual framework of the wider nationalist movement, and that he maintained an ambiguous relation to nationalist thought. Amongst Gandhi's main concerns were subjectivity, morality, conscience and spiritualism. Mobilising these traits, Gandhi attempted to speak for 'untouchables', the poor, peasants and women. The spiritual space of the home was Gandhi's domain and it is within his writings and speeches that one finds the home depicted as a site of agency, not just subjection.

Gandhi and Women

Mohandas Karamchand Gandhi (1869–1948) was born in Gujarat and trained in London and South Africa as a lawyer (Brown, 1989; Arnold, 2000). While defending the rights of Indians in Africa, he began to develop his thoughts on non-violent non-cooperation based around *satyagraha* ('truth force', non-violent resistance) and *ahimsa* ('love for all', action based on refusal to do harm). In 1919, back in India, he organised the mass protests against the Rowlatt Acts, followed by two years of national 'Non-Co-operation'. The suspension of the movement in 1922 was followed by eight years of 'constructive' work, aimed at the social uplift of 'untouchables', the poor, villagers and women, and encouraging Hindu–Muslim unity. Failure to achieve steps towards constitutional devolution or independence led to the Civil Disobedience Movements of 1930 and 1932. Their violent and effective suppression was followed by a further period of constructive work and limited participation in the ruling administration. The advent of the Second World War in 1939 led to further agitation, culminating in the Quit India movement of 1942, the largest and most diverse of all the movements and the last before independence in 1947. Each mass movement saw a widened support base and correlative decrease in control over the nature of each protest. From dramatically composed and disciplined processions and speeches in the early 1920s, Quit India saw widespread sabotage and violent protest.

However, there was also an evolution in Gandhi's thoughts and writings regarding women and political activity (Kasturi & Mazumdar, 1994; Kumar, 1994; Menon, 1999). Opposing Chatterjee's suggestion that women were resubjectivised during this period stands the commonplace nationalist imagery of Gandhi calling women out of the home and onto the street to protest against British rule. Gandhi idolised women as self-sacrificing and capable of infinite love (Kumar, 1993, p. 82) and encouraged them to be politically active (Mahan, 1999). However, Gandhi attempted to craft women in his image, using essentialist statements that contained 'a set of legitimisers that define[d] the concepts of femininity and womanhood which act[ed] as prescriptive frames of reference in his construction of women' (Patel, 1988, p. 378). Fox (1996) has stressed that although not representing another 'repressive regime', Gandhi's writings did not attempt to overturn the patriarchal ordering of domestic space.

While Gandhi insisted on female agency, he simultaneously insisted, especially in his early writings and speeches, that women's 'natural' space was the home. He warned that husbands should 'not, therefore, send your women out to work; protect their honour; if you have any manliness in you, it is for you to see that no one casts an evil eye on them' (Gandhi, 1920, pp. 43–44). To compensate, the home was emphasised as a political space in which women could be *active* (Thapar-Björkert, 1997, p. 494). Gandhi's essential means of targeting women and the home was that of *khadi* (homespun cloth). This had been mobilised as a political tool before, during the *swadeshi* (self made) movement of the early 1900s that aimed to replace cheap European imports with Indian-made goods. Gandhi took the idea of *swadeshi* and worked to integrate it into women's *dharma* (religious duty). In so doing, Gandhi simultaneously reimagined home and woman. As Kishwar (1986, p. 5) stated, 'Gandhi's relentless propaganda in favour of *charkha* spinning and the wearing of *khadi* was designed to bring the spirit of nationalism and freedom into every home, even in the remotest village. In this way, abstract political ideas, such as struggle against colonial rule, assumed concrete form for ordinary people'.

In addition to spinning, women's other duties in the home included educating their children to be proud Indians and soothing their husbands in the home after his trials in the outside world. During the 1920s Gandhi came to question the specific nature of women's hardships, rather than stressing that they too were oppressed by the British. In conjunction with his 'constructive' programmes against untouchability, sati and the treatment of widows came an attack on purdah. In an article entitled 'Tear down the purdah', Gandhi (1927, pp. 149–150) claimed that chastity 'cannot be protected by the surrounding wall of the *purdah*. It must grow from within, and to be worth anything it must be capable of withstanding every unsought temptation'. Yet, Gandhi would continue to stress that women's rightful place was in the home, crowning them 'Queens of the household' (Gandhi, 1928a, p. 195). Rather than this indicating, as Bald (2000, p. 82) has suggested, that women were secondary to Gandhi's concern for independence, this represents Gandhi's attempts to meet a compromise between the need for female agency and the forceful demands behind the emergent 'new patriarchy'.

With the onset of the Civil Disobedience Movement in 1930 women were discouraged from joining the Salt March to Dandi [2], but it was eventually conceded that their assumed patience made them better equipped to picket traders in liquor and foreign cloth (Patel, 1988, p. 379). It was from these injunctions that women could make legitimate claims for experiencing the outside world, and a female presence in community- or city-scale processions soon became an essential part of public performances. Yet the growing role of women in the public sphere should not be attributed simply to Gandhi. The role of journals (Orsini, 2002), education (Hancock, 2001), local Congress committees and women's associations (Basu & Ray, 1990) cannot be underestimated. By the 1942 movement, a minority of women were involved not only in spinning and public processions, but also in the organisation of underground movements, parallel governments and various other 'illegal activities' (Kumar, 1993, p. 93).

Despite both, to an extent, tying women to the home, Gandhi's writings about women and Chatterjee's interpretation of the women's question propose radically different visions of woman's place in society. Gandhi consistently strove to tear down purdah, but this 'veil' was one that cloaked women's self-confidence and dignity, rather than one that bound them to the home. Ashis Nandy (1983) has shown that the success of the colonial project rested on the colonised population internalising territorial conquest into the psychological realm. But this was also a process that occurred within the differentiated 'native' Other:

> This classic instance of the psychological defence of turning against self, by identifying with the aggressive male, draws attention to the way in which some social institutions have made woman herself a participant in her self repudiation and intra-aggression. The oppressive reality for women, one might suggest, is now only partly outside her. A part of that reality has been introjected through a long historical process of social learning. (Nandy, 1990, p. 149)

Women were not *just* doubly silenced by the oppressive structures of colonialism and patriarchy. They had to silence *themselves*, and it was this self-regulation that Gandhi targeted, encouraging Indians to reimagine a self without shame and encouraging women to conceive of domestic womanhood without purdah. Within a spatial framework, Gandhi's apparently contradictory essentialist statements about women make sense: they were superior within the home but not suitable for most tasks outside of it; women were to overcome purdah, yet their place was that of the home; women were to innovate and reorder their domestic lives, yet also to obey Gandhi's instructions on political-public activities. It is my contention that this paradox was the key to the partial success of Gandhi's vision of women, negotiating as it did the tension between women being, in Kandiyoti's (2000) terms, both representations of and participants in nationalism. The concept of spatially limited agency gained the support of men as a way of increasing the political workforce and bolstering Indian nationalists' claims to civility without threatening the security of their home lives. However, the indeterminacy that this paradox created around the role of the female could lead to abuse by men, in line with Chatterjee's argument, or abandonment by women to other, less prescriptive, nationalist regimes. A discussion of the literature regarding women and 'the home' will be strategically deployed within the interview analyses to help frame the investigation of the domestic scale. Before looking at the different ways women came to relate to the home in Delhi, the political-cultural context of the region will briefly be set out.

Delhi and Women

Unlike Calcutta, with its strong colonial influence and reforming middle classes, Delhi was an imperial city of tradition and custom. Independent of British rule until 1803 and only systematically reorganised along colonial lines after the 'Mutiny' of 1857, Delhi remained under the Muslim influence of the Mughal Emperors who had ruled there, in name, until the mid-nineteenth century. The tradition of purdah was strong throughout the female middle and upper classes, although the poor could rarely afford to have a potential earner inactive in the home (Jeffery, 1979; Raheja & Gold, 1994). Purdah was especially strong amongst the Delhi Municipality's substantial Muslim female population (Lateef, 1990, p. 203), which numbered 66,451 in 1931 and 100,816 by 1941, although this population section remained in both years at 19% of the total population and 46% of the female population. [3] Despite women's ability to adapt to these situations, by the late 1920s the fact still remained that many women were uneducated and had little consciousness of the outside world.

Delhi did not only lag behind in the broader debates on gender reform. Harsh retributions had followed the political disturbances of the 1850s and by the early 1900s Delhi was considered a political backwater compared to the turbulent political activity of Bengal and other urban centres in India. It is against this general backdrop that the appearance of women in the political public sphere was so shocking, as registered in the newspapers, government reports and nationalist self-assessments summarised below. As Forbes (1996, p. 148) commented, when women first came out of the home in protest in the 1930s, they gained attention in Delhi not because there were very many of them, but because the sight of them on the street was so rare. The activity within the home will be examined after a summary of women's public-political activity in Delhi.

During the 1930 Civil Disobedience movement the women of Delhi launched a successful offensive on the public sphere, not just in terms of urban space, but also of political and media space. Mary Campbell reported in the *Guardian* newspaper, back in England during June 1931, her surprise at the way Delhi women took up Gandhi's call to picket 'drink and drug shops'. She reported, 'I thought he had made a mistake this time that the Delhi women, so many of whom lived in *purdah*, could never undertake that task. But to my astonishment they came out, and picketed all the shops in Delhi' (quoted in Bala, 1986, pp. 24–25) The pages of the *Hindustan Times*, printed in Delhi and featuring local coverage, were filled with images and writings about women and their new roles as political activists. Women participated in liquor and foreign cloth picketing, but also in recitals of the informal national anthem, flag raisings and processions through the city.

One such procession to picket the Law Courts on 6 March 1930 was met by a police *lathi* (baton) charge [4], although the local administration's political report to the central government stressed that the women were 'shepherded to safety'. [5] These reports chart the growing awareness of female activity in the minds of the local government. On 6 June 1930 it was reported that 'An interesting, but troublesome feature during the fortnight was the processions of women shouting revolutionary slogans and carrying a national flag … These women of course rely on their sex for non-interference by the civil authorities, and their procedure is grossly mischievous'. [6] For the rest of the Civil Disobedience movement few major disturbances were reported without a mention of a significant female presence. Women remained involved in public sphere activities throughout the 1930s and played an equally important role in the Quit India movement of 1942. Beyond the major public processions, the Delhi District Congress Committee reported to the All India Congress Committee on 15 May 1930 that 'The successes of the campaign for the boycott of foreign cloth was solely due to the determined picketing of shops by ladies at very great personal inconvenience'. [7]

These are the events and characters that have featured in traditional accounts of women's contribution to the nationalist movement in Delhi. However, an examination of the biographies of some of the women who ventured out onto the streets blurs the boundary between public and private and forces a reconsideration of what constitutes a 'political' act. The accounts of the nationalist movement I received were filled with descriptions of Gandhi's influence and the sublime feelings of women who transgressed the public–private boundary within which they had previously been contained. These accounts included that of Sarla Sharma [8], whose mother paraded Chandni Chowk in the 1920s challenging those who 'cast obscenities' to give reasons why she should *not* be in the street. Mrs N. Chakravarty [9] spoke of the students at Indraprastha Girl's College running across the compound and climbing over the walls to join passing student processions in 1942, while Subhadra Joshi [10] recalled absconding from a local hospital, despite the repeated warnings of the doctor, in order to mediate a trade union dispute in the 1940s.

However, the physical movement onto the street should not always be equated with the achievement of a political act. Could this just be a parading of women as, to use the term of Yuval-Davis and Anthias (1989, p. 7), signifiers of the nation? While acknowledging the courage of such women, there can often be more evidence of political activity within the home than beyond it. Such issues were raised in an interview with Dr Gunmale Naulakha, whose father, Phool Chand Jain (1907–2000), was a prominent political activist in Delhi. [11] In 1932 Dr Naulakha's mother was asked by her husband if she was ready to participate in the Civil Disobedience movement. She prepared herself accordingly and was taken from the house by Phool Chand Jain to a local Congress worker who was organising political volunteers. Dr Naulakha recalled the conflict between her mother and grandmother at this point: 'My grandmother was so annoyed, so annoyed, that she just pushed her to the wall, she didn't beat her, she just said, 'What are you doing? Going out of the house? And you are wearing this khaddi sari? No, no, you are not going anywhere, you just go in the kitchen, sit there and do your thing. Yes, your brother-in-law can go and do this picketing, but no, nothing doing'. There was turmoil in the house when, leaving the home anyway, Phool Chand Jain's wife was arrested, on 6 February 1932.

As illustrated in the above quotation, part of the shock of this action was in the clothing used, a home-spun sari. This clothing was a political act, as Gandhi had outlined, and one which was repeated every day, rather than once in a lifetime, calling into being a gendered and nationalised subjectivity. Chand Jain's wife had long given up her silken clothes and silver jewellery and started wearing a *khadi* blouse and sari. As her daughter commented in the interview, for such an upper-class family, 'This was a revolution! … The whole concept of beauty had changed, beauty didn't mean using rich cotton and silks, no, beauty was that they were using khaddi and were still beautiful because of inner strength and beauty which was there in the women, it was amazing'.

After release from jail, Dr Naulakha's mother was never publicly active again, yet continued to wear *khadi* in the house.

While not denigrating the bravery of leaving the home, the above example illustrates that there may often be clearer evidence for 'political' activity within the home than outside of it. Leaving a house under the tutelage of one's husband, to be passed into the instruction of a political worker, did require courage. Yet, to alter one's daily routine and challenge the members of one's household on a daily basis required a firmly rooted determination. Those women who claimed the streets had to be backed up by legions of other women performing supportive roles in the home.

Women and the Home

The relationship of women to the home in a political context will here be separated into two interlinked groupings. The first is those homes of prominent, male politicians and the second is the homes of ordinary people who would not necessarily have identified themselves as political workers. Those women who have made it into the nationalist archive without mention of their activities on the street were the wives of well-known political activists. Especially after the declaration of the Congress committees as unlawful in January 1932 [12], Delhi's political headquarters had to take on a more peripatetic nature and flit between the households of high-ranking Congress workers.

Political Homes

One political home in Delhi was that of Dr M. A. Ansari (1880–1936), a prominent Muslim politician allied to the Congress in Delhi. His wife, Shamsunnisa (*Begum* Ansari [13]), had worked with the Delhi Khilafat Committee (Hasan, 1995, p. 15) in their protests against European post-war policy in Turkey and has been recorded as one of the few Muslim women to be politically active in 1920s Delhi (Mahan, 1999, p. 171). During Ansari's trips to Europe, Shamsunnisa ran the house and learned how to write Urdu in order to, amongst other things, keep registers of household expenditure. When her husband returned, she took to organising the house and accommodating the many guests that passed through this Congress meeting place. As Hasan commented, 'Confined as she was to the domestic role expected of a woman of her time and station, Shamsunnisa found an outlet for her interests and energy in the political career of her husband' (Hasan, 1995, p. 15).

Shamsunnisa maintained separate Muslim, vegetarian and European kitchens to ensure the satisfaction of her guests. Yet, as Ansari's adopted daughter, Zohra, later commented, 'she was a very excellent hostess and a very good organiser of the household, but she never came out because she was strictly observing *Parda*. It was always behind the scenes; her activities always remained behind the scenes whatever she did'. [14] E. M. Forster, who stayed with Ansari in 1912, complimented 'our host's wife—whose aimiable [*sic*] presence manifests itself through gifts of cigarettes, betel nuts, and scent, but not through the flesh' (Forster, 1912, p. 147). Gandhi himself stayed with Ansari during the 1931 Irwin talks (Slade, 1960, p. 120) and praised the organisation of the household in print following his stay. He wrote of a conversation with Dr Ansari in which Begum Ansari was praised for running the household; not just the kitchen but the finances and general organisation of the home (Gandhi, 1931).

In a dramatic re-enactment of his policy towards women, Gandhi was the only non-family male allowed to visit Shamsunnisa in purdah (Edib, 1937, p. 31). This, however, came after he had placed considerable pressure on Dr Ansari. In 1928 he claimed there was no point accepting Begum Ansari's invitation to stay because 'All the time I am there, they keep themselves at a safe distance hiding themselves behind the *purdah.* If they want me to be there, they will have to expose to view their superfluous bangles and other jewellery so that I can ease them of the superfluities and turn them to good account' (Gandhi, 1928b). In another step away from total purdah, she herself left the house in 1934, wearing burqa, to vote in the Congress elections, but this was viewed as very progressive under 'family rules'. [15]

Shamsunnisa represents a female form of participation without which one of the most important households in Delhi could have fallen into disarray. Yet, few houses were in positions of such prominence and few women received the direct tutorship of the spiritual leader of the nationalist movement. As Gandhi stated after praising Begum Ansari, 'It is not that there is only one such woman among the Hindus and Muslims, there must be countless numbers of them in every nook and corner. We do not know of them' (Gandhi, 1931).

A second home that depended on female household political labour was that of Desh Bandhu Gupta (1901–51), who had been involved in Congress politics, in addition to editing a local newspaper since the early 1920s, and was one of the leaders of the Delhi Congress throughout the 1930s and 1940s. Two of his sons still live in Delhi and were interviewed to get a sense of the home and the activities it contained. The sons recalled that amongst their earliest memories of the house were revolutionary shouts of '*Inquilab Zindabad*' (Long Live the Revolution). Ramesh Bandhu Gupta stated that 'even though it belonged to my Father, this was like a Congress house' while Vishwa Bandhu Gupta recalled that 'the entire psyche of the family was tuned to the freedom struggle in Delhi and also the newspaper, the gathering of news and following the movement through news'. Here, 'political issues' came to saturate the 'domestic', whether through memories of children wearing Quit India badges and armbands in their youth or through the walls of the house displaying Congress memorabilia and photographs of prominent nationalist politicians. A key figure in coordinating this saturation was Desh Bandhu Gupta's wife, Sona Devi.

Sona Devi had a range of duties to fulfil, from more active to more nurturing roles. She herself was imprisoned for a year and a half for her picketing activities, while in the home it was her duty to keep a boiler in the basement burning throughout the year so that if a police raid occurred then illegal literature could immediately be burnt. Yet she also had to contribute through organising the political household. Sona Devi cooked for volunteers on the street and when she herself left the house, other female members of the family assumed this role. Volunteers that arrived in Delhi to offer their services to the nationalist movement had to be fed and accommodated, often within the house that was located in New Delhi. The traditional hermetically sealed walls of purdah had imploded such that the house and the family, as Ramesh Gupta commented, 'was more like a banyan tree; everyone could shelter beneath it'.

When prominent members of the family were arrested then it was Sona Devi's role to make ends meet, while other Congress leaders would enter the house and make use of the facilities there. Yet Vishwa Bandhu Gupta stressed that these people never took the place of the Father figure; they were always regarded as guests because of the special duties Sona Devi maintained for her own children. These tasks, in line with those duties outlined by Gandhi, included reading educative books in the house and chastising the children when they did not wear *khadi.* Thus, while being politically and publicly 'active', Sona Devi was also active in the second form of participation identified by Yuval-Davis and Anthias (1989), the 'ideological' reproduction of the home itself as a political community.

Both the Bandhu Gupta and the Ansari houses show that even in overtly political houses, women had duties that were not 'public' yet were, in a variety of ways, political. These sorts of contributions invite ways of thinking that can open up households that have not previously been included, in detail, in accounts of the nationalist movement.

Homeplaces

Feminist academics have worked since the 1970s to illuminate the exploitative relationships and taken-for-granted assumptions that structure the household as a gendered space (Massey, 1994; Felski, 1995; McDowell, 1999). Young (1997, p. 134) has shown how de Beauvoir influentially portrayed the home as a supportive structure based on women's labour. Post-colonial feminists, such as Martin, Mohanty and de Lauretis, have also shown how the home as a secure place is often based on the exclusion of non-whites and the working classes (Pratt, 1998; George, 1999, p. 2; Young, 1997, p. 157). While acknowledging that the home can be oppressive and should not be viewed as the repository of a unified and secure identity, Young (1997, p. 156) stressed that the home also has liberating potential and can serve as a valuable base for personal and cultural identity.

In a similar vein, Honig (1994, p. 579) initially rejected the home as a 'safe place' and insisted on the search for 'coalitional politics' in unknown and risky places. However, she then went on to question the dividing lines between homes and the world, showing that coalitions relied on spaces of withdrawal, recovery and preparation and that homes themselves are 'invigorated' by political engagement (Honig, 1994, p. 583). Honig, like Young (1997, p. 160), turned to the work of bell hooks, who has stressed that in oppressive societies, homes can actually provide sites not only in which women are represented as nurturing, but in which active resistance can be crafted by women. As a site of memory and political reimagination, women's activities in the home have often gone unrecorded and unrecognised, just as their economic labour went unpaid and unvalued.

hooks (1991) has outlined a way of conceiving certain homes as 'homeplaces', cultural spaces which acted as havens in the face of white, male oppression in the outside world. Women constructed these spaces; they were 'their special domain, not as property, but as places where all that truly mattered in life took place—the warmth and comfort of shelter, the feeding of our bodies, the nurturing of our souls' (hooks, 1991, p. 41). hooks acknowledged that women may have been forced into these spaces by a form of oppression, but they took these spaces and made them into caring, revolutionary places. As such, hooks can be used as a corrective to mediate Chatterjee's statement that women were depoliticised and reconfined to the home in the twentieth century and to understand how women took up Gandhi's call and fashioned homes as sites of anti-colonial resistance.

So how were these homeplaces constructed and maintained? To begin with, they were grounded in *material* objects. hooks recalls that her grandmother 'was certain that the way we lived was shaped by objects, the way we looked at them, the way they were placed around us. She was certain that we were shaped by space' (hooks, 1991, p. 11). Nirmal Mukerji, whose father was Principal of St Stephen's College [16] and whose family were Christian, recalled in an interview that his childhood home contained a photograph of Jesus Christ above his mother's chair and, to its left, photographs of Jawaharlal Nehru and C. F. Andrews, both highly influential nationalists. These decorations were the duty and responsibility of the woman of the house. In the New Delhi home of her youth, Dr Sheila Uttam Singh recalled during an interview the traditional Indian drawings, images of Sikh gurus and photographs of Congress sessions the family had attended intermingling with family portraits, Western chairs, tables and cutlery. However, Dr Uttam Singh's grandmother refused to use the latter and insisted on eating from a traditional *thali* (a segmented Indian plate).

In addition to its hybrid material interior, the *discursive* space of this household was one heavily infused with nationalist sentiment and rhetoric. Dr Uttam Singh's father, an eye surgeon who had business relations with Europeans, would at home denounce British policies and go so far as to proclaim Hitler a saviour, on an 'enemy's enemy' basis. He would also read out newspaper articles that criticised the British, which often formed the basis for political discussions. Women were increasingly able to take part in discussions about politics due to increased education, often accessed through that of their children. Information was also transmitted as gossip through newly politicised yet traditional means of communication, such as discussions at temples, across rooftops in the dense centre of Old Delhi, in markets and through traders who came to the house. As such, the 'new woman' was not simply formed through elements of modernity selected by the new patriarchy, as Chatterjee suggested. Rather, they themselves actively sought out and procured the information that would allow them to ideologically reproduce the nation at the domestic scale.

The traditional 'purdah talk' (Minault, 1984) about relatives and family quarrels was replaced with discussions about the world events to which these women now had access and could educate their children. Rather than women leaving the home to enter the public world of politics, political information was infusing the homeplaces themselves. As Mrs Kesavan, a resident from the heart of Old Delhi, commented, 'Now the household knew what they were talking about, previously there had been such confined knowledge. Basically we were much better off'. Music also played an important role in these cultural spaces, as what Pandey (1995) referred to as minority languages within subaltern fragments of the nation. Many interviewees recalled singing *Bande Mataram* [17], with the opening line of 'Mother, we bow to thee', in the home with their family, as well as songs that referred to Gandhi as saviour, emperor or deliverer. Dr Uttam Singh only realised later in her life that a song she had picked up during her childhood had been part of the anti-Simon Commission agitation of 1928.

While these activities worked towards reimagining the nation and bolstering a gender performance that granted women political potential within the home, these houses were also sites in which Gandhi's *khadi* project took root. The appeal to spin was effective, especially in the early years, across communal and class lines (Tarlo, 1996). Kaushalya Lal recalled her sister, the Delhi leader Satyavati, giving away a finely embroidered *khadi* sari presented to her by leading female nationalist Sarojini Naidu because she came across a poor villager in a torn *dhoti* [18]. Kaushalya continued that 'A saree which I had embroidered with great pain for her also met the same fate.' [19] It was often the duty of the wife not only to spin, but to encourage the wearing of *khadi*. Nirmal Mukerji's mother burnt her husband's suits and encouraged him when his students suggested he wear *khadi*.

However, as Tarlo has commented, *khadi* did not succeed in overcoming all class, caste and community barriers (Tarlo, 1996, p. 91; Visweswaran, 1996). The uniformity of simple, *khadi* clothing and a 'Gandhi cap' were supposed to symbolically unite India against the 'divide and rule' tactics of the colonial government. However, personal taste was very often interweaved with the simple action of spinning. Dr Uttam Singh recalled her aunt creating a sari with the Congress tricolour around the hem, while as a baby she was presented with a box of gifts. Included amongst these was a piece of *khadi*, symbolically implanted within her life narrative, even if some of the family secretly considered *khadi* suitable only for Jats [20] and the poor.

Yet, the mutability of *khadi* also allowed it to be used in situations it otherwise might not have penetrated. Mrs Kesavan recalled that her father, a lawyer, was required to wear a suit in court and had thus acquired a *khadi* outfit. However, no *khadi* ties were available. As such, his daughters and wife unpicked a European tie to discover how it was made and produced a set of ties from Indian silk. These were received not only by Mr Kesavan but by the entire law firm. When he returned home he changed into a traditional Indian kurta. Even those people who could not wear *khadi* at work could change clothing, and identity, when they got home. As Mr Prithvis Chakravarty commented of his father in an interview, he 'would leave one personality at home and go and do work. When he got home, he would take off his blazer and become an Indian ... talk about speeches and maintain 100% support of the movement'. As Honig (1994, p. 583) suggested, this is a vision of the home 'invigorated' by politics and a site to reimagine one's subjectivity, rather than a space of denial or reluctance.

Despite the importance of these homeplaces, hooks stressed that a focus on the home in isolation was too nihilistic (hooks, 1991, p. 148), even if the home would remain a place one had to return to. The experiences of one's old life had to be combined with the search for new knowledges and experiences, exploiting the margins of society that involved personal risk and danger; combining the politics of the home with the politics of location (hooks, 1991, p. 145). Likewise, Honig has stated that 'If home is to be a positive force in politics, it must itself be recast in coalitional terms as a differentiated site of necessary, nurturing, but also strategic, conflicted, and temporary alliance' (1994, p. 583). As Gandhi discovered in 1930, women were not content to remain in the home, and gradually extended their influence outside of it, taking his prescriptions for action and adapting them to local circumstances. However, there were many different ways in which the influence of women was felt outside the home besides the picket and procession.

Prabhat Pheris were 'morning rounds' that paraded through the local community, singing songs to encourage people out of their homes. Even when women were not directly involved, children would often join in singing the songs they had practised at home and bring back new songs to be rehearsed with the family. There were also spaces that cannot easily be categorised as 'public' or 'private' that were politicised and appropriated by nationalist organisations. *Purdah Baghs* were walled, protected gardens in which women had traditionally met, that were used from the 1920s to discuss political issues. Such gardens were used as playgrounds for Indraprastha Girls' School and, Dr Kapila Vatsayan recalled in a personal interview, girls there were educated by local Congress activist Memobai and taught to sing songs such as 'Up, up Union Flag, down, down Union Jack'.

The home and the body mutually construct each other; the home is constituted by embodied practices while the body is nurtured and disciplined with the home. However, the body is mobile and deployable whereas the home is not. In wearing *khadi*, the home labour of women was displayed on the street. Yet women's influence could be felt in other ways also. For instance, Dr Uttam Singh spoke of the occasion when, after King George V's death on 20 January 1936 [21], the girls at her convent school were made to wear black armbands. When Kamala Nehru died, on 28 February, word was sent around between the children's families and they were sent to school wearing the same armbands. Likewise, associations were formed which presented bangles, traditional signs of effeminacy, to those who would not join the nationalist movement. Girls from Indraprastha Girls' School would picket the more conservative members of St Stephens College, although class issues again pervaded this practice. In reply to being presented with a bangle by the daughter of a high profile civil servant, Dr Sarup Singh suggested that she would do better to give it to her father. Therefore, the focus comes again to rest on public protests. However, the activities on the streets can now be seen as the accumulation of domestic activities and struggles. The power of the nationalist movement did not reside with the Congress elite or in the grand displays in the street. Instead, it circulated between the homeplaces through gossip, local political enterprises and community gatherings.

Radical Reinterpretations?

While we have seen that the incitations of Gandhi and the Congress were accepted and adapted by some women, there were many for whom these modes of womanhood were either not made available or rejected. At one extreme were those who accepted the calls for women's rights and respect, yet rejected Gandhi and the Congress's vision of what a new Indian woman should be. Throughout the nationalist movement, women collaborated with revolutionaries, harbouring absconders, smuggling guns and even working in bomb factories (see Thapar-Björkert & Ryan, 2002). Magan Kishore Johry was involved in bombing cinemas that attracted British audiences during the Quit India movement [22] while Aruna Asaf Ali famously turned away from the central Congress fold in 1942 when she went underground and organised Delhi's small-scale revolutionary activities (Kaul, 1991). [23]

At the other extreme, and more difficult to trace, were those women who met the fate Chatterjee envisaged; not unbounded female agency but a rebound subjection. Discussions of 'homeplaces' can often reify the object of which they speak and forget that such places were the sum of women's activities in the home. Drawing on Irigaray, Rose (1996, p. 69) has suggested that Woman represents the fixed and the stable, the source from which the subject draws 'his' reserves. While Man builds and constructs, Women are contained and enclosed. As Irigaray (1992, p. 47) spoke for Woman, 'What you intend for me is the place which is appropriate for the need you have of me. What you reveal to me is the place where you have positioned me, so that I remain available for your needs'.

Pratt (1998) is right to suggest that hooks's assumption of the dichotomous organisation of gendered space may work to reinforce these exclusions. If women *are* homeplaces, how can they draw sustenance from them? Many of the examples given of women's political activities in the house can just as easily be read as another chore added to their lists of duties by a publicly, politically active husband. Recalling the activities of his parents' generation, Sadiq-ur Rahman Kidwai suggested that women scraping together their resources to keep the family clothed and fed when a husband was sent to jail can be read as an essential contribution to supporting political workers. But did women, as wives, have any choice in these circumstances? Also, many women cited the education of children as a way of (re)producing an independent India, but was this a duty they had to fulfil regardless? And did this posit mothers as *representing* the rebirth of the nation, or actively *participating* in its (re)production? (Kandiyoti, 2000).

Some women may have been able to soothe and reassure their husbands, while

Sayeeda Kurshid [24] recalled that during periods of overt political activity many families were put aside from 'men's business'. These ambiguities were raised in many Muslim families, in which women found it harder to escape the tradition of purdah, yet it was these families that had more experience of dealing with secluded lives. Even Shamsunnisa Ansari, whose role in organising the political household was paramount, did not necessarily influence Dr Ansari's political thought. As Zohra Ansari commented, 'There was absolute understanding between the two, and she never objected to anything that he did in his political life and in his profession; he was a master of himself'. [25] Yet, Dr Naulakha stressed that even though her mother seemed subordinate to Phool Chand Jain, the 'Ties of the family were the binding force, because of love the women used to obey. This obedience was not suppression, these are two different things. Obedience means you believe and follow somebody, suppression is domination, there was harmony in our house'. While obedient, one could still craft and harbour one's own political agency. Suppression denotes the smothering of this agency within the film of representation and resubjectification.

All definitions of the 'political' must be arbitrary and subjective, but as Syeda Hameed stated in an interview, the distinction between 'oppression' and 'support' comes when a husband explains his actions and duties and imparts some sense of achievement to his wife. We must also resist the temptation to identify an internally motivated, humanist subject at the centre of these accounts (O'Hanlon, 2000, p. 74). At times women may have felt like the beating heart of the nationalist movement while at others they may have felt secluded and depressed. The aim of this work is to open up the domestic as a site for the consideration of female agency and to present the conceptual tools by which each individual and unique case can be analysed.

Conclusion

During the 1930s and 1940s Delhi witnessed an unprecedented upsurge in the activity and profile of its formally recluse, middle-class, female population. Women became involved in the realms of activity traditionally preserved for and by men, including the procession, the picket, writing in public journals and preaching in the street. While the activities of these women have been enshrined in the nationalist historiography, they have also served as a counterpoint against which those women who did not leave the home have been compared. Even subaltern historians have portrayed females who remained within the home as passive and contained.

This investigation has illustrated a number of ways in which women helped to politicise the home and assert their agency in a space often read as one of silence and subjection. While the wives and family members of high profile Congressmen helped to organise their homes as unofficial political headquarters, there were many more women who helped contribute to the anti-colonial struggle through interweaving the political with the domestic. In line with Gandhi's writings, women adopted their traditional nurturing roles, as well as transforming previous duties into new techniques of protest such as spinning, cooking, accommodating, clothing and singing. As such, political homeplaces were constructed that, while not eschewing the need for activism on the 'margins' and risk-filled spaces of society, provided discursive and material spaces of comfort and nourishment. Even when more women left the home, often in defiance of Gandhi's initial advice, there remained those to whom the private domain was the main site of political activity.

Gandhi had attempted to resolve the gender irony of nationalism using a spatial fix, delimiting a sphere in which women could participate politically without alienating the male population. As such, a formerly dormant political population was newly mobilised without levelling the gendered 'separate spheres'. The elasticity of Gandhi's concept of the political allowed a greater number of houses to be politicised. However, this openness allowed the rejection of these principles in favour of either a resubjectivised or a radically active womanhood. Accessing these spaces after such a length of time requires the parameters of academic study to be pushed and interrogated. Personal and family memory must be subject to distortion and error but, as the final repository of information about these domestic spaces, these risks must be accepted and countered using archival investigation and triangulation.

This research has relevance beyond the debate over the subalternist conception of female subjectivity or the nationalism of pre-independence Delhi. It provides evidence to back feminist assertions that women can participate, rather than just be represented, in political movements. Rather than just participating in the confrontational and traditionally masculine public sphere, this activity can take place in the private spaces of the home. As such, following Young, hooks and Honig, the home can be seen as an inventive and participatory space, rather than one of oppression and introspection. Finally, the porosity of the domestic boundary means that studies of urban political movements must attend to the trans-scalar flows of power and identity that link the nation and the 'public' to the locality and the 'private'; flows which are in many cases facilitated by the actions and movements of women, as well as men.

Acknowledgements

This article has benefited from comments made following presentations in Quebec, Cambridge and Los Angeles. I would also like to thank Jim and Nancy Duncan, Nicola Higgins, Gerry Kearns, Francesca Orsini, Linda Peake, Sarah Radcliffe, Simon Reid-Henry and the three anonymous referees for their useful advice and help. This research would have been impossible without the cooperation of the interviewees and the support of all those people upon whom I depended while in Delhi.

NOTES

[1] The Indian National Congress was established in 1885 and became India's largest nationalist organisation, providing the administrative machinery which linked negotiations with the Government to mass movements and social campaigning.

[2] To inaugurate the Civil Disobedience Movement Gandhi and a selection of devoted volunteers marched over 240 miles from Allahabad to Dandi on the west coast of India. Arriving on 5 April 1930, Gandhi made salt from the sea water, breaking the British monopoly on salt production and starting the mass movement.

[3] Hasan Khan, Khan Ahmad (1933) *Census of India 1931, Vol. XVI: Delhi Report and Tables* (Lahore, Civil and Military Press); Fazl-I-Ilahi (1942) *Census of India 1941, Vol. XVI: Delhi* (Simla, Government of India Press).

[4] Delhi Archives, Home Department File (Confidential) 1930 55B.

[5] Delhi Fortnightly Reports 1930, 6 May.

[6] Delhi Fortnightly Reports 1930, 6 June.

[7] Nehru Memorial Museum and Library (hereafter NMML) All India Congress Committee File G94/1930 Part 1.

[8] Sarla Sharma was a student and Communist supporter who was active in the 1940s and came from a politically supportive family.

[9] During the 1940s Mrs Chakravarty was a member of Indraprastha Girls' College, the staff and students of which were consistently involved in nationalist activity during the 1930s and 1940s.

[10] An active socialist worker in the 1940s.

[11] Government of India, Secret (1934) *Delhi Provincial Who's Who* (New Delhi, Government of India Press).

[12] Delhi Fortnightly Report 1932, 15 January.

[13] Begum is a term of respect of Muslim women of high rank.

[14] NMML Transcript 533, Smt Zohra Ansari, recorded by Dr Hari Dev Sharma on 3 November 3 1974, 26.

[15] NMML Transcript, Smt Zohra Ansari, 26.

[16] SN Mukarji became the sixth Principal of St Stephen's College, Delhi, in 1926, having joined the college in 1912 (Monk, 1935, p. 218). The college authorities felt that an Indian would be 'better fitted than a foreigner' to guide the university. This faith was rewarded during the civil disobedience of 1930, during which Mukarji sanctioned a debating forum named 'The Nationalist' in which 'inchoate patriotic fervour' was channelled into practical work such as the promotion of 'positive' swadeshi over 'negative' boycotts (Monk, 1935, p. 233).

[17] This song was popularised by Congress as a nationalist national anthem.

[18] A form of clothing that consisted of cloth wrapped around the waist that was associated with the lower classes.

[19] Kaushalya Lal (1985) 'My sister: my guru' in *Dilli ki Joan of Arc*.

[20] Jats are a diverse subgroup of the *Kshatriya* (warrior) caste, who are often agriculturalists although many were recruited to the British army. They have been condemned for not fully participating in the nationalist movement.

[21] The exact dates were not given in the interview but have been corroborated by archival research and are thus inserted here. For George V, see Rose (1983, p. 401) and Kamala Nehru, see Wolport (1996, p. 199).

[22] NMML Transcript Magan Kishore Zohry, recorded by Dr Hari Dev Sharma, 5.

[23] Asaf Ali found fame during the Quit India movement when she hoisted the Congress flag in Bombay during the mass arrests of August 1942. She returned to Delhi and eluded arrest until the end of the movement. She coordinated the student population as well as pushing for sabotage and bombings throughout the rural and urban realm. She became one of India's underground elite and refused the call of male, more orthodox Congress members in Delhi to return to non-violence.

[24] Sayeeda Kurshid was the daughter of Zakir Hussain, a prominent Delhi educationalist and politician.

[25] NMML Transcript 533 Smt Zohra Ansari, 27.

REFERENCES

AMIN, SHAHID (1995) *Event, Metaphor, Memory: Chauri Chaura 1922–1992* (Berkeley, CA, University of California Press).

ARNOLD, DAVID (2000) *Gandhi* (Harlow, Longman).

BALA, USHA (1986) *Indian Women Freedom Fighters 1857–1947* (New Delhi, Manohar).

BALD, SURESHT R. (2000) The politics of Gandhi's 'feminism': constructing 'Sitas' for swaraj, in: SITA RANCHOD-NILSSON & MARY ANN TÉTREAULT (Eds) *Women, States and Nationalism: at home in the nation?* (London, Routledge).

BANNERJI, HIMANI (2000) Projects of hegemony; towards a critique of Subaltern Studies, 'Resolution of the Women's Question', *Economic and Political Weekly*, 35, 11 March, pp. 902–920.

BASU, APARNA & BHARATI, RAY (1990) *Women's Struggle: a history of the All India Women's Conference 1927–1990* (New Delhi, Manohar).

BAXTER, JAMIE & EYLES, JOHN (1996) Evaluating qualitative research in social geography: establishing 'rigour' in interview analysis, *Transactions of the Institute of British Geographers*, 22, pp. 505–525.

BROWN, JUDITH (1989) *Gandhi: prisoner of hope* (New Haven, CT, Yale University Press).

BURTON, ANTOINETTE (1999) Introduction: The unfinished business of colonial modernities, in: ANTOINETTE BURTON (Ed.) *Gender, Sexuality and Colonial Modernities* (London, Routledge).

BUTALIA, URVASHI (2000) *The Other Side of Silence* (London, Hurst & Company).

CHAKRABARTY, DIPESH (1991) Trafficking in history and theory: subaltern studies, in: RUTHUEN KENNETH KNOWLES (Ed.) *Beyond the Disciplines: The New Humanities* (Canberra, Australian Academy of the Humanities).

CHAKRABARTY, DIPESH (1994) The difference-deferral of a colonial modernity; public debates on domesticity in British Bengal, in: DAVID ARNOLD & DAVID HARDIMAN (Eds) *Subaltern Studies VIII: Essays in Honour of Ranajit Guha* (Delhi, Oxford University Press).

CHAKRABARTY, DIPESH (1998) Minority histories, subaltern pasts, *Economic and Political Weekly*, 33, 28 February, pp. 473–477.

CONNEL, ROBERT WILLIAM (1990) The state, gender, sexual politics: theory and appraisal, *Theory and Society*, 19, pp. 507–544.

CHATTERJEE, PARTHA (1986) *Nationalist Thought and the Colonial World: a derivate discourse* (London, Zed for the United Nations University).

CHATTERJEE, PARTHA (1989) The nationalist resolution of the women's question, in: KUMKUM SANGARI & SUDESH VAID (Eds) *Recasting Women: essays in colonial history* (New Delhi, Kali for Women).

CHATTERJEE, PARTHA (1992) A religion of urban domesticity; Sri Ramakrishna and the Calcutta middle class, in PARTHA CHATTERJEE & GYANENDRA PANDEY (Eds) *Subaltern Studies: writings on South Asian history and society*, vol. 7 (Delhi, Oxford University Press).

CHATTERJEE, PARTHA (1993) *The Nation and its Fragments: colonial and postcolonial history* (Princeton, NJ, Princeton University Press).

EDIB, HALIDÉ (1937) *Inside India* (London, George Allen & Unwin).

FELSKI, RITA (1995) *The Gender of Modernity* (Cambridge, MA, Harvard University Press).

FORBES, GERALDINE (1996) *The New Cambridge History of India IV.2. Women in Modern India* (Cambridge, Cambridge University Press).

FORSTER, EDWARD MORGAN (1912) Letter to Florence Barger, in: *Selected Letters of E. M. Forster V.1: 1879–1920*, ed. Mary Lago & Philip Nicholas Furbank (London, Collins, 1983).

FOUCAULT, MICHEL (1980) Two Lectures: i) 7/1/76, in: MICHEL FOUCAULT *Power/Knowledge: selected interviews and other writings 1972–77*, ed and trans. Colin Gordon (Brighton, Harvester Press).

FOX, RICHARD (1996) Gandhi and feminized nationalism in India, in: BRACKETTE WILLIAMS (Ed.) *Women Out of Place: the gender of agency and the race of nationality* (London, Routledge).

GANDHI, MOHANDAS KARAMCHAND (1920) Speech at meeting of mill-hands, Ahmedabad, in: PUSHPA JOSHI (1988) *Gandhi on Women* (Ahmedabad, Navajivan Publishing House).

GANDHI, MOHANDAS KARAMCHAND (1927) Tear down the purdah, in: PUSHPA JOSHI (1988) *Gandhi on Women* (Ahmedabad, Navajivan Publishing House).

GANDHI, MOHANDAS KARAMCHAND (1928a) The function of women, in: PUSHPA JOSHI (1988) *Gandhi on Women* (Ahmedabad, Navajivan Publishing House).

GANDHI, MOHANDAS KARAMCHAND (1928b) Letter to Dr M. A. Ansari, in: *The Collected Works of Mahatma Gandhi, Volume 41* (Delhi, Publications Division, Ministry of Information and Broadcasting, Government of India).

GANDHI, MOHANDAS KARAMCHAND (1931) The Ansari Dharmasala, in: *The Collected Works of Mahatma Gandhi, Volume 51* (Delhi, Publications Division, Ministry of Information and Broadcasting, Government of India).

GEORGE, ROSEMARY MARANGOLY (1999) *The Politics of Home: postcolonial relocations and twentieth-century fiction* (Berkeley, CA, University California Press).

GREWAL, INDERPAL (1996) *Home and Harem: nation, gender, empire and the cultures of travel* (London, Leicestershire University Press).

HANCOCK, MARY (2001) Home science and the nationalization of domesticity in colonial India, *Modern Asia Studies*, 35, pp. 871–903.

HASAN, MUSHIRUL (1995) *M. A. Ansari* (New Delhi, Publications Division, Ministry of Information and Broadcasting, Government of India).

HONIG, BONNIE (1994) Difference, dilemmas, and the politics of the home, *Social Research*, 61, pp. 563–597.

HOOKS, BELL (1991) *Yearning; race, gender and cultural politics* (London, Turnaround).

IRIGARAY, LUCE (1992) *Elemental Passions*, trans. Joanne Collie & Judith Still (London, Athlone).

JEFFERY, PATRICIA (1979) *Frogs in a Well: Indian women in purdah* (London, Zed Press).

KANDIYOTI, DENIZ (1991) Identity and its discontents: women and the nation, *Millennium*, 20, pp. 429–443.

KANDIYOTI, DENIZ (2000) Guest Editor's introduction. The awkward relationship: gender and nationalism, *Nations and Nationalism*, 6, pp. 491–494.

KASTURI, LEELA & MAZUMDAR, VINA (1994) *Women and Indian Nationalism* (New Delhi, Vikhas Publishing House).

KAUL, TRILOKI NATH (1991) *Aruna Asaf Ali: a profile* (New Delhi, Lancer International).

KISHWAR, MADHU (1986) *Gandhi and Women* (New Delhi, Manushi Prakashan).

KUMAR, NITA (1994) Introduction, in: NITA KUMAR (1994) *Women as Subjects: South Asian histories* (Charlottesville, VA, University Press of Virginia).

KUMAR, RADHA (1993) *The History of Doing: an illustrated account of movements for women's rights and feminism in India* (London, Verso).

LATEEF, SHAHIDA (1990) *Muslim Women in India: political and private realities* (New Delhi, Kali for Women).

LUDDEN, DAVID (2001) *Reading Subaltern Studies: critical history, contested meaning, and the globalisation of South Asia* (Delhi, Permenant Black).

LUMMIS, TREVOR (1998) Structure and validity in oral evidence, in ROBERT PERKS & ALISTAIRE THOMSON (Eds) *The Oral History Reader* (London, Routledge).

MAHAN, RAJAN (1999) *Women in Indian National Congress* (Jaipur, Rawat Publications).

MASSEY, DOREEN (1994) *Space, Place and Gender* (Cambridge, Polity Press).

MAYER, TAMAR (2000) Gender ironies of nationalism: setting the stage, in: TAMAR MAYER (Ed.) *Gender Ironies of Nationalism: sexing the nation* (London, Routledge).

McDOWELL, LINDA (1999) *Gender, Identity and Place; understanding feminist geographies* (Cambridge, Polity Press).

MENON, NIVEDITA (1999) *Gender and Politics in India* (New Delhi, Oxford University Press).

MINAULT, GAIL (1984) Begmati Zuban: women's language and culture in nineteenth-century Delhi, *India International Quarterly*, 11, pp. 155–171.

MONK, F.F. (1935) *A History of St. Stephen's College, Delhi* (Calcutta, YMCA Publishing House).

NANDY, ASHIS (1983) *Intimate Enemy: loss and recovery of self under colonialism* (Delhi, Oxford University Press).

NANDY, ASHIS (1990) Woman versus womanliness: an essay in speculative psychology, in: BAL RAM NANDA (Ed.) *Indian Women; from purdah to modernity* (London, Sangam Books).

O'HANLON, ROSALIND (2000) Recovering the subject: subaltern studies and histories of resistance in colonial South Asia, in: VINAYAK CHATURVEDI (Ed.) *Mapping Subaltern Studies and the Postcolonial* (London, Verso).

ORSINI, FRANCESCA (2002) *The Hindi Public Sphere: language and literature in the age of nationalism* (Delhi, Oxford University Press).

PANDEY, GYANENDRA (1995) Voices from the edge: the struggle to write subaltern histories, *Ethos*, 60, pp. 223–242.

PATEL, SUJATA 1988 Construction and reconstruction of woman in Gandhi, *Economic and Political Weekly*, 23, pp. 377–387.

PERKS, ROBERT & THOMSON, ALISTAIRE (1998) *The Oral History Reader* (London, Routledge).

PRAKASH, GYAN (1990) Writing post-orientalist histories of the Third World: perspectives from Indian historiography, *Comparative Studies in Society and History*, 32, pp. 383–408.

PRATT, GERALDINE (1998) Geographic metaphors in feminist theory, in SUSAN HARDY AIKEN, ANN BRIGHMAN & SALLIE A.MARTSON (Eds) *Making Worlds: gender, metaphor, materiality* (Tucson, AZ, University Arizona Press).

RADCLIFFE, SARAH A. (1999) Embodying national identities: *mestizo* men and white women in Ecuadorian racial-national imaginaries, *Transactions of the Institute of British Geographers*, 24, pp. 213–225.

RAHEJA, G.G. & GOLD, A.G. (1994) *Listen to the Heron's Words: reimagining gender and kinship in North India* (Berkeley, CA, University of California Press).

RANCHOD-NILSSON, SITA & TÉTREAULT, MARY ANN (Eds) (2000) *Women, States and Nationalism: at home in the nation?* (London, Routledge).

ROSE, GILLIAN (1996) As if the mirrors had bled: masculine dwelling, masculinist theory and feminist masquerade, in NANCY DUNCAN (Ed.) *Bodyspace: destabilising geographies of gender and sexuality* (London, Routledge).

ROSE, KENNETH (1983) *King George V* (London, Weidenfeld & Nicolson).

SARKAR, TANIKA (1998) Orthodoxy, cultural nationalism and Hindutva violence: an overview of the gender ideology of the Hindu Right, in: RUTH ROACH PIERSON (Ed.) *Nation, Empire, Colony: historicizing gender and race* (Bloomington, IN, Indiana University Press).

SHARP, JOANNE P. (1996) Gendered nationhood: a feminist engagement, in: NANCY DUNCAN (Ed.) *Bodyspace: destabilising geographies of gender and sexuality* (London, Routledge).

SLADE, MADELINE (1960) *The Spirit's Pilgrimage* (London, Longman).

TARLO, EMMA (1996) *Clothing Matters: dress and identity in India* (London, Hurst & Company).

THAPAR-BJÖRKERT, SURUCHI (1997) The domestic sphere as a political site: a study of women in the Indian nationalist movement, *Women's Studies International Forum*, 20, pp. 493–504.

THAPAR-BJÖRKERT, SURUCHI & RYAN, LOUISE (2002) Mother India/Mother Ireland: comparative gendered dialogues of colonialism and nationalism in the early twentieth century, *Women's Studies International Forum*, 25, pp. 301–313.

VISWESWARAN, KAMALA (1996) Small speeches, subaltern gender: nationalist ideology and its historiography, in SHAHID AMIN & DIPESH CHAKRABARTY (Eds) *Subaltern Studies: writings on South Asian history and society*, vol. 9 (Delhi, Oxford University Press).

WOLPERT, STANLEY (1996) *Nehru: a tryst with destiny* (New York, Oxford University Press).

YOUNG, IRIS MARION (1997) *Intersecting Voices: dilemmas of gender, political philosophy, and policy* (Princeton, NJ, Princeton University Press).

YUVAL-DAVIS, NIRA & ANTHIAS, FLOYA (1989) *Woman–Nation–State* (London, Macmillan).

Part IV
Space, Imperialism and Postcolonialism

[12]

Between the book and the lamp: imaginative geographies of Egypt, 1849–50

Derek Gregory

This essay compares the imaginative geographies of Egypt produced by Florence Nightingale and Gustave Flaubert as they travelled up the Nile Valley and back to Cairo in 1849–50. Their experiences are used to emphasize the physicality (rather than merely the textuality) of travel writing. The differences between their imaginative geographies and in particular between their representations of landscape, space and people, illuminate the complex and fractured formation of Orientalism as a constellation of power, knowledge and spatiality, and its entanglements with patriarchy, sexuality and various colonialisms.

key words Egypt imaginative geographies Orientalism travel writing

Professor, Department of Geography, University of British Columbia, 217–1984 West Mall, Vancouver, British Columbia V6T 1Z2, Canada

revised manuscript received 13 June 1994

For the philosophy of history, what country stretches out its hands to press such facilities upon us as Egypt?
Florence Nightingale, *Letters from Egypt 1849–1850*

Let's not get lost in archaeology – a widespread and fatal tendency, I think, of the coming generation . . .
Gustave Flaubert to Louis Bouilhet, Cairo, 27 June 1850

Introduction

My title is indebted to two awkward companions: Michel Foucault and Edward Said. It was Foucault who posited the discovery of what he called 'a new imaginative space' in the middle of the nineteenth century. He suggested that conjuring up dreams of the fantastic was no longer confined to the stillness of the night but now took place in the hushed precincts of the modern library. Hence his claim that 'the imaginary now resides between the book and the lamp'.[1] This was a specifically European space, at least as Foucault described it, but when travellers ventured beyond Europe they took their pre-texts with them. In fact, of course, they had done so for centuries. The voyages of discovery and other expeditions and explorations all took place within a complex web of textualizations in which dreams of the fantastic were captured in intricate display. And it is that process of spinning, capturing and displaying, a process of inscription, uneven and unequal, that preoccupied Said when he urged the critical reading of what he called 'imaginative geographies'. These are figurations of place, space and landscape that dramatize distance and difference in such a way that 'our' space is divided and demarcated from 'their' space. For Said, imaginative geographies are discursive formations, tense constellations of power, knowledge and spatiality, that are centred on 'here' and projected towards 'there' so that 'the vacant or anonymous reaches of distance are converted into meaning for us here'. He derived the basic idea from Gaston Bachelard, but Said's particular purpose was to show how such a poetics of space – which was Bachelard's concern – is simultaneously a politics of space. This fed into his critique of Orientalism as a discourse that worked through a representation of space in which the Orient was constructed as a theatrical stage on which the Occident projected its own fantasies and desires.[2]

Although Said made no secret of his admiration for and indeed his debt to Foucault, the connections between the two of them have always seemed fractious. Critics have usually drawn attention to the difficulties of coupling Foucault's anti-humanism to Said's own humanism.[3] But I am more interested here in the ways in which they invoke spatiality and I want to make two comments that bear directly on what follows. First, Foucault insisted that his pre-occupation with space was more than metaphorical and that it enabled him to identify 'those points at which discourses were transformed in, through and on the basis of relations of power'.[4] In much the same way, Said has consistently remarked on the worldly nature of representations of space and on the relations of power in which they are implicated:

> Just as none of us is outside or beyond geography, none of us is completely free from the struggle over geography. That struggle is complex and interesting because it is not only about soldiers and cannons but also about ideas and forms, about images and imaginings.[5]

It is those images and imaginings, and their dense imbrications in the materialities of travel and colonialism, that are my main concern in the present essay. I want to move critical discussion of Orientalism beyond the library, to disrupt the usual distinctions between the text and the world, and recover the ways in which the physical passage of European travellers through other landscapes and other cultures marked the very process of their writing and their representations of those spaces. Secondly, both Foucault and Said offer accounts of the production of space that focus on what John Rajchman has called 'spaces of constructed visibility'. Their spatial analytics are not the same to be sure but they both focus on sites of enframing and envisioning, strategies of 'making seen' and 'making scene'.[6] My particular interest is in the production of imaginative geographies in the middle of the nineteenth century that brought Egypt within European horizons of intelligibility and visibility, and so in much the same way I will examine a series of discursive strategies that rendered Egypt as a text to be read or as the object of a gaze.

I accentuate the plural and the indefinite, however, because I want to resist the tendency in some versions of cultural theory and cultural geography to essentialize 'the' text or 'the' gaze. Instead, I insist that the processes of textualization and visualization that were involved in the construction and colonization of Egypt were complex, scored through multiple and often contradictory subject-positions. In saying this, I follow those critics who have found Said's construction of Orientalism to be too homogeneous. As he has since conceded, any account of its imaginative geographies must of necessity attend to the uneven topographies of these discursive formations.[7] For this reason I have found Lisa Lowe's reformulation of Said's project in more Foucauldian terms extremely helpful. Her use of Foucault's concept of *heterotopicality* is especially instructive in the present context because it conveys the multiple, interpenetrating and often antagonistic sites through which various constellations of power, knowledge and spatiality are put in place. She thus treats Orientalism as heterogeneous and contradictory, and makes two main claims:

> On the one hand, that Orientalism consist[s] of an uneven matrix of Orientalist situations across different cultural and historical sites, and on the other, that each of these Orientalisms is internally complex and unstable.[8]

My own view is that these multiple sites, instabilities and ambiguities assume a particular importance in the case of European travel writing in the nineteenth century, where constellations of class, culture and gender could be even more tense than they were at home, where the friction of distance between one site and another had an insistently physical dimension and where identities were often labile and subtly renegotiated in the course of the passage.[9]

With these differences in mind I return to my title because it also signals, somewhat more obliquely, two even less likely travelling companions: Florence Nightingale, whose nursing service with the British Army during the Crimean War (1854–6) earned her the soubriquet of the Lady with the Lamp – according to her (premature) obituary, the Victorian public pictured her carrying a lamp 'through miles of sick soldiers in the middle of the night' – and Gustave Flaubert, one of France's greatest novelists, whose writings were in fact the immediate inspiration for Foucault's 'fantasia of the library' to which I referred at the beginning. These are the principal subjects of my paper. They travelled through Egypt at more or less the same time, though quite independently of one another, when they were both in their late twenties. In their letters and other travel writings they mapped a series of imaginative geographies in exactly Said's sense of that phrase. As one might expect, given their ages, bourgeois origins and the cultural baggage of nineteenth-century Europe inevitably carried with them, they shared a number of assumptions and responses. But there were also significant differences between them – one a woman, one a man; one British, one French – that marked their constructions in other, equally salient ways. I hope that my emphasis on the physicality of their journeys helps to foreground their different subject-positions and the different degrees of freedom that framed their actions. In particular, there were experiences open to Flaubert which, by virtue of her gender, were denied to Nightingale. This bears directly on the intersections of Orientalism with other discursive formations, on those critical terrains in which, as Lowe says, class, 'race', gender and sexuality reinforce, cross-cut or contradict one another.

I will begin with Florence Nightingale and then move to Gustave Flaubert, following each of them from Alexandria to Cairo and then tracing their voyages up the Nile Valley and back (see Fig. 1). As my argument proceeds, however, I will also try to layer my account by making a number of comparisons between their readings. I will conclude with some general reflections on the relations between imaginative geographies and the production of colonial space.

A world turned upside down: Florence Nightingale in Egypt

'Cairo', wrote Eliot Warburton of his visit in 1843, 'is now the crowded thoroughfare of England and India; our flag has become as familiar to the Arabs of the Red Sea as to the people of Alexandria'. By the early 1840s there were regular P & O steamship services from Southampton to Alexandria and, by the end of that busy decade, four English steamers taking passengers up the Nile to Cairo; a string of post-houses dotted the desert between Cairo and Suez; and a regular steamship service made its way from Suez down the Red Sea to Bombay and Calcutta.[10] If most British travellers who visited Egypt in the first half of the nineteenth century were *en route* between Britain and India, however, an increasing number of them had come to see Egypt. One such was Florence Nightingale, who was then 29 years old. She had already caused her parents great consternation by announcing her intention to become a nurse. Like others of their class and generation in early Victorian Britain, they regarded the nursing profession as at best a rung above prostitution and pleaded with her to consider a career as a teacher or, better still, marriage to a suitable young man. But in vain. With her parents' approval – a sentiment which was not shared by other members of the family circle – Florence embarked on a tour to Egypt in the company of Charles and Selina Bracebridge. They were wealthy and accomplished travellers who had taken Florence to Italy with them a few years earlier, and all three of them had a keen interest in antiquity.

The Bracebridges, together with Florence and her maid 'Trout', arrived in Alexandria on 18 November 1849 and a week later the party set off by steamer for Cairo. They spent three weeks at the Hôtel de l'Europe on the Ezbekeiah, and then hired a dahabeeah (an Egyptian houseboat) for the Nile voyage. Although the boat had never carried Europeans before – apparently it had been built for a Bey's harem – it was much the same as the other houseboats hired by European travellers of means (Fig. 2). Its accommodation consisted of a day cabin, with a divan around the walls, and two sleeping cabins separated by a passage containing large closets. The crew lived on the open deck and meals were cooked in a small kitchen – little more than an open box – at the prow. The dahabeeah was chartered for three months at £30 per month. Provisions were stored in two chests on the deck and a basket of bread and two cages of oranges and meat hung overhead. These supplies were to be supplemented during the voyage: the crew baked bread in village ovens, Charles Bracebridge went out shooting, and milk, butter and other provisions were bought as circumstances allowed. The British flag was hoisted at the stern, Bracebridge's colours run up the rigging and Florence embroidered in Greek letters the name they had chosen for the dahabeeah – *Parthenope*, her sister's name[11] – on a blue pennant that fluttered at the yardarm. The party set off upriver on 4 December and, like other travellers, intended to make Aswan and the First Cataract as fast as possible. Such a plan almost always involved 'tracking' for much of the time, when the boat would be pulled on a long rope by the crew; Florence claimed this was because they had no idea of sailing or tacking. Most parties turned back at the First Cataract and sailed downriver with the current, inspecting the antiquities and sites at their leisure. But the Bracebridges and Florence were determined to travel up into Nubia and to get as far as Abu Simbel, which they made on 15 January 1850. They returned slowly and were not back in Cairo until

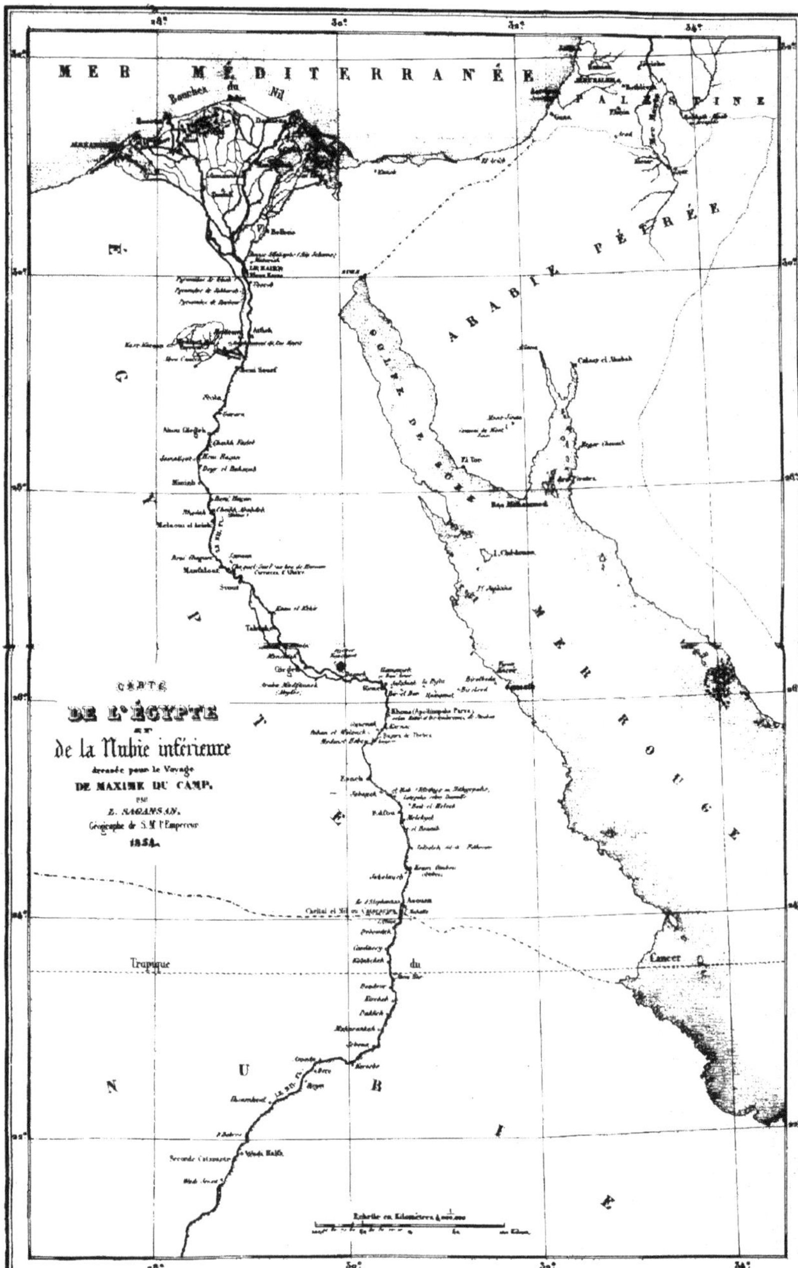

Figure 1. Egypt. This was the map produced to accompany Maxime Du Camp's version of his travels through Egypt with Gustave Flaubert

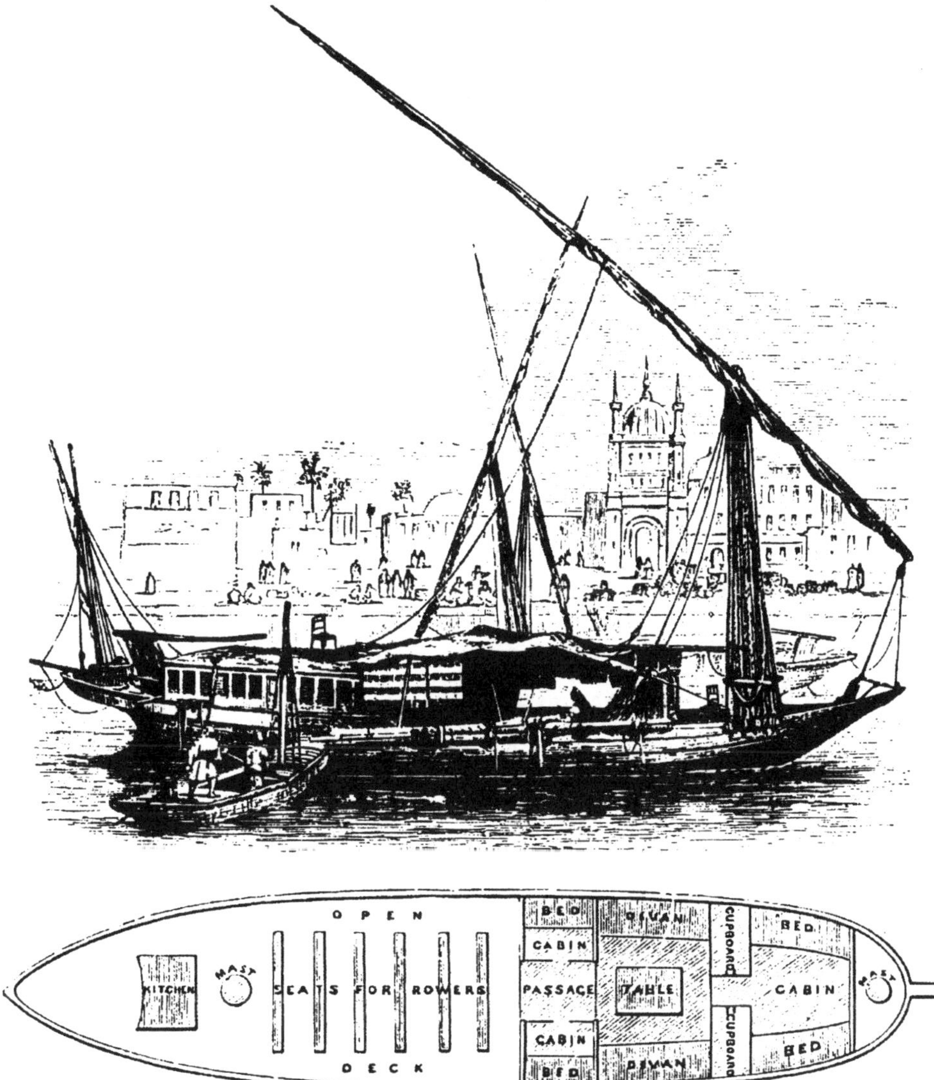

Figure 2. A dahabeeah (houseboat) on the Nile

mid-March. The following month the party left from Alexandria on their way home through continental Europe to England.

My account of Florence Nightingale's travels in Egypt is drawn from the long letters she sent home from Alexandria, Cairo and the dahabeeah. There is no record of the letters the party received while they were in Egypt, although Florence often refers to her pleasure at receiving them. The mails were, of necessity, haphazard beyond Alexandria and Cairo, and Florence always kept a letter 'ready and sealed' in case of a chance meeting with another boat. She excused her brief dispatch at one point by explaining that she was 'writing in the greatest possible haste for a steamer(!) which ... is going on to Cairo'. Still, making allowances for the uncertain progress of a Nile voyage, communication could be quite rapid. On 21 January, for example, Florence wrote: '*At last* I have a letter from you – dated 22nd November'. Two months must have seemed a long time to her, but it was really rather short, all things considered. All her letters were sent to her immediate family. She usually addressed them collectively ('My people') but sometimes the letters were addressed specifically to her mother; in any event, they were presumably read by the entire family because her sister edited and published them privately in 1854.[12]

It was by no means uncommon for such letters to be circulated in this way, but the prospect of publication was not uppermost in Nightingale's mind when she wrote them.[13] She looked upon the adventure as an occasion to gather her thoughts about her future. For many European visitors, however, the Nile voyage was fast becoming an extended house party with its own elaborate social conventions. For example:

> Every boat ascending the Nile hoists the flag of the country to which its proprietor belongs. Besides this, each traveller, before leaving Cairo, adopts a private flag, and registers it at the hotels with his own name and that of his boat. Thus, every stranger, on arriving at Cairo, learns who is 'up' the river, and for what flag to look.[14]

As this etiquette implies, many – perhaps most – of these European travellers sought out the company of their fellows. In Egypt, Nerval declared, 'any European becomes, in the eyes of another, a *Frank*, in other words a compatriot', and 'the map of our little Europe ... loses its fine distinctions'.[15] During the day their dahabeeahs would race one another and, if they found themselves moored together at night – as they usually did – invitations would be extended to the neighbours and, after tea, a game of bridge or even dinner, the day might end with an entertaining display of fireworks. The *Parthenope* was – inevitably and unavoidably – caught up in these trappings, the signs of European tourism and colonialism, but for the most part Florence and the Bracebridges avoided the company of other travellers. At Aswan, they stole away before sunrise to avoid the 'ruck' of English boats and, when the Hungarian count, the German professor and any one of a number of other visitors made their social calls at Thebes, Charles Bracebridge retired to bed at

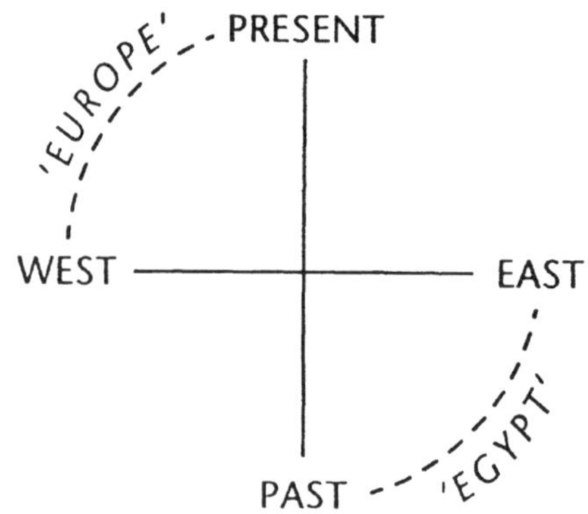

Figure 3. 'The two ends of Time and Space'

the first 'dreadful plash of oars', leaving his wife and Florence to 'unwillingly but nobly sacrifice ourselves to our duty' – though they soon learned never to serve tea, 'which has greatly limited these visitations'. As Nerval cautioned, 'the moment one meets one of these worthy travellers, all is lost. Society takes possession of you'. Florence and the Bracebridges understood that only too well: 'It is very hard to be all day by the deathbed of the greatest of your race', she wrote, 'and to come home and talk about quails or London'.[16]

The journey was a serious undertaking for all three of them, then, and for this reason Nightingale's letters also cast considerable light on her views of the Egyptian past and the European present. On several occasions she described Egypt and Europe as 'the two ends of Time and Space' and I want to read her letters through the grid which this suggests (Fig. 3).[17] The figure has several implications even in this dramatically simplified form but I will pay most attention to the way in which she sought on the one side to establish an essential distance and difference between Egypt and Europe while on the other side simultaneously moving to annul that separation.

Distance and difference: de-naturing the Egyptian landscape

In the first place, Florence Nightingale had immense difficulty in coming to terms with – in finding the terms for – the Egyptian landscape. 'All the colours of Africa are those of precious stones', she wrote, and the Nile itself appeared to her (as it did to countless other travellers) as a sheet of 'molten gold'. The sense of astonishment, of wonder, is unmistakable. But it was not to last long. These metaphors had a materiality that was profoundly disconcerting. She determined that the whole valley seemed

> so unlike nature – the descriptions of the gardens in the Arabian Nights, with the precious stones, seem no longer here fantastic or exaggerated – it is the description of the country.

And this 'un-natural nature' soon produced a feeling of unease, even of dislike:

> I cannot describe the unnatural colouring, a bright line of yellow green bordering the Nile, barley or lupins, the hard brown of the desert behind, a white ghastly Cairo in the background, dabs of Prussian-blue-and-gamboge trees stuck about. It looked as if a child had painted it and did not know how, and had made it unlike nature.[18]

Eventually, even these garish colours were bled from the landscape. 'The colourlessness of Egypt strikes one more than anything', she wrote. 'In Italy there are crimson lights and purple shadows; here there is nothing in earth, air, sky, or water, which one can compare in any way with Europe; but with regard to absence of colour, it is striking'.[19]

This Egypt was not only unlike any 'nature' she had ever known in its colours and compositions. In contrast to the familiar, changing landscapes of Europe, it seemed to her petrified, literally mortified. The desert was diabolical – 'an earth tumbled up and down; not as if Providence had made it so, but as if it had been created otherwise'; 'one almost fancies one hears the Devil laughing'[20] – and it was only the heavens that lifted her heart. Her letters accordingly made much of the opposition between the beauty and the animation of the sky and the deadness of the earth below.

> While the earth in our country is so rich and variegated with light and crowded with animation, the sky above contrasts by its deadness. Here, on the other hand, the sky is radiant, the light is living . . . One looks down, and the ungrateful earth lies there, hopeless and helpless, a dying, withered desert.[21]

Perhaps not surprisingly her imagery, as she passed through the nightmare landscape of tombs and temples, increasingly turned from the un-natural to the spectral. The European aesthetic literally deserted her. 'Travelling here is nothing like the tourist seeing sights of beautiful art and sunny landscape in Italy or the South of France', she wrote: 'It is like the ghoul haunting the tombs . . .'[22] She described the island of Elephantine thus:

> The island looks as if it were a world turned upside down, and then stirred up . . . It was such a world as might have been turned out of the cauldron of *Macbeth*'s Weird Sisters . . . It was as if a devil had been there, heaving underneath, upturning, tossing and tumbling it till everything was in atoms and confusion.[23]

Even here, of course, Europe was not far away, and Nightingale repeatedly called on Shakespeare's *Macbeth* to conjure up this strange and terrible world. But the image never fully satisfied her. 'It is useless to try to describe these things', she wrote at Syene, 'for European language has no words for them. How should it, when there is no such thing in Europe? All other nature raises one's thoughts to heaven: this sends them to hell'.[24] Although women travel-writers typically despaired of their attempts at description – Billie Melman suggests that they were 'painfully aware of their own inadequacy as writers' and repeatedly confessed the limitations of their powers of observation[25] – Nightingale's frustration turns not so much on the inadequacy of her descriptions as on the inadequacy of their *object*: the Egyptian landscape. It was, to her, a world turned upside down, an inversion of the ordered and Christian world of Europe: a Biblical landscape where the devil was now abroad.

The contemporary inhabitants of Egypt were identified with – and on occasion directly assimilated to – this diabolical un-nature. When Arab porters and hawkers boarded the ship at Alexandria, she was already persuaded that they were 'an intermediate race between the monkey and the man'. On shore, rattling through dusty streets to her hotel, the city – apart from 'the Frank Square' where the Europeans lived – reminded her of 'a vast settlement of white ants'. Upriver, she found huts in villages 'all tumbled together, up and down, as animals might build their nests, without regularity or plan'. In her eyes the city of Asyut, the capital of Upper Egypt, was no better: 'the sort of city the animals might have built when they had possession of the earth'. Sailing away from Luxor she was shocked at what she took to be the 'voluntary debasement' of the people. 'To see human beings choosing darkness rather than light', 'choosing to crawl upon the ground like reptiles', 'it seemed as if they did it on purpose to be as like beasts as they could': 'It felt as if one had trodden in a nest of reptiles'.[26] Then, at Durr, the capital of Nubia, the textual recoil, the shudder of revulsion becomes palpable:

> A sycamore by the river's shore, which was the coffee-house, was the only thing human – the white domes or bee-hives, the mud walls without windows, which enclosed a yard, in the corners of which were the lairs of families, the nests of little naked children, squatting between two stones (like nests of young foxes) – running away when you looked at them, and then baying like jackals after you . . . things about four months old, climbing about like lizards . . . Whether I was Gulliver or Captain Cook, I don't know, but certainly all this was as much out of our common habits of thought as if I had been either.[27]

A sycamore tree 'the only thing human', and human beings living in 'lairs' and 'nests', 'baying like jackals' and climbing 'like lizards': these were hideous commonplaces of colonial discourse.[28] In fact, Nightingale had prepared herself by reading the adventures of Mungo Park and James Bruce but she was now persuaded that their accounts completely failed to convey the 'debasement and misery' of African villages. At Elephantine she and her companions were greeted by what she called 'troops of South Sea savages' and, when she gazed down on six other houseboats moored at Mahatta, below the First Cataract, she described them as 'the English fleet' and thought it 'exactly like a wood-cut in one of Captain Cook's voyages – the savage scene, the neat English boats and flags in the little bay'.[29]

But there was a profound difference and one which required a second rhetorical movement. David Spurr has identified a rhetoric of negation as one of the characteristic tropes of the European colonial imaginary, and suggests that it involves both a 'negative space' and a 'negative history'.

> The discourse of negation denies history as well as place, constituting the past as absence, but also designating that absence as a negative presence: a people without history is one which exists only in a negative sense; like the bare earth, they can be transformed by history, but they cannot make history their own.[30]

But this could hardly be deployed very effectively in Egypt, which was the cradle of antiquity, of ancient civilization; or, at any rate, it required the metaphorical displacement of the present population. 'South Sea savages' were thus literally *out of place* in Egypt: they did not belong to its past and had no claim on its future. As I must now show, however, there was another, more general rhetorical strategy through which Nightingale articulated this discourse of negation.

Passages to antiquity: supernatural landscapes

In the second place, Nightingale effectively *depopulated* these landscapes of their human inhabitants and it was by this means that she was able to annul the distance between past and present. This was a familiar strategy too, of course, and Mary Louise Pratt has identified the effacement of human presence as a characteristic trope of colonial discourse. 'The residents of the country turn up in the narration mainly as traces on the landscape', she writes, as 'scratches on the face of the country'.[31] But this is given a peculiar power in Nightingale's letters by the metaphorical conversion of native people into beasts and reptiles, by the identity she presumes between them and 'un-natural nature', and also by the elective affinity between this 'dead nature' and what John Barrell has seen as a death-wish, a fantasy of extinction common in the writings of Victorian travellers in Egypt. Perhaps, as he concedes, this is only ill-considered hyperbole, a wish for silence and space in which to contemplate the sublime. Certainly, she often wished to be alone. And yet perhaps it also signals a real desire for a violent end to the horrors of the present: in Egypt, Nightingale declared, 'there is such a Past, no Present, and for a Future one can only hope for extinction!'[32]

Be that as it may, Nightingale simultaneously *repopulated* these landscapes with figures of the dead drawn from Egypt's ancient past. And she did so through a device that Walter Benjamin later made central to his vision of the Arcades Project: the dream. Nightingale's simple use of this dream motif bears little comparison with Benjamin's richly textured writings but there is something about his reflections on ruins, on dreamworlds and on wish-images that can be turned back on her writings. For Benjamin, modernity was itself a dreamworld but one whose dreams could be dispelled from within, so to speak, 'dissolving mythology into the space of history'. Nightingale's writings led in the opposite direction, dissolving history into the space of mythology, but her use of the dream motif can also be read (with Benjamin) as an attempt 'to overcome and transfigure the deficiencies of social reality'.[33] But *whose* social reality?

It should be said at once that 'dreaming' had an acutely personal significance for Florence

Nightingale. Throughout her late teens and into her twenties she had been consumed by 'the habit of dreaming'. Her diaries document how 'she fell into "trance-like" states in the midst of ordinary life' and how she gave way (and here one imagines Nightingale and her biographer shuddering together) 'with the shameful ecstasy of the drug-taker'. By 1849, Cecil Woodham-Smith maintains, she was drawing towards a crisis:

> 'Dreaming' became uncontrollable. She fell into trances in which hours were blotted out, she lost sense of time and place against her will. In daily life she moved like an automaton, could not remember what had been said or even where she had been. Agonies of guilt and self-reproach were intensified by the conviction that her worst fears were being realised and that she was going insane.
>
> Again and again she made resolutions to end dreaming, to 'tear the sin out', to 'stamp it out' – but they were always broken.[34]

Woodham-Smith attributes these 'dreams' to a spiritual crisis – to a 'call from God' whose exact nature remained undisclosed for years and which tormented Nightingale almost beyond endurance – but it would not be difficult to see a crisis of repressed sexuality in them too. In any event, Egypt pushed her resolve to the limit. Although her letters home gave no outward sign of her secret agonies, I suspect that her recourse to the dream motif may have been, in some measure, cathartic. It is important to distinguish between the dreams and the dream-images, of course, but there is a link between them. Woodham-Smith describes Nightingale's dreams as consoling visions, as withdrawals from the demands (and denials) of her everyday life, and in Egypt her dream-images gave her access to a parallel world that was, in its way, comforting and perhaps even consoling.

The dream-image was a commonplace of Victorian travel writing but it has a particular resonance in the case of Florence Nightingale in Egypt.[35] It suggests both her inner turmoil, her sense of disengagement and the disappointment she recorded in her small black notebook at the effect present-day Egypt was having on her, but also and, so to speak, redemptively, her proximity to Egyptian mythology and its intimations of (im)mortality.[36] Entering into that parallel world of the past was also perhaps a way of comprehending – of mapping? – her shifting subjectivity. Bénédicte Monicat has suggested that many women travellers in the nineteenth century came to question their identities, that they could 'neither locate themselves in the context of their "previous existence" nor [were] they certain of what they [had] become through their travels'. She argues that this process of estrangement was characteristically linked directly to a gendered identity – that it was an 'estrangement from the norm of the feminine' –and it is, I think, significant that both of her examples involve dream-images.[37] These suggestions might assume even greater significance here because it was during her travels in Egypt that Nightingale determined that she had been 'called' to nursing, which was itself coded, gendered and sexualized in ambiguous and complex ways.[38] But these are beads of speculation strung on a thread of autobiography and I am reluctant to attach too much weight to them. I am more interested in the consequences of the dream-image for the connections Nightingale saw between ancient Egypt and modern Europe.

To Nightingale, as to many other European visitors, Cairo was the locus of the fairy-tale world of the Arabian Nights.[39] Before she set off upriver she set down her first impressions.

> No one ever talks about the beauty of Cairo ever gives you the least idea of this surprising city. I thought it was a place to buy stores at and pass through on one's way to India, instead of its being the rose of cities, the garden of the desert, the pearl of Moorish architecture, the fairest, really the fairest place of earth below . . . Oh, could I but describe those Moorish streets, in red and white stripes of marble; the latticed balconies, with little octagonal shrines, also latticed, sticking out of them, for the ladies to look straight down through; the innumerable mosques and minarets; the arcades in the insides of houses you peep into, the first storeys meeting almost overhead, and yet the air with nothing but fragrance in it, in these narrowest of narrow wynds! But there are no words to describe an Arabian city, no European words at least . . .[40]

This is a far cry from her harsh impressions of the landscape of the Nile Valley but notice how her final sentence situates the city. She is indeed dissolving history into the space of mythology – into the space of the Arabian Nights – and she later underlines the distinction: 'Cairo is not Egyptian', she announced, 'it is Arabian'. Her language failed her here, so she seemed to be saying, not because the experience was un-natural but precisely because it was out of nature, beyond nature, virtually *super*-natural. When she returned to Cairo at the end of her voyage she remained convinced that it was impossible to describe it. 'You might as well try to record a dream', she wrote, 'and I do not know whether the waking dream of the living city within or the silent vision of the dead city without is most unreal'.[41] And she presented the journey upriver – which separated these two encounters – as a voyage into the supernatural.

> You feel, as you lie on the divan, and float slowly along, and the shores pass you gently by, as if you were being carried along some unknown river to some unknown shore, leaving for ever all you had known before – a mysterious feeling creeps over you, as if it were the passage to some other world . . . You lose all feeling of distance [and] all feeling of identity too, and everything becomes supernatural . . .[42]

Indeed, 'when in the evening a spectral bark with its glassy white sails comes standing in silently like a phantom ship upon this molten sea, you fancy yourself anywhere but in this world'. She and her companions were being 'carried away in a phantom-ship to Jinnee countries' and, later that night, when one of the crew roused himself from the deck, 'scattering fire upon the waves from his hand', he looked 'like anything certainly but a man'.[43]

Like other travellers, Nightingale saw the Nile passage opening the European present directly into the Egyptian past. Six years earlier Warburton had felt that 'as you recede from Europe further and further on, towards the silent regions of the Past, you live more and more in that Past'.[44] Similarly, for Nightingale and no doubt for many other Europeans, crossing that threshold was, in its way, a limit-experience 'perfectly distinct from that in any of our *living* countries' (my emphasis) and yet at the same time only accessible to travellers *from* those places.

> It is like going into the Sun, and finding there is not one living being left; but strewed about, as if they had just been used, all the work, books, furniture, all the learning, poetry, religion of the race, all the marks fitted to give one an idea of their mind, heart, soul and imagination, to make one feel perfectly acquainted with their thoughts, feelings and ideas, much more so than with those of many of one's own kin . . . But still, the star is a deserted one . . . [and] that Egypt is all but uninhabited. The present race no more disturbs this impression than would a race of lizards, scrambling over the broken monuments of such a star. You would not call *them* inhabitants, no more do you these.[45]

Brushing away the contemporary inhabitants of Egypt thus, she wondered at how she and her companions had 'become so completely inhabitants of another age' and 'how completely we were living in the time of 4000 years ago'. What made this possible was partly the ruins themselves and partly her educated sensibility. She marvelled at the paintings in the tombs and their delineation of 'all the details of everyday life'; the sheer prosaicness of it all afforded what she saw as a 'magic lantern glimpse into the domestic economy of 4000 years ago'. At the same time, she realized, 'every monument is its own interpreter' and the decipherment of the hieroglyphs had made it possible for educated European travellers – but not, by implication, the barely human Arabs who lived among the ruins – to appreciate Egypt's ancient past and hence to claim it as their own.[46] She herself had consulted with eminent Egyptologists, read widely in Orientalist literatures and, according to her sister, had set off for Egypt laden with learned books. And what did she find? At Beni Hassan she declared:

> All that one wants to know is that on this soil nearly 4000 or 5000 years ago men stood who felt and thought like us . . . I think the Egyptian must have been very much like some of the English clergy wives of the present day, who preach out of the Old Testament and make muslin curtains.[47]

Less extravagantly, but otherwise similarly, she found it 'astonishing how alike the human heart is in all periods and climates'; at Thebes she saw 'the same feelings we have in every sculpture and tomb and temple here'.[48]

Ambivalence and the philosophy of history

Yet this double strategy of estrangement/effacement, however consoling its vision may have been in personal terms, was shot through with a mesmerizing ambivalence. Nightingale was frankly appalled by most of what she saw of Egypt's present and not only the grinding poverty and misery of ordinary people. She also hated the physical violence, the resort to the cudgel and the bastinado, and was taken aback by 'what it is to be a woman in these countries'. And she realized that this could not all be laid at the door of the common people, however 'un-natural' she supposed them to be and no matter how 'voluntary' their 'debasement'. She dismissed the successive rulers of Egypt – Muhammad 'Ali, Ibrahim and Abbas – as being as bad as one another, presiding over 'the kingdom of the devil', but she reserved her most withering contempt for the present Pasha, Abbas, whom she deemed 'below all sentiment, either for the glories of his temples or the miseries of his people'. On occasion, however, and in the margins of her accounts, she conceded that these indignities and horrors were present in Egypt's past too. She disliked the Pyramids intensely for this very reason: they were a record of 'the most hideous oppression in the world', a monument to tyranny, and so she was not in the least surprised that her 'first impression of them should be nothing but repulsive'.[49]

If these continuities could be held in suspense most of the time, so that she could find solace of some kind in antiquity, it was much more difficult to negotiate the identities and commonalities she presumed between Egypt as the cradle of ancient civilization and Europe – and most particularly Britain – as the climax of modern civilization. For they surely called into question the very notion of 'progress' inscribed within European conceptions of modernity:

> It is good for a man [*sic*] to be here – good for British pride to think, here was a nation more powerful than we are, and almost as civilised, 4000 years ago – for 2000 years already they have been a nation of slaves – in 2000 years where shall we be? – Shall we be like them?[50]

Not surprisingly, this admonition soon turned into trepidation:

> It cannot be a law that all nations shall fall after a certain number of years. God does not work in that sort of way: they must have broken some law of nature which has caused them to fall. But are all nations to sink in that way? . . . And will England turn into Picts again . . . as Egypt has turned into Arabs?[51]

One might expect such concerns to have had a special resonance in the wake of the revolutions of 1848; that famous 'springtime of peoples' must have seemed quite otherwise to many members of the bourgeoisie. Even those separated from continental Europe by the English Channel had been rudely awakened by the spirited Chartist campaigns of the 1840s and, after the repeal of the Corn Laws in 1846, the middle class was probably even more alienated from the British working-class movement.

But the system of identities and oppositions set out in Figure 3 had another consequence for Nightingale's philosophy of history. Egypt and Europe, antiquity and modernity, were opposing poles and this made it virtually impossible for her to recognize 'modernizing Egypt' as it came into being at the centre of this force-field of tension. She was able to get her bearings on the Orient by using European constructions of Nature and History as sighting devices, but she deployed them in such a way that the idea of 'modern Egypt' simply could not be thought within the compass of her conceptual grid. Certainly, at mid-century it was a blurred and jagged image. If Muhammad 'Ali had been the modernizing autocrat, opening Egypt to Europe, introducing new cash crops and new agricultural practices, dredging and deepening irrigation canals, setting up new factories with modern machinery and planning improvement schemes for Cairo, then his grandson, Abbas, was a reactionary despot. During his brief rule, 1848–54, he set his face against the incursions of European modernity and abandoned many of his grandfather's innovations.[52] And yet, in so many ways, it was too late. In 1843 it had taken Warburton four days to travel between Alexandria and Cairo; in 1849 Nightingale was able to make the same trip on an English steamer, the *Marchioness of Breadalbane*, in just 24 hours. Where others would have seen this as a sign of the fragility of the Orient, of the fading of its imaginative geography, and mourned its disappearance, overwritten by the bolder and brasher strokes of modernity, Nightingale simply saw the Orient crumbling under the burden of its own unchanging present.[53] At the end of her Nile voyage there is a moment which might serve as a metaphor for her philosophy of history. She is in her cabin as dusk falls, writing one of her last letters from Egypt, with the khamsin raging outside and sand blowing in through the shutters. 'I could write much more easily on the table with my finger', she tells her family, 'than on the paper with my pen'.[54] Is it unduly fanciful to see in this an image of Egypt disappearing in the dust, the inscriptions of Europe struggling to make themselves visible, only to be overwhelmed by the deadening embrace of the desert?

A fantasy of escape: Gustave Flaubert in Egypt

Gustave Flaubert's writings about Egypt are textually more complicated than Florence Nightingale's because their construction is so multiform. Flaubert kept two notebooks during his travels and he used them to provide the raw materials for his letters home. The surviving correspondence consists of 31 letters from Flaubert and, since there is no trace of the 30 or so letters he received while he was in Egypt, we have direct access to only one side of these often intimate conversations. Each letter presents his travels to a different audience and is thus composed from a different position. Although he often copied and reworked passages from one letter to another, there is a vivid contrast between the letters to his mother and those to his close friend, Louis Bouilhet. He wrote to Bouilhet much less often but at much greater length, returning to descriptions drafted in letters to his mother and elaborating them, by which I mean not only giving more detail – including erotic encounters which he kept from his mother – but also polishing and perfecting his prose. This is of the utmost importance, for this correspondence mattered greatly to Flaubert. He was comforted by news from his family and friends and was always worried when he did not hear from them but he had other reasons too: he tormented himself with the thought that some of his own letters might have been lost in the post not least, I think, because he intended to make use of them himself. He had promised to incorporate extracts from the letters into articles for the *Revue de l'Orient et d'Algérie*, the bulletin of the Société Orientale de France, but soon after he arrived in Egypt he decided 'to publish nothing for a long time yet'. When Flaubert returned to France he re-read all the letters he had sent to his mother (he had instructed her to number them in sequence and keep them safe), probably those to Bouilhet and perhaps even the notes taken by his travelling companion, and then used them in conjunction with his own notes to compose his *Voyage en Egypte*. As he had said, however, this was not intended for publication. It was a purely private project in which Flaubert tried to lay down his experiences like a young wine to mature for his literary future: 'to make the Orient into a place of personal memory, a sort of immense treasury of images'.[55]

Flaubert left France in the fall of 1849 in the company of Maxime Du Camp. Both men were in their late twenties and had been friends since university. They had long been fascinated by the Orient but it was Du Camp who took the initiative in making arrangements for the expedition. He persuaded the French government to entrust each of them with a diplomatic mission that he hoped would open various doors *en route*. The Ministry of Public Instruction endorsed his own proposal to photograph ancient monuments and their inscriptions, while the Ministry of Agriculture and Commerce charged Flaubert with collecting commercial information. Du Camp took his task very seriously – Flaubert often describes his companion busy with plates and negatives while he sat around talking or writing – and he eventually published a remarkable series of photographs.[56] But Flaubert was quite uninterested in his own mission. He enjoyed the status it conferred upon him but in Cairo, at the very beginning of the adventure, his ministerial instructions already 'seem[ed] to be waiting patiently for the day I'll use them as toilet paper'. They simply got in the way of what he wanted to do. Towards the end of the journey, he confessed:

> I pay no more attention to my mission than to the King of Prussia. To 'discharge my duties' properly I should have had to give up my journey – it would have been absurd . . . Can't you see me in every town, informing myself about crops, about production, about consumption? 'How much oil do you shit here? How many potatoes do you stuff into yourselves?'[57]

Counter-discourse and contaminated geographies

Flaubert's motives were complex. Like many other European writers and artists of the time, he was haunted by the desire to escape the conventions and confinements of bourgeois Europe. 'If France is in the same state as it was when I left', he wrote towards the end of his journey, 'if the bourgeoisie is still as ferociously absurd and public opinion as base as it was, in a word, if the whole kettle of fish smells as bad then I don't miss any of it'.[58] Perhaps he hoped to find in the counter-experience of Egypt the terms of what Richard Terdiman calls a *counter-discourse* 'from which the *ennui* and the banality characteristic of Flaubert's life and his consciousness for a long time would have been banished'.[59] Like Florence Nightingale, there were also intensely personal reasons for Flaubert wanting to get away from Europe – the death of a close friend, the mental illness of his brother-in-law – but, unlike her, he had no desire to efface the distance between modernity and antiquity. Terdiman compares him to Jules in the first *Education sentimentale*: 'Modern life began to seem too restricted to him and he returned to antiquity to find subjects for pleasure and objects of desire'. In Terdiman's eyes, Flaubert hoped to hold the two apart, to leave the one and lose himself in the other, and realize a dream

> not so much of naturalizing within his own writing (in the form of some sort of curative antidote) the discourse of the Orient, but much more radically of absorbing himself – almost *obliterating himself* – in this other discourse.[60]

The preoccupation with antiquity was scarcely novel and neither was its particular identification with the Orient but in Flaubert's case these associations assumed a special form. I have mentioned their presence in the first *Education sentimentale*, which was completed in 1845; they are also centrally involved in Flaubert's *Conte Oriental*, which was his next project and which he abandoned in order to work on *La tentation de Saint-Antoine*. This extraordinary text was set in Egypt and Flaubert completed a first version immediately before his departure: the stinging criticism it received from Du Camp and Bouilhet provided another reason for him to get away.[61] But this was more than an escape into the exotic. Eugenio Donato argues that Flaubert's 'nostalgia for antiquity' (if that is what it was) was a longing for 'an absolutely original, unmediated form of writing' and that his Orient was constituted within a complex metaphorical network that, among other functions, staged his own act of writing: that it was, in effect, 'a spatial metaphor for an absolute difference which would permit a textually unmediated representation'.[62]

And yet, as Donato also remarks, such was not (and could not be) the case for the Egypt of *La tentation* or the Carthage of *Salammbô*, the first novel in which Flaubert drew directly on his travels in the Orient:

> Between them and Flaubert [stood] the archaeological museum and the library of erudition, and it is only through the Museum and the Library that the modern writer [could] have access to them.[63]

Foucault captured these textual mediations in his striking reflection on the fantasies that fill *La tentation*, to which I referred earlier:

> Possibly, Flaubert was responding to an experience of the fantastic which was singularly modern and relatively unknown before his time, to the discovery of a new imaginative space in the nineteenth century. This domain of phantasms is no longer the night, the sleep of reason, or the uncertain void that stands before desire, but, on the contrary, wakefulness, untiring attention, zealous erudition, and constant vigilance. Henceforth, the visionary experience arises from the black and white surface of printed signs, from the closed and dusty volume that opens with a flight of forgotten words; fantasies are carefully deployed in the hushed library, with its columns of books, with its titles aligned on shelves to form a tight enclosure, but within confines that also liberate impossible worlds. The imaginary now resides between the book and the lamp.[64]

One might say the same of Flaubert's travel writings. He read widely in the literature of Orientalism between 1845 and 1849, and these 'pre-texts' enframed his own texts.[65] It is thus scarcely suprising that, as Terdiman claims,

> Despite the plenitude which stems from a constant reference to places, monuments, terms and mores which were exotic for [Flaubert], these texts really represent an *absence of rupture* from the protocols of the dominant discourse. On the contrary: these protocols are unexpectedly inscribed at the center of a mode of writing whose mission was precisely to exclude them.[66]

In Egypt, Flaubert found a modernity always present in his antiquity and an antiquity always present in his modernity. But – and this is central to my own argument – this was more than a matter of poetics, of textual enframing. It had the most material of dimensions too:

> The West, having preceded Flaubert to the East, is obliging him to record it there, like an unanticipated referent which forces its representation upon the text whose function had nominally been to evade it . . . Thus it is that Flaubert's text registers its contamination by Europe as a systematic and more or less conscious critique of this mode of intervention and domination. It organizes itself in the form of a network of tiny points of European infection, distributed over its pages as they seem to have been over the map of the Orient itself. Such contaminated geography it takes as its model.[67]

Terdiman's reading seems to me largely persuasive, and I will make use of it in what follows, but it is hardly inclusive. Lowe draws attention to the complex, often contradictory foliations of Flaubert's texts: to the preoccupation with defining a coherent national identity, an 'imagined community' in Benedict Anderson's sense of that phrase, at a time when it had been called into question by the revolutions of 1848; to the eroticization and sexualization of the Orient as female Other, different from and desired by a masculine European subject; and even, in his later writings, to the ironic counter-commentary on Orientalism as 'sentimentality'.[68] My sense is that these themes are all present in his appropriations of Egypt.

Landscapes and ruins

Flaubert and Du Camp arrived in Alexandria on 15 November 1849, accompanied by their Corsican servant, Sassetti. The trio left for Cairo 10 days later – perhaps travelling on the same steamer as the Bracebridges and Florence Nightingale[69] – and stayed first at the Hôtel d'Orient and then at the Hôtel du Nil, where they remained for over two months. They then hired a houseboat, a blue-painted *cange* which was somewhat smaller than the *Parthenope*. Flaubert wrote to his mother,

> For quarters we have a room with two little divans facing each other . . . a large room with two beds, on one side of which there is a kind of alcove for our baggage and on the other an English-type head; and finally a third room where Sassetti will sleep and which will serve as a store-room as well.[70]

Still, it must have been capacious enough; Flaubert had more than 1200 pounds of baggage. It was certainly well-equipped and came with a crew of nine.

They set off up the Nile on 5 February 1850 and, although they were hit by a succession of storms and the crew had to track for long distances, they reached the First Cataract on 11 March. They travelled up into Nubia as far as Wadi Halfa which they reached on 22 March. As was the custom, their return was more leisurely but it was made even longer by a 10-day excursion overland across the desert from Keneh to the Red Sea at Koseir. They then resumed their voyage and returned to Cairo on 25 June, where they stayed at the Hôtel du Nil for a week or so before travelling back to Alexandria. They spent a fortnight there and left Alexandria (for Beirut) on 17 July 1850.

For the most part, Flaubert affected not to be greatly interested in either Egypt's landscape or the ruins of its past civilizations. But this apparent neglect needs to be treated carefully. Soon after his arrival in Cairo Flaubert wrote to Louis Bouilhet that he was 'very little impressed by nature here – i.e. landscape, sky, desert (except the mirages)'. And certainly he rarely presented panoramas in his notes and letters or his novels, preferring to focus on images of particular moments; but these had a visual density so vivid, a *focalisation* so intense that passages in his travel writings have reminded several critics of an artist's sketchpad rather than a writer's notebook.[71] One evening at Denderah, for example, when Flaubert was strolling under the palms, gazing at the purple-red of the distant mountains, the blue of the Nile, the ultramarine of the sky and the livid green of the vegetation, nothing moving, he imagined that the scene was like 'a painted countryside, an immense stage set made expressly for us'.[72] But this was not the artistic conceit of Florence Nightingale, who thought the landscape looked 'as if a child had painted it and did not know how'. Not only did Flaubert exult in these landscapes; *he* 'painted' them and he *did* know how: in fact, one commentator regards these as the great pages of his *Voyage*.[73] Flaubert's descriptions were often acutely physical in their ability to convey changing and intersecting arcs of sensation in which vision virtually fades into sound and back again and in their almost physical recomposition of the scene itself. Consider this passage, written after Flaubert had climbed to the top of one of the Pyramids to wait for the sunrise:

> The sun was rising just opposite; the whole valley of the Nile, bathed in mist, seemed to be a still white sea; and the desert behind us, with its hillocks of sand, another ocean, deep purple, its waves all petrified. But as the sun climbed behind the Arabian chain the mist was torn into great shreds of filmy gauze . . . Everything between the horizon and us is all white and looks like an ocean; this recedes and lifts. The sun, it seems, is moving fast and climbing above oblong clouds that look like swan's down, of an inexpressible softness; the trees in the groves around the villages . . . seem to be in the sky itself, for the entire perspective is perpendicular; . . . behind us, when we turn around, is the desert – purple waves of sand, a purple ocean.[74]

The passage begins with an image of arresting stillness but the scene – the canvas? – is suddenly cut by the sun and its composition thrust into motion: even the trees are rotated into the sky. Nightingale was struck by the colourlessness of Egypt but here waves of colour wash over sky and land, and one can almost hear them crashing into the desert sand.

It is necessary to be equally cautious about Flaubert and antiquity. At Abu Simbel he flatly declared that 'the Egyptian temples bore me profoundly' and his notes on them as he returned downriver seem to be increasingly perfunctory; there is certainly nothing in his writings that approaches the endless descriptions of ruins that filled Florence Nightingale's letters. Yet what bored him, I think, was not so much the ruins as their incorporation within the increasingly elaborate textual apparatus of Orientalism and, in particular, within the discourses of archaeology and tourism that policed their appropriation. 'Are they going to become like the churches in Brittany, the waterfalls of the Pyrenees?' he demanded.

> Oh, necessity! To do what you are supposed to do; to be always what, according to the circumstances (and despite the aversion of the moment) . . . a young man, or a tourist, or an artist . . . is supposed to be![75]

Flaubert was genuinely moved at a number of sites, most notably at Giza, Luxor and Thebes, and he had an eye for detail so acute – sometimes much more so than Du Camp – that at least one commentator is sceptical about his show of disinterest. He was drawn to the figures shown on the frescoes and, while he sometimes chaffed at the rigidity of Egyptian art, he paid particular attention to its representation of the human body – to the lines of naked captives displaying their hands and uncircumcised penises to the king, to the naked women and the 'deliberately lubricious aspect of the[ir] thighs, with the knee deeply inset' – as a confirmation of what de Biasi calls 'the unchangeable modernity of desire and eroticism'. Flaubert's own account was more prosaic: 'So dirty pictures existed even so far back in antiquity?'[76] But the point is, I think, essentially the same. Far from being unmoved by the traces of antiquity, Flaubert found in their very physicality – at least in his response to and his rendering of that physicality – the possibility of bringing that past into his present. In much the same way, death was not contained by the lost worlds of the temples and tombs. It was constantly all around Flaubert and reached out to touch him: thus dogs tearing at the remains of a donkey; the muzzles of hunting-dogs purple with clotted blood caked in the sun; camels dying in the desert while vultures wheel in the sky; the shrill cries of an Englishwoman in her death agony; the dessicated corpse of an Arab, 'the mouth screaming with every ounce of human strength'.[77]

When Flaubert wrote that landscape and architecture 'seem the work of the same hand', therefore, this was not to render them as flat and immobile surfaces but to situate their geometries within a multidimensional, almost kinetic representation of space that was intensely corporeal. If Flaubert's space is one in which places are crisscrossed by glances in a geometric *quadrillage*, those glances and their geometries are not disembodied.[78] On the contrary, Michel Butor suggests that Flaubert's travel writings were shot through with a critique of what one might call the body-culture of the bourgeoisie. For Flaubert.

> The contemporary epoch, that of the bourgeoisie, destroys the human body by covering it with clothes and cutting it out. Generalised castration.[79]

When Flaubert declares that 'everything in Egypt seems made for architecture', therefore, 'the planes of the fields, the vegetation, the human anatomy, the horizon lines', his 'architecture' is not a formal exteriority but rather an organic interiority.[80]

Physicality, passivity and representations of space

I have made this point in relation to landscape and antiquity, but it can be sharpened most acutely in those other fields that Flaubert made intersect with them. And it needs to be sharpened in a particular way, as Butor's invocation of castration – and, by implication, of desire – implies. There can be no doubt that, as Flaubert said himself, he was 'enormously excited by the cities and the people' and that his appropriation of them was a thoroughly sensual one.[81] Like Florence Nightingale, Flaubert landed in Alexandria 'amid the most deafening uproar imaginable'; unlike her, however, when he recorded himself 'gulp[ing] down a whole bellyful of colours, like a donkey filling himself with hay', he applied the zoological metaphor to *himself*.[82] He soon tired of Alexandria – what Nightingale saw as 'a vast settlement of white ants', Flaubert saw as a city adulterated by Europe – and he roundly declared that 'the Orient begins at Cairo'. He arrived in the capital at the end of November and found that city 'inexhaustible; the longer I stay here, the more we find to discover'.[83] In mid-January he was still

> scarcely over the initial bedazzlement. It is like being hurled while still asleep into the midst of a Beethoven symphony, with the brasses at their most ear-splitting, the basses rumbling and the flutes sighing away; each detail reaches out to grip you; it pinches you; and the more you concentrate on it the less you grasp the

> whole. Then gradually all this becomes harmonious and the pieces fall into place of themselves, in accordance with the laws of perspective. But the first days, by God, it is such a bewildering chaos of colours that your poor imagination is dazzled as though by continuous fireworks . . .[84]

As the physicality of the prose suggests – once again, sound fading into vision and the details of the scene reaching out, gripping and pinching – Flaubert's responses were more than mental exercises. 'I try to *take hold* of everything I see', he wrote to his mother, and he was greatly exercised by the sheer physicality and corporeality of his encounters.[85]

This was true in the most literal of senses. Where Florence Nightingale was sickened by the physical brutality that she took to be part of everyday life in Egypt, Flaubert was entranced by it. 'You would scarcely believe the important role played by the cudgel in this part of the world', he wrote to his mother. 'Buffets are distributed with a sublime prodigality, always accompanied by loud cries; it's the most genuine kind of local colour you can think of'. He was amused by the assaults he witnessed – on one occasion his own servants forced another man into the ocean with their sticks and he and Du Camp laughed 'like fools' – and he and Du Camp questioned the Nazir of Ibrim about the bastinado with something approaching relish.[86] But physicality entered still more intimately into Flaubert's images. A wedding procession with 'two naked wrestlers, oiled and wearing leather shorts, but not wrestling, just striking poses'; a performance by 'dancers [whose] gauze ripples on the hips like a transparent wave with every move they make'; a visit to the baths, alone in the hot room, 'very voluptuous . . . while the naked *kellaas* call out to one another as they massage you'; a night with the courtesan Kuchuk Hanem, 'feeling of her stomach against my buttocks . . . her mound warmer than her stomach, heated me like a hot iron'; the clown of the crew dancing 'a naked, lascivious dance that consisted of an attempt to bugger himself', driving off mendicant monks who swam to the side of the boat by 'show[ing] them his prick and his arse and pretend[ing] to piss and shit on their heads'.[87]

This is not quite the passivity that Sartre finds in Flaubert and, although I cannot discuss his reading in the detail it deserves, I do need to establish the distance between our two Flauberts. Sartre's interpretation is derived from a mix of Marxism and psychoanalysis. Of particular relevance is his account of Flaubert's sexuality, which he uses to characterize Flaubert's being-in-the-world as essentially passive. Through his sexual encounters Flaubert tries to 'recuperate' his being by constantly placing himself in the hands of others – literally so – and 'turning himself, through complaisant submission, into a fascinating object for his executioners and simultaneously for himself'. The object of this recuperation, so Sartre contends, is 'his passivity itself'; in another formulation, Flaubert's passivity 'desires to become flesh under the manipulations of others'. For Sartre this is indicative of 'a secret femininity' of Flaubert's vision of himself as an 'imaginary woman', and his sexual encounters in Egypt and elsewhere reveal a desire to submit 'in order to coincide in orgasm with his objective being'.[88] These are, I think, tendentious claims, and for several reasons. Less preoccupied with diagnosing Flaubert's sexuality, Said none the less seems at least half-convinced by Sartre's analysis and translates 'passivity' into Flaubert's penchant for the theatrical: at its best, he writes 'this provides a spectacular form, but it remains barred to the Westerner's full participation in it'. In the Orient, so Said claims, Flaubert – 'like so many others – felt his detached powerlessness, perhaps also his self-induced unwillingness, to enter and become part of what he saw'.[89]

If one accepts these interpretations, then Flaubert in Cairo, dressing up 'in a large white cotton Nubian shirt trimmed with little pompoms', shaving his head 'except for one lock at the occiput (by which Mohammed lifts you up on Judgement Day)', adorning himself with 'a tarboosh which is of a screaming red', was doing exactly that: dressing up (Fig. 4).[90] 'We look quite the pair of Orientals', he wrote to Bouilhet, and that was precisely the point. They were playing a part; they could not be – and neither, I think, did they wish to become – 'Orientals'. Towards the end of their journey on the Nile they entertained themselves by *faisant les sheikhs*: play-acting as sheikhs and asking portentous questions about literacy, railways and the spread of socialism. That this was, in part, a satire on Flaubert's neglected mission and, perhaps, on a 'modernization' that made the inhabitants strangers in their own country, does not detract from their ridicule of the 'informants' but the fact remains that this was not how Flaubert and Du Camp conducted themselves; their impressions did not rely on formal interrogatories.[91] I think that de Biasi claims too much when he asserts that Flaubert was somehow

Figure 4. Gustave Flaubert in the garden of the Hôtel du Nil, Cairo, 1850 (photograph: Maxime Du Camp)

able 'to represent the other in his or her alterity', since, as I will show shortly, the hermeneutic circles in which Flaubert's narrative moved were uneven and unequal.[92] Nevertheless, neither Flaubert nor Du Camp sought to insulate themselves from local society and there are several passages in which Flaubert records his contempt for those who travelled through Egypt without making any concessions to their surroundings (most notably – and one hopes imaginatively – an Englishman travelling in Syria who insisted on eating four meals of roast beef a day and taking regular cold baths).[93]

Equally, Flaubert's homoerotic adventures in the baths in Cairo were frissons of arousal and anticipation but not, if we are to believe Sartre, consummations.[94] Although here too Flaubert satirized his mission – 'Travelling as we are for educational purposes, and charged with a mission by the government, we have considered it our duty to indulge in this form of ejaculation' – this was, I think, more than play-acting on his part. Flaubert was not merely making a gesture to the symbolic homoeroticism that some contemporary critics now see in the colonial European encounter with the exotic East.[95] He seemed to be remarkably well-informed about the practical arrangements:

> You reserve the bath for yourself . . . and skewer the lad in one of the rooms . . . The final masseurs, the ones who come to rub you when the rest is done, are usually quite nice young boys. We had our eye on one in an establishment near our hotel. I reserved the bath exclusively for myself. I went, and the rascal was away that day!

Flaubert's first encounter was unsuccessful – 'a man in his fifties, ignoble, disgusting' – but he later told Bouilhet that he did indeed consummate 'that business at the baths' and that he was determined to do it again.[96] His admission may have been perfunctory but, when these passages are seen in the light of the highly mobile, intensely physical sexuality that is so vividly present in Flaubert's writings, it becomes difficult (for me, anyway) to treat his appropriations of Egypt as essentially and intrinsically passive.

That said, there was one sense in which Flaubert's passage through Egypt was as passive as Florence Nightingale's. At the start of her voyage she declared that she was 'no dahabeeah bird, no divan incumbent' and that she longed 'to be wandering in the desert by myself, poking my own nose into all the villages and running hither and thither, making acquaintances *où bon me semble*'.[97] But it quickly became apparent that, for all their excursions, she and her companions constituted a sort of collective moving eye on which successive sights/sites were registered.[98] In much the same way, Flaubert boasted that he and Du Camp lived

> in the grossest idleness, stretched out all day on our divans watching everything that goes by: camels, herds of oxen from the Sennaar, boats floating to Cairo laden with negresses and elephants' tusks.

Du Camp went much further and claimed that his companion

> would have liked to travel, if he could, stretched out on a sofa and not stirring, watching landscapes, ruins and cities pass before him like the screen of a mechanical panorama.[99]

But Du Camp was being less than fair (as he was in a number of other places). For there is also a real sense of Flaubert plunging into Egypt's present that is almost wholly absent from Nightingale's letters and which is the product of more than a difference in style. I say 'almost wholly absent' for good reason; the exceptions are revealing. Soon after their arrival in Alexandria, Nightingale and Selina Bracebridge disguised themselves as Egyptian women in order to visit a mosque during the mid-day prayers. They had to confine themselves to the womens' gallery, but 'the people' (whether men or women is unclear) crowded round them, laughing and pointing. 'That quarter of an hour seemed to reveal to one what it is to be a woman in these countries', Nightingale recalled: 'God save them, for it is a hopeless life'.[100] The sense of hopelessness was at once lessened and heightened by her visits to the sisters of St Vincent de Paul, who ran a hospital and a school for the poor in Alexandria. Her engagements were mediated by a (feminized) charitable apparatus that not only, I imagine, helped harden her resolve to become a nurse: as the sisters of the order intervened in the everyday life of the city, so they also amply confirmed her horror at Egypt's present, at what she took to be its culture of improvidence and indifference.[101] Nightingale made no secret of the fact that 'without the past' she conceived Egypt 'to be utterly uninhabitable', and she exulted in the 'absolute solitude' of the Great Temple at Abu Simbel precisely because it provided 'the absence of a present, of any of one's fellow creatures who contrast the past with that horrible Egyptian present'.[102] For Flaubert, however, it was an inhabited version of the present – and, in particular, the complex temporalities of past and present, the sometimes grotesque, sometimes dazzling irruptions of the past into the present – that was so exciting.

And if one thinks about Flaubert's *topoi*, the places in which these experiences *take* place and the representation of space within which they are convened, then his experiences and encounters are scarcely passive. In a letter to his mother Flaubert wrote that when he galloped across the desert with Du Camp 'we were *devouring* space' and the image captures the sense of physical engagement that I want to invoke more generally here.[103] It is quite different from the way he has his mother figure the space of Egypt. He often suggested that she follow his progress on the map he had sent her from Paris, and at Philae he imagined her tear-filled eyes looking on 'that map that only represents an empty space where your son is lost'.[104] Yet Flaubert's own space was not empty; the image of his mother and her map functions as a contrast that has the effect of confirming that his Egypt was not an abstract cartography but rather a physical space constituted as sensation and plenitude.

Sexuality and representations of space

This must be pressed further, however, because this space is constituted in part through Flaubert's writing, and there are filiations between his sexualization of writing and his representation of space. The first is surely unexceptional. 'Literature!' Flaubert exclaimed to Bouilhet, in a vividly graphic transposition of the two,

> That old whore! We must try to dose her with mercury and pills and clean her out from top to bottom, she has been so ultra-screwed by filthy pricks![105]

The connections between sexuality and writing – which are scarcely foreign to French critical theory in general or criticism of Flaubert in particular – are of course considerably more complicated than such a casual, scatalogical image can convey.[106] But if these connections are granted, then it becomes all the more important to ask about the gendering and sexualization of Flaubert's imaginary geographies. Butor regards Flaubert's European travels as sketches for his expedition to the Orient and argues that his descriptions of the Pyrenees, Corsica and parts of western France feminized those landscapes. Of most direct relevance to my own discussion is the suggestion that, in those early descriptions, 'light circulates through the interior of this feminine landscape . . . and the spectator desires to become a moment in that circulation'.[107] These are clearly heterosexual readings and one needs to recognize (and make space for) the ambivalence of Flaubert's sexuality but, seen like this, his view from the Pyramids – or, rather, his act of viewing – becomes not only aestheticized but also eroticized. Flaubert represents the sky and its play of light as 'a medium that envelops, penetrates and transforms the body and subjectivity'.[108] Throughout his writings about Egypt and, for that matter, the corpus of his work as a whole, this sense of transubstantiation is usually achieved through metaphors of fluidity and liquidity. Gazing down on the city of Cairo from a minaret, for example, across to the desert, the Nile and the Pyramids beyond, Flaubert remarked how 'the liquid light seems to penetrate the surface of things and enter into them'.[109] The metaphor is precise; for Flaubert, water is almost always associated with sexuality. In Cairo, the voluptuousness of the baths is blended with and conveyed through their humidity; in the Red Sea, Flaubert 'lolled in its waters as though I were lying on a thousand liquid breasts that were caressing my entire body'.[110] What is particularly arresting is that Flaubert invokes water to convey not just the clouds floating in the sky, the waves in the sea or the flowing currents of the Nile but the desert too. Like many travellers he was fascinated by the mirage, by those

> great stretches of water with trees reflected in them, and at their farthest limit, where they seem to touch the sky, a gray vapour that appears to be moving in a rush, like a train

but his whole sense of the desert, of its oceans of sand and waves of colour, was a liquid one. Riding back from the Pyramids in the early evening, Flaubert describes the desert pavement thus:

> The small stones covering the ground glitter, literally bathed in purple light; it is as though one were looking at them through water so transparent as to be invisible.

During their excursion to the Red Sea, Flaubert and Du Camp were enveloped in a dust-cloud as a caravan passed them from the opposite direction. It seemed to him

> that the camels were not touching the ground, that they were breasting ahead with a ship-like movement, that inside the dust-cloud they were raised high above the ground, as though they were wading belly-deep in clouds.[111]

Whatever one makes of these images, they are quite dramatically different from Florence Nightingale's representations of Egypt: her landscapes are hard, angular, dessicated, mortified; she maintains a scrupulous distance from them. I do not mean to read her as somehow representative of all European women travellers but it is true that, by and large, women writers did not eroticize the landscapes through which they travelled and there are very few passages in which Nightingale explicitly engendered or sexualized the landscape. She constantly referred to the Nile in the masculine but then so did most European travellers; she described Cairo as 'a bride adorned for the marriage' and Egyptian 'nature' as 'an oriental queen in gorgeous jewellery' but – as she more or less said herself – these were common clichés of Orientalism.[112] If many women travellers adopted what Sara Mills has termed a 'confessional' mode of writing, and Nightingale was no exception, her letters (like her experiences) were far removed from the licentious confidences Flaubert eagerly shared with his young male friends back in France.[113] Her account of her own visit to the baths in Alexandria provides a vivid contrast to Flaubert's 'haunting' of the baths in Cairo. She dwells on the marble interior and the shafts of sunlight, 'the whole like an Arabian Night's description', and likens the 'enchanted gardens' outside to 'the Chatsworth conservatory', whereas Flaubert's sense of enchantment derived from more corporeal architecture.[114]

And yet: if Flaubert was able to disrupt at least some of the conventions of Orientalism – to free himself from the abstracted gaze that enframed Nightingale's accounts and to construct a series of imaginative geographies in which the inhabitants of Egypt were indelibly *present*, not as traces or imprints but as *bodies* and *subjects* – he was able to do so, in large part, by virtue of the privileges conferred upon him by a particular constellation of masculinity and patriarchy. Not only did he have considerably more freedom of action than Nightingale – he was a young man travelling with a close male friend, not a young woman chaperoned by friends of the family – but the powers inhered in masculinity and patriarchy enabled him to make visible the spaces that remained opaque ('impenetrable') to Nightingale.[115] This makes it impossible to disentangle his evident pleasure in Egypt's present from his eroticization of the Orient and there is a sense, I think, in which Flaubert's plunging into Egypt's present was, at the same time, an invasion and even an exploitation that Nightingale's horrified recoil, grounded as it was in an unalloyed racism, denied to her. Flaubert found a place for bodies and subjects in his imaginative geographies but they were bodies and subjects in the most literal of ways.

Under the sign of Europe

This is an unsettling reading and I need to make three further, closely connected remarks. First, I hope it will be clear that – vivid and engaged as Flaubert's account is – I do not wish to imply that it was somehow more 'real' than Nightingale's. His were carefully wrought presentations in which the Orient was *scripted*, and I mean by this more than an artful re-working of his notes and letters. There is something artful about the way in which Flaubert set up many of his encounters in the first place. Julian Barnes provides a witty and erudite gloss on Flaubert's view from the top of the pyramid. Gazing down, Flaubert saw a small business card inscribed 'HUMBERT FROTTEUR' and he realized that Maxime Du Camp had climbed up ahead of him and left it there for him to find. And yet a few lines later it turns out that this 'sublime surprise', this ironic eruption of the everyday into the extraordinary, had been planned by Flaubert himself, who had deliberately brought the card with him from Croisset.

> So, ever stranger: Flaubert, when he left home, was already preparing the special effects which would later appear entirely characteristic of how he perceived the world. Ironies breed; realities recede.[116]

Or again, he writes to Bouilhet of his visit to 'the whores' quarter' at Kena, where he walked back and forth through the streets, 'giving *baksheesh* to all the women, letting them call me and catch hold of me'. But, he continues:

> I abstained deliberately, in order to preserve the sweet sadness of the scene and engrave it deeply in my memory. In this way I went away dazzled, and have remained so. There is nothing more beautiful than these women calling you. If I had gone with any of

> them, a second picture would have been superimposed on the first and dimmed its splendour.[117]

These were experiences constructed in Egypt from within a European imaginary and with a European sensibility in mind.

Secondly, I have made much of Flaubert's physical appropriation of the Orient – 'gulping down a bellyful of colours' at Alexandria, 'devouring space' in the desert and, I might have added, 'filling [himself] with sun and light' and 'inhaling the vastness' at Bulak[118] – but this metaphoric of consumption had an insistently, intrusively material dimension too. Not only did Du Camp and Flaubert inscribe Egypt and carry it home with them as the trophies and records of an imaginative geography; their stay in Cairo and their passage on the Nile was freighted with many of the assumptions and usages of colonial consumption. The most striking image is one that I have already cited without comment. When Flaubert watches 'boats floating down to Cairo laden with negresses and elephants' tusks', it is surely not his idleness that is so unsettling but rather the casual elision between the objects of his gaze: between black women – who, it seems from descriptions elsewhere in his letters, were slaves – and animal trophies.[119] They were reduced to the same level, objectified and commodified, and these elisions entered into a complex chain of colonial privileges and appropriations. Let me list just some of them: in Alexandria, soldiers were supplied to hold back the crowds when Du Camp set up his photographic apparatus; at Rosetta, the pasha entertained them to a dinner at which they were waited on by 10 negro servants in silver jackets; prostitutes, 'their hair spangled with gold piastres', were hired to dance for them at a party on the river; in Cairo, hospital patients stood on their beds and, at a sign from the doctor, displayed the chancres on their anuses to the entranced visitors; the French consul suggested they ask their guide to have a black woman strip for them at the slave market; Du Camp bargained with another woman for her coral necklace; Flaubert treated their donkey-drivers to a prostitute by the aqueduct for 60 paras; upriver, men dived into the cataract to entertain them; Flaubert had tresses and ornaments cut from the heads of two women for 10 piastres apiece; and, climactically, Flaubert and Du Camp were entertained by Kuchuk-Hanem, who performed an erotic dance and had sex with them.[120] Terdiman's reading of these appropriations is uncompromising:

> The text and the appropriative dynamic which energizes it moves from spectacles and entertainments to women's tresses to whole women: everything is made the Occidental's lawful prize in that accelerating absorption of the foreign which structures the *Voyage*.[121]

As I have said, I accept that these are invasions and so, in his way, did Flaubert. 'One curious thing here', he wrote to his mother, 'is the respect, or rather the terror, that everyone displays in the presence of "Franks", as they call Europeans'.[122] But the practices through which these incursions were put in place were more complicated than Terdiman seems prepared to acknowledge. The slaves and the ivory were not the exclusive objects of a European economy, and Flaubert and Du Camp were not simply travelling under the sign of an all-powerful European colonialism. If the ties between Egypt and the Ottoman Empire were frayed by the middle of the nineteenth century, Egypt was none the less still a colonized country, ruled and policed by an Albanian-Turkish élite under the nominal authority of the Porte. And there were other grids of power – of class, of religion, of gender – that scored Arab society too. The practices of colonial power in which Flaubert and Du Camp were imbricated were thus multiple and interleaved with intricate local grids of power: thus local dignitaries supplied the soldiers and arranged dinners for them, doctors and donkey-drivers evidently had no compunction about using their power over patients and prostitutes and, while there is no way of knowing what the other people involved in these transactions made of them, it seems unlikely that these exchanges – however unequal – were all one-way affairs between a dominant Occident and a submissive Orient. Flaubert's encounter with Kuchuk-Hanem in particular was highly ambivalent, at once disavowing and fetishizing, and it did not leave him unmoved.[123]

Thirdly, and in some degree against the grain of the previous paragraph, Europe was never far from Flaubert's construction of the Orient. 'You ask me whether the Orient is up to what I imagined it to be', he replied to his mother.

> Yes, it is; and more than that, it extends far beyond the narrow idea I had of it. I have found, clearly delineated, everything that was hazy in my mind. Facts have taken the place of suppositions – so excellently so, that it is often as though I were suddenly coming upon old forgotten dreams.[124]

And later:

> Anyone who is a little attentive *re*discovers here much more than he discovers. The seeds of a thousand notions that one carried within oneself grow and become more definite, like so many refreshed memories.[125]

Flaubert constructed his Egypt in advance, in Europe: he even wrote a description of the view from the top of one of the Pyramids before he left France. He arrived in Egypt with dreams of a timeless Orient, of a place which, precisely because it had escaped the compulsions and conventions of nineteenth-century Europe, had somehow stolen a march on time itself. His voyage was more than an *aide-mémoire*, of course, and his travel writings more than a record of his dreams turned to sand and stone. But once he arrived in Egypt, his idealized, exoticized Orient did not disappear; he was still enough of a Romantic to recognize (and even be entranced by) its shimmering images. And he was already enough of a realist to see that these imaginative geographies were terribly vulnerable. For many critics, this oneiric precariousness, if I may so call it, was the paradoxical product of an Orientalism that both valued the region 'because it could be imagined as unknown' and yet 'impregnated [it] by a textual network so dense that it threatened to exhaust its own referent completely', to make it strangely familiar, the place of Flaubert's 'forgotten dreams' and 'refreshed memories'.[126]

But nineteenth-century Orientalism was not self-sufficient; it was articulated with the ideological grids of capitalist modernity and one might argue that it was, in part, the immanent disappearance of the exotic – the fading of the dream – that made its revival so desperately necessary. For the Orient to become a commodity like any other in the age of commodities, its 'fictive otherness' had to be reproduced, those 'worked landscapes of the imagination' assiduously, artfully reworked.[127] Even this seems inadequate, however, because the presence of Europe in Egypt was more than textual conceit: or, rather, those textualizations were also acutely material. Flaubert complained endlessly about the graffiti left by previous European travellers on ancient monuments but these were not only marks of their 'unshakeable stupidity', as he wrote to his uncle Parain, they were also petty inscriptions of the advent of capitalist modernity.[128] 'It's time to hurry', he wrote to Gautier:

> In a little while the Orient will have ceased to exist. We may be its last witnesses. You can't imagine how much it has already been ruined. I've seen harems go by on steam boats.[129]

Conclusion

That final image from Flaubert juxtaposes with great economy a timeless, traditional and thoroughly exoticized Orient with a changing, modern and increasingly familiar Egypt. By the middle of the nineteenth century the transition from one to the other, or more accurately the coincidence between them, was beginning to be registered through a shift in the European spatial imaginary. But it was a complicated, fractured ideological landscape. Florence Nightingale and Gustave Flaubert shared the same viewpoint – the commanding heights of European modernity – but their prospects were vastly different. Where Nightingale looked back to the ruins of ancient Egypt, Flaubert anticipated the ruins of present-day Egypt; where Nightingale dreamed of the past, Flaubert saw the West as the 'executioner of its own dreams'.[130] This makes it difficult, perhaps even foolhardy to generalize but I want to make three final observations about the European construction of imaginative geographies of Egypt at mid-century.

In the first place, the accounts that I have considered here rendered Egypt as a text to be read and as the object of a gaze. These strategies overlap and I do not think they correspond to any simple distinction between imaginative geographies produced by writers and those produced by artists. Neither Florence Nightingale nor Gustave Flaubert spent time sketching in Egypt but there is an intensely visual quality to their images. They were both caught up in what Linda Nochlin calls a 'politics of vision' which, as she shows in one of her own essays, allows Said's critique of Orientalism in its textual forms to be brought to bear on the visual arts.[131] But I want to emphasize that, for both Nightingale and Flaubert, this was a politics of *reading* or *looking* rather than listening and in this they were not alone. Most Europeans had little choice but to read the landscape because they could not speak the language. Many of the dignitaries they met spoke French or English, and it is perfectly true that Flaubert made some attempt at learning Arabic script and that Nightingale had a lesson in Arabic from the English Consul-General in Cairo; presumably both of them picked up the odd word or phrase in Arabic, perhaps in Turkish too. But neither of them could conduct a conversation with ordinary people: 'We are only sorry we don't get on with our Arabic', declared an exasperated Nightingale, and a bemused Flaubert wrote of 'love-making by interpreter'.[132]

Like other travellers, they had to rely on a *dragoman* – a cicerone, an interpreter-guide – to act as an intermediary, and even the best *dragoman* could not overcome the one-sidedness of these exchanges. He invariably took pride in his superiority over the common people, and often over his European charges too, but he also redoubled the different expectations of Nightingale as a woman and Flaubert as a man. As I have repeatedly emphasized, the conventions of bourgeois Europe allowed Flaubert considerably more freedom of manoeuvre than Nightingale but her 'priceless cicerone' was also keenly aware of the proprieties to be observed by a woman in Egypt. He 'is so careful of me that he won't let anybody come near me', she wrote, and 'if they do, he utters some dreadful form of words, which I don't understand, and they instantly fall back'.[133] How different from Flaubert, dressing after a night in a Cairo brothel, who records his *dragoman* 'Joseph's expression amid all this'.[134]

In the second place, the textualization of Egypt occurred in a number of different, interlocking registers. By the middle of the nineteenth century, European visitors could invoke a dense web of textual associations which constituted a collective 'fantasia of the library' in something like Foucault's sense of the phrase. As Said has shown, Orientalism was inherently citationary: it re-presented the Orient less as a place rooted in history and geography than as a chain of references embedded in the library.[135] It was in this spirit that Nightingale called Egypt 'a vast library, the finest, the Alexandrian library of the world' and sprinkled her letters with references to Champollion, Lane and others. Similarly, Flaubert often suggested that his mother might consult a book to get a better idea of what he was describing and several times drew her attention to the plates in the *Description de l'Égypte* and to Lane's *Modern Egyptians*.[136] But these references were not only helpful injunctions to the audience at home. They were also sighting-devices for the visitors themselves, a discursive apparatus that enabled them to make (their) sense of what they saw.

Murray's *Handbook* recommended an essential library of some 30 scholarly books to accompany the independent traveller to Egypt and most seem to have taken the advice to heart. When Warburton recalled reclining on cushions under a thick awning, gazing at the landscape, 'our books and maps . . . beside us, ever ready to explain or illustrate what we saw', he was describing a common experience – and *expectation* – of the Nile voyage that shaped the way in which it was appropriated.[137] European travellers read their texts, wrote their journals and letters and, in the very act of doing so, enframed Egypt itself as at once a text to be transcribed and translated and as a picture to be composed and exhibited. Thus, when Flaubert described the village of Mahatta as being 'like an engraving, an oriental scene in a book', he was invoking one of the conventions of Orientalism's imaginative geographies.[138] But it was a convention that, as I have implied, gained much of its power from the incursions and even corruptions of European modernity. Nochlin suggests that the very notion of the picturesque is premissed on its (creative) destruction: 'Only on the brink of destruction, in the course of incipient modification and cultural dilution, are customs, costumes and religious rituals of the dominated finally seen as picturesque'.[139] It is in this light that I peer over Flaubert's shoulder, as he and Du Camp 'have translations of songs, stories and traditions made for [them] – everything that is most folkloric and oriental'.[140] Conversely, Nightingale welcomed the destruction of Egypt's 'horrible present', not by modernity but certainly from its privileged position, and even she succumbed to the picturesque in Moorish Cairo. 'In riding home by moonlight', she wrote, with 'the Turk sitting cross-legged smoking under a low vaulted arch, there is not a corner that is not a picture': at least, not to a sophisticated European traveller who had seen such sights in books or galleries.[141]

In the third place, these imaginative geographies cannot be severed from the tangled but tightening ties between Europe and Egypt. It was always a complicated relationship, riven by political and economic rivalries between England and France and made all the more uncertain by the geopolitical ambitions of imperial Russia and the shifting sands of the Ottoman Empire. 'At the first sign of trouble in Europe', Flaubert predicted, 'England will take Egypt, Russia will take Constantinople, and we, in retaliation, will get ourselves massacred in the mountains of Syria'.[142] It did not come out quite like that, though several years later Nightingale found herself in the thick of the Crimean War. But I want to emphasize that the tottering balance of power between Europe and the Orient cannot be measured on the scales of profit and loss or in the currency of influence and intrigue alone. The readings that I have been describing here were moments in the installation of what Timothy Mitchell calls 'the world-as-exhibition'. He suggests that by the closing decades of the nineteenth century, it was a characteristic of European ways of knowing to render things as objects to be viewed and to 'set the world up as a picture . . . [and arrange] it before an audience as an object on display, to be viewed, investigated and experienced'. This machinery of representation was not confined to the zoo, the museum and the exhibition – all European icons of the nineteenth century and all of them involved in the appropriation and re-presentation of the Orient – but was, so Mitchell claims, constitutive of European modernity at large. The colonial process, he argues, re-ordered Egypt

> to appear as a world enframed. Egypt was to be ordered up as something object-like . . . it was to be made picture-like and legible, rendered available to political and economic calculation. Colonial power required the country to become readable like a book, in our own sense of such a term . . .[143]

In other words, reading Egypt was never an innocent metaphor. As I have tried to show in this essay, it was implicated – and continues to implicate us – in constellations of power, knowledge and spatiality.

Acknowledgements

The research for this essay has been supported by a grant from the Social Science and Humanities Research Council. Early versions were presented to audiences at the University of British Columbia, the University of California at Berkeley, Rutgers University and the University of Washington, and I am grateful to them for their comments and suggestions. I am also grateful to Trevor Barnes, Alison Blunt, David Ley, Geraldine Pratt, Michael Smith, Bruce Willems-Braun and to two anonymous referees for improving successive drafts of the essay.

Notes

1. **Foucault M** 1977 Fantasia of the library in **Bouchard Donald F** ed *Language, counter-memory, practice: selected essays and inteviews* Cornell University Press, Ithaca 87–109; the quotation is from p. 90
2. **Said E** 1978 *Orientalism* London, Penguin 54–5; **Bachelard G** 1969 *The poetics of space* Beacon Press, Boston (first published in French 1958)
3. For example, **Clifford J** 1988 On Orientalism in *The predicament of culture: twentieth-century ethonograpy, literature and art* Harvard University Press, Cambridge, MA 255–76; **Young R** Disorienting Orientalism in *White mythologies: writing history and the West* Routledge, London 119–40
4. **Foucault M** Questions on geography in **Gordon C** ed *Power/knowledge: selected interviews and other writings* Harvester, Brighton 63–77; the quotation is from p. 70
5. **Said E** 1993 *Culture and imperialism* Alfred Knopf, New York 7
6. **Rajchman J** 1991 Foucault's art of seeing in *Philosophical events: essays of the 80s* Columbia University Press, New York 69–102
7. **Said** *Culture and imperialism op cit.* xxiv
8. **Lowe L** 1992 *Critical terrains: French and British Orientalisms* Cornell University Press, Ithaca and London 3, 15. Foucault defines a heterotopia as both a juxtaposition of 'several sites that are in themselves incompatible' and as a space of compensation, 'a space that is [radically] other': **Foucault M** 1986 Of other spaces *Diacritics* 16 22–7. He goes on to define this other space as 'another real space, as perfect, as meticulous, as well arranged as ours is messy', which might seem out of joint with Orientalism but it gestures towards the Orient as a counter-site, a space of ambivalence in which both the superiorities *and* the anxieties of the Occident are inscribed. Foucault's own acount is not unproblematic, however, and (in its closing paragraphs in particular) seems to shade into its own Orientalism
9. My understanding of travel writing is indebted to **Blunt A** 1994 *Travel gender and imperialism: Mary Kingsley and west Africa* Guilford, New York; **Mills S** 1991 *Discourses of difference: an analysis of women's travel writing and colonialism* Routledge, London; **Pratt M L** 1992 *Imperial eyes: travel writing and transculturation* Routledge, London
10. **Warburton E** 1908 *The crescent and the cross: romance and realities of eastern travel* Maclaren, London (first published in 1845) 196; see also **Sattin A** 1988 *Lifting the veil: British society in Egypt* J M Dent, London 44–9
11. 'Parthenope' had been given the Greek name for the city of Naples, where she was born, and in this sense I suppose the Bracebridges and Florence were literally travelling under the sign of classical antiquity
12. **Nightingale F** 1854 *Letters from Egypt* A and G A Spottiswoode. All my quotations are drawn from the modern edition selected by **Sattin A** 1987

Letters from Egypt: a journey on the Nile 1849–1850 Weidenfeld and Nicolson, New York. Details of the dahabeeah in the previous paragraph come from pp. 45–6, 60, 68; details of mail deliveries in this paragraph come from pp. 49, 106, 150

13. In fact, one of her biographers says that she was not pleased at their publication, 'but acquiesced and corrected the proofs': **Cook E** 1913 *The life of Florence Nightingale Vol. 1: 1820–1861* Macmillan, London 95
14. **Warburton** *Crescent and cross op cit.* 87
15. **De Nerval G** 1927 *Voyage en Orient* Le Divan, Paris (first published in 1851) Vol. 1 389. De Nerval exempted the English from these remarks, however, 'for they stay on an island apart'.
16. **Sattin** *Letters* 120, 137, 139; **de Nerval** *Voyage en Orient op cit.* 254
17. See, for example, **Sattin** *Letters op cit.* 123
18. *Ibid.* 45–7
19. *Ibid.* 81
20. *Ibid.* 34, 49
21. *Ibid.* 49
22. *Ibid.* 72. The contrast between 'travel' and 'tourism' was a common one: see **Buzard J** 1993 *The beaten track: European tourism, literature and the ways to 'culture', 1800–1918* Oxford University Press, Oxford. But I am not sure of its purchase here. Nightingale had visited Italy with the Bracebridges in 1847–8, and she was surely not implying that it was a less intellectually serious destination than Egypt. See **Keeler M** ed. *Florence Nightingale in Rome* American Philosophical Society, Philadelphia
23. **Sattin** *Letters op cit.* 86
24. *Ibid.* 87
25. **Melman B** 1992 *Women's Orient: English women and the Middle East, 1718–1918. Sexuality, religion and work* Macmillan, London 213
26. **Sattin** *Letters op cit.* 22, 24, 44, 61, 81
27. *Ibid.* 111–12
28. John Barrell suggests that this reached its ugly climax in **Martineau H** 1848 *Eastern life past and present* Edward Moxon, London, in which the inhabitants of Egypt were likened to apes, ants, beavers, bees, camels, deer, frogs, pigs, rabbits and sheep: **Barrell J** 1991 Death on the Nile: fantasy and the literature of tourism 1840–60 *Essays in Criticism* 41 97–127, see p. 27
29. **Sattin** *Letters op cit.* 61, 86, 120
30. **Spurr D** 1993 *The rhetoric of empire: colonial discourse in journalism, travel writing and imperial administration* Duke University Press, Durham 98
31. **Pratt** *Imperial eyes op cit.* 58–9
32. **Barrell** Death on the Nile *op cit.* 115; **Sattin** *Letters op cit.* 74
33. **Buck-Morss S** 1989 *The dialectics of seeing: Walter Benjamin and the Arcades Projects* MIT Press, Cambridge 253–84; **McCole J** 1993 *Walter Benjamin and the antinomies of tradition* Cornell University Press, Ithaca 292–5
34. **Woodham-Smith C** 1950 *Florence Nightingale 1820–1910* Constable, London 46, 76
35. On the dream-image and travel writing see **Buzard** *Beaten track op cit.* 181–2
36. **Woodham-Smith** *Florence Nightingale op cit.* 79. Her entries in her notebook are tantalizingly brief, which is why I have had to rely on her letters
37. **Monicat B** 1994 Autobiography in women's travel writings in nineteenth-century France: journeys through self-representation *Gender, Place and Culture* 1 61–70; the quotation is from p. 68
38. **Poovey M** 1988 A housewifely woman: the social construction of Florence Nightingale in *Uneven developments: the ideological work of gender in mid-Victorian Britain* University of Chicago Press, Chicago 164–98
39. This world was first introduced to a European audience in the early eighteenth century with the publication of Antoine Galland's French translation, which was almost immediately translated into English and enjoyed considerable popular success. But Nightingale's imagery was almost certainly drawn from Edward Lane's illustrated translation from the Arabic, *The Arabian Nights' entertainment* 1839–41. All of these translations were highly expurgated, however, and in accentuating the fantastic and the fabulous they established the collection as a classic of the nursery and the drawing room. It was not until Richard Burton's translation of 1885–6 that the eroticism of the original was available to a wide (and often scandalized) English audience
40. **Sattin** *Letters op cit.* 32
41. *Ibid.* 187, 200. Other European travellers, both men and women, also used the dream-image to evoke Cairo as the city of the Arabian Nights. For example, 'It seems as though one were travelling in a dream through a city of the past inhabited only by phantoms': **de Nerval** *Voyage en Orient op cit.* 163
42. **Sattin** *Letters op cit.* 48. The loss of 'all feeling of identity' may be an example of the 'estrangement from the norm of the feminine' identified more generally by **Monicat** Autobiography *op cit.* 68. But it was not a uniquely female experience. 'All nature seems so tranced', wrote Warburton of his Nile voyage, 'and all the world in such a dream, that we can scarcely realise our own identity': **Warburton** *Crescent and cross op cit.* 79
43. **Sattin** *Letters op cit.* 65–8
44. **Warburton** *Crescent and cross op cit.* 77
45. **Sattin** *Letters op cit.* 187–8

46. *Ibid.* 54–5, 161. Champollion had deciphered the hieroglyphs in 1822, but it was not until mid-century that his successors were able to read more than the odd cartouche
47. *Ibid.* 54–5. There is an interesting shift in this passage from the masculine – 'men who felt and thought like us' – to the feminine – 'like English clergy wives of the present day' – which is common in the writings of Victorian women travellers: see **Blunt** *Travel, gender and imperialism op cit.* 72–8. This may also speak to the 'estrangement from the norm of the feminine' proposed by **Monicat** Autobiography *op cit.* 68
48. **Sattin** *Letters* 131
49. *Ibid.* 26–8, 39, 80, 84; see also pp. 170, 178–82
50. *Ibid.* 63–4
51. *Ibid.* 74. Later she confessed that 'the desire of the mind to find some law, to learn some reason, for this rise and fall of nations, is almost painful in Egypt' (p. 85)
52. See **Al-Sayyid Marsot A L** 1984 *Egypt in the reign of Muhammad Ali* Cambridge University Press, Cambridge; **Vatikotis P J** 1991 *The history of modern Egypt: from Muhammed Ali to Mubarak* 4th edn Weidenfeld and Nicolson, London. A fuller account would have to nuance the contrast I make here. Muhammad 'Ali's impact on the countryside was perhaps less revolutionary than this implies and some historians would prefer to describe Abbas as merely 'a typical conservative Ottoman governor': see **Cuno K** 1992 *The Pasha's peasants: land, society and economy in Lower Egypt, 1740–1858* Cambridge University Press, Cambridge; **Toledano E** 1990 *State and society in mid-nineteenth century Egypt* Cambridge University Press, Cambridge. But Vatikiotis insists that Abbas was indeed a 'reactionary despot' if seen 'in the context of the move inaugurated by the Muhammed 'Ali dynasty to develop a "modern state" in nineteenth-century Egypt'. The momentum of modernity was resumed with the accession of Said in 1854 and especially Ismail, 'the impatient Europeanizer', in 1863: **Vatikiotis** *History of modern Egypt op cit.* 70, 508
53. Those others included Warburton – who asked the 'Naiads of the Nile' whether their 'deified stream must now be harrowed up by a greasy, grunting steamship, like the *parvenues* rivers of Europe' – and, as I will show later, Flaubert. See **Warburton** *Crescent and cross op cit.* 29, 183n; **Sattin** *Letters op cit.* 29–31
54. **Sattin** *Letters op cit.* 176
55. I have used the critical edition of Flaubert's letters edited by **Naaman A Y** 1965 *Les lettres d'Egypte de Gustave Flaubert* Nizet, Paris; this is more complete than the versions contained in *Oeuvres complètes de Gustave Flaubert: correspondence, deuxième série (1847–1957)* Louis Conard, Paris (first published 1926). I have also used the critical edition of **Flaubert G** 1991 *Voyage en Egypte* edited by **de Biasi P M** Grasset, Paris; the remark about the Orient as a place of personal memory is de Biasi's and appears on p. 13. Extracts from both of them have been edited by **Steegmuller F** 1972 as *Flaubert in Egypt: a sensibility on tour* Bodley Head, London, and since this is so widely available I have cited it wherever possible. All other translations are my own
56. See **Dewachter M and Oster D** eds 1987 *Un voyageur en Egypte vers 1850: 'Le Nil' de Maxime du Camp* Sand/Conti, Paris
57. **Steegmuller** *Flaubert in Egypt op cit.* 41, 199. For a discussion of their missions, see **Starkie E** 1967 *Flaubert: the making of the master* Weidenfeld and Nicolson, London 167–9
58. **Naaman** *Lettres d'Egypte op cit.* 264
59. **Terdiman R** 1985 Ideological voyages: on a Flaubertian dis-Orient-ation in *Discourse/counter-discourse: the theory and practice of symbolic resistance in nineteenth-century France* Cornell University Press, Ithaca and London 227–57; the quotation is from p. 232
60. *Ibid.* 237
61. Flaubert told them that if they did not 'roar with enthusiasm, it will be because nothing is capable of moving you!' Four days later they delivered their awful verdict: 'We think you should throw it in the fire and never speak of it again'. See **Troyat H** 1992 *Flaubert* Viking, New York 92–3
62. **Donato E** 1993 Flaubert and the question of history: the Orient in *The script of decadence: essays on the fictions of Flaubert and the poetics of Romanticism* Oxford University Press, New York 35–55
63. *Ibid.* 41
64. **Foucault** Fantasia of the library *op cit.* 90
65. **Bruneau J** 1973 *Le 'Conte Oriental' de Flaubert* Denoël, Paris 63–70
66. **Terdiman** Ideological voyages *op cit.* 234
67. *Ibid.* 242–3
68. **Lowe** *Critical terrains op cit.* 75–101
69. **Sattin** *Lifting the veil op cit.* 84 suggests that the Bracebridges and Florence were the 'English family, hideous' that Flaubert records in his travel notes, but there is no way of knowing. Even more intriguing, given their itineraries, is Flaubert being taken to the rock-tombs of old Lycopolis: 'Our guide takes us by the hand and leads us with an air of mystery – to show us the print of a woman's shoe in the sand! She was an Englishwoman, there a few days ago'. This was at the end of February; Florence would have been there in December. But it may be that their guide's sense of time – or their understanding of what he said – was uncertain. See **Steegmuller** *Flaubert in Egypt op cit.* 35, 109–10
70. **Steegmuller** *Flaubert in Egypt op cit.* 92, 106; Du Camp describes the crew on pp. 223–8.

Contemporaries would not have regarded Flaubert's baggage as excessive. **Warburton** *Crescent and cross op cit.* 220 explained that 'the Oriental traveller' had to take 'an assortment of goods like those of an upholsterer, comprising every article his various exigencies may require, from a tent to a toasting-fork'. The desire to recreate the familiar in the midst of the exotic extended beyond these various accoutrements, of course, and was reflected in etiquette and conduct: the hoisting of flags, the invitations to bridge or tea, the shooting parties, and a host of other social practices that helped to bring a comforting Europe into an otherwise disconcerting Egypt

71. **Steegmuller** *Flaubert in Egypt op cit.* 42; **de Biasi P-M** Introduction to *Voyage en Egypte op cit.* 64; see also **Bottineau L** 1984 La représentation de l'espace dans *Salambbô* in **Masson B** ed. *Gustave Flaubert 1 (Lettres modernes)* Minard, Paris
72. **Naaman** *Lettres d'Égypte op cit.* 246
73. **Sattin** *Letters op cit.* 47; **Bart B** 1967 *Flaubert* Syracuse University Press, Syracuse 194; see also **Bart B** 1956 *Flaubert's landscape descriptions* University of Michigan Press, Ann Arbor, MI 7–11. I ought to qualify these remarks, because there is a sense in which scripting the voyage through Egypt was a process of Flaubert's *learning* to 'paint'; but even his preliminary sketches and his experimental designs were far from rough or untutored
74. **Steegmuller** *Flaubert in Egypt op cit.* 52. This passage is taken from **de Biasi** *Voyage en Egypte* 210–11 and closely follows a letter to Bouilhet: **Naaman** *Lettres d'Egypte* 179–80
75. **Steegmuller** *Flaubert in Egypt op cit.* 142. Flaubert was running ahead of himself. Although the Nile voyage was already popular among wealthy European travellers, and Sir John Gardner Wilkinson's *Modern Egypt and Thebes* – the forerunner to Murray's *Handbook* – had appeared in 1843, it took another 20 years or more for Egypt to become scripted as a site for popular tourism. See **Gregory D** (forthcoming) Scripting the Orient: possession, memory and tourism in late nineteenth-century Egypt
76. **Steegmuller** *Flaubert in Egypt op cit.* 142, 173; **de Biasi** *Voyage en Egypte op cit.* 386: **Dewachter and Oster** *Voyageur en Egypte op cit.* 35, **de Biasi** Introduction *op cit.* 60. Flaubert made a similar comparison in a letter to Bouilhet, where he described finding in the rock-tombs at Thebes 'a bordello scene from a dirty Palais Royal picture of 1816'. He continued: 'That gave us a good laugh and made with dizzy . . . It's all so modern you would almost think condoms must have been known at the time of Sesostris'. See **Naaman** *Lettres d'Egypte op cit.* 287–8
77. **Steegmuller** *Flaubert in Egypt op cit.* 63–4, 65, 75–6, 206
78. *Ibid.* 170; **Bottineau** Répresentation de l'espace *op cit.* 89. Bottineau's remark about places and glances refers to *Salammbô*, the first novel in which Flaubert drew directly on his Egyptian travels, but the observation has a wider purchase
79. **Butor M** 1984 A propos des 'Voyages' in *Improvisations sur Flaubert* Le Sphinx, Paris 43–77; the quotation is from p. 51. My English translation does not convey the French word-play: *tailler* means both to tailor ('cut-out') and to 'cut': hence 'castration'
80. **Steegmuller** *Flaubert in Egypt op cit.* 58
81. *Ibid.* 42
82. *Ibid.* 29
83. **Sattin** *Letters* 24; **Naaman** *Lettres d'Egypte op cit.* 143, 200, 307
84. **Steegmuller** *Flaubert in Egypt op cit.* 79
85. *Ibid.* 160 (my emphasis)
86. *Ibid.* 33–4, 145–6
87. *Ibid.* 39, 69, 85, 114–18, 127
88. **Sartre J-P** 1987 *The family idiot: Gustave Flaubert 1821–1857* University of Chicago Press, Chicago Vol. 2 32–43 (first published in France in 1971): For a commentary by a critic persuaded by Sartre's reading, see **Barnes H** 1981 *Sartre and Flaubert* University of Chicago Press, Chicago 77–82. Others are less well disposed. In a thoughtful commentary Ronald Aronson notes that the first two volumes of Sartre's *Family idiot* are 'an endless spiral of speculation' in which interiority and the imaginary 'have become the pure element of his activity'. See **Aronson R** 1980 *Jean-Paul Sartre: philosophy in the world* Verso, London 325–54. Only in the second two volumes does Sartre address socio-historical formations, but he largely fails to integrate interiority and exteriority. Bourdieu has offered a counter-reading to Sartre, but it is desperately in need of one of Said's contrapuntal readings because it is almost hermetically sealed around Flaubert's Paris. See **Bourdieu P** 1992 *Les règles de l'art: genèse et structure du champ littéraire* Ed. du Seuil, Paris
89. **Said** *Orientalism op cit.* 188–9
90. **Steegmuller** *Flaubert in Egypt op cit.* 41–2
91. **Naaman** *Lettres d'Egypte op cit.* 297–8
92. **De Biasi** Introduction *op cit.* 64
93. **Naaman** *Lettres d'Egypte op cit.* 142, 162
94. **Sartre** *Family idiot op cit.* 39–40
95. See, in a different context, **Suleri S** 1992 *The rhetoric of English India* University of Chicago Press, Chicago
96. **Steegmuller** *Flaubert in Egypt op cit.* 84–5; **Naaman** *Lettres d'Egypte op cit.* 293
97. **Sattin** *Letters op cit.* 42
98. I have borrowed this imagery from **Pratt** *Imperial eyes op cit.* 59

99. **Steegmuller** *Flaubert in Egypt op cit.* 126, 140
100. **Sattin** *Letters op cit.* 26
101. *Ibid.* 24, 204
102. *Ibid.* 99, 139
103. **Naaman** *Lettres d'Egypte op cit.* 151 (my emphasis). Whereas Florence Nightingale thought the desert an abomination, Flaubert was entranced by it. 'His' Orient, he declared to Bouilhet, was not Byron's Orient, 'the Turkish Orient, the Orient of the curved sword, the Albanian costume and the barred window looking on the blue sea' – which was probably also in large measure Nightingale's Orient –but rather 'the baked Orient of the Bedouin and the desert, the vermilion depths of Africa, the crocodile, the camel and the giraffe'. See **Sattin** *Letters op cit.* 34, 63; **Naaman** *Oeuvres complètes op cit.* 255
104. **Naaman** *Lettres d'Égypte op cit.* 255
105. **Steegmuller** *Flaubert in Egypt op cit.* 213
106. See, for example, **Bem J** 1979 *Désir et savoir dans l'oeuvre de Flaubert* Éd. de la Baconnière, Neuchâtel
107. **Butor** Improvisations *op cit.* 48
108. **De Biasi** Introduction *op cit.* 66
109. **Steegmuller** *Flaubert in Egypt op cit.* 47
110. **Naaman** *Lettres d'Egypte op cit.* 291. For a fuller discussion of the significance of water in Flaubert's work, see **Richard J-P** 1954 *Littérature et sensation* Éd. du Seuil, Paris 143–7 and *passim*
111. **Steegmuller** *Flaubert in Egypt op cit.* 56, 182
112. **Sattin** *Letters op cit.* 41, 47
113. **Mills** *Discourses of difference op cit.* 82
114. **Sattin** *Letters op cit.* 22–3
115. Gillian Rose describes a masculinist conception of space that is 'absolutely knowable', in which nowhere is too threatening or too different to be made known and in which space is rendered as transparent: **Rose G** 1993 *Feminism and geography: the limits of geographical knowledge* Polity Press, Cambridge 38–9. This is precisely the effect that Flaubert achieves, I think, but he does not do so through any 'masculinist denial of the male body', and the connections between gender, sexuality and space seem to me more complex than Rose's discussion allows
116. **Barnes J** 1985 *Flaubert's parrot* Picador, London 69–70. The incident is described in **Naaman** *Lettres d'Egypte op cit.* 153. Humbert was a polisher who had a shop in Rouen and was part of the comic mythology of Flaubert and his circle
117. **Steegmuller** *Flaubert in Egypt op cit.* 128–9
118. **Naaman** *Lettres d'Egypte op cit.* 151; **Steegmuller** *Flaubert in Egypt op cit.* 74, 89
119. **Steegmuller** *Flaubert in Egypt op cit.* 126. Flaubert encountered the slave trade in several places. In Cairo, he and Du Camp visited the slave market; at Asyut they met caravans dragging slaves behind them for sale to the *gellabs* (traders); and above Durr, the two Frenchmen boarded boats belonging to slave traders *en route* to Cairo – 'the black women are packed in, in all kinds of positions' – and bought belts and amulets from them: *ibid.* 67, 109, 131. Their seeming indifference was neither uniquely French nor peculiarly masculine. Nightingale saw the slave traders but the plight of their victims excited little comment: 'The Ethiopian slaves are sold by their parents willingly for a couple of handkerchiefs, or a little box, and are often exposed and picked up. We passed a boatful yesterday, crammed together, all women, half-naked . . . Our guide poked one with his stick, when it was sitting down, as if it were a frog': **Sattin** *Letters op cit.* 88
120. **Steegmuller** *Flaubert in Egypt op cit.* 30, 43, 47, 65, 67, 76, 113–19, 146
121. **Terdiman** Ideological voyages *op cit.* 247–50
122. **Steegmuller** *Flaubert in Egypt op cit.* 29
123. **Lowe** *Critical terrains op cit.* 86–7
124. **Steegmuller** *Flaubert in Egypt op cit.* 75. Cf. **de Nerval** *Voyage en Orient op cit.* 407: 'I had seen it so many times in the dreams of my youth . . . It seemed to me that I was treading in the footprints I had made before; as I went along I said to myself: "When I get past this wall, when I go through this gate, I shall see such a thing . . ." and there the thing was, ruined but real'
125. **Steegmuller** *Flaubert in Egypt op cit.* 81
126. **Terdiman** Ideological voyages *op cit.* 243
127. **Gilsenan M** 1986 *Imagined cities of the East* An inaugural lecture delivered before the University of Oxford, May 1985 Clarendon Press, Oxford 5
128. **Naaman** *Oeuvres complètes op cit.* 243–4; **Steegmuller** *Flaubert in Egypt op cit.* 54
129. **Terdiman** Ideological voyages *op cit.* 243
130. The phrase belongs to **Hentsch T** 1992 *Imagining the Middle East* Black Rose Books, Montreal 151. Although this captures Flaubert's sensibility with marvellous precision, Hentsch was in fact referring to **de Nerval** *Voyage en Orient op cit.* 408: 'Let us think no more of it. That Cairo lies beneath ashes and dust; the spirit and progress of modernity have triumphed over it like death. In a few more months, the European streets will have cut right through the old city, dusty and silent, that now crumbles peacefully upon its poor inhabitants. The part that is flourishing, glittering and growing is the quarter of the Franks, the town of the Italians, Provençaux and Maltese, the future emporium of British India. The Orient of former days is wearing out its old costumes, its old palaces, its old customs, but it is in its last days . . .'
131. **Nochlin L** 1989 The imaginary Orient in *The politics of vision: essays on nineteenth-century art and society* Harper and Row, New York 33–59. See also **Stevens M A** ed. 1984 *The Orientalists: Delacroix to*

Matisse. European painters in North Africa and the Near East Royal Academy of Arts, London

132. See **Naaman** *Lettres d'Egypte op cit.* 201 for Flaubert's attempts at learning Arabic script; **Sattin** *Letters op cit.* 68 for Nightingale's frustration at her lack of Arabic; **Steegmuller** *Flaubert in Egypt op cit.* 40, for Flaubert's amorous interpreter
133. **Sattin** *Letters op cit.* 44. Whether she would have seen much *point* in understanding them is a different question: given her interest in the remote past and her belief that present-day Egyptians had no interest in, knowledge of or claim to the ancient civilization whose ruins they inhabited, it seems unlikely
134. **Steegmuller** *Flaubert in Egypt op cit.* 40
135. **Said** *Orientalism op cit.* 176–7
136. **Sattin** *Letters op cit.* 183; **Naaman** *Lettres d'Egypte op cit.* 229, 261
137. **Warburton** *Crescent and cross op cit.* 83
138. **Steegmuller** *Flaubert in Egypt op cit.* 124
139. **Nochlin** Imaginary Orient *op cit.* 50
140. **Steegmuller** *Flaubert in Egypt op cit.* 86
141. **Sattin** *Letters op cit.* 39
142. **Steegmuller** *Flaubert in Egypt op cit.* 81
143. **Mitchell T** 1988 *Colonising Egypt* Cambridge University Press, Cambridge 33

[13]

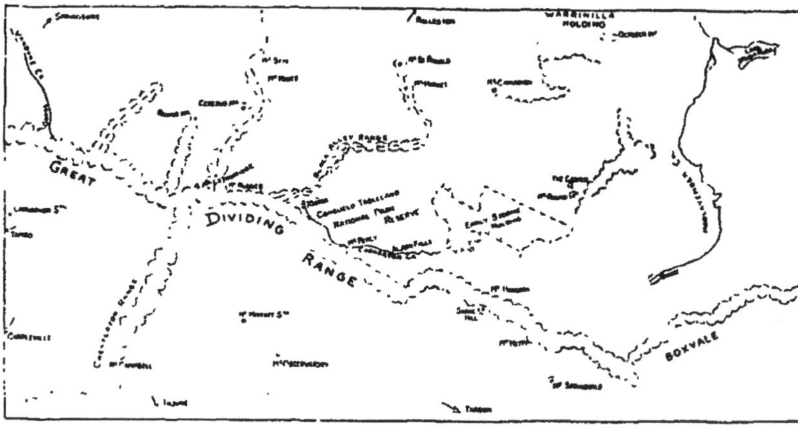

Source O'Brien, 'The Carnarvon Ranges'

Sketch map of the Carnarvon Ranges. The map shows that the early Carnarvon Gorge 'expeditions' were entering country which was far from unknown or un-named.

Kooramindanjie: Place and the Postcolonial

by Jackie Huggins, Rita Huggins and Jane M. Jacobs

Touring Aboriginal Land

In Carnarvon Gorge National Park, Central Queensland, a rock art trail guides visitors to an Aboriginal place in the past. The interpretative sign suggests that if the visitors had arrived a little earlier – some 10,000 years, it says – then they may have 'followed a family group of Aboriginals across the dry plain to this oasis gorge'.[1] *It is April 1993 and I* (Jane Jacobs) *am following 'a family group of Aboriginals' (Jackie Huggins, her mother 'Auntie Rita' and her son, young John Henry), across the plain, to the oasis gorge and down this very trail. I do not have to project into the past to move through this country in the manner recommended by the National Parks' sign.*

This journey was the result of a collaborative writing project between myself and historian Jackie Huggins whose family is from the Carnarvon Gorge area. During 1992 Jackie and I were colleagues at the same academic institution. We discussed our work and in the course of those discussions realized that we had both made our first and, at that time, only visits to Carnarvon Gorge in 1985. Jackie had gone to the Gorge to see her mother's country. On that visit Jackie never made it into the Gorge itself because of torrential rains. My visit was a result of my involvement in a research project for the Australian Heritage Commission on the impact of tourism on

Aboriginal rock art sites. The weather was good. I saw all the main art sites and came away with at least 200 colour slides. It was in part through the ironic unevenness of our experience of this place that we came to be in Carnarvon Gorge together in April 1993. At Jackie's invitation, I joined her on a visit to the Gorge in the company of her son and mother. Our objective was to write about Carnarvon Gorge in a way which might render visible the complex history of colonialism as enacted at one point in 1985 when our paths almost crossed in a place Jackie thinks of as home (but on that first visit never got to see) and I think of as a tourist destination (and saw on my first visit all that the tourist sees, with photographs to prove it).

Each of us involved in this project sought to draw upon the specific histories which shaped our associations with this place. It is these differing histories of place which are presented here. Our intention is not simply to show that places have many pasts. Rather we seek to politicize the notion of a 'sense of place', to reveal the power relations which not only shape meanings of place but also the very way in which we might dwell (or not dwell) in place. We set conditions. Jackie did not want me to be the *ethnographer, the expert on her culture. But then I did not want to be* her *ethnographer, I wanted to think and write about* my *place here.*

My land

Kooramindanjie (Carnarvon Gorge) is 600 kilometres north west of Brisbane in the sandstone tablelands of the Great Dividing Range of Australia. It is an oasis in the desert plains and Aboriginal people would travel many hundreds of miles to visit this place for water, food, shelter and ceremonies. It is the country of Rita my mother and my maternal grandmothers, it is Bidjara country.

Our people lived in this area for over 19,000 years (archaeologically speaking) in the maze of gorges, ranges and tablelands. Other tribal groups living around the area were Kairi, Nuri, Karingbal, Longabulla, Jiman and Wadja. However since the government removals policy, descendants from these tribal groups are now scattered over wide areas of Queensland and northern New South Wales.

Non-Aboriginals must understand their cultural and ethical limitations in studying, researching and writing about Blacks. They must learn to step back in areas where they are not welcome, although they often think and presume they are, where they are intruders not accomplices. Otherwise, they can do great damage to Aboriginal people and our struggles, adding to the burden rather than alleviating it. Binney suggests that we cannot translate others' histories into our own – we can merely juxtapose them.[2] Translation is detrimental to the integrity of one or the other, or to both historical traditions. Rather than trying to understand the past on its own terms, some academics have sought to explain Aboriginal pasts in terms of a contemporary ethnographical present which they confront but do not fully comprehend.

There are different experiences of the world, different bases of experience. If we begin our understanding as we actually experience the world, it is at least possible to see how we are located. What is known of the 'other' is conditional upon that 'other's' relative location. Whites must not ignore this by taking advantage of their privileged speaking positions to construct an external version of 'us' which may pass for our 'reality'. There must be limits to the ways our worlds are re-written or placed in conceptual frameworks which are not our own.

I don't have to say to Jane that I didn't want her to be the big white expert in this exercise. In fact she knows well my feelings on that topic. But I can see she's itching to get to that spiral notepad. In fact my mother asks me 'Why isn't she taking notes?'. She told everyone about Jane when we returned home, referring to her as another writer, like myself, but she was bewildered at Jane's refusal to take notes.

Ambiguous senses of place

The interpretative sign on the walking trail at Carnarvan Gorge directs the visitor to 'pause . . . reflect and wonder at the ways of those who wandered this land before us'. The Huggins family group and I – their trailing tourist – thought more about the ambiguous status of those of us who wander the land now. We discussed the history of violence and dispossession that resulted in large numbers of Aboriginals in the area being massacred late last century and those who were left being relocated earlier this century. As we walked the trail, Jackie and I discussed the fact that this visit is one of only a few ever made to the area by Auntie Rita – even though it is her homeland. I talked of my first visit to the Gorge in 1985 as an 'expert' on the management of rock art sites suffering the pressure of tourism. I wondered uncomfortably if my recommendations had manifested themselves on the ground in the form of this time-tunnel interpretative trail. Many attempts to narrate a sense of place presume the necessity of embeddedness, permanence, presence. In contrast, our presence in Carnarvon Gorge is surely constituted by our intersecting senses of displacement.

> Within 50 years of the first visit by a white man to Carnarvon, few traces survived of the culture of the Aboriginals who had inhabited this land. Only a sketchy picture of their lives and long history in the area remains.[3]

Auntie Rita's born country[4]

I was only a small child when we were taken from my born country. I only remember a little of those times there but my memories are very precious to me. Most of my life has been spent away from my country . . . but I remember about the land I come from. It will always be home, the place I belong to.

My born country is the land of the Bidjara-Pitjara people, and is known now as Carnarvon Gorge. . . . There were huge cliffs and rocks, riddled with caves where many of my people's paintings were. Most caves and rock faces showed my people's stencilled hands, weapons and tools, and there were engravings here, too. Fertility symbols and the giant serpent tells us of the spiritual significance of the place. This place is old. My people and their art were here long before the whiteman came.

The caves were cool places in summer and warm places in winter, and offered shelter when the days were windy or when there was rain. They offered a safe place for the women bringing new life into the world. As had happened for my mother and her mother before her, going back generation after generation, I was born in the sanctuary of one of those caves. My mother would tell us how my grandmother would wash my mother's newborn babies in the nearby creek, place them in a cooliman and carry them back to suckle on my exhausted mother's breast. . . .

My mother, Rose, had a Bidjara-Pitjara mother known as Lucy Conway from the Maranoa River and a white father who was never married to her mother. I never knew who her father was. I don't know much about the contact my mother had with whites. She had a whiteman's name, but she also had a tribal name, Gylma, and she spoke language and knew the old ways. My father, Albert Holt, was the son of a Yurri woman known as Maggie Bundle and a white man, the owner of Wealwandangie Station. My father was named after that man. . . . Dadda was brought up on the station, away from his mother's people. When he grew up he wanted to be with Aboriginal people, and started visiting the camps. He saw my mother there and wanted to marry her. After that, he stayed in the camp with her, and then the children started coming.

One winter's night, troopers came riding through our camp. My father went to see what was happening, and my mother stayed with her children to try to stop us from being so frightened. One trooper I remember clearly. Perhaps he was sorry for what he was doing, because he gave me some fruit – a banana, something as unknown to me as the whiteman who offered it. My mother saw and cried out to me, 'Barjun! Barjun!'

Dadda and some of the older men were shouting angrily at the officials. We were being taken away from our lands. We didn't know why, nor imagined what place we would be taken to. I saw the distressed look on my parents' faces and knew something was terribly wrong. We never had time to gather up any belongings. Our camp was turned into a scattered mess – the fire embers still burning.

What was to appear next out of the bush took us all by surprise and we nearly turned white with fright. It was a huge cage with four round things on it which, when moved by the man in the cabin in front, made a deafening sound, shifting the ground and flattening the grass, stones and twigs beneath it. We had never seen a cattle truck before. A strong smell surrounded us when we entered the truck and we saw brown stains on the wooden floor. . . .

The truck went on, travelling for two terrible days, going further south. As if in a funeral procession, we were loud in our silence. We were all in mourning. I can't remember what we had to eat or drink, or where we stopped on the journey, and by the time we reached our destination we were numb with cold, tiredness and hunger. And this new country was so different from our country – flat, no hills, and valleys, arid and cleared of trees.

It was Barambah Reserve (renamed Cherbourg in 1932). . . . Here we were separated from each other into rough houses – buildings that seemed so strange to me then, with their walls so straight. Each family was fenced off from the other into their own two little rooms where you ate and slept. The houses were little cells all next to each other in little rows. A prison. . . . The place in fact had its own gaol. A prison in a prison. There were white and Aboriginal areas.

No one had the right to remove us from our traditional lands and to do what they did to us. We were once the proud custodians of our land and now our way of life became controlled by insensitive people who knew nothing about us but thought they knew everything. They even chose how and where we could live. We had to stay in one place now while the whiteman could roam free.

Dis-place-ment

During the late 1920s Rita and her family were rounded up by the troopers and sent on the back of a cattle truck to the then Barambah Aboriginal Reserve, later known as Cherbourg. Rita reminisces about the time of traditional Aboriginal existence in a manner which most people would find beautiful. Her stories are glimpses of a pristine past prior to the white invasion. She tells me:

> My mother would make soap from the leaves of a tree and unfortunately for us there was no excuse not to take a bogey. Goanna fat was used for cuts and scratches on bare feet and limbs as well as soothing treatment for aches and pains, eucalyptus leaves for coughs and bark for rashes and open wounds. Witchetty grubs for babies' teething while charcoal was used for cleaning teeth. Bush tucker also thrived in this environment and we were never left with empty bellies. The men would go hunting for kangaroos, goannas, lizards, snakes, porcupines with their spears, boomerangs and nulla nullas while the women gathered berries, grubs, yams, edible roots, wild plums, honey and waterlilies with digging sticks. Children always accompanied the women as there was less

likelihood that animals would run away when disturbed. The creeks supplied an abundant and rich source of fish – jew, yellow belly, perches and eels.[5]

While my mother recalls an idyllic Carnarvon Gorge her memories efface a history of an opposite kind; the bloody massacres which occurred not only here but right across the continent. They were acts of violence intended to demonstrate white superiority and power.

The perpetrators were seen as heroic pioneers who had entered into the place they saw as 'wild' but which we called home. Their quest was to tame the 'natives' and possess the land. Their greed cost my people dearly. There was only one thing they could do with us – get rid of us completely so they would not have anybody else to consider. Then our tranquil home could become their fortresses and their recreational areas. They did this through force and theft, by poisoning flour and waterholes, burying Blackfellas alive, tying them to trees for shooting practice, and much more. The scarcity of white women in colonial times meant that the colonizers sexually exploited Aboriginal women. The result was rape and 'half-caste' children who were usually disowned by their white 'fathers'. The conquerors killed Aboriginal dogs and game, they dispossessed us of our hunting grounds and destroyed our sacred sites.

In 1857 the Jiman people whose country bordered Carnarvon Gorge retaliated against the invaders.[6] They attacked and killed those who lived and worked on the Fraser homestead of Hornet Bank Station. This acted as the catalyst for what became known as the six months 'little war' which was waged against the Jiman and other Aboriginals in the Central Queensland area by white vigilantes. Many Aboriginal people – men, women and children – in the area were shot down as punishment for the killing of the Fraser family. The reprisals developed into an uncontrollable rage and resulted in the wholesale slaughter of innocent human lives. No effort was ever made to bring the white murderers to court. The killings and mutual hatred created during this period went on long after, and all over Australia.

The people of my mother's generation display a profound lack of bitterness about their lot, something which I find both frustrating and amazing. This trait has often polarized old and young Aboriginals. It has encouraged so many of my generation to become active about the continuing injustices occurring to Aboriginal people. We were once the proud custodians of our nation and then our way of life was irreversibly shattered by the controlling hand of colonialism. Aboriginal people could not choose how or where to live again. Under the powers of the Queensland Aborigines Protection and Restriction of Opium Act 1897, we were dictated to by government officials and told to stay in one place. If we wanted to go off the reserves set aside for us we had to get clearance from the reserve officials. This clearance was known as a permit and the system was standard practice in Queensland from late last century until the 1980s. It was only through the permit system that we were allowed 'freedom' from the reserves. Aboriginal people had to stay put now while the whiteman could roam free. We became like prisoners or occasional tourists in our own country.

Geography and imperialism

The discipline of Geography has a problematic relationship with the historical process of colonisation. Geographical knowledges of 'other' lands were crucial tools in the building of empires and provided practical guides for dispossessing 'others' of their space. The colonial history of nation-states such as Australia testifies to the symbolic and practical possession enacted through the map. Contemporary geographers and historians have begun to re-read the map in this historic context.[7] *Paul Carter tells us that the naming of places and the making of maps, created a colonized place out of a pre-modern Aboriginal space. In Carter's terms this is the process whereby the 'haze' of space is turned into the 'clear outlines' of place.*[8]

For the map and the act of mapping/naming to be positioned as the *signifier of colonialism is a most seductive notion for a geographer. But in the critical appraisal of the map as a 'state of geographical knowledge' it is important that it does not remain 'alone on stage'. De Certeau argues that cartographic knowledge dis-places other knowledges and practices which are essential to the very production of maps. He makes a plea for the inclusion of the 'tour describers', an insertion of the 'itineraries' of occupation which give rise to and make use of the map.*[9] *This concern with 'tour describers' returns us to Carter's spatial history which foregrounds the ambiguous itineraries of colonisation – the journeyings of explorers, travel writers and others, and their encounters with Aboriginals. Such journeyings led at times to the making of maps, at other times depended upon the knowledges of maps and yet at other times darted away from the structures of knowledge and power held within the map. What is central here is the idea of 'travel' of movements into and out of places, what the anthropologist James Clifford describes as the processes of dwelling and travel.*[10] *In the Australian nation dwelling and travel have an ironic resonance. For many Aboriginal Australians the experience of being under colonialism has been one of dispossession and forced movement, that is, dwelling out of place. For non-Aboriginal Australians, occupation is in part constituted out of travel: the early travel to and through the interior, but also the transient engagements most contemporary coastal-hugging Australians have with the interior of the nation as tourists.*

To think about dwelling and travel allows the relationship between identity and place to be understood as more than an unproblemized, unruptured notion of 'sense of place' (perhaps derived from the notion of being-in-place)[11] *which is frequently said to be undermined by the alienating power of, say, commodification (including tourism).*[12] *A more productive notion of identity and place might attempt to reconcile these two possibilities, and to recognise a certain co-habitation between the global and the local, the commodified and the vernacular, the mediated and the 'real'.*[13] *Doreen Massey's 'progressive sense of place', for example, argues against a sense of place which is artificially bounded and localized and for a sense of place constituted out of the relationship between the local and the non-local.*[14] *But this 'friendly' co-habitation may not be deconstructive enough to accommodate the often ambiguous 'senses of place' which are generated in contemporary settler Australia and which are forced into a political space by the arbitrations of Aboriginal eligibility to land rights.*

The land rights provisions of this nation, including the most recent Native Titles provisions, still privilege Aboriginal claims which are supported by traditional place-specific knowledges and often uninterrupted occupation. Yet many contemporary Aboriginal Australians have an uncanny sense of place. Colonialism has rendered their country, their home, unfamiliar to them.[15] *In 1993 Jackie, Auntie Rita, John Henry and I were guided through Carnarvon Gorge by paths and signs made by the Queensland National Parks and Wildlife Service. We practiced a map, followed an itinerary, which had been devised by the National Parks authority as a means of presenting this place – as a scene – to the visitor. Carnarvon Gorge may well have been Jackie and Auntie Rita's country, but the path of their return was made for viewers other than themselves. A truly postcolonial land rights provision might accommodate such estrangements and presume its goal to be the provision of the means for places to once again be rendered familiar. This is an active and inventive land rights, not a compensatory land rights of loss.*

Lost in place

I felt out of place at Carnarvon Gorge. Jane and I actually thought we were lost in the area at one stage. I felt I should know this place but could not get us out of the mess we had found ourselves in; we did not know where we were as we drove along the dirt track in the darkness of night. I could sense that Jane was too frightened to offer directions – directions I had expected, her being a geographer and all.

On the first of our many walks to various cultural sites a certain confusion arose. It was a journey which began with the wrong name. We set out for a site Auntie Rita had referred to as 'what I think they call "Hemisphere"'. Having been to the Gorge before, and being a geographer and reading my maps, I knew there was no site called 'Hemisphere' but that there was one called 'The Amphitheatre'. Close enough. We set out the next day to the place on the map called 'The Amphitheatre'. When we arrived at the site it was not the place Auntie Rita had visited before and wanted us to see. This place had no rock art. Auntie Rita was greatly distressed, thinking this was the place but that it no longer had any art. With my geography and my translation we had come to the wrong place.

Indigenous touring

The 'uncanny' might also describe contemporary non-Aboriginal experiences of this place. Non-Aboriginal tourists were clearly unsettled when they confronted 'indigenous tourists' walking along the Carnarvon Gorge trails – even more so when Aunty Rita struck up a conversation and announced 'I was born here, this is my country'. This is not to say that non-Aboriginal tourists do not expect or want to meet Aboriginal people when out in the bush. Most do, and throughout Australia National Park authorities and mainstream tourist ventures are busily 're-Aboriginalizing' place in the service of the tourist industry. Aboriginal communities are often actively engaged in this process, producing arts and crafts for the tourist, running their own 'indigenous Australia' tours, and in some areas owning and managing the areas open to the tourist.

But in Carnarvon Gorge the Huggins family seems 'out of place'. This is a place rendered in popular literatures, including current park brochures, as devoid of living Aboriginal people but rich in Aboriginal 'heritage'. Official brochures mention the 'passing' of local Aboriginals but not the local massacres. Aboriginality is presented in Carnarvon Gorge through the rock art which, safe and static, hugs the walls of the Gorge. The Huggins family's claims of belonging and ownership, along with their presence as 'tourists' in the Gorge, are rendered unbelievable by the myth of a place thoroughly cleared of Aboriginals.

> Their passing is marked only by the fragile art on the sandstone cliffs. This is damaged if the fine dust at the base of the rock wall is disturbed. To enable you to view the art, boardwalks have been constructed at the three art sites. . . . Keep to these to protect a valuable and irreplaceable part of our heritage. . . . Please look without touching. The artists are gone and it would be a further tragedy to lose the heritage they left. By keeping on the boardwalk you will see the art to the best possible advantage.[16]

Carnarvon Gorge is a place occupied by tourists, shaped by tourism. There is a transitory but continual occupation of the Gorge by tourists. In 1993 some 400 visitors each weekend confirmed this is no longer Bidjara country, but a public space, a National Park. The National Parks sign we confronted on our first walk together through the Gorge directed the tourist to gaze upon an imaginary Aboriginal time, in what is now an ambiguously Aboriginal place. Imagine if the National Parks interpretation sign directed the visitor's gaze not to an Aboriginal pre-history but to a more recent colonial history. Here we might see the imaginative projections which have ensured that this place became a tourist space, serving the transient in search of an authentic experiences of inland Australia.

When I set out to write about Carnarvon Gorge from my own non-Aboriginal positioning, I never imagined that I'd find a colonial history so close to my professional identity as a geographer. For Carnarvon Gorge, tourist destination, was created through the vision of an amateur geographer, Danny O'Brien, who was a member (and later Director and General Secretary) of the Royal Geographical Society of Australasia (Queensland Branch) from the 1930s through to the mid 1960s.

The first expedition by the Royal Geographical Society of Australasia (Queensland Branch) to the Carnarvon Ranges left Brisbane on July 6, 1937 with 11 members to 'explore that vast area known as the Carnarvon Ranges'.[17] *It was led by the enthusiastic but undoubtedly eccentric society member, Danny O'Brien. It was the first of at least fourteen expeditions to the sandstone tablelands between 1937 and 1958. O'Brien was a controversial figure in the Queensland Society. He is remembered today less as an heroic advocate of the Carnarvons, the maker of an inland service industry resource, than as a 'abrasive' and 'self-indulgent' personality who seriously damaged the reputation of the Society among scientists. Indeed, the government is reported to have complained that under his leadership the Society was no longer serving its function of providing geographical information for the administration of Queensland.*[18]

O'Brien's expeditions were representative of a modern Australia, a modernising Australia. The group travelled not by horseback, but by train and car. Trails were not carved through unknown lands. The groups travelled with local guides from nearby pastoral leases and frequently arranged ritualistic ceremonies of governmentality, various civic welcomes and farewells, to mark their passage through the area. By this stage in the settlement and development of modern Australia, this place at best could be described by O'Brien as 'little known land'.[19]

In this history 'the geographer' is not simply the maker of maps; the area had already been surveyed and mapped by the time the Society came to Carnarvon Gorge, no frontier cartography was necessary. Less than ten years after the first expedition, the Carnarvons was captured on an aerial photograph run. All major natural features were already named, even renamed, as O'Brien's own account testifies:

> *We stood on Consuelo Tableland . . . and viewed the great valley of Lethbridges' Pocket with its deep walls. . . . We saw in the distance, Tyson's Nugget, Mount Lethbridge (also known as Keniff's Lookout), Vandyke Creek, veering away towards Springsure and Mount Sugarloaf.*[20]

All that was left to name for this mimic explorer were overlooked geographical features or newly-constructed engineering works. O'Brien immodestly named after himself a road he had lobbied local authorities to grade and a makeshift causeway made by the members of one of his expeditions.

O'Brien did not just lead expeditions to these areas; he wrote about, photographed and filmed them. These 'explorers' did not harbour the reluctance for photography Paul Carter noted in their predecessors.[21] *O'Brien's writings appeared in special Society publications and his visual records of the Gorge were presented in lantern slide and film showings in Brisbane and regional centres throughout Queensland. O'Brien's writings on his visits to Carnarvon Gorge are not part of that genre of information which might help us reconstruct the 'real' geography of Carnarvon Gorge, but they are important in reconstructing an imaginative geography of desire. His writings are more aptly described as travelogues than scientific treatises. They are fledging promotional documents with descriptions of place more romantic than systematic. O'Brien's activities on behalf of the Royal Geographical Society during the 1930s and 1940s thus mark a moment when the geographical imagination shifts from the cartographic to the scenic, in the quest to invent a tourist destination.*

O'Brien's 'pioneer expedition' was considered a triumph less because of the new knowledges it established than the verification that the gorge could be reached by car over level country, and that, if an all-weather road was built, the Gorge could be reached in a 'comfortable' two-day motor trip from Brisbane. In his writings O'Brien was self consciously constructing a place which was to have a particular role in the on-going development of Australia and the state of Queensland. His 1942 publication entitled 'Carnarvons' was written 'for the development of the state' and copies were duly posted to President F. D. Roosevelt, Henry Ford, David Lloyd George and Prime Minister Menzies.[22] *He was, by his own description, 'the honorary advocate of Carnarvon range development' and his rewards were 'the knowledge that there now exists a tourist traffic to Carnarvon National Park'.*[23]

In creating this tourist destination, O'Brien's depictions of Carnarvon Gorge sought to reinvent a frontier land, an unoccupied land open to 'discovery'. In relation to Aboriginal occupation, O'Brien did not have to rely on imaginative constructs like terra nullius *(land unoccupied), which was the founding myth of the colonization of Australia. In this part of Queensland the fiction of land unoccupied had (almost) become fact through the massacres in the area and the forced removal of Aboriginal families to reserve areas. When O'Brien described his 1937 visit to the Gorge he referred to the 'old Dead Myall race'.*[24] *But the area was not empty of Aboriginals: there were families that still worked and lived on nearby pastoral leases and there was still a small community camped on the edge of Springsure. On one of the early expeditions, O'Brien discovered that a local Aboriginal population did exist, but noted that it was confined to a camp on the edge of the nearby town of Springsure and that the group was mainly 'mixed blood'. Under the supervision of a police officer the expedition visited the camp. Although it is very probable there were quite a large number of Aboriginals at this camp, O'Brien was only interested in photographing a man he refered to as 'the Last of the Tribe group of Aboriginals'. King Chookey's photograph was to appear in the travelogues to be issued by the Society.*[25] *In the context of the travelogue this photograph functioned as an exotic curiosity. But King Chookey, 'last' of the tribal Aboriginals, would also provide an assurance to would-be travellers that Aboriginals of this kind were few and far between in this place. Prospective travellers were told in one promotional booklet that they might look forward to camping in an 'Aboriginal cave' but were assured that the Aboriginal inhabitants had disappeared 70 years ago.*[26] *O'Brien's erasure of a 'threatening' Aboriginal population was premised on the distinction he made between 'full blood' and 'mixed blood' Aborigines. The former he relegated to the 'relic', the latter he simply discounted. In almost all of his publications Aboriginality is presented in more 'comfortable' and aesthetically pleasing ways. It is the visually impressive rock art which lines the walls of Carnarvon Gorge which is to provide visitors with a calm and scenic experience of a pacified and naturalized Aboriginality.*

In O'Brien's quest to develop this place as a tourist destination for those seeking an experience of the interior it was also necessary to construct this place as untouched by non-Aboriginal settlers. O'Brien reaches Carnarvon Gorge through modern technology and engineering and relies on non-Aboriginal guides from the local pastoral leases. Yet he depicts the place as untouched by settlement – where 'the footprints of the white man are still missing from many of its most secluded places'.[27] *Discursively depopulated, O'Brien could label his first trip to the Gorge a 'pioneer expedition' and speak convincingly of the 'insurmountable difficulties' he faced in traversing 'new country'. His writings not only made a place but also render him the pioneer hero. But he was an explorer who, like many before him, followed in the paths of others.*

> *. . . we passed over a sloping razor back with only a few feet of even surface, and although but a few hundred yards in extent, it proved to be the most thrilling experience of our whole journey. There was a precipice on each side, hundreds of feet down without a stop. The expedition moved in single file. Our guide went on before us in a casual and unconcerned way, and we followed his lead.*[28]

His heroic possession of the 'undiscovered' Carnarvons were indeed a fiction. By the end of the nineteenth century the Carnarvon Range areas had been divided between the pastoral leases of Mt Moffatt, Carnarvon, Consuelo and Bandana. By 1932 the Gorge itself had been designated by the state authorities as National Park. His desire to re-invent the masculinist project of pioneering exploration was perhaps most seriously unsettled by the presence of a small travellers cottage at the mouth of the Gorge, which was built by the local Country Women's Association around the time of the Society's first expedition to the area.

O'Brien was beset with a nostalgia for an earlier age of exploration when geographical knowledges of description and mapping were on more solid ground. O'Brien and the Royal Geographical Society attempted to consolidate the precarious masculinity of their expeditions through reference to science and technology. The very description 'expedition', from the noun 'expedite', evokes an authorized, systematic venture with an intent of clearing difficulties, or hastening progress. O'Brien consciously brought modernity to the Gorge. His journeys were marked by orchestrated technological rituals which were later reported to the Society. He claimed to have made the first chartered air flight over the Gorge and the first wireless transmission from the Gorge. This performance of progress ultimately gave way to the pursuit of pleasure.

By 1956 a circular letter encouraging people to come on a Royal Geographical Society trip to the Gorge was still referring to the ventures as 'expeditions'. But the practice of 'expedition' was giving way other notions. The brochure which promoted the Society tours describes what travellers might expect:

> *'An outstanding feature of recent expeditions is that though we are not now attracting so many scientists, we are getting a great many intelligent younger people. Thus the real aims of our Society are being accomplished, and instead of geology, anthropology, etc., just being laborious subjects in the lecture room, they are now 'picnic' subjects – education by pleasure. Our students don't get nervous collapses caused by study, though some have collapses caused by Cupid. . . . This is Queensland's biggest picnic trip'.*[29]

O'Brien's dream was being realized: the Society's visits to Carnarvon Gorge might well be seen as infant forms of niche tourism: an adventure tour, a cultural tour, an eco-tour, perhaps even Australia's first version of ClubMed.

Despite O'Brien's avid advocacy of Carnarvon Gorge and the early interest by the Queensland Tourism Bureau in the area this place, Carnarvon Gorge was received ambivalently. A post-war (1949) visit to the area by tourism officials and the press led to reports that the area was 'over-rated as a tourist attraction', had 'disappointing' scenery and poor accessibility. At best, it concluded, the area might serve as a cheap holiday destination for 'low income western residents'.[30]

Carnarvon Gorge has remained a relatively inaccessible place, a destination only for the keen long distance traveller or the local. Tourist numbers remain relatively modest when compared with the massive levels of tourism experienced by the more famous and spectacular National Parks of Kakadu or Uluru. The main road to the Gorge is still unsurfaced and there is no five-star hotel resort. The pace of recent tourist development of the area has been largely regulated by a local non-Aboriginal, a self appointed gatekeeper of Carnarvon, who works for the National Parks Service. O'Brien saw the opening up of the Gorge as dependent upon the intervention of engineering, the construction of roads, the building of resorts. The current manager of this site has used engineering and technology to different ends – to contain visitors to direct their paths through the Gorge. In the early 1980s military helicopters airlifted timber into the Gorge so that massive boardwalks, the first of their kind at art sites in Australia, might be built. The boardwalks were designed to keep tourists away from the art, they protect 'heritage'. But the boardwalks also opened the art to more precise and regulated scrutiny, directing where visitors go, drawing the visitor's attention to specific details, interpreting the art in terms of technique and meaning and presenting the signature of 'heritage' on the landscape of Carnarvon Gorge. These structures, built to protect the art, inscribe the on-going possession of this place by tourism.

> The true meaning of the gorge to the Aboriginal people may never be explained. The Aborigines coexisted successfully with the environment in this area before the pyramids of Egypt were built, but they left no written record.[31]

Reclaiming Place

Returning to my mother's 'born country' complemented my own sense of identity and belonging, and my pride in this. It was important to me that we made this trip together as she had been insisting for quite some time, pining for her homelands. We shared a special furthering of our mother-daughter bond during this time, although we argued incessantly about nothing as usual or, as she calls it, 'fighting with our tongues'. I began to gain an insight into and understanding of her obvious attachment and relationship to her country and how our people cared for this place long before the Royal Geographical Society and park rangers ever clapped eyes on it. The way my mother moved around, kissed the earth and said her prayers will have a lasting effect on my soul and memory because she was paying homage and respect to her ancestors who had passed on long ago but whose presence we could both intensely feel.

This was our place, my sense of becoming. The land of my mother and my maternal grandmother is my land too. It will be passed down to my children and successive generations, spiritually, in the manner that has been carried out for thousands of years. As Rita's daughter, I not only share the celebration and the pain of her experience but also the land from which we were created.

Like most Aboriginal people it is my spiritual and religious belief that we come from this land hence the term 'the land my mother'. This land is our birthing place, our 'cradle'; it offers us connection with the creatures, the trees, the mountains, the rivers and all living things. This is the place of my dreaming. There are no stories of migration in our Dreamtime stories. Our creation stories link us intrinsically to the earth. This is why place and land is so important to us, it does not matter where and when we were born.

The High Court Mabo decision of 1992 and the subsequent Native Titles Act will attempt to rectify some of these injustices through the new land claim process. Claims are already being prepared on behalf of the Bidjara people for the Carnarvon Gorge area. Five other clan groups may also lay claim to it, as this place was used by a number of groups for ceremonial purposes. Tensions are mounting between the Aboriginal traditional owners and interest groups, pastoralists, National Parks and Wildlife Service and of course the white experts. Decisions made by the Aboriginal Land Claims Tribunal will need to consider the concerns of all interested parties and not just Aboriginal concerns. Any claims we Bidjara people make to this country under the provisions of the new Native Titles Act will have to counter the myths that there are no Aboriginal people for this place. The claims we as Aboriginal owners of this land can make are tainted by dispossession and its violence on our memories.

Rita and other elders will now be pressured to remember their childhood memories of their place to help prove our association with this country. Not so long ago the protectionist and assimilationist policies of the day pressured our elders to deny and forget. How ironic it is that now every detail must be recalled and told and that this remembering and this telling is now honourably encouraged and documented and forms the basis for land claims being won or lost. History finds a way of reinventing itself for Aboriginal people, for better or for worse.

In conclusion, I want to return to the early tourists. Their desire for tourist pleasure was always shadowed by the memory of violence. Mrs Hamlyn-Harris was one of the first women to go on a Royal Geographical Society Carnarvons expedition. She wrote a series of poems about this place as a tribute to the intrepid leader Danny O'Brien. In one of her poems, simply entitled 'Carnarvon Gorge', she contemplated the history that led to her coming to my place.

> Yes, what have we done?
> The past is dead by our own hand. The olden race is gone.
> They are but ghosts, those spirits of the wind and trees,
> That sigh such doleful threnodies!
> We of To-day by duty bound must wake
> And silence them with Joy's sweet laughter ringing,
> And children's voices singing!
> We must arise, and of this primal playground make
> A resting place for toiling families of the drought stricken west
> Needful of Nature's healing and of rest.
> Thus shall we at last
> Make humble expiation for the past.[32]

Turning Kooramindanjie Place into Queensland's 'biggest picnic ground' has not made 'humble expiation' for the past.

NOTES

1 Walking trail interpretation sign produced by Queensland National Parks and Wildlife Service, Baloon Cave, Carnarvon Gorge National Park (n d.)

2 Judith Binney, 'Maori oral narratives, Pakeha written texts· two forms of telling history', *New Zealand Journal of History*, 21(1), 1987, pp. 30–32.

3 Queensland National Parks and Wildlife Service, *Carnarvon National Park· Park Guide*, QNPWS, 1991. By the QNPWS estimates, the area would have been clear of Aboriginals by the 1880s Auntie Rita and her family were taken from her country in the 1920s.

4 The following section is reprinted in Rita Huggins and Jackie Huggins, *Auntie Rita*, Aboriginal Studies Press, for the Australian Institute of Aboriginal and Torres Strait Islander Studies, 1994, pp. 7–12.

5 Reprinted in Huggins and Huggins, *Auntie Rita*, pp. 7–8.

6 Gordon Reid, *A Nest of Hornets. The Massacre of the Fraser Family at Hornet Bank Station, Central Queensland 1857 and related events* Oxford University Press, 1982; and Raymond Evans, Kay Saunders and Kathryn Cronin, *Race Relations in Colonial Queensland: A History of Exclusion, Exploitation and Extermination*, University of Queensland Press, St. Lucia, 1988.

7 See as examples J. Brian Harley, 'Maps, knowledge and power' in Denis Cosgrove and Steven Daniels (eds) *Iconography of Landscape*, Cambridge University Press, 1988, pp 277–312: David N. Livingstone, *The Geographical Tradition*, Blackwell, 1992.

8 See for the more general point of mapping and naming Paul Carter, *The Road to Botany Bay*, Faber and Faber, 1987, pp. 320–352: Paul Carter, *Living In A New Country*, Faber and Faber, 1992, pp. 9–27 and p. 123–124

9 Michel de Certeau, *The Practice of Everyday Life*, University of California Press, 1984, p. 121.

10 James Clifford, 'Traveling Cultures', in Lawrence Grossberg, Cary Nelson, Paula Treichler (eds), *Cultural Studies*, 1992, pp. 104–5.

11 Martin Heidegger, 'Building, Dwelling, Thinking', in Martin Heidegger, *Basic Writings*, Harper San Francisco, 1977, pp 319–340.

12 Sharon Zukin, *Landscapes of Power From Detroit to Disney World*, University of California Press, Berkeley, 1991.

13 David Harvey, 'From place to space and back again: Reflections on the condition of postmodernity', in Jon Bird, Barry Curtis, Tim Putnam, George Robertson and Lisa Tickner (eds), *Mapping the Futures· Local Cultures, Global Change*, Routledge, 1993, p 11.

14 Doreen Massey, 'Power-geometry and a progressive sense of place' in Jon Bird, Barry Curtis, Tim Putnam, George Robertson and Lisa Tickner (eds), *Mapping the Futures: Local Cultures, Global Change*, Routledge, 1993, p. 64.

15 See Ken Gelder and Jane M. Jacobs, 'Uncanny Australia', *Ecumene*, forthcoming

16 Queensland National Parks and Wildlife Service, *Carnarvon National Park Park Guide*, QNPWS, 1991.

17 Danny O'Brien, *Carnarvons* Published under the auspices of the Carnarvon Range Association, Brisbane. Records of the RGSA (Qld), 1942, p 2.

18 P. D Griggs, *A Dream in Trust The Centenary of the Royal Geographical Society of Australasia, Queensland, Incorporated*. Special issue: *Queensland Geographical Journal* 8, July 1985, pp. 82–84.

19 O'Brien, *Carnarvons*, p 8.

20 Danny O'Brien, *The Carnarvon Ranges. Expedition of 31 all told, 18.9.'50 to 3.11.'50.* Unpublished manuscript. Records of the RGSA (Qld), 1950, p 2

21 Paul Carter *Living in a New Country*, Faber and Faber, 1992, p 31. Carter argues that there is an odd absence of photographical technologies in early exploration and geographical travels.

22 O'Brien, *Carnarvons*, p. 8.

23 O'Brien, *Carnarvons*, p. 36.

24 O'Brien, *Carnarvons*, p 21

25 O'Brien, *Carnarvon Pictorial*, p 6

26 Danny O'Brien, *The Carnarvon Ranges*, RGSA (Q), Records of the RGSA (Q), 1947, p. 12.

27 Danny O'Brien, quoted in Anon , Manuscript in Records of the RGSA (Q), n d., p. 1.

28 O'Brien, *Carnarvons*, p. 5.

29 Circular letter from Peter Grant and Tom Denning, Joint Organisers of the 1956 RGSA (Qld) Carnarvon Expedition, to prospective participants, Records of the RGSA (Q), February 1956.

30 *Charleville Times*, 'The Carnarvon Ranges Fails as Tourist Attraction', 22 December, 1949.

31 Queensland National Parks and Wildlife Service, *Carnarvon National Park: Park Guide*, QNPWS, 1991.

32 M. Hamlyn-Harris, 'Carnarvon Journey and other Verses of the Series "Mid Highways and Byways of Queensland"'. Tribute to D. A. O'Brien from Friends RGSA. RGSA (Qld) Records, n.d.

[14]

Paradise or pandemonium? West African landscapes in the travel accounts of Victorian women

Cheryl McEwan

Recent attempts to include women in histories of geography have often drawn on the work of women travel writers in the British empire. Research to date has tended to focus on discourse theory, and the spatial context of the journey has been somewhat neglected. In their efforts to assert that travel writing is an essentially gendered form of representation, several theorists have failed to recognize the complexities in the varying responses to and depictions of overseas territories by women travellers. This paper emphasizes the importance of the regional context of the journey by analysing landscape descriptions in the narratives of Victorian women travellers in West Africa. It explores how women travel writers contributed to the popular geographies of West Africa in Victorian Britain and, more specifically, investigates the extent to which the images contained in their narratives were informed by, or challenged, Britain's myth of the "Dark Continent" during the nineteenth and early-twentieth centuries. It also analyses how the temporal context of the journey, the evolving nature of British imperial culture, and contemporaneous ideas about romanticism, "wilderness" and "sanctuary", were important factors in the landscape descriptions of West Africa by women travellers.

Recent years have witnessed a renewed interest, particularly in cultural studies and literary theory, in travel writing.[1] The renewed interest has also captured the attention of geographers, particularly with regard to imperial travel writing, and several authors have made the case for interpreting travel writing, both contemporary and historical, as a form of geography.[2]

Increasingly, gender has also become an important concern in analyses of travel narratives. Much of the research into women travel writers has taken place in disciplines outside geography, in particular, anthropology and history, and within discourse theory.[3] However, feminist geographers have begun to produce theoretical critiques of both travel writing and imperialism. Mona Domosh refers to women travel writers in her contention that historiographies of geography should incorporate feminist approaches. However, by claiming women travel writers in a disciplinary sense she ignores the fact that these women did not consider themselves geographers and fails to address the imperialist implications of both their travels and their narratives. Alison Blunt's exploration of the narratives of Mary Kingsley addresses this very point. Blunt offers a critical reading of the social, cultural and intellectual context of British imperialism and of the complex relationships between white women travellers and the empire. Her thesis is a persuasive argument about the place of women in historiographies of geography.[4] Nineteenth-century women travellers cannot be claimed in a disciplinary sense since they did not consider themselves geographers and, ultimately, did not contribute to the establishment of geography as an academic discipline. However, in publishing accounts of their travels at a time when travel writing was extremely popular, and also an important medium for the dissemination of "knowledge" about overseas territories, women travellers were important in the production of *popular* geographies. The term popular geographies, therefore, refers here to the creation of images of West Africa and its peoples outside academic and scientific discourse, but available for mass public consumption. They were especially important because the vast majority of people in Victorian Britain did not have the opportunity to experience overseas territory first-hand. Popular geographies were created through a variety of media, including advertising, colonial fiction and travel writing,[5] and women's contributions to these popular geographies merits investigation, particularly in light of recent critical readings of travel narratives by women.

Early critiques of women travellers and their narratives tended to treat both as a monolithic category from which conclusions about an essential "feminine" *genre* of travel writing or experience of empire can be drawn.[6] More recent theoretical critiques have sought to address this problem. For example, Alison Blunt confronts issues of subjectivity and produces an in-depth reading of one narrative by a single women; Sara Mills suggests that travel writing by men and women is not always "fundamentally and necessarily different", but that the gendered context of imperialism always makes a difference to the construction and reception of the text; Billie Melman argues that gender is important as an analytical concept only when it is considered together with other relationships of power such as race and class.[7] It is the contention of this paper that further studies would benefit from a *regional* focus, and an analysis of travel narratives about specific *places*. Such an approach facilitates the exploration of differences within travelogues written by women, and avoids notions concerning an essential feminine experience and description of empire.[8]

The regional focus of this paper is West Africa;[9] the analysis centres on the construction of a particularly powerful vision of Africa, focused very much on West Africa, during the nineteenth century—the vision of the "Dark Continent"—and how women travel writers responded and contributed to this. Several writers have analysed the creation of this "myth" of Africa and its imperialist connotations through a variety of sources, including exploration literature, colonial fiction, imperial art and scientific treatises;[10] this paper explores the vision of the "Dark Continent" in relatively under-researched sources, namely the travel narratives of Victorian women. It focuses particularly on the representation of the physical environments of West Africa in these narratives.[11] Imperial strategy was characterized by a tendency to homogenize the "other",[12] but West Africa was not a homogeneous entity; clearly the location for the narrative, both spatial and temporal, was an important factor influencing the landscape descriptions of women travellers. By exploring the differences in their representations of West African landscapes, it is possible to highlight some of the complexities and contradictions in nineteenth-century notions of the "Dark Continent".

All the women of this study drew on references to "the Dark Continent", "the Land of Death", and "the White Man's Grave", but their landscape descriptions were often strikingly different. The creation of a particular narrative was informed and influenced by the spatial and temporal context of the journey, and by the individual circumstances of the author concerned. This paper analyses the publications of four women who, together, highlight the complexities of any reading of imperial imagery in West Africa.[13] Elizabeth Melville was from an aristocratic background.[14] She married a colonial judge and lived and travelled with her husband in Sierra Leone between 1840 and 1846. Her account of her experiences was entitled *A Residence at Sierra Leone* (London 1968, first published 1849), and contained a strong anti-slavery message which was typical of the time. Zélie Colvile was the wife of the Acting Commissioner of Uganda and travelled in the Oil Rivers area of Nigeria in the early 1890s, immediately prior to the onset of "high imperialism" in West Africa, as part of a coastal journey around the continent. Like Melville she had an aristocratic background,[15] and the fact that she travelled purely as a tourist makes her unique to this study. She published an account of her journey entitled *Round the Black Man's Garden* (Edinburgh 1893). Mary Kingsley was a traveller/explorer who journeyed throughout tropical West Africa between 1892 and 1895 and gained the reputation as being *the* expert on West Africa.[16] She published three books about West Africa and numerous articles; her major publications were *Travels in West Africa* (London 1895) and *West African Studies* (London 1899). Constance Larymore was the wife of a military administrator who travelled with her husband in Southern and Northern Nigeria between 1902 and 1908, at the beginning of the colonial phase of Britain's involvement there. She published *A Resident's Wife in Nigeria* (London 1908).[17]

West Africa and the myth of the "Dark Continent"

Depictions of the physical environment were important in establishing enduring images of West Africa in nineteenth-century Britain. Patrick B. Rantlinger has argued powerfully that the vision of Africa in the imaginations of many Victorians, which was informed by scientific theory, exploration literature, fiction and art, evolved with the changing nature of British cultural, economic and political imperialism in the region. The anti-slavery supporters at the time of Melville's travels fostered an image of an Edenic West Africa, an unspoilt wilderness that would be perfectly peaceable were it not for the intrusions of the slave trade. However, with the failure to convert West Africans to Christianity, and as a result of the high mortality rate among Europeans travelling to West Africa, this image was soon transformed into a more negative portrayal. West Africa became exotic, unfamiliar and frightening. The harsh climate and prevalence of disease were used in arguments in favour of Britain's abandonment of West Africa, and such portrayals held sway throughout the middle of the nineteenth century. By the time of the revival of Britain's interest in West Africa in the 1880s, prior to the journeys of Colvile and Kingsley, the image of West African landscapes had become one of absolute pandemonium, sadly lacking the order of "civilization". This depiction of the physical environment coincided with Britain's desire to extend its empire in Africa at the time of the Berlin Conference, and persisted into the twentieth century when Larymore travelled in West Africa. Africa became the "Dark Continent", and West Africa was epitomized as the centre of this "Dark Continent".[18] Of course, Africa was the "Dark Continent" only in the tautological sense that many people in Britain defined it as such, the necessary mirror-image confirming the superiority of an enlightened and ordered Europe.[19]

The descriptions of the physical environment by women travellers thus need to be framed within this broader picture of British imperialism, and the changing popular images of West Africa throughout the nineteenth century. As stated previously, most of these women were influenced by negative images of the physical environment of West Africa. Zélie Colvile's chapter on West Africa was entitled, rather sensationally, "The Land of Death". Mary Kingsley described how friends had warned her not to go there because of the threat of disease, that West Africa was the "white man's grave" and the "deadliest spot on earth", but she claimed that she chose to travel there precisely because it was coloured black on Scheele's map of the distribution of disease.[20] Kingsley also described the psychological malaise that could affect Europeans in West Africa, and although she felt at home in the forests, she believed that for those who could not identify with their surroundings, life there was extremely difficult.[21] Elizabeth Melville was constantly ill during her years in West Africa, and on her first furlough home to Britain she decided to leave her children rather than take them back to Sierra Leone. Although Melville's descriptions of the hills outside Freetown perhaps came closest to evoking Edenic images of the landscape consistent with her anti-slavery views, she was convinced that disease and death were the price to be paid for the exquisite beauty of West Africa.[22] This attitude was consistent with contemporary opinion of Sierra Leone as "The White Man's Grave". The climate of West Africa, which was believed during the nineteenth century to be the cause of fever, disease and death, was the one aspect of the physical environment which drew the darkest descriptions from women travellers. However, if one looks more closely at the texts there were differences and complexities in the imagery created by these women, and often these images challenged contemporary visions of the "Dark Continent".

British women travellers in West African landscapes

Landscape descriptions by women travellers contributed to Britain's popular geographies of West Africa. As a consequence of the lack of available information, such as publication figures and sizes of lecture audiences, it is difficult to assess the extent of the audience to which these women had access,[23] but it can be supposed that, because of the popularity of the literature of empire during the nineteenth century, the imagery that they created could be consumed by potentially large audiences. Colvile broadened her audience by publishing an article in *Blackwood's Magazine*,[24] and she also publicized her book by giving lectures.[25] As with Melville, her narrative was aimed as light reading for the general enthusiast, and the potential for these works was considerable. There existed a fascination with empire among various leisured middle- and upper-class women in Britain, which was an extension of the desire among many Victorians for titillation by tales of adventure in far-off, "exotic" locations. The tales of "the daughters of Albion" establishing little Englands around the world were extremely popular.[26]

There is more evidence regarding the popularity of Mary Kingsley's work. Her audience was a varied one, incorporating those interested in travel narratives from the empire, and those engaged in commerce, science and politics. Kingsley's readership was also quite substantial. The first edition of *Travels in West Africa* ran to 1500 copies and was reprinted eight times up to 1904, the reprint runs totalling 7500. *West African Studies* was published in 1899 in an edition of 2000 copies. It sold 1200 copies the week it was published.[27] It was reprinted in 1901 in an edition of 1500 copies.[28] Both books were reviewed very widely by both the national and provincial press, and were also reviewed as far afield as France, Australia and the United States. Kingsley also published articles in the influential periodicals of the time (*The Spectator*, *The National Review*, *The Fortnightly Review*, *Manchester Chamber of Commerce Monthly Record*, *British Empire Review*, and other learned journals such as the *Transactions of the Liverpool Geographical Society* and the *Scottish Geographical Magazine*). Kingsley broadened her audience further by embarking upon an exhausting series of lecture tours around Britain, addressing most notably the Scottish and the Liverpool Geographical Societies, but also various audiences in many locations around Britain. She would often address over 2000 people at these lectures.[29] Thus the popular geographies created by women travellers had potentially large audiences, and the expectations of the audience may have played a part in shaping the production of these narratives.

The conventions of the time exercised some influence on the language employed in landscape descriptions by women travellers. For the most part, they did not employ in these descriptions the sexual metaphors often found in travel accounts by men at this time;[30] the language of penetration, conquest and domination was a language generally not employed by women, although not exclusive of them. Instead they considered themselves to be observing and describing the physical environment, rather than confronting and doing battle with their surroundings in an effort to establish dominion over them. However, the very acts of observing and describing have implications for the relative positions of power attained by white women abroad; observation and description were also means of mastering and appropriating the landscapes of West Africa.[31]

Romanticism was a common factor in the women's responses to the natural environment of West Africa, and this had its roots in the wider philosophical trend taking place in Victorian Britain. As suggested, within nineteenth-century exploration literature not only were the landscapes of West Africa depicted as wild and chaotic in opposition to the ordered and controlled landscapes of Europe, they were also among the few places that could still be referred to as "the country" in contrast to an increasingly urbanized Europe.[32] However, it was in the works of Mary Kingsley that this romanticism was most strongly represented. Kingsley's beliefs were more akin to an eighteenth-century romanticism, which was characterized by "the delight in, and wonder at, the beauty and beneficence of nature", and "the ascendence of emotion and intuitive perception over reason",[33] than they were to late-Victorian romanticism. Her responses to the landscapes were on a deeply personal and aesthetic level. Her claims of objective observation of West Africa were undermined by the subjectivity of her responses.[34] However, as Alison Blunt argues, Kingsley "claims subjective authority by stressing the individuality of her response".[35] She allowed the landscapes to exercise their almost magical powers over her rather than seek to impose herself upon her surroundings. Kingsley located herself within the forests and attempted to blend with them instead of "mastering" them. She was emotionally connected to the landscapes through which she travelled, and insisted, "I am more comfortable there than in England".[36] Although raised by agnostics, her experiences in West Africa had instilled within her a deep private faith in pantheism, and she derived her spiritual beliefs from the works of Spinoza.[37] For Kingsley, as with Spinoza, God was the immanent, not the transcendent, cause of all things; God was not the creator of things, but the things themselves.[38] After her journeys, Kingsley identified these beliefs as "firm African", and she maintained that she was happiest when among the "non-human things" of nature. She wrote of the Ogowé rapids (present-day Gabon):

> The majesty of the scene fascinated me, and I stood leaning with my back against a rock pinnacle watching it. Do not imagine it gave rise, in what I am pleased to call my mind, to those complicated, poetical reflections natural beauty seems to bring out in other people's minds. It never works that way with me; I just lose all sense of human individuality, all memory of human life, with its grief and worry and doubt, and become part of the atmosphere. If I have a heaven, that will be mine.[39]

As Alison Blunt argues, Kingsley's descriptions were less likely to "establish the landscape as a stage for the exercise of the imperial viewers' power" than they were to prompt self-questioning, "further highlighting her personal and reflexive sensitivity to landscape".[40] For example, Kingsley wrote that while she was sat on her verandah overlooking Victoria and taking in the scenery beneath her, she thought, "Why did I come to Africa? ... Why! who would not come to its twin brother hell itself for all the beauty and charm of it"[41] The extent to which she identified with the landscapes of West Africa was revealed in a letter written shortly before she died, where she wrote:

> I am no more a human being than a gust of wind ... My people are mangroves, swamps, rivers, and the sea and so on—we understand each other. They never give me the dazzles with their goings on, like human beings do by theirs repeatedly.[42]

In contrast, a sense of dislocation from the physical environment is evident in Larymore's narrative. She emphasized her position as the first white women to travel from the southern forests to the deserts in the north, and thus was able to depict herself as discover/explorer. In this guise, the "monarch-of-all-I-survey" trope, which Mary Louise Pratt suggests is an essentially masculine *genre* of travel writing,[43] became accessible to Larymore. The context for her journey was crucial to her attitude; as part of a military expedition to suppress indigenous populations and delineate the boundaries of British territory, she was also aware of her relative position of authority within Nigeria. Her sense of her own superiority and of her presence on the imperial frontier are apparent in her landscape descriptions. For example, she described the view at Egga:

> I had a glorious and uninterrupted view of mile upon mile of grassland, flanked in the distance by the curious flat-topped hills at Padda. The distance was marked only by the "wire road", the telegraph line leaving Egga and disappearing into the pearly iridescent Harmattan mists in an ever diminishing perspective—the one link with civilization.[44]

Pratt argues that within the "monarch-of-all-I-survey" trope three conventions can be identified which create qualitative and quantitative value for the explorer's achievement. These conventions are recognizable in Larymore's landscape descriptions. Firstly, the landscape is "aestheticized", the view is seen as a painting and the description is ordered in terms of background and foreground, perspective and colouration. Secondly, "density of meaning" is sought, particularly through the modification of nouns. Larymore's "pearly" mists [and on other occasions, "pearly blue haze" (p. 71), and "emerald green" plains (p. 112)] tie the landscape to her home culture, "sprinkling it with little bits of England". Finally, the metaphor of the painting is important in terms of Larymore's "mastery" of the scene. Larymore became a verbal painter reproducing a scene for others; what she saw was all there was. The landscape was ordered with reference to her vantage point. The "ever-diminishing perspective" to which she referred privileged her position as observer. As Pratt argues, the observer has the power "if not to possess, at least to evaluate this scene".[45] These conventions are repeated throughout Larymore's narrative. The panorama was a device for seeing the country as a future colonised country, and thus by including panoramic scenes Larymore arrogated to herself the power of the colonizer.[46] This was very different to the response of Kingsley who, although an imperialist, was anti-colonialist and travelled in southern Nigeria before the implementation of Lugard's colonial policies. Unlike the savanna lands of the north, the panoramic view was almost impossible in the southern forests.

Despite Larymore's adoption of a colonial "gaze", there is a certain tension within her landscape descriptions; she was obviously captivated by the landscapes through which she travelled, but she struggled simultaneously to maintain objective, even scientific, rationality in her descriptions. As Gillian Rose (1992) argues, many explorers and travellers, in their quest to produce geographical knowledge, had to resist what J. K. Wright referred to as "the sirens of *terrae incognitae*".[47] Taking pleasure in viewing the landscape was often seen as a threat to the scientific validity of the observations contained in the narrative. This tension is apparent in Larymore's landscape descriptions. For example, she wrote:

> autumn is almost the best time of the year to "see the country"; in the farms the guinea-corn was just beginning to ripen and droop its massive plumes of grain, . . . and we got magnificent views of miles of wooded hill and plain unrolling themselves into the dim blue distance.[48]

Here, Larymore provided knowledge on when to acquire the best view of the landscape, and detail on the nature of the landscapes of Northern Nigeria, but she succumbed to the inherent pleasures of gazing upon the landscape. In the colonial context, Larymore's position as author-colonizer allowed her to adopt this "gaze", which both Rose and Pratt argue is essentially a masculine gaze, and her descriptions thus contain the same tensions found in exploration literature between scientific narrative and the sense of almost voyeuristic pleasure the observer has in viewing the landscape.

In addition to her adoption of the colonial gaze, Larymore also used other conventions of exploration narrative which, because of the nature of social relations at the time, were perhaps not as readily available to those women who travelled in the nineteenth century. She described herself surmounting geographical barriers such as the rapids on the Upper Niger:

> At the worst point, where the whole face of the river appeared to be barred with a rush of falling waters, and no smallest passage was visible amidst the tumbling foam, the canoes were hauled under the steep bank, and their entire contents bundled out thereon, we, the passengers, clambering, by the aid of roots and branches, to a place of some security, where we sat on the warm sand and watched the manoeuvres down below.[49]

Here Larymore confronted the landscape, the rapids were obstacles which she must overcome in order to progress on her journey. On other occasions, she described herself as "marching" into the towns of Northern Nigeria, identifying herself with the British military expedition that had facilitated her travels. Furthermore, whilst in Kabba she wrote, "Outside the town, there is a little stretch of forest belt, and, as no one has ever disputed its possession with me, I am pleased to consider it exclusively my own property".[50] She referred to this forest as "my 'kingdom'". Significantly, Larymore "claimed" this land not for Britain but for herself; she claimed it not for its potential wealth, but because of the beauties of both flora and fauna that were contained within the forest. Thus the convention of laying claim to territory, which Pratt views as a masculine convention, was tempered by her delight in the physical environment; her desire to possess was rooted in the aesthetic beauty of the landscape she encountered. However, this metaphorical claiming of land indicated how Larymore placed herself in a position of power and authority in relation to the landscapes that she described, and is not found in the narratives of those women travelling before the end of the nineteenth century.

The imperial strategies contained in Larymore's narrative seem contrary to Kingsley's personal identification with the landscapes of West Africa, and whereas the former privileged vision, Kingsley's descriptions included other sensual responses such as sound. However, in common with most colonial texts, there is a certain amount of ambivalence in both narratives.[51] For example, despite her emotional connection to the landscapes of West Africa, Kingsley sought a panoramic view in her ascent of Mungo Mah Lobeh. However, her ability to identify with the imperialist trope of the panoramic view was undermined by the vista being obscured by mist.[52] Disappointed with the lack of view from the top, Kingsley nonetheless appreciated the aesthetic effects of the mist which was "exceedingly becoming to the forest's beauty".[53] The landscape was feminized, "but its attraction lay in it being veiled rather than being unveiled".[54] Similarly, despite the presence of imperial strategies in Larymore's narrative, she also wrote:

> Picture to yourself a green—truly emerald green—plain, holding an area of, roughly, ten square miles, dotted with palm trees, their tall slender stems crowned with crests of graceful drooping plumes, and bearing a respectable fortune in the palm-oil contained in the closely clustering bunches of nuts on each tree. Hundreds of acres are under cultivation ... Away, beyond, rise the blue hills, in a huge circle, jealously shutting in this little green paradise from the tiresome world of restless white folks, who would take count of time, make roads, try to introduce sanitation and otherwise employ themselves in fruitless and unnecessary works to the dire discomfort of the peaceful denizens of peaceful places![55]

Larymore's ambivalence is clear in passages such as this. As the wife of a military administrator in Nigeria she acquired freedoms and a degree of authority that would not have been available to her in Britain, yet she expressed her worries about the impact of British economic and cultural imperialism, and the colonial administration her husband was helping to impose, upon the very landscapes and peoples she was observing. Larymore was clearly a romantic and wished to see the landscapes she encountered remain unchanged, yet she realized that British imperialism, in which she played a part, would have a drastic effect upon the pace of life in Nigeria.

Confronting "chaotic" West Africa

A major feature of landscape descriptions by British women in West Africa was their recognition of order in the natural environment, and this is particularly evident in the narratives of Mary Kingsley. For Kingsley, West Africa was not a homogeneous confusion of nature, but a series of ordered physical environments—coastal, riverine, montane and forest. The horizontal bands of colour in the landscape when viewed from the coast—the blue of the sea, the white of the surf upon the yellow of the sand, the brown and green of the mangroves, the deep green of the rainforest, and the purple of the distant mountains—were evidence of natural order rather than of pandemonium. Other women travellers remarked upon these bands of colour which formed their first impressions of West Africa from the Atlantic Ocean, and, although bereft of the poeticism of Kingsley's descriptions, still conveyed the impression of order. Colvile noticed the "long low line of palms, the strip of sand, and the fringe of breakers which are characteristic of this part of the African coast",[56] Larymore described the "sunshine, sapphire water, the fringe of low grey coast-line ... [T]he huge swell swept shorewards, to break in its thundering surf, away by the grey palm-trees and the yellow sand".[57] The common perception was of order rather than chaos.

It seems that it was the natural order and organization of the physical environment that Kingsley found most alluring about West Africa, and it could thus be argued that she employed a scientific, ecological "trope" in her landscape descriptions. This can be observed in her descriptions of Corisco Island, off the coast of Angola. For Kingsley, Corisco Island was an exquisitely ordered physical environment. She wrote:

> [Leaving the beach] we clamber up the bank and turn inland, still ankle deep in sand, and go through this museum of physical geography. First, a specimen of grassland,

> then along a lane of thickly pleached bush, then down into a wood with a little (at present) nearly dried up swamp in its recesses; then up out onto an open heath which has recently been burnt and is covered with dead bracken and scorched oil palms ... There is such an elaborate completeness about this museum, and we have not even commenced the glacier or river departments.[58]

Similar descriptions are found in the narratives of other women travellers. Although Melville's landscape descriptions frequently alluded to "wilderness", and she sometimes found the physical environment of West Africa threatening, she portrayed the landscapes around Sierra Leone as inherently ordered. She wrote:

> There are three distinct phases of landscape here. The first is hill and dale, clothed in all their original exuberance of stately forest, and appearing in their primeval grandeur, as it were, fresh from the hands of their Maker; the second is the first denuded and laid waste by fire and hatchet ... and that is the scenery I would gladly see changed; the third is the second rich in partial cultivation, and which, with the first, constitutes the peculiar beauty of the tropics, and in it I certainly desire no variety.[59]

Melville perceived those landscapes that remained in their natural state, and those that were only partially cultivated, as the most aesthetically pleasing; those landscapes which remained untouched, or "wildernesses", were equally ordered as those under cultivation.

Familiarizing "exotic" landscapes

A further method employed by women travellers to dispel the myth of pandemonium was their attempt to acquaint their readers with unfamiliar landscapes. This was by no means exclusive to women writers, but it is a theme that seems to run through many of the narratives by women travellers in West Africa. They alluded to familiar aspects of Western art and literature in creating their imagery. For example, Kingsley wrote of the scenery of the forest of the Ogowé: "It is as full of life and beauty as any symphony Beethoven ever wrote; the parts changing, interweaving and returning".[60] Her publications and lectures abounded with such descriptions of the forest, "a vast, seemingly limitless cathedral with its countless columns covered, nay, composed of the most exquisite dark-green, large-fronded moss, with here and there a delicate fern embedded in it as an extra decoration".[61] Here climbing plants became "coverlets" spread over the forest to keep it dry; bush-ropes and lianas became "some Homeric battle of serpents that at its height had been fixed forever by some magic spell";[62] sunshine lit up the red sand-banks which then glowed like "Nibelungen gold" and the sun played with the mist in a vision of "Turneresque" beauty.[63] These frequent literary and artistic allusions provided a frame of reference through which her readers could familiarize themselves with the landscapes of West Africa. Kingsley referred to popular authors of the time (Dickens, Twain, Dr Johnson, Stevenson, Goethe, Kipling), and made unfamiliar landscapes recognizable by comparing them to the works of Turner and the Pre-Raphaelites, re-emphasizing her own romanticism. She also compared the sounds of the forest to symphonies by Beethoven and Handel.

Similarly, Melville compared the "soft quiet beauty" of the beach landscape to paintings by Claude and Poussin,[54] and an approaching tornado was described as "great fleecy clouds rising above Mount Oriel; their curled outlines forming many a Hogarth-like portrait against the sky".[65] As with Kingsley, Melville sought to familiarize the sounds of the forest to her readers. Not all of the associations were pleasant or pastoral: "When darkness sets in, the hum of millions of insects arose—and a very unsentimental memory it brought along with it, being exactly like the noise of a large manufactory where spinning machines are constantly in motion".[66] Women travellers also used the technique of drawing analogies between British and West African environments in order to render them more familiar. Melville compared the deforested hills around Sierra Leone to those in Scotland, "where green herbage, heather, and furze appear in alternate patches".[67] Colvile described a scene approaching Bonny Town in a similar manner: "Two minute's walk brought us to the margin of a deep pool surrounded by gigantic cotton trees, whose heavy shade and buttressed roots would almost have made one imagine one's self in some early English cloister".[68] Larymore described a scene at Jebba in Northern Nigeria which:

> ... changed abruptly from low-lying grassy marsh land and warm sand-banks, where the wild duck and geese were wont to gather, to great beetling cliffs and walls of rock, which rose sheer from the still water, seemingly shutting in the river altogether, and giving the impression of one end of a highland loch.[69]

By painting their more familiar pictures of West African landscapes and making them recognizable to their readers, women travellers challenged the image of the exotic and frightening pandemonium which characterized many descriptions of West Africa during the nineteenth century. The landscapes in the narratives of these women were ordered and, to some extent, demystified.

The pastoral myth and West Africa as "garden"

If there were some similarities in the ways in which women travellers represented the physical environments of West Africa as ordered and familiar, there were also great differences in their designs upon their immediate surroundings. Kingsley attempted to blend in with the surrounding landscapes, and related to them in a personal and subjective way. In contrast, Melville and Larymore perceived West Africa as a potential garden that must be cultivated. In Victorian Britain suburbs were developing rapidly, and gardens came to be perceived as refuges amidst the enormity of urban sprawl, "bowers and oases in a desert of brick and mortar".[70] In West Africa, the garden was a retreat from the surrounding "wilderness", a small area of "civilization" on the imperial frontier. If the urban gardens of England could only be imagined in relation to the city, which shaped and defined them by surrounding and threatening them, then the gardens in West Africa could only be imagined in relation to the surrounding "wilderness". Both survived under siege, "a passive persistence emphasized and symbolized, here, by closure and concentration".[71] Furthermore, to quote Helen Callaway, "The garden served as a rich metaphorical link; English order and productivity amid the tropical chaos of overgrown jungle or scrubby savannah".[72] Sections within the narratives of Larymore and Melville were illustrative of attempts by white women to impose their own personalities as well as British order upon the landscapes of West Africa. They were both impressed by the creation of gardens and farmland in the forests of West Africa, and remarked upon the contrast between these ordered features and the surrounding wilderness. Melville created a vision of West African landscapes which was consistent with anti-slavery depictions of an earthly paradise. She particularly delighted at the gardens in Freetown which, despite her exoticism of them, reminded her of home:

> I was enchanted by the luxuriance of the trees ... Innumerable blossoms shone in all directions; one resembling a branch of red coral; another still more gorgeous, with its festoons of orange and scarlet, reminded me of the feathers of the bird of Paradise; while the pale lilac colours of a third recalled the image of more northern gardens, and

> claimed a kindly remembrance of old familiar flowers, although the perfume of orange and lime trees was all around us.[73]

Constance Larymore had a different attitude which was reflective of her presence in Nigeria at the beginning of the colonial period. She took the theme of the garden a step further than Melville, insisting that it was the duty of every Englishwoman resident in Africa to attempt to grow a garden as a constant reminder of England amid this "wilderness". She wrote at Lokoja, "I then and there took to heart the lesson which I have tried to practise ever since—the absolute *duty* of planting trees everywhere for the benefit of one's successors".[74] Larymore attempted to create a "sanctuary for an idealized domesticity" on the imperial frontier of Nigeria.[75] She devoted an entire section of her book to advising her readers how to grow a flower garden, a verandah garden, a vegetable garden, a lawn, and trees and shrubs in West Africa. The imposition of English order upon the physical environment appealed to Larymore. Her intention was to create "little Englands" wherever she went in Nigeria, and believed it was the duty of the wife of the Resident to ensure that this task was undertaken, both for the increased comfort of their husbands, and for the wider benefit to British imperialism in Africa. She perceived the garden to be an extension of the domestic sphere, a private sanctuary to be managed and maintained by the wife for the good of the household.[76] In her chapter on creating a lawn, she wrote:

> It is said to be very dear to the heart of every Englishman to own a lawn, and it certainly should be doubly so to John Bull in exile; in a tropical country well-kept turf is much to be desired, there is nothing as cool and refreshing to tired eyes dazzled with the glare of the sunshine and baked earth, and perhaps, nothing that gives such a home-like and cared-for look to a West African compound.[77]

Such responses to the landscape were in marked contrast to the more harmonious relationships sought by Kingsley. Although both responses were imperialist they were very different, and they highlight both the complexities within imperialist responses to West Africa and the importance of the spatial and temporal context of the journey.

Conclusion

White women travellers were exposed to images of the "Dark Continent" before they travelled to West Africa through various media, including colonial fiction, exploration literature, late-nineteenth century photography, imperial art and scientific treatises, and these often informed their descriptions of the physical environments of West Africa. This is particularly evident in their comments on climatic conditions, and the supposed links between climate and disease. However, their visions of West Africa were complex. This paper has illustrated that various literary traditions in Victorian Britain influenced the ways in which women travellers wrote about the physical environment of West Africa. Victorian romanticism, ideas about the imperial frontier, and notions of the "wilderness" and "sanctuary" all exerted powerful influences upon the narratives of these women, and it could be argued that these were standard ways of representing the physical environment in travel literature at the time. The ambivalence apparent in these narratives was common to many colonial texts. However, to a certain extent the descriptions of the physical environment by the women in this study were different from standard Victorian travel texts. Their romanticism was overtly expressed because they were constrained to emphasize their landscape descriptions rather than to engage in scientific observations or political comments. As Sara Mills suggests, the scientific narrator figure was not unavailable to women travel writers, but was constantly undermined by other elements within their texts.[78] Nineteenth-century convention held that women's travel narratives were not supposed to be "scientific" and authoritative, but, rather, supposed to be amateurish. It is for this reason that travel narratives by women were often prefaced with disclaimers which denied "any scientific, academic, literary or other merit".[79] Furthermore, an emphasis on landscape description was a means of maintaining this amateurish style. This is recognizable as a common characteristic in travel literature by women travellers in West Africa, as are their attempts to familiarize landscapes rather than exoticize them.

This paper has suggested that the spatial and temporal context of the journey, particularly with regard to contemporary British imperial culture and popular images of West Africa, was an important influence on the landscape descriptions of women travellers. The differences in their descriptions are revealing of the women's differing attitudes towards British imperialism in West Africa. Elizabeth Melville's attitude was typical of the 1840s, and her landscape descriptions were heavily influenced by philanthropic perspectives on West Africa, anti-slavery concerns, and the anti-colonialism in Britain at the time. She portrayed West Africa as almost Edenic; it was a land of gardens and cultivation, of flowers, blossoms and perfumes. However, it was not Europe's Eden, but Africa's. She believed West Africa unfit for European colonization, primarily because of climatic conditions; the presence of Britons would be unnecessary once the slave trade had been abolished. Abolishing slavery would ensure that West Africans could be left to live peaceably in their own paradise. Constance Larymore's descriptions of Nigerian landscapes were rather different. Larymore was unique among the women of this study in that she adopted the "monarch-of-all-I-survey" position to describe the landscapes she encountered. She claimed territory as her own in her writings. She was able to do so because, unlike the other women, she travelled through a part of West Africa which had been conquered by the British and which had been formally incorporated into the British empire. In effect, she was surveying aesthetically territories that were about to be colonized and surveyed for military and economic purposes. Mary Kingsley's landscape descriptions were also unique, not only in their great abundance and poeticism, but also because they challenged many of the images of West Africa which prevailed during the 1890s, defying the conventions of the time. Unlike Zélie Colvile, who portrayed West Africa as the "Land of Death", Kingsley drew on earlier romantic depictions in order to challenge negative visions of West Africa, and to raise interest in the possibilities for British imperialism there. She believed that if she could convince the British public that West Africa was not a land of darkness, it would encourage the Government to increase Britain's trading interests in the region.

Despite their misgivings about the climate, very often the visions of West Africa created by women travellers offered up challenges to the image of the "Dark Continent". For example, the depiction of West Africa as a "garden" in travel texts written after 1840 was unique to women travellers such as Melville, for by mid-century West Africa had come to be popularly perceived within Britain, and particularly by explorers, travellers and the formulators of imperial policy, as the heart of a dark and dangerous continent. While the vagaries of the climate steered women travellers away from Arcadian visions of West Africa, they also tended to avoid depictions of West Africa as pandemonium, and relied more upon evoking the intricate beauties of their surroundings. However, in the words of Annette Kolodny, although they "shied away from paradisal projection, they nonetheless seemed eager to tend the garden".[80] Therefore, although they often portrayed West African landscapes in standard ways, white women travellers made important contributions to the popular geographies of West Africa. Their narratives are revealing of the complexities inherent in the metaphor of the "Dark Continent", and provide an important archive which helps elucidate the complexities in Britain's response to, and perception of, West Africa during the nineteenth century.

School of Geography
University of Birmingham
Birmingham B15 2TT

Acknowledgements

I would like to thank Pyrs Gruffudd and Michael Heffernan for the helpful comments on an earlier draft of this paper, and two anonymous referees for their insightful comments. I am also grateful to the British Academy for funding the original research which forms the basis for this paper.

Notes

[1] For example, James Clifford, 'Travelling cultures' in Lawrence Grossberg *et al.* (Eds) *Cultural Studies* (London 1992) 96–116; M. Kowalewski, *Temperamental Journeys: Essays on the Modern Literature of Travel* (Athens, Georgia 1992); Mary Louise Pratt, *Imperial Eyes: Travel Writing and Transculturation* (London 1992); George Robertson *et al.* (Eds), *Travellers' Tales. Narratives of Home and Displacement* (London 1994)

[2] For example, Trevor Barnes and James Duncan (Eds), *Writing Worlds. Discourse, Text and Metaphor in the Representation of Language* (London 1992); Peter Bishop, *The Myth of Shangri-La* (London 1989); Jonathon Crush, Grazing on Apartheid: post-colonial travel narratives of the Golden City, in Paul Simpson-Housley and Peter Preston (Eds), *Writing the City* (London 1994); Derek Gregory, Between the book and the lamp: imaginative geographies of mid-nineteenth century Egypt *Transactions of the Institute of British Geographers* **20** (1995). The inspiration for the recognition of both the enduring appeal of travel writing as a mode of representation, and of the overlaps between geography as a discipline and travel-writing as an imaginative and sometimes "scientific" *genre* of literature was Edward Said's *Orientalism* (London 1978); see also Sara Mills, *Discourses of Difference. An Analysis of Women's Travel Writing and Colonialism* (London 1991); Felix Driver, Geography's empire: histories of geographical knowledge *Environment and Planning D: Society and Space* **10** (1992) 26; Pratt, *ibid.* (*passim*)

[3] See Helen Callaway, *Gender, Culture and Empire. European Women in Colonial Nigeria* (London 1987); Shirley Foster, *Across New Worlds: Nineteenth Century Women Travellers and Their Writings* (London 1990); Billie Melman, *Women's Orients: English Women and the Middle East 1718–1918* (London 1992); Mills, *ibid.*; Pratt, *ibid.*

[4] Mona Domosh, Towards a feminist historiography of geography *Transactions of the Institute of British Geographers* **16** (1991) 95–104; Alison Blunt, *Travel, Gender, and Imperialism* (New York 1994); see also Alison Blunt and Gillian Rose (Eds), *Writing Women and Space. Colonial and Postcolonial Geographies* (New York 1994)

[5] On popular images of empire see Felix Driver *ibid.*; John MacKenzie, *Propaganda and Empire* (Manchester 1984)

[6] See, for example, Catherine Barnes Stevenson, *Victorian Women Travel Writers in Africa* (Boston 1982); Foster, *ibid.*; Susan L. Blake, What difference does gender make? *Women's Studies International Forum* **13** (1990) 347–355; Callaway, *ibid.*

[7] Blunt, *Travel, Gender and Imperialism*; Sara Mills, Knowledge, gender and empire, in Blunt and Rose, *ibid.* 30; Melman, *ibid.*

[8] For examples of recent studies which have adopted this approach, see Cheryl McEwan, How the "seraphic" became "geographic": women travellers in West Africa, 1840–1915 (unpublished Ph.D. thesis, Loughborough University, 1995); Morag Bell, Citizenship not Charity: Violet Markham on nature, society and the state in Britain and South Africa, in Morag Bell, Robin Butlin and Michael Heffernan (Eds), *Geography and Imperialism* (Manchester 1995, 189–220)

[9] For ease of reference I refer to West Africa in the light of the Victorian delineation of this

region. The term is certainly reductive and contestable; as illustrated by Kingsley's narratives, references to "West" Africa during the nineteenth century included areas of equatorial Africa (modern-day Zaire, Gabon and the Republic of Congo). The spellings of place names appear in this paper as they do in the original narratives

[10] See, for example, Patrick Brantlinger, Victorians and Africans: the genealogy of the myth of the Dark Continent *Critical Inquiry* **12** (1985) 166–203; Philip Curtin, *The Image of Africa. British Ideas and Action 1780–1850* (London 1965); Dorothy Hammond and Alta Jablow, *The Myth of Africa* (New York 1977). Also of relevance is Abdul R. JanMohammed, The economy of Manichean allegory: the function of racial difference in colonialist literature *Critical Inquiry* **12** (1985) 59–87

[11] The focus here is on landscapes primarily because of the profusion of landscape description in these narratives. Although political and scientific commentary were not exclusive of women, most chose to devote large sections of their narratives to descriptions of the physical environment. One could also focus upon the depictions of Africans, African cultures and customs, and so on, by women travellers (see McEwan *op. cit.*)

[12] See Said, *op. cit.*

[13] Between the years 1840 and 1915 many British women travelled in West Africa in various roles, but an exploration of bibliographical and archival sources (for example, the *Dictionary of National Bibliography*; Jane Robinson, *Wayward Women. A Guide to Women Travellers* (London 1991) and the collections on women Members and Fellows at the Royal Geographical Society) has revealed only seven who published accounts of their travels (see McEwan *op. cit.*). The four women who provide the focus of this study were chosen because their travels span the period from the 1840s to the early 1900s and three broad phases of British relationships with West Africa (exploration, anti-colonialism and high imperialism), and because landscape description was prominent in their narratives

[14] Edward Walford, *The County Families of the United Kingdom* (London 1874) 162

[15] Walford, *ibid.* 704

[16] For a comprehensive bibliography of Kingsley, see Deborah Birkett, *Mary Kingsley. Imperial Adventuress* (London 1992)

[17] In addition to printed books, this paper also draws upon the various journal articles by Kingsley about West Africa, and her personal correspondence

[18] Joseph Conrad popularized this image of tropical Africa in his novel, *Heart of Darkness* but, as Brantlinger argues, the "darkening" of West Africa had begun several decades before the publication of Conrad's novel in 1899 (see Brantlinger, Victorians and Africans: the genealogy of the myth of the Dark Continent *Critical Inquiry* **12** (1985) 166–203)

[19] There are a number of reasons why the myth of the "Dark Continent" was focused on West Africa as opposed to other parts of the continent, including the prevalence of disease, the non-colonial nature of Britain's involvement there (in comparison to eastern and southern Africa, which were considered far more suitable for settlement), the image of the mysterious, impenetrable forest and its strange creatures (the gorilla was "discovered" by du Chaillu only in the 1860s), and stories of cannibalism and strange cultural practices. Obviously, the "darkness" of West Africa was more a product of British ignorance of its geography during the nineteenth century than a true reflection of the state of the region

[20] Mary Kingsley, *Travels in West Africa* (London 1982) 2–3

[21] Kingsley wrote, "A more horrible life than a life in such a region for a man who never takes to it, it is impossible to conceive ... [I]t is a living death," *West African Studies* 39

[22] Elizabeth Melville, *A Residence at Sierra Leone* (London 1968) 8

[23] Melville's *A Residence at Sierra Leone* was reprinted in 1968 at around 400–500 copies, but there is no information for the 1849 edition (Frank Cass Publications, personal correspondence). Similarly there are no records of publication figures for the books of Larymore and Colvile

[24] Zélie Colvile, Ten days on an Oil River *Blackwood's Magazine* March (1893) 372–382

[25] She wrote, "I have hopes there may be a demand for *Round the Black Man's Garden* which I have been advertising by giving a lecture on Madagascar at Liverpool G. D. Society. My first attempt met with a most enthusiastic reception—and they found my lecture too short ..." (National Library of Scotland, Blackwood Papers 1893–1900, Colvile to Blackwood 16/12/98)

[26] Jane Robinson, *ibid.* 200

[27] Deborah Birkett, *ibid.* 135

[28] Information courtesy of Macmillan Publishers Limited. Both books were reissued in the 1960s by Cass, and *Travels in West Africa* was reprinted in 1972, 1982, and 1987, the latter an abridged version
[29] Kingsley to John Holt, 19/11/99, *John Holt Papers*, Mss. Afr. s. 1525, Box 23, Rhodes House Library, Oxford
[30] See, for example, Rebecca Stott, The Dark Continent: Africa as female body in Rider Haggard's adventure fiction *Feminist Review* **32** (1989) 69–89
[31] Mary Louise Pratt, *ibid.* 119–143
[32] Raymond Williams, *The Country and the City* (London 1973) 281–3
[33] Donald Koster, *Transcendentalism in America* (Boston 1975) 8
[34] Alison Blunt, Mapping authorship and authority: reading Mary Kingsley's landscape descriptions, in Alison Blunt and Gillian Rose (Eds), *Writing Women and Space* (New York 1994) 51–72
[35] Blunt, *Travel, Gender and Imperialism*, 95
[36] Kingsley, *Travels in West Africa*, xiii. Kingsley never felt comfortable in England. The circumstances of her birth were partly to blame for this; her father, George, was a member of the illustrious Kingsley family; her mother, Mary Bailey, was the daughter of an inn-keeper whom George married 4 days before the birth of his daughter (see Deborah Birkett, *ibid.*). From Kingsley's correspondence it seems that she was shunned by the Kingsley side of the family, and her working class mother precluded her involvement in the rounds of polite socializing which occupied the time of other young middle-class women (correspondence with Alice Stopford Green, National Library of Ireland, Dublin). West Africa represented a welcome escape
[37] Mary Kingsley, West Africa from an ethnologist's point of view *Transactions of Liverpool Geographical Society* (1897) 63
[38] On pantheism, see Thomas McFarland, *Coleridge and the Pantheist Tradition* (Oxford 1969) 63
[39] Kingsley, *Travels in West Africa* 178
[40] Blunt, *Travel, Gender, and Imperialism* 97
[41] Kingsley, *Travels in West Africa* 608
[42] Kingsley to Matthew Nathan, 12/3/99, Nathan Papers, Bodleian Library, Oxford
[43] Pratt, *Imperial Eyes* 213
[44] Larymore, *ibid.* 37–38
[45] Pratt, *ibid.* 204–5
[46] See also Timothy Mitchell, *Colonising Egypt* (Cambridge 1988), for more on the "imperial gaze"
[47] Rose explores the geographer's "gaze" at nature in terms of the distinction between the feminisation of the landscape, which requires interpretation, and the masculine, scientific gaze which produces this interpretation (see Gillian Rose, Geography as a Science of Observation: the Landscape, the Gaze and Masculinity, in Felix Driver and Gillian Rose (Eds), *Nature and Science: Essays in the History of Geographical Knowledge* (London 1992) 8–18
[48] Larymore, *ibid.* 69
[49] Larymore, *ibid.* 159
[50] Larymore, *ibid.* 118
[51] Homi Bhabha argues that ambivalence is a marked feature of many colonial texts (see Bhabha, *The Location of Culture* (London 1994), esp. 85*ff*); it was not, therefore, exclusive to women travel writers
[52] Alison Blunt discusses ambivalence in Kingsley's narratives. She writes that Kingsley's account of her ascent "illustrates the ambiguities of being constructed as inside and outside and moving between patriarchial and imperial discourses", *Travel, Gender, and Imperialism* 101
[53] Kingsley, *Travels in West Africa* 570
[54] Blunt, *Travel, Gender, and Imperialism* 102
[55] Larymore, *ibid.* 112–3
[56] Colvile, *ibid.* 285
[57] Larymore, *ibid.* 4
[58] Kingsley, *Travels in West Africa* 385–6
[59] Melville, *ibid.* 225

[60] In Cecil Howard, *Mary Kingsley* (London 1957) 86
[61] Kingsley, *Travels in West Africa* 570
[62] Mary Kingsley, The ascent of Cameroons Peak and travels in French Congo *Transactions of the Liverpool Geographical Society* (1896) 272
[63] Kingsley, *Travels in West Africa* 239
[64] Melville, *ibid.* 6–7
[65] Melville, *ibid.* 91
[66] Melville, *ibid.* 8
[67] Melville, *ibid.* 6
[68] Colvile, *ibid.* 303
[69] Larymore, *ibid.* 153
[70] Andrew Griffin, The interior garden and John Stuart Mill, in V. C. Knoepflmacher and G. B. Tennyson (Eds), *Nature and the Victorian Imagination* (Berkeley 1977) 171
[71] Griffin, *ibid.* 171
[72] Callaway, *ibid.* 178–9
[73] Melville, *ibid.* 8
[74] Larymore, *ibid.* 7
[75] Annette Kolodny, *The Land Before Her. Fantasy and Experience of the American Frontier, 1630–1860* (London 1984) xiii. Certain parallels can be drawn between those women, such as Larymore, living on the imperial frontier in West Africa and the experiences of women living on the North American frontier. Kolodny writes:

> "After initial reluctance at finding themselves on the wooded frontiers . . . , women quite literally set about planting gardens in these wilderness places. Avoiding for a time male assertions of rediscovered Eden, women claimed the frontiers as a potential sanctuary for an idealized domesticity. Massive exploitation and alteration of the continent do not seem to have been part of women's fantasies. They dreamed, more modestly, of locating a home and a familial human community within a cultivated garden"

[76] There is an analogy here with the notion of Victorian domesticity and the suburban garden (see Leonore Davidoff and Catherine Hall, *Family Fortunes: Men and Women of the English Middle Class 1780–1850* (London 1987) 370–375)
[77] Larymore, *ibid.* 224
[78] Mills, *Discourses of Difference* 77–8
[79] Mills, *ibid.* 83
[80] Kolodny, *ibid.* 47

[15]

Embodying colonialism? Domination and resistence in nineteenth-century Ceylonese coffee plantations

James S. Duncan

This paper examines plantations as laboratories of modernity in the colonial tropics during the middle decades of the nineteenth century. Plantations can be conceived of as modern technologies for the reconfiguration of space, tools, scientific instruments and other material resources, bringing together culturally heterogeneous populations, stripping them of their former social attachments and reconstituting them as workers through the use of space-time strategies of monitoring and control. I argue that coffee plantations in Ceylon were intended to be sites of what Henri Lefebvre has termed abstract space: the commodification and bureaucratisation of everyday life. As an ideal, abstract space requires the construction of 'abstract bodies' to conform to it. This ideal remained largely unattained, however. In the first part of the paper, the strategies and technologies adopted by British planters as they attempted to produce and regulate abstract space and bodies are examined. Disciplinarity as internalised control, was rarely achieved and consequently direct surveillance and the threat of physical force was applied. The second section explores the way in which the Tamil plantation workers countered the planters technologies of control by engaging in tactics of resistance. In the final section, the strategies planters adopted in order to oppose the workers' tactics and the middle space of Tamil overseers are examined.

Introduction

Europe's tropical colonies in the nineteenth century can be seen as laboratories of modernity, and nowhere was this clearer than on plantations.[1] Plantations can be conceived of as modern technologies for the reconfiguration of space, tools, scientific instruments and other material resources, bringing together culturally heterogeneous populations, stripping them of their former social attachments and reconstituting them as workers through the use of space-time strategies of monitoring and control.[2] The development of plantation production in Asia was intimately associated with the industrial revolution in Europe, as plantations became sources of cheap raw materials for industrial production and bourgeois consumption. The rise of what Courtenay has termed "industrial plantations" occurred nearly simultaneously before mid-century in British South Asia, the Dutch East Indies, French Indochina and the Philippines.[3] These plantations developed labour regimes markedly different from earlier slave plantations in the Americas.[4] They were characterised by migrant (indentured or free) labour, rather than fixed slave labour, and management practices and technologies that attempted to maximise the use of labour over a short period of time (often a five-month contract) before sending labourers back to their villages. Planters had a shorter-term interest in the physical well-being and enculturation of their labourers than did slave owners. Because plantations operated on relatively slim profit margins, they exerted a high degree of control over labour and other production costs, striving for efficiency of production.[5] In this regard at least, these plantations achieved only limited success as early experiments in developing a rationalised workforce in Asia. As plantation owners found that they had to resort to pre-modern carceral techniques of coercion alongside their more modern methods, the industrial plantation was a hybrid entity, rather than the successful experiment in modernity that was envisioned. As Tinker put it "the world of slavery still survived; the plantation was a world apart, on its own, subject to laws—or whims—of those in charge: the overseers and the manager or the proprietor."[6]

The plantation was constructed to be a site of what Henri Lefebvre has termed abstract space: the commodification and bureaucratisation of everyday life.[7] As an ideal, abstract space requires the construction of 'abstract bodies' to conform to it.[8] Abstract bodies, are bodies that are made docile, useful, disciplined, rationalised, normalised, and controlled sexually; such bodies are seen as economic resources to be protected and utilised to capacity.[9] The successful production of abstract bodies in a place like Ceylon would have required the creation of a re-cultured worker.[10] The workers in question were poor Tamil villagers displaced from southern India, and consequently torn from their traditional social networks: their social, judicial, religious, and medical support systems. However, the attempt to produce rationalised bodies for plantation space was undermined because elements of the workers' networks and ways of life remained intact and served as important bases for resistance. Just as the harsh material realities of tropical nature strongly resisted rationalisation, so the partially rent Tamil networks lent support to the workers' resistance to attempts to re-culture them.[11] Stoler argues that plantations rarely transformed peasants into full-fledged proletarians.[12] Rather the plantation system allowed and even encouraged some elements of peasant life, such as the cultivation of private garden plots. As such, the modern plantation can be understood as a site of heterogeneous networks, radically reconstituted into a new socio-material configuration, with some degree of economic stability and success achieved at a very high social cost, borne principally by the Tamil workers but also by the planters themselves.

I will first examine some of the strategies and technologies adopted by European planters as they attempted to produce and regulate abstract space and bodies on coffee plantations. I will argue that planters never achieved the condition of disciplinarity, as internalised control, that is necessary for the production of docile bodies.[13] In part this was due to the fact that plantation owners and managers, saw workers as short-term, expendable, racialised, commodities that were still so thoroughly embodied by their raw 'animal' nature and lack of exposure to European civilising forces that self-discipline was barely possible.[14] Hence rather than educating workers through any formal English schooling, the most that could be hoped for was an appreciation of 'the dignity of labour.' Most worker discipline was imposed through direct surveillance by Europeans and through the threat of physical force, based on the common environmental determinist view of race which assumed that coercion was required to get tropical peoples' to labour.[15] The refusal of workers to discipline themselves was not, however, a passive refusal. Rather they actively engaged in what Scott has termed "the weapons of the weak" a set of tactics of resistance to the demands placed upon them within the experimental space of the plantation.[16] Unable to consistently instil discipline through anything other than constant surveillance and punishment, the planters continually sought more effective ways of producing more complete discipline, in part by involving the colonial state in the disciplining process. This not only brought the planters into conflict with the workers and the Tamil supervisors' (kanganies') networks of power and authority, but also with other bureaucratic forces such as British government agencies and philanthropic groups who campaigned against the planters' continual resort to force.

There have been a number of previous studies of domination and resistance on colonial plantations. One of the earliest and best known is Genovese's discussion of everyday resistance on slave plantations.[17] But trying to discern covert resistance from colonial records is extremely difficult. Ramasamy outlines some forms of resistance on coffee plantations in late nineteenth-century Malaya and concludes that "lack of evidence prevents us from providing a detailed picture of individual forms of resistance among Indian labourers on plantations."[18] Perhaps Warner has undertaken the most sustained set of studies of Indian plantation labour.[19] Her study of the plantation industry in Mauritius in the nineteenth century provides a detailed understanding of the workings of many aspects of plantations and the recruitment of labour. However, she has relatively little to say about labourer resistance, presumably for the same reasons as Ramasamy.[20] More generally, Adas argues that non-confrontational resistance on plantations can be divided into what he terms everyday resistance, exit strategies such as desertion, and retribution such as damaging crops and equipment.[21] He claims that everyday resistance is simultaneously the most common form and the most difficult to study, in part because it involves individual rather than collective action and in part because it entails 'calculated errors' and incompetence rather than sustained protest. One of the principal goals of my paper is to provide a detailed account of the forms of everyday resistance employed by plantation workers.

Barnett[22] and McEwan[23] have rightly suggested that academics have overstressed 'voice' as a key element of resistance on the part of the subaltern. This is very much in line with the findings of Scott, whose theory of the weapons of the weak is premised on the fact that the oppressed usually cannot afford to let their voice be heard. All three conclude that a focus upon practices in understanding resistance may prove more productive. This paper explores resistance as a set of embodied practices. Butler has argued that the body is a "site of incorporated history."[24] The Tamil workers on coffee plantations can be seen as sites of bodily memory and Hindu beliefs and practices brought with them from their villages in India. In fact, Hinduism is especially focused on the body, having many bodily prescriptions and prohibitions associated with the practices of caste. The plantations, then, were sites in which the incorporated histories and bodily orders of Indian villagers were violently confronted by modern European disciplinary regimes.[25] Although the planters had little expectation of producing fully modern individuals with interiorised self-control and an alert public sense of civil vigilance, they did hope to accomplish more limited goals through disassociating the labourers from their Hindu cultural incorporation. Through modern work discipline, it was hoped that positive culture change could be accomplished.

The coffee planting enterprise

Arab traders introduced coffee to Ceylon and until the early nineteenth century it remained a peasant crop.[26] The first European plantations on the island were created in the 1820s by the Governor, Sir Edward Barnes, and Lt Colonel George Bird, an English businessman.[27] Governor Barnes freely mixed his personal financial interests with what he took to be those of the colony by abolishing export duties on coffee, waiving the tax on coffee lands for twelve years,[28] building a road from the coast to the coffee-producing region in the highlands, and relocating the Government botanical research station from the capital in Colombo, to a site adjacent to his plantation.[29] But in spite of these measures, coffee in the late 1820s and early 1830s remained largely a peasant crop. The shift from peasant production to plantations occurred suddenly with the passage of two key pieces of legislation. In 1835 the duty on West Indian and Ceylonese coffee was equalised in Britain and in 1840 the Crown Lands Encroachment Ordinance Number 12 was passed, seizing for the British Crown all lands that were not continuously under cultivation and for which no deed or tax receipt could be shown.[30] This latter ordinance was disastrous for the local Kandyan peasantry, few of whom had satisfactory documentation; it meant that they lost much of the lands they used for shifting cultivation, a vital supplement to their rice lands.[31] The influence of this land alienation as an external force impacting the plantation system can be seen from the fact that in one day alone in 1840, 13 275 acres of land were transferred to British officials at five shilling an acre.[32] The first investors in these expropriated lands were government officials, clergy and businessmen. During the early 1840s production skyrocketed but then crashed in the middle of the decade when a general depression in Europe coincided with the removal of the preferential duty on colonial coffee. The price for coffee plummeted from 130 shillings per hundred weight in 1840, to 43 shillings in 1847. Consequently, 90 per cent of investors from the 1830s and 1840s were bankrupted.[33] When the recovery came in the 1850s, these first planters were replaced by more professional planters and by agency houses. By 1870 the price of coffee had climbed to 55 shillings per hundred weight, a return which allowed a modest profit, but by this time Ceylon coffee was coming under attack by a devastating disease, (hemileia vastatrix), and by the mid-1880s, production was nearly wiped out.[34] Another round of bankruptcies followed, and the only plantations which survived were those that switched to tea.[35]

The recruitment process

The first planters assumed that they would be able to recruit the Kandyan villagers who lived around the plantations to work as agricultural labourers. Much to their frustration, the Kandyans for the most part refused. There was a difference of opinion amongst the British on the reason for their unwillingness to work. For someone like Governor Barnes, who owned one of the first plantations in the island, the answer was clear. The Kandyans were 'innately prone to idleness.'[36] *The Ceylon Observer*, which was often sympathetic to 'native' interests noted more neutrally that the Kandyans were "generally above labouring for hire."[37] One major reason for this was that the Kandyans were small peasant farmers who owned sufficient land to support their families, and their own agricultural season coincided with the picking season (from November to March) on the plantations. In addition, the Kandyans considered the rate of pay to be too low. Some colonial administrators defended the Kandyans against charges of idleness and lay the blame for their unwillingness to work squarely on the European planters themselves. None less than the Colonial Secretary, Emerson Tennant, in a despatch to the Colonial Office stated that whilst the Kandyans had for a time laboured on the plantations, they had given up, because of "the want of good faith on the part of their employers by breach of engagements and the unkindness of general treatment."[38] In that same year, it was reported to the Colonial Office, by the Police Magistrate in Gampola, an important town in the planting districts, that village headmen had told him that their villagers had ceased working on the plantations as they were often cheated of their wages.[39]

However, whilst unwilling to plant, the Kandyans and lowland Sinhalese did work on short-term contracts as artificers, carters, and traders, and on longer-term contracts as domestics.[40]

Given that the Kandyans could not be persuaded to labour, the planters turned first to more distant sources of workers; the lowland Sinhalese, and finally to South Indian Tamils from across the Palk Strait in Madras.[41] The lowland Sinhalese, like the Kandyans, had little interest in labouring on the plantations, and planters complained that they were unreliable, often heading home for two or three months to fulfil 'social obligations' at times when they were needed for planting or picking the crop.[42] The planters then turned to the Indian Tamils. Governor Barnes and Lt Colonel Bird were the first to import Tamil labourers, bringing 150 over to work their plantations at 15 shillings per man per month. However, the experiment failed and all returned to India within the year.[43] Nevertheless, South Indian Tamils remained an attractive option to the planters, and systematic recruitment began in 1839 coinciding with the beginning of the boom in coffee production.[44] The Tamils were willing to come for a number of reasons, some directly related to the British control over Madras. Firstly, the local population grew as the result of a British-imposed peace in the area producing an increase in the number of landless labourers. Secondly, poverty in the region was exacerbated by the decline of village weaving, as Indian cloth manufacture could not compete with the British cotton that had flooded the market. Thirdly these problems were compounded by periodic famines in the region.[45] As a consequence, there was a persistent wage differential between Madras and Ceylon throughout the coffee growing period.[46] Difficult as conditions were for Tamil labourers in Ceylon, they appeared to be better than in their home villages. As one might expect, the labourers were drawn from the lower castes.[47] The planter Sabonadiere noted that low caste labourers were preferred as they were thought to be harder workers.[48] The planters adopted a system of seasonal employment, using the kangany (South Indian headman) system of recruitment. The kangany was paid a head fee to obtain labourers from the south of India and to feed them on the march to the plantations. This cost was deducted from the labourer's wages after he arrived. After delivering them to the plantation, the kangany remained on as gang boss. Heidemann divides the kangany system in Ceylon into two periods, pre- and post-1850.[49] During the 1830s and 1840s, kanganies were more likely to come from the labouring classes, to recruit from their home village and bring their own relatives as labourers. Consequently, it is reasonable to assume that during these decades they felt somewhat more solidarity with their labourers. After 1850, with the takeover of plantations by institutional investors, the recruitment of labour changed as well. Kanganies increasingly filled the growing demand for labour by recruiting outside their own villages. As a result the relations between kangany and labourer became increasingly divorced from kinship ties.

This system had many advantages from the point of view of the planters. First, it delegated recruiting to a third party. Second, given the seasonal demands of coffee planting, workers were only hired when needed.[50] Third, workers paid their own way to the plantations. Fourth, the labourers, being far away from home, disconnected from all of their support systems, placed into the highly circumscribed carceral space of the plantation, were easier to control than the low country Sinhalese, who could easily return to their home villages.[51] Once the majority of the labour force was Tamil, it became even more difficult to recruit Kandyan and lowland Sinhalese workers. The local press at the time noted with regret that, they refused to work with low caste Tamils and were especially reluctant to work under the alien regime of Tamil kangany overseers.[52]

The production of abstract space

During the early 1840s, newly expropriated forestland was cut and burned at a feverish pace to make way for coffee bushes. This newly cleared land was rationalised according to modern models of plantation management developing in India, Java, Jamaica and elsewhere.[53] This putting into practice in Ceylon of spatial models of plantation management and worker control developed elsewhere and learned from texts rather than practical local experience was problematic for a number of reasons. Firstly, Ceylon's environment (soil, weather, and pests) differed from other coffee growing regions, and secondly, and more importantly, most of the owners, supervisors, and labourers on the new plantations had never grown coffee before. With the price of coffee high in the early 1840s and with a few plantations making quick profits, speculators rushed in buying land at inflated prices and hiring inexperienced staff. When the price of coffee collapsed in 1847, the ensuing debts were enormous. This collective failure was put down not only to a depression in Europe but also to greed on the part of speculators. The Governor Lord Torrington summarised his view of the reasons for the collapse to the Colonial Office thus:

> Soldiers whose discharges were purchased from the ranks were sent up to the interior to manage plantations on salaries of £300 to £400 per annum; houses for their use were purchased by the agents at excessive prices, and their style of living, wines, and their expenditures of every description were on a scale of the most absurd proportions, while the proprietors were mortgaging every security and raising money at 9 or 10 per cent to support this expenditure, buoyed up by the confident expectation that the first golden harvest would reimburse every outlay and leave them in possession of a splendid and permanent income.[54]

While the aristocrat Torrington, put the problem down to the wastefulness of the lower classes who put on airs and took advantage of their superior's greed, a planter of more modest background writing in the *Ceylon Examiner* explained it as a loss of rationality which he compared to addiction: "an infatuation appears to have possessed them [the European planters], and they slumbered on like the narcotised opium smoker who having filled his lungs with the fumes of the pernicious drug, throws himself back on a couch."[55] The implication of both critiques is that the early planters failed to behave like proper British imperialists.[56] Rationalised, textualised, abstract space and European modernity could not be produced under such conditions.[57] What was needed to make planting a success were institutions that could collect and communicate locally specific information, and a class of self-disciplined, modern, professional managers who could more effectively control the workforce.[58] During the early 1850s just such an institutionalisation began to be attempted. Links were strengthened with the Botanical Gardens, which opened a second research centre in the high mountains above Kandy. The Planters' Association was formed in 1854 to represent planter interests to the government and collect and diffuse technical information useful to planting. Increasingly during these years, Agency Houses purchased plantations and rationalised business by taking over sales and export and by sending agents to advise planters on the latest agricultural technologies and plantation management strategies. Central to this modernisation was a significant increase in locally produced handbooks, memoirs, newspaper articles, and the *Proceedings of the Planters' Association*. Scientific papers were prepared on everything from weather, soil, plants, and disease, to plantation management. This literature was a mix of technical advice and moral judgement, all framed to produce a new, improved rationalised space, tailored to work in this *particular* environment.

The average plantation covered 200 acres and employed 140 workers (see Figure 1) specifications were developed for cutting and burning the forest, and staking, holing, and

Figure 1. Peacock Hill plantation, 1864. Painted by Captain O'Brien.

the orderly planting of coffee trees.[59] Trees were to be planted in straight lines in order that supervisors might better observe the workers (see Figure 2).[60] The pulping house, where the coffee beans were separated from the fruit, was typically located on a stream, with the stores and drying ground nearby. The lines, as housing for workers was called, were located near the pulper and stores. The planter's bungalow was ideally situated between the buildings and fields to further facilitate supervision.[61] The standard size of a line was 6–10 rooms, each of which averaged about 150 square feet and contained 10–20 people.[62] It was believed that relatively small lines gave a sense of belonging, and decreased both fighting and plotting against the planter.[63] In most plantations, there were individual garden plots staked out adjacent to the rooms, but this was controversial as well and there were debates in the *Proceedings of the Planters' Association* about whether gardens improved the labourers' health and kept them on the plantation, or encouraged them to waste time and steal from each other.[64] All of these calculations were aimed at producing rationalised spatial arrangements most conducive to producing a well-ordered labour force.

The production of the abstract body

The un-cultivated body

Microtechnologies of control were employed in order to attempt to transform Tamils into abstract bodies. Labourers, exhausted from their trip from Madras to northern Ceylon, were forced to march a further 150 miles in six days before reaching the plantations. Often the kanganies who recruited them provided insufficient food for the trip. Consequently, they arrived weak and sick at the plantations.[65] The Tamils were figured by the British as base humans, more animal than human in many ways.

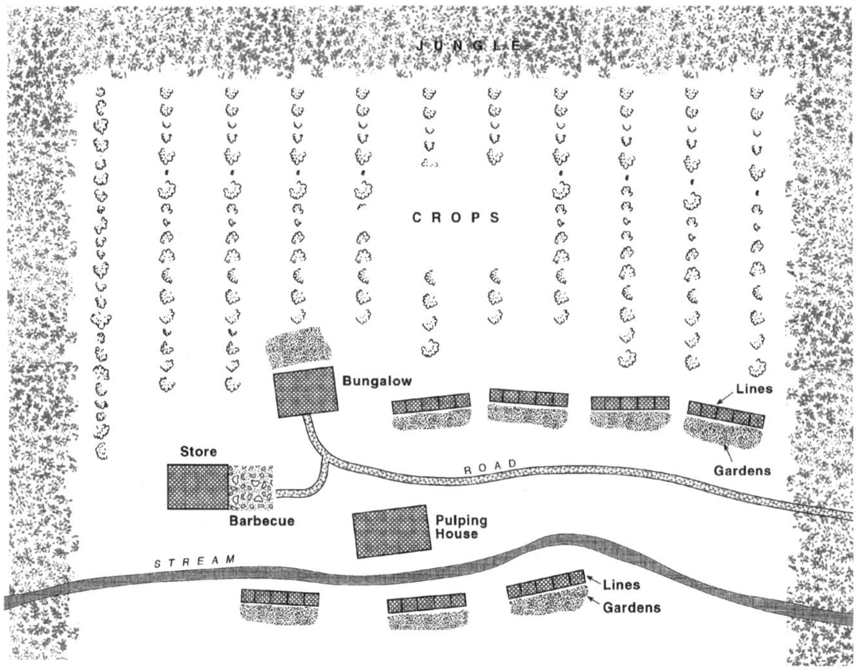

Figure 2. Schematic diagram of a coffee plantation, mid-nineteenth century.

Dr Kynsey, the Principal Civil Medical Officer said of the labourer, or 'coolie' as he was called, he "is naturally lazy, indolent and docile... He has strongly developed animal passions..."[66] But it was thought the Tamil worker could be transformed through the modernising effects of the plantation system. For the character of the Tamil was considered to be human nature "in its original, uncultivated state."[67] As Governor Robinson wrote in a despatch to the Colonial Office, "The Tamil coolie is perhaps the simplest, as he is certainly the most capricious, of all Orientals with whom we have to deal in Ceylon. He is like a child requiring the sense of the strong arm of power. He must know that he is subject to parental authority."[68]

Disciplining the Tamil body

Respect for authority was impressed through the routines of abstract space. Just as the Tamil worker would cultivate the soil, so work discipline was seen to transform his or her uncultivated nature. The principal way in which this transformation was to take place was through work discipline. The time worked, tasks assigned, and pay, were based on calculations of the relation between the "flawed but improving" Tamil body and the abstract body, based on a Europeanised ideal. Time and space were organised on this basis. For example, there was a strict work schedule on plantations. The tom–tom for reveille was beaten at 4 a.m. Labourers took their morning meal at 5–6 a.m., and then worked from 6 a.m. to 4 p.m. Adult males were expected to pick two bushels per day as the norm,[69] and were not to carry more than 40 pounds for more than 25 miles in

24 hours.[70] The lists of tasks, the time necessary and the cost of each job were calculated.[71] Similar formulas were calculated for women and children.[72] Wages were paid according to the type of body, age and gender. In the 1860s men were paid nine pence per day, women seven pence, and children paid by height.[73]

Health and the body

As Arnold has argued, "colonial rule built up an enormous battery of texts and discursive practices that concerned themselves with the physical being of the colonised."[74] However, such knowledge can best be understood in terms of the economic and enculturating goals of the colonisers. Hygiene and medical theories were specifically tailored to theories of racial difference and tropical environment.[75] Arnold states, "the history of colonial medicine, and of epidemic diseases with which it was closely entwined, serves to illustrate the more general nature of colonial power and knowledge, and illuminate its hegemonic as well as coercive processes."[76] "Colonialism," he says, "used, or attempted to use, the body as a site for the construction of its own authority, legitimacy and control."[77]

Dr Kinsey, the chief medical officer wrote of the Tamil labourer, he "is filthy . . . with all this, he is capable of undergoing severe labour and much privation, but when once attacked with illness he speedily sinks unless treated with great care."[78] The Tamils were thought to be sickly by nature and consequently short-lived. They developed quickly and had to be used quickly. As the planter Jenkins put it, "in the East . . . a youth of 9 or 10 is mentally as precocious as our boys at double that age."[79] Such attitudes dovetailed with the short term requirements of labour on the plantations. They also justified and naturalised high death rates. But there was dispute about what exactly the death rate was, and what constituted an acceptable level of death among the Tamils. Planters were sensitive to charges of high death rates among Tamils because there were powerful forces both in Britain and India who took a great interest in the welfare of Indian migrant labourers.[80] The Ceylon Observer estimated that between 1841 and 1848, 70 000 workers, or 25 per cent of those who came to the island died.[81] Ten years later in 1859, A. M. Ferguson writing in the *Ceylon Plantation Gazetteer* argued that 25 per cent was an accurate figure.[82]

There were also debates over how much medicine workers should have. It was assumed that they did not need the same level of care as Europeans, but how much was unclear. How much death was acceptable? Should European medicine be forced on them?[83] There was a constant battle with the government over these issues and who was to blame for the unusually high death rate. The planters argued that the Tamils' unsanitary habits, their refusal of western medicine, the exploitation by the kanganies on the trip to the plantation, were responsible.[84] However, Government officials challenged the planters' assumptions about what constituted adequate health care. As early as 1846, the Governor, Sir Colin Campbell, wrote to the Secretary of State for the Colonies arguing that if the planters did not take care of their workers then a direct assessment might be necessary.[85] Planters fought this and consequently it was not until 1873 that Medical Ordinance No. 10 was passed allowing the Government to recover the cost from planters of treating sick workers. The planters appear to have made an economic calculation about the value of labour and how much it was worth spending to keep workers alive versus the cost of finding new labourers. As there was nearly always a ready supply of labour available, it seemed uneconomic to spend too much on medicine. Similar calculations were made regarding food. Healthy bodies meant healthy workers. But again there was a fine balance to be achieved between the cost of food and the return on labour. Planters debated the question of how much food a worker needed and how much and how often they should consume hot food.[86] Such rationalisation of bodies may have been extreme in its lack of empathy with the plight of workers. However, the broader tradition of 'political anatomy' and medicine, as it was developing in Britain at the time, likewise entailed the regularisation, surveillance and technical control of bodies abstracted from their socio-cultural context.

Transformed bodies

It was believed that plantation technologies of control had a degree of success. As the Government Agent for the Kandyan Province stated in his annual report for 1867 "the coolie after a residence of a few years is a very different being from what he was when he came; instead of being a weak, stupid, ill-fed, badly clad person, he, from being well housed, well cared for, well fed, and employed at regular and sufficient work, returns to his country a strong, intelligent, lusty labourer, to bring back others willing to try their fortune also."[87] Dr Van Dort, the Assistant Colonial Surgeon in charge of the Badulla hospital, echoes this. It is "genius of labour," he said, which transforms the labourer. His countenance, "mirrors the newly awakened soul, and the consciousness of powers and capabilities, hitherto dormant and unused, stamps the physiognomy with an expression of manliness and intelligence which is never seen in the raw, uncivilised, newly landed coolie."[88] The planters and government officials in the planting districts argued against their critics in Britain that the plantations were not exploitative but were positively transformative.

Resistance in everyday life

Covert resistance

There are many different tactics used by workers to resist the planters' attempts to discipline and transform them, some covert and some overt. In situations of greatly unequal power, covert resistance—'the weapons of the weak'[89]—are most important. These are tactics used by people without the power to directly challenge those in authority. They, therefore, "poach" to use de Certeau's term, within the spaces of power.[90] The planters tried to wring as much labour from the workers as possible for as small a wage as possible. The workers countered this by minimising their labour output.

There were numerous ways of doing this. One was to try and establish the local standards by testing a new master to see how little work they could get away with and whether he was a strict disciplinarian.[91] They would seek out any 'cracks' in the imposition of abstract space. Within these early-established frameworks of demands and consequences on each plantation, workers developed tactics for avoiding work. One tactic was to feign sickness, as planters were obliged by law to feed their sick workers. Another was to return to the lines after the 6 a.m. muster on a pretext of having forgotten something, and then to hide there for the rest of the day.[92] If the planter searched the lines he often found it difficult to see workers hiding in the dark, smoky interiors. The darkness and smoke gave the worker time to scale the partitions between the rooms in a line and return to the fields.[93] But even while in the fields, the workers could use their knowledge of the landscape as a tactic. For example, they would hide behind trees and coffee bushes or slip into the jungle at the edge of the plantation for a short rest.[94] Tired workers would lighten their load by cutting open and pouring out a small portion of coffee or rice from the heavy bags they were carrying.[95] They were also known to abandon the bags in the underbrush where enraged planters would find them rotting months later.[96] During the harvest, one of the onerous tasks was separating the beans from the coffee fruit. Having to continuously feed heavy baskets of fruit into the pulper, workers would periodically slip the belt on the machine in order to catch a few moments rest.[97]

Another tactic was to pretend to fill one's daily quota. Such a tactic depended on laxness of supervision. For example, workers on James Taylor's plantation in the 1850s had to dig 25 holes for young coffee bushes per day or they would lose half a day's pay. Taylor writes, "they are the greatest cheaters that ever existed, for they must be called back to every hole they make two or three times before it is large enough... they will come asking for tickets for 25 holes though they have only done 15."[98] Other ways of quickly filling ones quota was to strip green coffee off the bushes in order to gather ripe coffee more quickly,[99] or to steal coffee from the store at night and bring it in the next day to fill their quota.[100] Workers would also supplement their income by accepting bribes from bazaar keepers to drop ripe coffee under the bushes, as the bazaar keepers had contracts to glean coffee that had been dropped in the fields,[101] or by stealing coffee beans or rice from the store or from bags during transportation to sell in the bazaars at the edge of the plantations.[102]

All of these forms of resistance used fragments of space and time. Workers discovered ways to escape the planters and overseers' panoptical procedures; to discover places that could not be seen; to learn how to take advantage of moments when supervision was lax; to manoeuvre, as de Certeau put it, within an enemy field of vision.[103] A key role in the counter-surveillance was played by household servants in the planter's bungalow and Tamil women in the lines. They would report, for example, if the planter was drunk and consequently staying at home that day, or if he had left the plantation.[104] Once it was established that the planter or supervisor was not coming to the fields, then work would slow down or cease. Flows of information were incompletely controlled. The planter and the workers played a cat and mouse game with each trying to adjust to and counter the other's moves. But the game was a dangerous one for the workers, as their covert resistance, once detected was punished with public humiliation, beatings and loss of food.

Open resistance

While covert resistance was constantly practised, open resistance could rarely be risked as the potential costs of such behaviour nearly inevitably outweighed the benefits.[105] There were, however, a variety of different types of open resistance that workers adopted. Perhaps the most common form was insubordination, which could take the form of insolence or openly disobeying an order. George Wall, the President of the Planters' Association wrote, "the coolies are getting a great deal too much of their own way... [and, unless they are controlled]... they will soon be virtually the masters." He went on to observe that on plantations where workers were given contracts to weed, the "independence and insolence of coolies was rampant."[106] Millie, writing in the late 1870s felt that "the general submissive character of the coolies has very much changed for the worse," and went on to add, "in a country like Ceylon, uncalled for disrespect or contempt for an employer, even though shown in a very quiet way, should meet with prompt and immediate punishment."[107] Insubordination was seen as such a serious challenge to the authority of the planter, that it was punished by flogging.[108]

Another form of open resistance was to go to court to attempt to collect unpaid wages. Sir Emerson Tennant reported to the Colonial Office during the crash of 1847 that planters were not paying wages and were responding to worker demands to be paid with "blows."[109] The following year the Governor in a Despatch to the Colonial Office wrote that a portion of the crop was lost on certain estates as workers refused to pick until back wages were paid.[110] In theory, the Service Contracts Ordinance #5 of 1841 allowed labourers to sue planters for non-payment of wages.[111] However, as the Police Magistrate at Kandy pointed out, "the coolies are nearly all ignorant that any law exists for regulating agreements."[112] Having said that, during the crisis years of 1847–1848 there were 2584 complaints for non-payment in the Kandy Courts.[113] Part of the problem was that a number of the Justices of the Peace in the planting districts were themselves planters who were quite understandably reluctant to prosecute their fellow planters. The 1841 law was strengthened in 1865 by the passing of Ordinance 11, which allowed workers to terminate employment with two days notice if wages had not been paid in a month. The Planters' Association opposed this ordinance.[114] The first plantation strike was in 1854 when workers successfully struck for one penny extra to match to pay of road labourers.[115] After this, it was not uncommon for workers to engage in brief strikes over working conditions. But they were short lived and restricted to single plantations.

Desertion

Workers were tied to their plantations by contracts and it was a crime to leave before they expired. But many apparently found conditions on individual plantations so intolerable that they fled. Such a decision was not undertaken lightly, for workers ran the risk of being attacked and robbed by Sinhalese villagers who were anti Tamil or being hunted and jailed by planters and the police.[116] In light of this, it is quite remarkable that a survey in 1862 revealed that seven per cent of all Tamil immigrants had been arrested for desertion. How many escaped it is not known.

There were a number of reasons for desertion. Perhaps the most common was the accumulation of an unbearable level of debt. From the time of their arrival on the plantations, workers were in debt to planters for the cost of their passage from India. Furthermore, they were paid so little that most remained in debt throughout their stay. Consequently, they often borrowed heavily from kanganies at 120 per cent interest on average and if they were unable to borrow from them, they turned to the chetties, Indian bazaar keepers, outside the plantations at the rate of 200 per cent interest.[117] Surveys conducted by the planters in the 1870s revealed that the average worker was one to two years' income in debt.[118] While the planters outwardly deplored this level of debt, it was the wage structure that produced it. Planters were, in fact, ambivalent about debt, realising that on the one hand it helped tie the workers to the plantation, while on the other it transferred authority away from them to the kanganies and bazaar keepers.

Workers sometimes deserted when they were physically mistreated on their plantation, or because a kangany from a neighbouring plantation lured them away with promises of better wages. At times, although this was rare, they even deserted out of fear of the British. For example in 1871, several hundred workers fled the plantations after a rumour started that the British were about to offer a blood sacrifice of a 1000 Tamils to the gods to mark the successful completion of the railroad to the planting districts.[119] Earlier that year, Governor Robinson wrote the Colonial Office that, "coolies are fleeing back in hot haste to their own country on an absurd rumour that war had broken out in China and that they were to be forced into the service of the Government as camp followers and soldiers."[120] The ultimate form of open resistance was violence, including murder. But this was rarely seen as a viable option as retribution by the British was invariably swift and severe.[121]

Resistance on the part of the workers to the appalling conditions in which they found themselves was made more difficult for a number of reasons. First, their poor health made them submissive to authority. Second, they were surrounded by Sinhalese villagers who were hostile to them. Third, their distance from home in the south of India cut them off from much support.

Strategies of domination

The workers' tactics of resistance were countered by a number of strategies of domination by the planters. These were premised on planters' notions of the Tamil workers as thoroughly embodied and therefore, lacking in what the British referred to as honour and character.[122] Of particular concern was malingering, since Indians were thought to avoid labour, by nature. Those who claimed to be too sick to work were often sent to the fields.[123] Only if they dropped in the fields were they judged to be truly sick. The planter W. A. Swan told the Annual Meeting of the Planters' Association how he deterred malingerers: he went "to muster on Tuesday morning, picked out and sent to police court, one or two men who had not come to work on Monday; the beneficial effect on the following day's turnout has been most marked."[124] The usual punishment for workers who did not work hard enough or made mistakes was a beating. James Taylor when a young assistant wrote that the superintendent on his estate "thumped a coolie for nearly half an hour till he was nearly dead, then threw him off the estate."[125] Beatings were also administered at times to discipline the "moral life" of workers. For example, the planter R. B. Tytler caned one of his kanganies for having an affair. He also caned two women workers whom he suspected of infidelity to their husbands.[126] Millie, in arguing a more liberal line on punishment, demonstrates just how embodied the workers were thought to be by the majority of the planters. He urges that planters beat a labourer "without hurting his body; but . . . [to] hurt his feelings..yes, hurt the feelings of a coolie! . . . in this way. All his comrades jeered at him so that he suffered a great humiliation."[127] Others, such as Sabonadiere, argued that a pay cut was even more effective than a beating for people earning so little.[128] The kangany Carpen acknowledged that both policies were used: "An angry master beats us on the back. A quiet master beats us on the belly."[129] The potential problem with both of these types of punishments was that the workers were often in such fragile health that the punishments impaired their ability to work well. As such, planters had to consider that the punishment might be more counterproductive than the behaviour it was trying to change.

From the beginning, planters argued that what was good for the plantations was good for the colony more generally. Consequently they fostered close ties with the colonial government wherever possible. While in the early years planters disciplined their workforces as they saw fit, increasingly pressures were brought to bear on them, by the Government of Ceylon, the Colonial Office, and the Government of India to restrict illegal beatings of workers. In response, planters pressured the Government of Ceylon, through their representative on the Legislative Council, to put in place laws that would close the cracks that the workers' tactics opened up in abstract space. Among these were the Labour Ordinance of 1841 and 1865, which punished non-work with a short prison sentence and desertion with a term of up to three months of hard labour. So prevalent was desertion, and so effective was the state in aiding the planters, that in the words of the Queen's Advocate Richard Morgan, "the jails are crammed with scores and hundreds of men, women, and children arrested on warrants of desertion."[130] As prison did not appear to sufficiently deter desertion, a number of planters urged the Government to allow flogging for desertion, but the Government refused.[131] There was a continual tension between the enforcement practices of the planters and Government attempts to protect the rights of workers. At the end of the coffee period the Committee of the Legislative Councillors ruled that there was systematic abuse of warrants to recapture deserters.[132]

Another strategy of domination was to attempt to control the margins of the plantation to produce a space of disciplinary monotony and confinement. One way in which this was done was by trying to exert control over the bazaars around the plantation. This strategy, however, had limited success because the planters could not offer a viable alternative and furthermore could not stop workers from leaving the plantation for short periods of time. Attempts to set up bazaars on the plantations were seen as highly problematic because it was felt that this would simply transfer more power to the kangany. Unsuccessful attempts were also made to close down arrack taverns at the edge of plantations. Planters opposed these, not out of opposition to drink *per se*, but because they generated debt and consequently encouraged stealing. However, the planters were never able to close down the taverns because the government generated a large amount of revenue from the sale of arrack, the local liquor.

Planters became increasingly concerned that it was extremely difficult to stop the theft of coffee. There were simply too many places on the plantations where the workers could hide coffee in the jungle. The Sinhalese cart men also stole coffee once it left the plantation. The problem in all of these cases was that stolen coffee could not be identified as such, as Sinhalese villagers grew coffee themselves. The Planters' Association became particularly concerned when large numbers of pulpers and boutiques dealing in coffee sprang up around plantations.[133] As they could not ban these establishments, they tried to make it difficult for them to procure and sell coffee. To this end, the planters argued that 'natives' with coffee should be made to account for where it came from. After years of arguing their case before the Government, the Coffee Stealing Ordinance Number 13 of 1876 was passed, which judged individuals to be guilty of theft if they could not prove where their green coffee came from. The punishment was raised from a five shilling fine to flogging. Once again, we see the ideal of purified, abstract plantation space breaking down, in this case due to the impossibility of securing watertight boundaries. What some South Asia scholars find true in India can equally be said of Ceylon: that ultimately, the British did not have "much capacity to regulate or transform indigenous society and that... British rule appears as a 'limited raj'—weak, alien, ill informed, often blunderingly ineffectual—and hence an improbable candidate for Gramscian hegemony."[134]

The middle space of the Kanganies

While it is tempting to paint a picture of native solidarity against an imperialist master, the kanganies, in fact, occupied a middle space between the planters and the workers and showed limited solidarity with either.[135] Millie sums up the kangany's position when he writes, "if he attended to the master's interest and worked the coolies, he displeased and lost his interest with them. If he humoured the coolies too much, the master was down on him and threatened him with the sack."[136] Kanganies, from the 1840s on, were professional recruiters who collected labour from different villages in South India.

Far from feeling solidarity with their fellow Tamils, such was the rapaciousness of many kanganies that they misappropriated the part of the head fee intended to feed the labourers on their trip to the plantations, turning the march south into a race against starvation, where the weak and sick were abandoned along the roadside to die.[137] Such callousness was exacerbated by the fact that the kanganies were often of higher caste and class than their workers. In spite of this, or perhaps because of it, kanganies exercised great control over workers. This was understandable, as it was the kangany who hired the workers. They were his labour gang and often were in debt to him. Furthermore, the kanganies acted as crucial intermediaries between the planter and the labourers, for, as most planters spoke little Tamil, the kanganies occupied the powerful position of translators. The planters feared that the kangany and worker would collude against them, by the latter accepting a small bribe from workers to check them off as attending muster.[138] The centre of power for the kangany was the separate house he had at the end of each line. It was here workers would come for advice, loans, justice, or simply in order to pay deference.[139] On large estates where there was more than one kangany, the head kangany often tried to consolidate his power by offering sub-kangany posts to members of his extended family or patronage network.[140] Whilst this strategy was usually effective, there were cases recorded of sub-kanganies defrauding head kanganies.[141] Likewise, whilst planters were nervous that their kangany might desert with his labourers and keep part of the coast advance, or procure a better deal for himself on another plantation, kanganies also fought among themselves over deserters.[142] For example, the Police Report from Badulla district on 29 March 1867 stated that an estate kangany was murdered by five other kanganies and coolies, during a quarrel about deserters.[143]

The planters understandably had a constant fear of lack of control over their kanganies. For although they were "natives", they were powerful ones inserted into the very heart of this particular project of modernity. The planter R. J. Corbet bemoaned the dependence on kanganies saying that it was not "compatible with the high character that an Englishman should bear that we should go on as we are doing, begging, borrowing, hiring, crimping coolies from each other, according to our respective notions of right or wrong, stooping to acts we should be ashamed of."[144] As Governor Gregory put it, "The kangany system is no doubt a bad one. There is hardly a planter in Ceylon who would not abolish it if he could. But it is impossible to do so. It is custom and anyone connected with the East knows what a barrier that word is to any innovation however palpably beneficial."[145] The kangany was a threat to the planter because he represented the traces of an often harsh village authority, which intruded onto the abstract space of the plantation. These traces of older social and spatial practices, it was quite rightly felt, challenged white authority. But the planters could also use this middle space which the kanganies occupied in order to shift blame from themselves. For example, the very high death rates mentioned earlier were continually blamed, not on the plantation system itself, but on the viciousness of kanganies.[146] For the labourers the kanganies were ambivalent figures as well, for although they could at times pit the kanganies and planters against one another, more often than not the former were caught between two sets of exploiters.

Conclusion

Because Ceylonese workers strongly resisted attempts to transform them into abstract, docile bodies, planters resorted to surveillance and force, justifying such coercion on the grounds that the workers were degraded, and using their acts of resistance as confirmation of their moral failure. To resist the demands of the disciplined body was considered laziness; to reject the space-time organisation of abstract space was called dishonesty and lack of self-discipline; escaping from intolerable conditions was called desertion, a lack of loyalty. Although Tamil workers were forced to adapt to the highly routinised plantation regime, they cannot be said to have ever fully embodied colonial European modernism. Hegemony was never accomplished; self-discipline made little sense within the network of limited coloniser expectations and colonised ambitions of the forced Tamil insertion into the heterogeneous, fissured space of the plantation.

The planters' strategies of power and their desire to record in minute detail all of their observations and theories, their rationalisation of space and time, have provided me with my data; perhaps not the voices, but at least the practices of the workers.[147] Those who have no voice must rely on their oppressors to speak for them. Through a close reading of government reports, and British planters' pamphlets and manuals warning each other of workers' tactics of resistance, one can begin to develop a detailed record of the spatial and temporal practices which take place within and in opposition to the plantation owners' attempts to create a rationalised, space and disciplined, docile bodies. Thus, even the workers' place in the historical record is highly constrained within this bureaucratised, textualised and commercialised space. As the workers resisted their domination through inhabiting the marginal spaces within the purified, abstract space of the plantation, so likewise, their resistance now occupies the margins of the colonial documents which themselves were part of the socio-technical apparatus of domination. It is out of the margins of the discourse of domination that an historical geography of the voiceless will probably have to be written.[148]

Emmanuel College
Cambridge CB2 3AP
UK

Notes

[1] T. R. Metcalf, *Ideologies of the Raj* (Cambridge 1995) 29, argues that India was a laboratory for the creation of the liberal administrative state in the mid-nineteenth century. For a discussion of Dutch colonialism as an experiment in modernity see, A. L. Stoler, *Race and the Education of Desire* (Durham 1995) 15. For the French case see, P. Rabinow, *French Modern: Norms and Forms of the Social Environment* (London 1989).

[2] Here, I draw language and inspiration from Bruno Latour and actor-network theory. See, B. Latour, *Science in Action: How to Follow Scientists and Engineers* (Cambridge, MA 1987); Idem., *We Have Never Been Modern* (Hemel Hempstead 1993); J. Murdoch, Towards a geography of heterogeneous associations, *Progress in Human Geography* **21** (1997) 321–337; J. Law and J. Hassard, *Actor-Network Theory and After* (Oxford 1999). While the focus of this paper is the coming together of heterogeneous human systems, in the larger research project of which this paper is a part I bring together a wider range of human and non-human associations. I will, however, be using this language of the hybridity of nature and culture with some trepidation and care to avoid the implied complicity in using language with which the British colonisers felt comfortable. Popular nineteenth-century concepts of race entailed the ambiguous conception of humanity and civilisation with some races seen as not fully human. The power to define what was human was seen as the coloniser's prerogative. The assertion of equality among humans based on a clear nature/culture dualism is an achievement to be rejected only with extreme caution. Also see R. Young, *Colonial Desire: Hybridity in Theory, Culture and Race* (London 1995) for his different but related concerns about the use of the concept of cultural hybridity which he sees as possibly carrying some unwanted baggage from its "Victorian extreme right" connections.

[3] P. P. Courtenay, *Plantation Agriculture* (London 1980) 55–56.

[4] For a discussion of slave plantations, see P. Curtin, *The Rise and Fall of the Plantation Complex: Essays in Atlantic History* (Cambridge 1990).
[5] R. Kurian, *State, Capital and Labour in the Plantation Industry in Sri Lanka 1834–1984* (Amsterdam 1989) 1–12.
[6] H. Tinker, *A New System of Slavery: The Export of Indian Labour Overseas 1830–1920* (Oxford 1974) 177.
[7] H. Lefebvre, *The Production of Space* (Oxford 1991). Although Lefebvre applies the notion of abstract space to capitalism in Europe, I would argue that it was equally a goal in industrial plantations in tropical colonies. That it was crosscut by race and older forms of pre-capitalist non-European social relations is the subject of this paper.
[8] M. Landzelius, personal communication. See also M. Poovey, *Making a Social Body: British Cultural Formation, 1830–1864* (Chicago 1995).
[9] M. Foucault, *Discipline and Punish: The Birth of the Prison* (New York 1979) 135–169.
[10] Planters, as we shall see, viewed 'natives' as the product of a degrading climate and moral condition, which, it was believed, European capitalism could begin to transform. For a discussion of the discourse of the degraded 'native,' see D. Arnold, *Colonising the Body: State Medicine and Epidemic Disease in Nineteenth Century India* (Berkeley 1993); Idem., *The Problem of Nature: Environment, Culture and European Expansion* (Oxford 1996); D. N. Livingstone, The moral discourse of climate: historical considerations on race, place, and virtue, *Journal of Historical Geography* **17** (1991) 413–434; Idem., Tropical climate and moral hygiene: the anatomy of a Victorian debate, *British Journal of the History of Science* **32** (1999) 93–110; J. S. Duncan, The struggle to be temperate: climate and 'moral masculinity' in mid-nineteenth century Ceylon, *Singapore Journal of Tropical Geography* **21** (2000) 34–47; For a working out of some of these ideas in Britain, see F. Mort, *Dangerous Sexualities: Medico-Moral Politics in England Since 1830* (London 1987) 25.
[11] Duncan, *The struggle.*
[12] A. L. Stoler, Plantation politics and protest on Sumatra's east coast, *Journal of Peasant Studies* **13** (1986) 124. E. Wolf, Specific aspects of plantation systems in the new world, in V. Rubin (Ed.), *Plantation Systems of the New World* (Washington 1959) 43, says that such plantation workers live "double lives," while M. Tausig, *The Devil and Commodity Fetishism in South America* (Chapel Hill 1980), 92, 103, 113, says they are "limnal beings."
[13] Foucault, *op. cit.* L. Stewart, Louisiana subjects: power, space and the slave body, *Ecumene* **2** (1995) 227–245 explores New World slave plantations in much the same terms.
[14] The struggle that white male planters waged against their own embodiment in the tropics is the subject of Duncan, The struggle.
[15] D. Arnold, *The Problem of Nature: Environment, Culture and European Expansion* (Oxford 1996) 160.
[16] J. Scott, *Weapons of the Weak: Everyday Forms of Peasant Resistance* (New Haven 1985). Like Steward, *op. cit.*, I am keen on de Certeau's use of the term tactics, to refer to the manner in which the oppressed 'poach' in the space of their oppressors, but like her I am uneasy with his claim that the oppressed can only resist through controlling time and not place. See M. de Certeau's, *The Practice of Everyday Life* (Berkeley 1984). Having said this, I will use de Certeau's notions of tactics to refer to the workers' resistance and strategies to refer to the planters attempts at domination.
[17] E. Genovese, *Roll Jordan Roll: The World the Slaves Made* (New York 1972) 285–324, 599–657. A more recent and differently theorised exploration slave plantations is to be found in Stewart, *Louisiana.*
[18] P. Ramasamy, Labour control and labour resistance in the plantations of colonial Malaya, in E. V. Daniel, H. Bernstein and T. Brass (Eds) *Plantations, Proletarians and Peasants in Colonial Asia* (London 1992) 100.
[19] M. Carter, Strategies of labour mobilisation in colonial India: the recruitment of indentured workers for Mauritius, in Daniel, Bernstein and Brass, *op. cit.*, 229–245; Idem., *Servants, Sirdars and Settlers: Indians in Mauritius 1843–1874* (Delhi 1995).
[20] Carter, *Servants*, 225–226.
[21] M. Adas, From foot dragging to flight: the evasive history of peasant avoidance protest in South and Southeast Asia, *Journal of Peasant Studies* **13** (1986) 64–86.
[22] C. Barnett, Sing along with the common people: politics, postcolonialism and other figures, *Environment and Planning D: Society and Space* **15** (1997) 137–154.

[23] C. McEwan, Cutting power lines within the palace: countering paternity and Eurocentrism in the geographical tradition, *Transactions, Institute of British Geographers* **23** (1998) 371–384.
[24] J. Butler, Performativity's social magic, in T. R. Schatzki and W. Natter (Eds), *The Social and Political Body* (New York 1996) 30. For a similar point with regard to social memory see, P. Connerton, *How Societies Remember* (Cambridge 1989).
[25] The concept of embodiment valorised in contemporary social thought and seen to transgress entrenched dualisms was quite familiar to racist, imperial modes of thinking. The crucial difference today lies in the belief that all humans are thoroughly embodied whereas the imperial view saw only natives, women and the lower classes as embodied. Nevertheless, I would wish to use this concept with great care.
[26] D. M. Forrest, *A Hundred Years of Ceylon Tea, 1867–1967* (London 1967).
[27] See I. H. Van Den Driesen, Coffee cultivation in Ceylon (1), *The Ceylon Historical Journal* **3** (1953); Idem., Coffee cultivation in Ceylon (2), *The Ceylon Historical Journal* **3** (1953); A. C. L. Ameer Ali, Peasant coffee in Ceylon during the nineteenth century, *Ceylon Journal of History and Social Science* **2** (1972) 50; D. Moldrich, *Bitter Berry Bondage* (Colombo 1989); E. Meyer, Enclave plantations, hemmed-in villages and dualistic representations in colonial Ceylon, *Journal of Peasant Studies* **19** (1992) 199–228.
[28] See Concessions for the Cultivation of Certain agricultural Products, Regulation number 4 of 21 September 1829, in G. C. Mendis (Ed.), *The Colebrooke Cameron Papers: Documents on British Colonial Policy in Ceylon 1796–1833*. vol. 2. (Oxford 1956) 279.
[29] L. A. Mills, *Ceylon under British Rule, 1795–1932* (London 1933) 224; Kurian, *State*, 47.
[30] During the nineteenth century, England imported coffee from the West Indies, Java, Ceylon, and Brazil.
[31] For studies of the political economy of the Kandyan Highlands, see A. Bandarage, *Colonialism in Sri Lanka: The Political Economy of the Kandyan Highlands 1833–1886* (Berlin 1983); Kurian, *State*; N. Perera, *Society and Space: Colonialism, Nationalism, and Postcolonial Identity in Sri Lanka* (Boulder 1998).
[32] R. E. Lewis, *Coffee Planting in Ceylon, Past and Present* (Colombo 1855).
[33] *Ceylon Examiner* 16 June 1852.
[34] Coffee blight began in plantations to the south and west of Kandy in 1845. See Lewis, *Coffee Planting*, 29.
[35] According to Kurian, *State*, 53, the average area planted to coffee rose from 4000 ha in the early 1840s to a high of 125 000 ha between 1875–1879, and dropped to 45 000 ha by 1886. By the late 1880s, 400 of the island's 1700 coffee planters had left the island and most of the others had abandoned coffee production. See Moldrich, *Bitter Berry*, 10.
[36] C. R. De Silva, *Ceylon Under the British Occupation 1795–1833* (Colombo 1962) vol. 2, 152.
[37] *Ceylon Observer* 3 June 1843.
[38] Sri Lanka National Archive (SLNA) Despatches 5/34, 19 April 1847.
[39] Colonial Office (CO) 54/235.
[40] *Ceylon Examiner* 16 June 1852.
[41] *Ceylon Observer* 10 July 1843. An early attempt was also made to bring in labourers from China but this was quickly abandoned on the grounds that they were too expensive and not sufficiently docile.
[42] *Ceylon Observer* 3 June 1843.
[43] C. R. De Silva, *Ceylon*, 484.
[44] The average number of annual arrivals for 1839–1842 was 5300. This rose to 115 416 between 1873 and 1879 and then dropped to 46 152 between 1880 and 1986 when coffee was in its final decline. See Bandarage, *Colonialism*, 202.
[45] D. Snodgrass, *Ceylon: an Export Economy in Transition* (Illinois 1966) 24; Tinker, *A New System* 119.
[46] Wages in Madras remained at three pence per day until the 1850s, while a male labourer in Ceylon could earn four pence per day in the 1840s and between six and nine pence per day in the 1850s. See *Ceylon Observer* 20 April 1843; *Ceylon Examiner* 16 June 1852. For a general discussion of Indian labour on plantations, see M. Roberts, Indian estate labour in Ceylon during the coffee period 1830–1880, *The Indian Economic and Social History Review* **3** (1966) 3.

[47] *Proceedings of the Planters' Association* (Kandy 1871–1872), 162–163. The Planters' Association conducted a survey in the early 1870s and found that 72 per cent of their workers were lower caste.
[48] W. Sabonadiere, *The Coffee Planter of Ceylon* (Guernsey 1866) 85.
[49] F. Heidemann, *Kanganies in Sri Lanka and Malaysia: Tamil Recruiter-cum-Foreman as a Sociological Category* (Munich 1992) 12–13.
[50] For a discussion of similar employment practices in twentieth-century Californian agriculture see D. Mitchell, *The Lie of the Land* (Minneapolis 1996).
[51] E. Meyer, The plantation system and village structure in rural South Asia, in P. Robb (Ed.), *Linkages, Change and Development* (London 1983) 40.
[52] *Ceylon Observer* 3 June 1843; *Ceylon Observer* 5 January 1852.
[53] A. Brown, *The Coffee Planter's Manual* (Colombo 1880). As noted previously, Governor Barnes moved the Botanical Gardens from Colombo to Peradeniya adjacent to his newly opened plantation, precisely so that he could avail himself of the latest scientific advice on coffee production. The Ceylon Agricultural Society was formed in 1842. Its stated purpose was to improve agriculture, organise the introduction of Tamil labourers, form a library and correspond with other colonies about agricultural matters. The Society's first proceedings published a list of plantations and the number and wages of workers employed. *Proceedings of the Ceylon Agricultural Society* (Kandy 1842). In the following year, the Society communicated with the Calcutta Agricultural Society, and with societies in Egypt and the Cape. *Proceedings of the Ceylon Agricultural Society* (Kandy 1843). Some of the early planters such as R. B. Tytler had previous experience planting coffee in Jamaica (J. Ferguson, *Pioneers of the Planting Enterprise in Ceylon* vol. 1 (Colombo 1894)) 14.
[54] SLNA Despatches 5/35 4 July 1848 Lord Torrington to Earl Grey.
[55] *Ceylon Examiner* 16 June 1852.
[56] On the importance of the notion of rationality as a sign distinguishing European imperialists from other peoples, see M. Adas, *Machines as the Measure of Men: Science, Technology, and Ideologies of Western Dominance* (Ithaca 1989).
[57] This worry about the unreliability of certain types of imperialists is discussed at length by Stoler with particular reference to the Dutch East Indies in the nineteenth century. See A. L. Stoler, Rethinking colonial categories: European communities and the boundaries of rule, in N. B. Dirks (Ed.), *Colonialism and Culture* (Ann Arbor 1991) 119–152; Stoler, *Race*.
[58] For a discussion of the anxiety surrounding self discipline in nineteenth century Ceylon see Duncan, "The struggle."
[59] Lewis, *Coffee Planting*, 37–44; Sabonadiere, *The Coffee Planter*; T. C. Owen, *First Year's Work on a Coffee Plantation: Oonoogalla estate, Madulkelle (being the essay which received the second prize from the Ceylon Planters' Association in 1877)* (Colombo 1877); Brown, *The Coffee*.
[60] Brown, *The Coffee*, 13.
[61] Lewis, *Coffee Planting*. 44.
[62] W. Sabonadiere, *op. cit.* 65–66; Brown, *The Coffee* 20.
[63] *Proceedings of the Planters' Association* (Kandy 1876–1877) 195.
[64] *Ibid.* Stoler, *Plantation politics* 139, argues that plantation gardens in Sumatra were an important part of the plantation system in that they made it cheaper to feed workers. But this also, as the Ceylon planters were well aware, came at the cost of retarding the proletarianisation of the work force.
[65] K. M. De Silva, *Social Policy and Missionary Organisations in Ceylon 1840–1855* (London 1965) 264–265.
[66] Dr Kynsey, Minutes of Evidence, Medical Aid Commission. Papers Laid Before the Legislative Council of Ceylon (Colombo 1879) 168.
[67] P. D. Millie, *Thirty Years Ago: or Reminiscences of the Early Days of Coffee Planting in Ceylon* (Colombo 1878) ch. 15.
[68] SLNA Despatches 5/53 27 May 1866.
[69] Brown, *The Coffee* 42.
[70] Labour Law of 1864. This was communicated to planters in the *Proceedings of the Planters' Association* (Kandy 1864–1865).
[71] Brown, *The Coffee*, 95.

[72] During the 1840s, the ratio of women to men labourers was 1:38. By the 1860s, it had increased to 1:6 as it was felt by the planters that access to more women would make their male labourers more manageable.
[73] SLNA Despatches 5/54. 6 March 1867.
[74] Arnold, *Colonising the Body*, 8.
[75] Arnold makes the point that Western medicine was more than simply a tool of empire. European doctors working in South Asia found it to be an excellent laboratory for the study of variation in disease. Western medicine in India had to be adapted to local conditions, "it had to negotiate its passage." See Arnold, *Colonising the Body, 292–293.*
[76] *Ibid.*
[77] *Ibid.*
[78] Dr Kynsey, *Minutes*.
[79] A Planter (Richard Wade Jenkins) *Ceylon in the Fifties and the Eighties, a Retrospect and Contrast of the Vicissitudes of the Planting Enterprise During a Period of Thirty Years and of Life and Work in Ceylon* (Colombo 1886), 12.
[80] During the coffee period, Colonial Office officials such as the Colonial Secretary Lord Glenelg and the Permanent Under-secretary James Stephen were evangelical and much interested in the morality question. Another influential group who took an interest in the Tamils was the Aboriginal Protection Society. Likewise, the Indian Government, concerned about the treatment of indentured labourers in places like Mauritius and the West Indies, passed the Indian Act Number 14 of 1839 prohibiting the employ of Indians abroad. See D. Northrup, *Indentured Labour in the Age of Imperialism 1834–1922* (Cambridge 1995). While emigration was allowed to Ceylon, the planters were very aware that the Indian Government could cut off the labour supply if it thought that workers were being seriously mistreated. For a discussion of abolitionist and liberal influences on British policy in India, see Metcalf, *Ideologies*.
[81] *The Ceylon Observer* 4 October 1849.
[82] *Ceylon Plantation Gazetteer* (Colombo 1859).
[83] Debates about Tamil prejudice against European medicine, high caste prejudice against hospitals, and how consequently the planter's own home remedies were best were frequently aired in the *Proceedings of the Planters Association*. See, for example, *Proceedings of the Planters Association* (Kandy 1871–1872,) 102–103; (1875–1876) 250.
[84] *Proceedings of the Planters' Association* (Kandy 1869–1870) 55–115 reprints Dr Van Dort's charge that plantation workers receive inadequate health care, and letters in reply from planters which debate the level of care which Tamil labourers need and shift the responsibility for the high death rate to the kanganies or the workers themselves.
[85] SLNA Despatches 11 November 1846.
[86] See *Proceedings of the Planters' Association* (Kandy 1876–1877) for a survey of workers' consumption patterns.
[87] R. C. Pole, Assistant Government Agent, Matale, *Administrative Report* (Colombo 1867) 40.
[88] Van Dort *Proceedings, Planters' Association*, (Kandy 1869–1870) 55.
[89] J. Scott, *Weapons*.
[90] De Certeau, *The Practice*.
[91] Sabonadiere, *The Coffee Planter*, ff. 34.
[92] W. A. Swan, *Proceedings of the Planters' Association* (Kandy 1869–1870) 23.
[93] Millie, *Thirty Years Ago*, ch. 5.
[94] Brown, *The Coffee* 43.
[95] Millie, *Thirty Years Ago*, ch. 3.
[96] *Ibid.*, ch. 5.
[97] *Ibid.*, ch. 6.
[98] J. Taylor, 23 June 1852. Papers of James Taylor (1851–1892) His Correspondence, etc. MSS 15908-10. Department of Manuscripts, National Library of Scotland.
[99] Millie, *Thirty Years Ago*, ch. 6.
[100] *Ibid.*, ch. 5.
[101] *Ibid.*, ch. 6.
[102] *Proceedings of the Planters' Association* (Kandy 1880–1881), xxii,xxvi.
[103] De Certeau, *The Practice*.
[104] Millie, *Thirty Years Ago*, ch. 34.

[105] Scott, *Weapons* makes this same point with reference to Malaysian peasant farmers.
[106] G. Wall, *Proceeding of the Planters' Association* (Kandy 1865–1866) 80.
[107] Millie, *Thirty Years Ago*, ch. 38.
[108] Ferguson's Ceylon Directory (Colombo 1865–1866) 127. The other was coffee stealing.
[109] SLNA Despatches 5/34 21 April 1847.
[110] SLNA Despatches 5/35 11 December 1848.
[111] M. Roberts, The master servant laws of 1841 and the 1860s and immigrant labour in Ceylon, *Ceylon Journal of Historical and Social Studies* **8** (1965).
[112] K. M De Silva, *Social Policy*.
[113] *Ibid*., 247.
[114] *Proceeding of the Planters' Association* (Kandy 1865–1866) 76–78.
[115] C. T. Arasaratnam, A brief history of the development of labour relations in Ceylon. *Ceylon Labour Gazette* April 1970.
[116] SLNA Despatches 5/29 22 December 1842. For a discussion of the persistence of Sinhalese violence against Tamil workers throughout the coffee period, see G. K. Pippet, *A History of the Ceylon Police*. vol. 1 1795–1870 (Colombo 1938) 156; B. Basliampillai, The South Indian immigrants' trek to Ceylon in the mid nineteenth century, *Sri Lankan Journal of Social Science* 7 (1984) 52–53.
[117] Moldrich, *Bitter Berry* 83–84.
[118] *Proceedings of the Planters' Association* (Kandy 1876).
[119] *Times of Ceylon*, 29 September 1871.
[120] SLNA Despatches 5/58, 2 February 1871.
[121] In 1866 a planter was murdered by a worker and in the early 1880s three more were, all for being strict disciplinarians. In each case the killer was hung. See Prippet, *op. cit*., 231; Administration Report (Colombo 1882) 23c.
[122] Sabonadiere, *The Coffee Planter*, 86 writes "untruthfulness comes as naturally to a Tamil as mother's milk. No dependence whatever can be placed upon the statements of the coolies as a class." As Metcalf, *op. cit*. 24, points out, the British portrayed Indians as uniquely predisposed to corruption and mendacity, which had the effect of reaffirming their own moral superiority and right to rule. For a discussion of the importance of 'character' in colonialist thinking, see B. Anderson, *Imagined Communities* (London 1983), 137; Stoler, *Race*.
[123] Swan, *Proceedings*, 23.
[124] *Proceeding of the Planters' Association* (Kandy 1869–1870) 24.
[125] J. Taylor, 23 June 1852. *Papers of James Taylor (1851–1892) His Correspondence, etc*. MSS 15908-10. Department of Manuscripts, National Library of Scotland.
[126] W. Digby, *Life of Sir Richard Morgan* (Madras 1879) vol. 2, 49–50. It appears that men and women labourers were punished in similar fashion. Sexual relations between planters and women Tamil labourers were frowned upon as it was thought to cause trouble with the workforce. It was common, however, for planters to take a Sinhalese mistress from a neighbouring village. For a discussion of this , see Duncan, *The struggle*.
[127] Millie, *Thirty Years Ago*, ch. 38.
[128] Sabonadiere *The Coffee Planter*, 8; see also Swan *Proceedings*, 23.
[129] Carpen, *The Diary*, 88.
[130] Digby, *Life*.
[131] *Proceedings of the Planters' Association* (Kandy 1864–1865) 163.
[132] Sessional Papers 33, 1884.
[133] *Proceeding of the Planters Association* (Kandy 1864–1865) 6, 7.
[134] Arnold, *op. cit*. (1993) 243.
[135] Heidemann, *op. cit*.
[136] Millie, *Thirty Years Ago*, ch. 4.
[137] K. M. De Silva, *Social Policy*, 264–265.
[138] *Ibid*.
[139] Millie, *Thirty Years Ago*, ch. 4.
[140] *Proceedings 1873/74*, p. 206.
[141] *Ibid*.
[142] *Proceedings 1870–1871*, p. 13.
[143] Administration Reports. Ceylon 1867 (Colombo 1868) 208.
[144] *Proceedings of the Planters' Association* (Kandy 1878–1879) 28–29.
[145] SLNA despatches 5/59, 9 July 1872.

[146] *Proceedings of the Planters' Association* (Kandy 1869–1870) 55–115.
[147] See Barnett, *op. cit.*; McEwan, *op. cit.*
[148] On rereading colonial archives in order to recuperate subaltern voices, see R. Guja, *Elementary Aspects of Peasant Insurgency in Colonial India* (Delhi 1992); V. Das, Subaltern as perspective, in R. Guja (Ed.) *Subaltern Studies VI: Writings on South Asian History and Society* (New Delhi 1994) 310–324; J. S. Duncan, Complicity and resistance in the colonial archive: some issues of method and theory in historical geography, *Historical Geography* **27** (1999) 119–128.

[16]

Colonialism's afterlife: vision and visuality on the Northwest Coast

Bruce Braun

Department of Geography, University of Minnesota, Minneapolis

This paper explores the relationships between landscape and power, colonialism and its aftermaths, and state territoriality and its contestation, in the work of two popular Northwest Coast landscape painters: Emily Carr and Lawrence Paul Yuxweluptun. The work of both artists is explored in terms of their representation of relations between indigenous peoples, physical landscapes, state power, and modernity, and in the context of ongoing political struggles over land, resources and the environment between First Nations and the Canadian government. The paper also calls attention to the multiple and fractured nature of postcolonial visualities, to the discursive, social, technological and institutional relations that shape how landscapes are experienced and represented, and, ultimately, to the trace of colonial pasts in the environmental and political imaginaries of a postcolonial present.

The Northwest Coast will always remain in our minds as Carr saw it (Ruth Appelhof)[1]

If it could be said that there is an observer . . . it is only as an *effect* of a heterogeneous network of discursive, social, technological and institutional relations. There is no observer prior to this continually shifting field (Jonathan Crary)[2]

The relevance of postcolonial studies to the Americas has been widely debated, both within and outside geography. For some, its theoretical approaches have offered useful concepts that help elucidate the operations of power, historical narratives, forms of subaltern knowledge and consciousness, and national modernities, that emerged in the region in the wake of European exploration and settlement.[3] For others, its concepts are too rooted in colonial experiences elsewhere – India and South Africa, for instance – to have much more than general use in the Western Hemisphere. There have been other reasons for reticence too. In Latin America, Spanish and Portuguese control was markedly uneven, and ended early, such that the present seems much less directly shaped by past colonial projects than other post-Independence events. In the United States, colonialism has often seemed irrelevant, since it is widely considered a solely European matter. The United States may have become an imperial power in the twentieth century, but, the argument goes, it has never sought to rule foreign lands in a manner similar to European powers in the nineteenth and early twentieth centuries. Further, the applicability of postcolonial theory *internally* has been occluded through ideologies of manifest destiny and the naturalization of the nation in the ongoing drama of the frontier, narrated as the nation forging its unique character in the encounter with, and overcoming of, natural forces, of which 'Indians' were merely a part. That this 'part' has never disappeared, and that the United States found itself in the position of governing other nations and peoples within the territory it claimed as its own, is often separated out from the story of its founding and removed from questions of 'domestic' politics. This is further elided today through the rhetoric of multiculturalism, and pushed to the margins of academic work by the appeal to other frameworks of analysis: diaspora studies, transnationalism, ethnic studies. Within the Americas, it is perhaps in Canada, where the sovereign territoriality of the state has been consistently placed in question, that the relevance of postcolonial theory has been most vigorously asserted, often through direct comparison with Australia and New Zealand.

Even in such a brief survey, it is clear that the significance of colonialism, and the experience of postcoloniality across the Americas, is highly differentiated.[4] Thus, *thinking* about postcoloniality must necessarily attend to spatial and temporal specificity; it must account for the unique configuration of colonialisms and their aftermaths as they occur in particular places, as well as the contingent and open-ended nature of specific postcolonial relations.[5] It is with this sort of specificity in mind that this essay examines postcolonial visual cultures in the Northwest Coast of North America, a region today constituted as Canadian territory. While the focus in what follows rests primarily on the politics of *vision* and *visuality* in relation to struggles over land and environment – and the continued hold of colonial ways of seeing – what is at stake is ultimately the sovereign territorial claims of the Canadian nation-state. That social and political relations in the region have come to be subsumed within a *national* discourse must be understood as a provisional achievement. Nor is this solely the result of political agreements such as the Oregon Treaty of 1846 that divided the Pacific Northwest between two colonial powers – the United States and England – while ignoring the territorial claims of aboriginal peoples.[6] Rather, the continued force of this political geography today – in the face of ongoing and serious challenges by First Nations – is achieved in part through the reiteration of notions of sovereign Canadian territory in cultural construction and practices, such as landscape painting, which can work to consolidate or contest political geographies even as it purports to merely 'represent' peoples and places. As in other white settler societies, the representation of land, nature and environment on the Northwest Coast is politically charged.

This paper traces the intertwining of vision and politics, colonial past and postcolonial present, the nation-state and its disavowals, through the work of two landscape artists who have taken Northwest Coast landscapes as their visual object. While I am concerned foremost with the matter of colonialism's 'afterlife' – the ways in which colonialism remains an ineradicable trace in the social relations and visual cultures of the present – this question is never far removed from the matter of sovereignty, and the practices by which this corner of the Americas has been assigned its political identity. It is perhaps a first step forward thinking North American geographies 'postnationally'.[7]

Carr/Yuxweluptun: vision, land and politics

In 1997 the Vancouver Art Gallery (VAG) hosted a small exhibition of two landscape painters: Emily Carr and Laurence Paul Yuxweluptun. Carr (1871–1945), the daughter of English immigrants, had long been an iconic figure in Canadian culture, revered throughout the country for her images of West Coast Indians and rainforests (see Figures 1 and 2). Yuxweluptun (b. 1957), a Coast Salish artist, gained prominence in the 1990s for his striking re-visioning of West Coast natural and cultural landscapes (see Figure 3). Although a modest exhibition it raised important political and theoretical questions about vision and visuality 'after colonialism' on the Northwest Coast.[8] On the one hand it

FIGURE 1 Emily Carr, *Blunden Harbour* (circa 1930). Oil on canvas, No. 4285, National Gallery of Canada, Ottawa; purchased in 1937. Reprinted with permission of the National Gallery of Canada

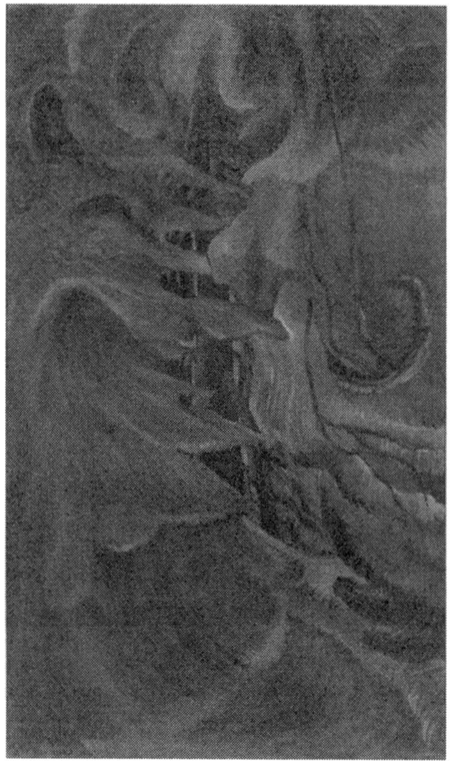

FIGURE 2 Emily Carr, *Cedar* (1942). Oil on canvas, Vancouver Art Gallery, Emily Carr Trust, VAG 42.3.28. Photograph by Trevor Mills. Reprinted with permission of the Vancouver Art Gallery

questioned the central place accorded Carr's work in Canadian culture and challenged assumptions about its political neutrality. The decision to place Yuxweluptun's *Thou Shall Not Steal* (three painted wooden panels emblazoned with the words NATIVE LAND CLAIMS) amid Carr's famous images of Native villages and verdant rainforests provocatively identified the visual arts – and not only politics and law – as complicit with colonial power. On the other hand, the exhibit raised anew a set of questions about the incorporation of Native art within Canadian cultural institutions. The show was one among many exhibits of contemporary First Nations' artists staged in Canada during the 1990s, a decade in which the 'indigene' became fashionable again, especially if indigeneity was articulated in a way that brought European and non-European aesthetic traditions and cultural identities into tension, displaying hybrid forms that troubled distinctions between Europe and its others, or between the categories of art and artifact. Yuxweluptun's work – with its self-conscious mixing of Northwest Coast native motifs and European modernisms – fitted this slot better than most, and was added to the National

FIGURE 3 Lawrence Paul Yuxweluptun, *Scorched Earth, Clear-cut Logging on Native Sovereign Lands, Shaman Coming to Fix* (1991). Acrylic on canvas; no. 36950, National Gallery of Canada, Ottawa. Reproduced with permission of artist

Gallery's permanent collection in the first half of the decade.[9]

Such incorporations are double-edged – a sign of increased visibility for aboriginal peoples, but containing echoes of earlier incorporations, especially in the 1920s and 1930s, when native peoples and artifacts were romanticized and incorporated as the nation's 'heritage', even as the dispossession of Native lands continued unabated.[10] It would be overly cynical, though, to dismiss these efforts entirely. Today, Native artists are often involved in planning the exhibits (or at least writing commentaries for them), and the staging of Native cultural production as integral components of the contemporary 'multicultural' nation – although frequently in ways that reaffirm the primacy of the nation – is a marked departure from earlier practices that positioned aboriginal peoples as residual elements of the nation's pre-history. Indeed, as we will see in the case of Yuxweluptun, incorporations of the modern indigene within the artistic canons of white settler societies are decidedly ambivalent, since they disturb the narratives and sovereign claims of the singular nation-state, even as the indigene is brought within its institutional frames.[11]

As important as these issues are, this paper limits its focus to an analysis of competing constructions of Northwest Coast landscapes at a time in which vision and visuality are politically charged. Fierce struggles over land and environment have engulfed the region in the past 20 years. On the one hand, First Nations have sought to regain a measure of sovereign control over lands they have never ceded to the Canadian state.[12] Incarcerated within a system of reserves in the late nineteenth century, and subject to the paternalism of the federal government throughout the twentieth, Native groups have pursued land rights across a range of arenas – from constitutional conferences and legal challenges, to cultural productions and international media campaigns. On the other hand, local, national and international environmental groups, at times in concert with First Nations, have fought to reduce, even halt, the further extension of industrial forestry, which frequently occurs on lands claimed by First Nations as their traditional territories.[13] In such a context, W.J.T. Mitchell's assertion that landscape and power are simultaneous takes on added significance.[14] What cultural work is done by images of land and environment? How do representational practices lend support to, or authorize, particular social, political, ecological and national projects? What histories of seeing have informed the environmental imaginaries of Canadians, and how have they been questioned, contested or reconfigured? How have different visual practices framed the Northwest Coast as a landscape of fantasy and desire, and with what effects?

Placed alongside these struggles over land, resources and environment, matters of vision and visuality can be seen to be paramount rather than marginal, part of the way that a particular Northwest Coast postcoloniality has been set in place and made to evolve. The work of Carr and Yuxweluptun do more than provide a valuable window onto these matters, they are *constitutive* elements of the region's social, spatial and political relations. Both take as their 'object' the region's temperate rainforests and Native communities. Both are also immensely popular. This is especially true of Carr, who almost 60 years after her death remains a central figure in Canadian culture. Her art is displayed routinely in the nation's galleries and museums. It is reproduced endlessly on postcards and calendars. Books of nature photography explicitly try to emulate her vision. Poets find inspiration in her struggles to 'capture' the coast, while playwrights bring her life to the stage. Posters of Carr's work hang on the office walls of environmental organizations, realtors and travel agents. The only permanent gallery in the VAG, located at the center of its majestic building, is devoted to Carr's works. All other exhibits are temporary, situated around the periphery of the gallery's space. Even scholarly books on British Columbia by historians and economists with no stated interest in aesthetics or landscape place Carr's work on their covers, usually without additional comment. For many Canadians, Carr *is* BC, her ubiquitous presence needs no explanation, no supplementary text.

Borrowing from Homi Bhabha, I will suggest that Carr's work plays an exemplary pedagogical role in Canadian culture, educating Canadians on how properly to 'see' Northwest Coast landscapes, and how to see the them as 'properly' Canadian.[15] This leads to a series of additional questions. What does it tell us about the 'postcolonial' that Carr's work is so widely disseminated today, almost 60 years after her death? What does this suggest about colonialism and its (multiple) temporalities? Her continued popularity raises other questions: what did Carr 'see' when she traveled the inlets and passages of the coast in the early decades of the twentieth century? What social, spatial, political and ideological relations enabled and limited her gaze? And, crucially, how are these relations carried into the present? Here Carr's popularity must itself be subject to inquiry. How do we account for the metonymic relation between 'Carr' and the 'West Coast' within Canadian culture? How did *her* vision – and not others – come to stand in for the region as a whole? It is crucial to note that the VAG exhibit – with its juxtaposition of Carr and Yuxweluptun – would never have occurred had Carr merely become a footnote in Canadian art history. To fully understand postcolonial visual cultures on the Northwest Coast, then, we must ask how Carr's work came to occupy the position it does – a story that is as important as the story of her art.

Although in what follows I focus on writing historical-geographies of the present, and thus focus primarily on Carr, I return to Yuxweluptun's work at the end in order to ask how it compares with or disrupts Carr's enframing of Northwest Coast landscapes. What is the significance of his self-styled 'salvation art'? What does it tell us about postcolonial visual cultures, their shifting aesthetic and political coordinates, and their implication in questions of land, nature and nation? I also ask what it means to *juxtapose* Yuxweluptun and Carr. As I discuss later, it is tempting to understand Yuxweluptun in terms of difference, and subsequently imagine his work speaking into, and disrupting colonialist visual regimes from a place *external* to colonial power. Indeed, this was implied to some extent in the VAG exhibit, which despite the stated intentions of the curator, positioned Yuxweluptun as an oppositional artist. But does the language of 'oppositionality' or 'alterity' fit when political, cultural and visual fields are not only multiple and fractured, but folded together? Or, do we perhaps need to imagine the 'anti-colonial' differently?

'Artist of the Canadian West Coast': framing Carr

Emily Carr, Stephanie Walker insists, is intrinsic to the Canadian imagination.[16] It does not follow, however, that her art alone explains her appeal. National artists are made, not born. This simple statement reminds us that we need to look elsewhere than the artist and her work in order to understand the popularity of Carr in the present, and her close association with the Northwest Coast.

Carr's central place in Canadian culture is arguably as much a historical accident as a product of artistic excellence, due in part to her work fitting well into an ideological slot available within an anxious Canadian nationalism in the years immediately preceding the Second World War. Although she had already painted for more than 30 years, Carr was relatively unknown until the late 1920s and 1930s when she was 'discovered' by eastern critics, and came to be discussed by critics as a Western Canadian counterpart to the Group of Seven, a loosely organized circle of artists in Eastern Canada. These eastern artists were celebrated for forging a Canadian School of landscape painting at a time when nation-states were seeking to establish unique cultural traditions.[17] In Canada, this meant finding cultural forms that were different from those in Europe and the United States. What was thought to set the Group of Seven apart from more derivative painters was their immersion in the landscape. From this they were said to obtain their inspiration and their uniquely *Canadian* style.[18]

From 1927 onwards, Carr's work was viewed in similar terms, although perhaps more than any Group of Seven painter, her work came to be associated with a specific region. Even before her death in 1945, she was described as the artist who had the most intense and open response to the Northwest Coast, capturing its unique character better than any of her peers, and distilling its landscapes into two defining elements: a monumental, albeit vanishing, native culture, and a powerful, primeval nature. As Gerta Moray has shown, the broad outlines of this view were set in place by critics and art historians like Mortimer Lamb and Ira Dilworth in the 1930s and 1940s.[19] But it is perhaps Doris Shadbolt who did the most to reiterate and thus stabilize this perception of Carr in the present. Her influential 1979 book-length study and catalogue of Carr's work, for instance, begins with a simple heading: 'Artist of the Canadian West Coast'.[20] The double meaning of the 'of' was not, I suspect, entirely accidental (nor the adjective 'Canadian'). In Shadbolt's eyes, Carr was at once an artist who *portrayed* West Coast landscapes, and an artist who was *of* the West Coast. For Shadbolt the two had a necessary relation: one could truly be 'an artist of the West Coast' only if one was from the coast and had an intimate attachment to it. Immediately following this heading a short paragraph established what is now a widely accepted understanding of Carr and her art.

> Emily Carr was born in Victoria on Vancouver Island 13 December 1871, died there 2 March 1945, and lived most of her life within a few blocks of the house where she was born in the James Bay district of that city. Her genius throve in the island's isolation from mainland British Columbia and in the province's isolation from Canada and the world. The two great themes of her work derived from the most characteristic features of that region – a unique and vanishing Indian culture, and a powerful coastal nature. It is logical to think of Carr and the Canadian West Coast at the same time, for her paintings and her writings bear the indelible imprint of her long attachments to the place where she was born and where she chose to remain.[21]

Here 'seeing' is conjoined with 'proximity' to form a powerful rhetoric of unmediated vision. For Shadbolt, Carr's art was organic, a product of a long and sustained encounter with Northwest Coast landscapes. Because of her intimate attachment, these landscapes left their 'indelible imprint'; if one remained *in place* one's vision would remain unpolluted by outside influences – the filters of culture and ideology would fall away, and things would become present in their true form.

Today, notions of a pure space outside global circuits of capital, commodities, ideas and information seem rather quaint, even naïve.[22] So, also, do suggestions that vision is unmediated.[23] Indeed, it now goes without saying that critical analysis must interrogate the *situatedness* of vision in order to draw out its enabling conditions.[24] Accordingly, some of Carr's recent biographers have placed much greater emphasis on the social, psychological and political conditions that may have shaped her life and influenced her work.[25] It is now well known, for instance, that despite living much of her life in Victoria, Carr was thoroughly enmeshed in a web of extra-local social, political and aesthetic relations that linked her 'isolated' world in Victoria with artists in Eastern Canada, New York, Paris and elsewhere (even before her so-called discovery by eastern critics). She was in close contact throughout her career – through books, friends and numerous trips – with developments in Western culture and aesthetics, spending many years studying abroad (San Francisco 1890–93; London 1899–1903; Paris 1910–12). Downplayed in

Shadbolt's rhetoric of unmediated vision were her acquaintances with ethnologists, her contacts with members of the Group of Seven, her appreciation of Fauvism, Expressionism and Cubism among other streams of post-impressionist modernisms, and her trips to places like Toronto and New York, where she had contact with such prominent artists as Lawren Harris and even Georgia O'Keefe. Nor does such rhetoric allow for a full acknowledgement of what it meant that she read Walt Whitman religiously, or that for a period in the early 1930s she struggled to come to terms with, and incorporate into her art, the spiritual vision of theosophy.

Accounts that assume that Carr captured the essential truths of Northwest Coast landscapes turn on notions of unmediated vision, yet Carr's art (and journals) show a wealth of references to the work and writings of others. Her work was decidedly *intertextual*, in the sense that it obtained its meaning through its relation to other works. This intertexuality troubles Shadbolt's image of Carr as an organic artist, since it places her work in a complex web of cultural relations, and throws into question the immediacy of Carr's 'seeing'. This was evident even to Shadbolt, who at points interrupts her narrative of 'isolation' to acknowledge Carr's extra-local connections. Yet, in the last instance she sustains her reading of the purity of Carr's vision by appealing to a well-worn discourse of art as an expression of an inner journey.

> Emily Carr was an individual who did not sit comfortably in close company. Like any artist, she drew on whatever art sources were available to her, but she was never truly in step with any group, movement, or trend. She was inspired by the strong fresh look and the vigorous nationalist sentiment of the Group of Seven. She did not, however, adopt their painting conventions and she was fundamentally different from them in the degree to which she finally made her work the vehicle of feeling, and in the strong mystical tendency she shared with [Lawren] Harris alone . . . Because of the quality of Carr's painting and its originality . . . affinities are bound to suggest themselves and to hint at a larger relationship with other artists whose strength, like hers, grew out of isolation, particularly an *inner isolation*.[26]

Shadbolt was far from the first to describe Carr in these terms. In 1939 Eric Newton opined that Carr 'belonged to no school' and that her inspiration was 'derived from within herself'.[27] Nor did this assumption of an unmediate seeing end with Shadbolt's 1979 study; a recent 1990 National Gallery exhibit of Carr's work repeated this message, and the VAG continues to encourage visitors to understand Carr in this way.

If Carr is intrinsic to the Canadian imagination, we are now in a position to suggest why: in part it is a result of a national 'longing for form' that was particularly acute in mid-twentieth-century Canadian culture.[28] It is also in part a result of an art historical discourse that carefully policed the boundaries of art and politics, finding in the artist a unique response to landscape that was derived from 'inside', a result of 'getting clear' of politics and ideology, rather than enabled by both. Thus, not only was Carr made to be a representative of the Northwest Coast, she was seen to express its very essence.

The story of Carr's reception needs to be further complicated, however, for the rhetoric of unmediated vision is commonly accompanied by a 'developmental' narrative. Curators, critics and art historians have regularly framed Carr's life in terms of a progressive chronological and artistic development, proceeding from her 'early accomplishments' to her 'mature years' (those that followed after her trip to Eastern

Canada in 1927). Typically her career is divided into three stages: an initial documentary period (1907–13) during which Carr produced over 200 sketches and paintings of West Coast totem poles and Native villages; a second period of relative inactivity (1914–27) following her unsuccessful attempt to sell these sketches and paintings to the newly established Provincial Museum;[29] and a third so-called mature period (1928–45) during which Carr is said to have discovered a 'deeper vision' that took her to the essential spirituality and formal aesthetic properties found in Native artifacts and West Coast landscapes. As explained by Shadbolt,

> It was not until she was in her late fifties that she came to understand the Indian's carvings as the expression of his experience in a primordial environment and was led in turn to a bolder perception of the forests and coastal landscape. When she finally was able to identify the primal energy she found in nature with the spiritual energy she was seeking as the manifestation of God, the elements in her art became integrated at a higher level . . . At this point her art reached its full range of effectiveness.[30]

This developmental story has effectively devalued work Carr did before the 1920s (when she was in her mid-50s), and elevated work she produced in the last decades of her life.[31] In Shadbolt's 1979 biography and catalogue, for instance, fully four-fifths of the text and plates focus on the period after 1927, even though Carr had previously produced hundreds of canvasses and sketches. Likewise, since only work from the latter period is cloaked with the mantle of artistic excellence, it is paintings like *Blunden Harbour* (Figure 1) and *Cedar* (Figure 2) that have come to represent the most sophisticated expressions of 'BC seeing'.

Revisiting Carr

This developmental narrative is far from innocent, for it shapes the archive of Northwest Coast images that Canadians have ready-at-hand. Carr may be closely associated with the Northwest Coast, but only certain of her works circulate widely, while others are rarely seen. To return to an earlier question, this partly determines *how* the past circulates in the present, giving shape to what Stuart Hall has described as the uneven and differentiated temporal rhythms of postcoloniality.[32]

This uneven reception may be important in other ways too. Gerta Moray has argued that in her early years Carr brought to bear on BC landscapes a decidedly anti-colonial optic, one that was much diminished in her more popular later work.[33] This argument merits further examination, not only because it forces us to consider the implications of 'dividing' Carr, but because it necessarily leads us to Carr's work, and the manner in which she framed BC landscapes.

Appropriations or interventions? Debating Carr's legacy

Moray's argument was in part a response to a brief yet sharp critique of Carr's work written by the Haida/Tsimshian art critic Marcia Crosby in the early 1990s.[34] In her paper, Crosby examined Carr's work alongside a number of other nonaboriginal writers and artists who had 'appreciated and supported indigenous art forms and culture at a time when no one else did'.[35] Far from celebrating this, however, Crosby showed its paradoxical nature, writing that such work 'ha[d] more to do with the observers' own values' than with the situation of First Nations peoples, or the changing character of First Nations sites visited and depicted.[36] In Crosby's view – drawing heavily on Robert Berkhofer and Edward Said – the 'Indians' that Euro-Canadian artists such as Carr sought to represent were always, and could only ever be, imaginary Indians since these artists were never able to step outside their own cultural frames.[37]

In Carr's case, Crosby argued that her writings and paintings were notable for their produced authenticity, portraying only a pre-contact or uncontaminated native life. While Carr paid tribute to the Indians she 'loved' in her journals and short stories, Crosby suggested that they were nostalgic figures rather than the people who took her in gasoline-powered boats to the abandoned villages that she painted. She finds the same to be true of Carr's paintings and sketches, which 'intimate that the authentic Indians who made them existed only in the past, and that all the changes that occurred afterwards provide evidence of racial contamination, and cultural and moral deterioration'.[38] Such portrayals, Crosby argued, devalued *present-day* native cultures in the guise of celebrating their past.

It is worth noting that Crosby's interventions occurred concurrent with two important political and intellectual developments: the (re)assertion of First Nations rights (political and territorial) in the 1980s and 1990s; and the 'culture wars' that brought into view the relation between representation and power and issues surrounding cultural appropriation.[39] Writing into these controversies, Crosby sought to map the connections between cultural production and (neo)colonialism. By showing how apparently sympathetic portrayals of native peoples were not free from colonialist power/knowledge relations, Crosby radically undermined claims for a disinterested literary or artistic expression, or, for that matter, for the possibility of a 'love' for Indians independent from the cultural discourses that shaped how Native peoples could be viewed and represented. For Crosby (in contrast to Shadbolt) representation was always already implicated in cultural and political dynamics that exceeded the stated intentions of artists. To represent Natives was therefore to enclose them within particular frames of reference, which claimed Natives aesthetically, and, by extension, politically, for the nation-state.[40]

Moray, on the other hand, cautions that determining whether an artist's work was colonialist is actually far more complicated. According to Moray, critics such as Crosby make two errors. First, they reduce *political consciousness* to *social location*, and therefore fail to situate artistic work – and its reception – with the fragmented (post)colonial cultural and political fields within which it is produced and received. To support her argument, Moray provides a careful reading of the social worlds of early twentieth century BC, suggesting not only that settler society was *heterogeneous* in terms of ethnic, class and religious differences, but also in terms of competing political opinions. Within this context, Moray argues, Carr's paintings cannot be reduced to a singular and seamless colonial discourse. This is especially true, she continues, in the case of Carr's *early* years before her discovery by eastern critics, when her effort to document native villages may have deliberately challenged the racist views held by the

FIGURE 4 Emily Carr, ***Memalilaqua, Knight Inlet*** (circa 1912). Oil on canvas, no. 5043, National Gallery of Canada, Ottawa; purchased in 1950. Reprinted with permission of the National Gallery of Canada

British gentry, missionaries and state administrators (see Figure 4). By this view, the second error of Carr's critics is that they accept unquestioned the image of Carr provided by art critics and historians, who, swayed by the developmental narrative, have largely overlooked this earlier work.

It is worth noting that the differences between Crosby and Moray are symptomatic of a shift in the early 1990s from the study of colonial discourse (often presented in Said's totalizing terms) to an analysis of colonialism as a fractured and contested field in which the operation of power is found to be more ambivalent and uncertain.[41] Moray, for instance, places great stress on the existence of *multiple* and *overlapping* social, political and discursive fields, and notes the presence of vastly different ideological positions regarding the 'Indian question' during Carr's lifetime, explaining that these could be held simultaneously and in contradictory ways by individuals.[42] Significant contrasts could be found in attitudes towards 'Indians' held by white settlers, Anglican and Catholic missionaries, Hudson Bay Company employees, and the British gentry who dominated political and economic activity in the province.[43] Carr moved within and between these groups, and was keenly aware of these differences. It is well known that she preferred to see herself as 'colonial' or 'Canadian' rather than as part of the British gentry to which her family belonged, and her writings often harshly criticized the efforts of missionaries, even as she relied heavily on them for her travels and contacts. Likewise, Carr strained against the protocols of Victorian femininity. Her career as a painter departed significantly from other white female artists who lived and worked in the region but who remained more closely bound to notions of the 'picturesque' and to subject material drawn from gardens, or town life.[44]

This fractured field suggests to Moray the possibility that oppositional positions were available *within* non-Native society. As Moray puts it, since 'images are generated within a *field* of political and psychological interests, not all such imagery is equally closed, unsympathetic or reductive'.[45] Rather than write totalizing critiques based on identity, she argues, Carr's work must be studied for its 'local levels of engagement and meaning' and evaluated in terms of 'the local circumstances of her work's making'.[46] Moray's point has considerable merit. Many of Carr's contemporaries considered native peoples to be culturally inferior, and destined to wither away in the presence of a far superior European civilization. In important ways Carr's work contested this narrative, emphasizing the *vigor* and *scale* of native life, and evaluating settler society in far less favorable terms. On this basis Moray concludes that Carr's early paintings not only ventured into subject matter that most women painters (and many men too) were either unwilling or unable to depict, but were also intended to stand as a 'vindication' of Native traditions against accounts by missionaries and administrators who condemned what they saw as the Indian's dark, heathen past. 'Through her choice of images, her style of representation and her textual exploration of her works', Moray writes, 'Carr took quite specific issue with certain views of native culture being propagated at the time by missionaries, government agents, anthropologists and the press'.[47]

I wish to retain Moray's insight that colonial cultures in early twentieth-century British Columbia were heterogeneous, and thus that Carr's work would have allowed for interpretations that departed from the assumptions of much of the British gentry. Yet, I think it is important to question recent efforts to defend Carr from critics such as Crosby. Invariably, Carr's defenders have appealed to her *writings* in order to support their assertion that Carr was sympathetic to the predicament of Native peoples. Douglas Cole, for instance, turns to Carr's autobiographical *Klee Wyck*, in which she described many of her travels to Native communities. Here he finds a portrait 'of a people living in a transient present, caught in conflicts between old and new, between traditional and colonized modern, who are making their way as best they can'.[48] Carr's reflections, Cole writes, were of everyday life, not an imaginary authenticity. Likewise, despite her insightful comments about not all images being equally open or closed, Moray frequently evaluates Carr's work in terms of statements she made *about* her painting, or about the cultures she portrayed. In journals published posthumously, Carr explained that her project in these early years had been to 'picture totem poles in their own village settings, as complete a collection as I could'.[49] This statement, along with comments from a 1913 lecture in which she discussed the 'mythology' of a number of the poles she painted, and other comments disparaging missionaries (among others) for their attempts to disrupt Native life, have been invoked as evidence that Carr sought to vindicate native traditions in the face of settler and missionary hostility.

I have no doubt that Carr was sympathetic to the situation of Northwest Coast Natives. Where these efforts go awry, however, is in their attempt to interpret Carr's *visual* work through her *written* work. This is to make two related mistakes. First, it recuperates a notion of intentionality where Carr's statements about her work (often written decades later) are interpreted as the truth about her painting. Secondly, it posits Carr as a *unified* subject, whereby her writings and her paintings are interpreted as the product of a single and singular vision. While intuitively this seems reasonable (if she *wrote* sympathetically of native life, why would she not have done the same in her paintings?), it fails to recognize that writing and painting are different genres governed by very different codes. In their own ways, Crosby, Moray and Cole each fail to recognize that Carr the writer, and Carr the visual artist, need not be the same. Here we need to introduce different questions. Rather than turning to Carr's statements about her art, we must ask what it meant to 'observe' Indians on Canada's West Coast in the first decades of the twentieth century. What shaped 'seeing' on the coast? What did it mean to produce a *visual record* of Indians and, more to the point, of their totem poles? Why these objects and not others? To what extent could recording native villages be considered an anti-colonial gesture?

Observing natives

To answer these questions is to take seriously Moray's call to attend to the 'local circumstances' of Carr's art. Between 1907 and 1912 Carr made several trips along the coast, documenting totem poles and village sites, eventually producing over 200 canvases (see Figures 5 and 6). Moray argues that this work was framed by scientific and political interests more than aesthetic ones, setting it apart from her later paintings (precisely why Shadbolt disparages them as inferior). By comparing Carr's paintings with photographs of the villages taken during the same period, for instance, Moray demonstrates how Carr strove for accuracy and detail. From points outside each village, Carr painted panoramas designed to place each pole in relation to the others (Figure 5), and in relation to the village as a whole. From vantage points inside the villages, Carr painted detailed studies of individual poles (Figure 6). In short, Carr produced a detailed *geography* of all of each village's totem poles, and an equally detailed *iconography* of each pole.

There is some merit in Moray's argument that this work contested prevailing images of Native peoples. Carr's exhaustive record would have reminded white settlers that the coast was never an empty land without prior residents.[50] Further, Carr's direct engagement with 'Indian' themes diverged from other contemporary female artists who, by avoiding 'Indian' content altogether, merely reinforced white settler society as

FIGURE 5 Emily Carr, *Tanoo* (1913). Oil on canvas. Province of British Columbia Archives, PDP02145. Reprinted with permission of BC Archives

normative. There is also some basis for arguments that a political impulse in this early work may have been blunted over time. Carr's pre- and post-1927 work diverged in important ways. In comparison with her later work, Carr's documentary project gave far more attention to the *details* of individual poles and their village contexts (compare Figure 1 with Figure 6). People were also often present (although limited in terms of how they appeared – usually only in the guise of 'traditional' culture, see Figure 4). While Carr painted similar scenes after 1927, they became increasingly stylized in accordance with modernist conventions, most notably expressionism and cubism, and also in terms of the primitivism that pervaded both.

While it is important to remember that neither colonial culture nor visuality in early twentieth century BC was singular, it is also important to investigate what it meant to be an *observer* of peoples and landscapes on the West Coast in the first decades of the twentieth century. How was the desire to 'see' channeled along certain lines? What social, technological, discursive and institutional relations shaped the visual fields through which Carr moved? To document – with meticulous attention to 'detail' and 'accuracy' – is, after all, not in itself a progressive politics; it is equally as important to ask why some things – such as totem poles – showed up as objects of interest and not others. Carr's vision may have been more conventional than her defenders allow. Further, the language of rupture used to separate an 'early' from 'later' period may be as problematic when used to identify an 'anti-colonial' moment in Carr's work as it is when used to establish a

FIGURE 6 Emily Carr, *Skidegate* (1912). Oil on card on board, Vancouver Art Gallery, Emily Carr Trust, VAG 42.3.46. Photograph by Trevor Mills. Reprinted with permission of the Vancouver Art Gallery

discourse of artistic 'excellence'. Indeed, by attending to the details of Carr's seeing during her early period we may be able to also think more carefully and critically about the later images that circulate so widely today.

What, then, did it mean to be an observer of nature and natives on the Northwest Coast at the beginning of the twentieth century? In *Growing pains*, Carr traced the origins of her documentary project to a trip through the Inside Passage to Alaska that she took with her sister in 1907. Traveling the popular tourist route, Carr would have encountered for the first time many of the native artifacts that were to become her all-encompassing passion. This would lead to additional trips over the next decades.

Typically, these trips have been cited as evidence that Carr, more than any other artist, had an intimate relation to the land and captured the essence of the coast, since they entailed considerable effort and hardship and allegedly took her away from metropolitan cultures to the domain of the objects themselves.[51]

Such accounts studiously ignore the conditions that enabled and shaped possibilities of seeing for white settlers and travelers at the beginning of the twentieth century.[52] If, as Jonathan Crary argues, the observer is an effect of material and discursive practices, rather than an autonomous and intentional actor, it becomes important to ask how not only the objects seen, but seeing itself, has material histories.[53] It is also important to bring the *body* of the observer into view, since vision must always be anchored in material practices. In his brilliant *Techniques of the observer*, for instance, Crary documents the disciplining of modern observers in the nineteenth century, and the visual technologies and scientific practices that were central to this.[54] Crary's central insight that seeing is an effect of heterogenous relations can be extended to the Northwest Coast at the last *fin-de-siècle* in order to consider the technological, social, discursive and spatial relations within which subjects were constituted as observers. Crucially, any account of vision and visuality in BC in this period must take on the question of movement. Here we might follow Nicholas Green, who in his superb study of nineteenth-century Paris traces the role new transportation technologies played in the reconfiguration of both vision and landscape in the city and its environs.[55] The increased appeal to nature and the search for the picturesque, he explains, was not merely a reaction to industrial modernity, but enabled by new forms of mobility that allowed the bourgeoisie to experience city and country as distinct domains. Indeed, Green intimates that this mobility – and constructions of 'city' and 'country' that they allowed – was not merely something available to the bourgeois subject, but constitutive of emerging bourgeois subjectivities.

Similar dynamics were present in British Columbia during the first half of Carr's life. Cole Harris has sketched the outlines of a system of transportation and communication that emerged in the province during the first decades of the twentieth century.[56] Drawing on the Canadian economic geographer Harold Innis, as well as Gilles Deleuze and Felix Guattari, he links these emerging networks with the spatial extension of imperial and colonial power.[57] His point is simple yet critical: technological and spatial transformations at the time were not culturally or politically innocent; they *repositioned* people and things. For instance, the time–space compression that came with these networks allowed for capital to colonize space in British Columbia in new ways. From wagon-roads to railroads, postal service to telegraphs, paths opened for people, capital and commodities that were not previously available, enabling new social, economic and cultural relations and drawing formerly distant places together. In short, these transformations reterritorialized the coast and ushered in new spaces of identity, culture and politics.

It is worth pausing here to emphasize a key point: there is not one history of 'space', another of 'society', a third of 'technology' and yet another of 'vision'; these must be thought together. This insight has considerable consequence for how we understand Carr's 'BC seeing'. To take only one example of this intertwining of material and visual cultures, in the last decades of the nineteenth century steamship service radically reordered both the geographies and the optics of social life in the region, a point made explicit by one of the many companies offering tours along the coast (Figure 7). Service began on the South Coast as early as 1836, and in the North by the 1860s, but the volume of travel along coastal routes accelerated most rapidly during and after the 1880s.[58] By then the regions outside the orbit of Vancouver and Victoria increasingly found themselves within the economic, political and cultural influence of these cities, as well as connected to East Coast centers of culture and commerce. This had a number of important effects. To begin, it helped produce the 'West Coast' as a distinct Canadian region, both in terms of a set of material relations, and also in terms of emerging identities and imaginative geographies. As well, increasing linkages along the coast had the paradoxical effect of strengthening distinctions *between* the city and its surrounding country, even as both were increasingly coupled within a unified space-economy. To say this differently, in the years before Carr's trips the 'countryside' of the Northwest Coast had become an important cultural and visual resource for emerging metropolitan cultures, and not only an economic one.[59] For bourgeois travelers the Inside Passage had become an important rite of passage.

FIGURE 7 Cover of Grafton's Tours guide to Alaska and the Inside Passage. Source: Grafton (1894)

The extension of regular and rapid service along the coast altered Native–White relations too. In earlier periods travel by Europeans had been dependent on establishing close, if unstable, relations with Natives. By the early 1900s, with a modernizing system of transportation and the spatial incarceration of Natives within a system of reserves, the movement of Europeans was increasingly distanced from direct encounters with Native peoples. No longer as reliant upon their knowledge and labor, the emerging space economies of the late nineteenth century on the coast often by-passed their villages. As Harris explains, 'Improvement in transportation and communication enabled the world economy to use BC's space not through Native intermediaries, as during the fur trade, but by distributing Western technologies, labour and settlers across the land'.[60] This allowed Native villages and people to be viewed in a new light. No longer essential to networks of social and economic exchange within which the recognition of coevality was of paramount importance, 'Indian' villages could now be viewed as anachronistic elements of a landscape that was passively viewed without direct contact.[61]

This brings us to a key point: the very way that Euro-Canadian travelers *looked* at landscape underwent significant transformations during the period. In his classic study of travel in the nineteenth century, Wolfgang Schivelbusch argued that new transportation technologies gave rise to 'panoramic travel'.[62] Before this period travelers belonged to the same space as perceived objects (what he calls the space of 'landscape'). New transportation technologies, however, resulted in travelers seeing objects through the apparatus that moved them through the world ('geographical' space). Modern transportation technologies, he wrote,

> . . . choreographed the landscape. The motion . . . shrank space, and thus displayed in immediate succession objects and pieces of scenery that in their original spatiality belonged to separate realms. The traveler . . . acquired a novel ability . . . to perceive the discrete . . . indiscriminately.[63]

This involved a degree of detachment. Objects were now 'viewed', and viewed from a distance. It also reorganized the hierarchy of senses. The relation of the traveler to the landscape was now predominantly visual, a development accentuated by an array of popular and widely available technological devices such as binoculars and cameras that accented the visual appropriation of sights.

Although much slower than trains, steamship service at the turn of the century also remade the coast as a succession of sites/sights. Travel along the coast was now self-contained in scenery-machines that moved along the region's passages and inlets. Detached from local intermediaries, and insulated from the hardship of travel, movement was rationalized into a regular itinerary of fixed stops and predictable arrivals. From the comfort of the ship deck one could view nature – and Natives – as a passing spectacle. The effect of such distancing was to evacuate specific meanings assigned to places and incorporate them instead into a mythological space that had little to do with the lived details of particular sites and much more to do with discourses of European modernity and national history. It is not insignificant that during this period Native villages were far more frequently appraised *aesthetically* than were non-Native settlements (which did not give themselves to such detachment) and much more prone to being situated *outside* history in contrast with white communities in which history was thought to unfold.[64] By the 1870s the route to Alaska had grown rapidly in popularity among tourists, many from the United States.[65] In 1890 alone, almost two decades before Carr headed north, over 5000 excursionists traveled the route. So extensive was the tourist trade that the itineraries of steamships were altered in order that the most spectacular scenery would be passed during daylight hours.[66]

It is important to situate Carr's movements within these dynamics. At the time she and her sister explored the region as many as 38 steamships were operating along the coast, with departures from Vancouver occurring almost daily. This was far from uncharted territory; it was equally far from an unmediated seeing. Indeed, measured on a per capita basis, travel along the coast by residents of Vancouver and Victoria was likely far more common in Carr's time than it is today, and by the 1890s numerous guidebooks and accounts of the Inside Passage were in circulation. These guidebooks outlined important sites/sights, scripting in advance an iconography of landscape that viewers would recognize as they passed through waters that were in important respects already known in advance.[67] As Badlam's *The wonders of Alaska* – already by 1891 in its third edition – explained to readers, they could 'follow these pages and be fully informed of all the principle points of interest along the Inland Sea, with its innumerable islands, the great resources of this wonderful country, its native villages, the grandeur of its scenery, the traditions of the Indians, the success of its mission schools and the extension of its civilization'.[68] Nor did these guides spare detail. Grafton's 1894 guide ran to 87 pages, its cover distilling the Inside Passage into two 'typical' scenes – glaciers and totem poles – anticipating Carr's doubled vision by several decades. The same company explained that its tours offered 'an excellent way in which to journey and see strange lands under auspices which will enable [travelers] to learn and see more than if they should go alone. Where to go and what to see has all been arranged ... [travelers] only have to keep their eyes open and look'.[69]

Clearly, this was a well-disciplined gaze that found certain objects and scenes to be of interest and others not. What travelers saw along the Inside Passage had as much to do with a series of discourses at home, and the organization of the view from on board, than it did with the land and the people themselves. This does not mean that all travelers approached Native communities in the same way (or that painting and 'kodaking' were identical). Carr, for instance, went to great lengths to distance herself from tourists, both rhetorically and physically. In her autobiographical writings she was cruelly disparaging of her fellow travelers, complaining about their fabricated experience and decrying their destructive impact on Native villages. Her description of Hazelton, BC, was typical:

> Hazelton was a tough little mining town. It had three hotels, rough and turbulent as the rivers ... The coming of rail eased travel and gentled Hazelton's hotels. Tourists came. The G.T.P. ran its line close to several of the Indian villages and the tourists looked curiously to see how our Aborigines lived. They did not see real Indian life, the original villages, or the grand old totem poles because, flattered and boastful, the Indians tore down their crude but grandly simple old community houses, built white man's houses – shoddy, cheap frame buildings. They turned the totem poles that had faced the river, welcoming visitors who had come by canoe up Skeena: they turned them to face the railroad by which visiting tourists now came. They loaded garish commercial paint over the mellow sincerely carved old poles till all their meaning and beauty

> were lost under gaudy, bragging show off . . . The Indians hurriedly made baskets and carving too – careless, shoddy things to catch the tourist's eye – which brought in a few dollars but lowered the standard of their handicraft . . . Everything was changed, cheapened.[70]

It is tempting to take these statements as evidence of a counter-vision. Yet, such statements must be approached cautiously. While fault lines like these within cultures of colonialism are important, they do not add up to a position of exteriority. Beyond the fact that Carr's statements were often written long after the events she described (and in which she willingly participated) it is important to remember that criticism of 'mere tourism' is often framed in the very terms that structure the tourist experience in the first place: authenticity. Indeed, criticism of tourism was at the time as common as tourism itself and as W. Jackson Rushing has shown for the United States during the period, Carr's disparaging attitudes were widely shared by other artists who sought to capture the 'real' Indian.'[71]

For the most part Carr traveled the same routes as tourists. During part of the time she worked on her documentary project she was paid by the Grand Trunk Pacific Railway to travel on their trains and steamships.[72] Likewise, her departures were necessarily shaped in relation to these routes. One stop on Carr's initial journey to Alaska is illustrative. The town of Sitka was an important station along the popular route Carr followed in 1907. In the years immediately preceding Carr's trip, a 'totem pole walk' had been established for tourists in the forest behind the town. This consisted of poles that had been removed from Native villages along the coast. Indeed, many of these were as well-traveled as the visitors who came to see them. Prior to their display at Sitka they had been exhibited at the Louisiana Purchase Exposition in St. Louis and then again at the Lewis and Clark Exposition in Portland.[73] The irony – that these poles were 'repatriated' not by local natives, but by the white population of Sitka, and then only to be set up again as objects for tourists – was not lost on Carr. In a later collection of stories Carr expressed profound disdain for this walk.

> When in delicious remembering my mind runs back to Sitka, what does it see? Not the . . . Totem-Pole Walk winding through trees beside a stream named Indian River, named for the pleasing of tourists but really the farthest possible distance from the village of Indian shanties – a walk ornamented at advantageous spots with out-of-setting totem poles, transplanted from the rightful place in front of an Indian chief's house in his home village, poles now loaded with commercial paint to make curiosity for see-it-all tourists, termed by them 'grotesque', 'monstrous', 'heathenish'! . . . Tourists came back from their twenty-minute stopover at Sitka and lectured on Indian totem poles and Indian *un*culture, having seen a few old Indian squaws who spread shawls on the end planks of the wharf and sold curios . . . Civilization cheapened her Indian.[74]

Whether Carr held these opinions at the time of her travels is unclear. We know that Carr proceeded to paint the scene, suggesting either that she had not yet developed her virulent anti-tourist rhetoric, or that she did not see her own activities through the same lens.[75] What I wish to stress is the close affinity between the tourist who travels to Alaska to 'see' the poles (even if through an experience of 'produced authenticity') and the artist who seeks to return these poles to their 'rightful place'. What appears as Carr's negation of the romantic tourist gaze was perhaps merely its most pure form!

Significantly, it was not the attention paid by the tourist to *totem poles* (rather than modern Native culture) that Carr decried, but rather the displacement of these poles from their 'rightful place' and the application of 'commercial paint'.

If this is an anti-colonialism, it is a peculiar one at best. Despite what has been read into Carr's statements, her *visual* interest lay not in the everyday life of villages, but in the cultural artifact, often in ways remarkably similar to the abstraction and commodification of the object that she had experienced – and decried – at Sitka. Conceiving of her work as a documentary project, Carr went to great lengths to be exhaustive and representative, locating the poles in their 'correct' places and faithfully reproducing their detail, but her paintings ignored Native life, much as did state policy at the time. While these paintings *do* depict some of the details of houses and villages, this aspect remains secondary to the documentation of the artifact, much like dioramas were used in the same period to authenticate museum displays.[76] When people are present they are often little more than placeholders in a background scene; we learn almost nothing about their activity. The village setting functions merely to inform the viewer that Carr, in contrast to the tourist, traffics in the 'real'. In short, as much as she hated the 'inauthenticity' of the tourist spectacle, her 'love' of Indians was channeled along remarkably similar lines. Like the tourist, Carr found that these 'awkward objects' had attached to them a 'sort of fascination'.[77]

Salvage anxieties and primitivist fantasies

Tourists were not the only travelers to follow these newly opened paths. By the end of the nineteenth century the Northwest Coast had become the site of intense activity by anthropologists and collectors. In the 1880s Franz Boas was already studying the Kwakiutl settlements that 20 years later would provide Carr with a wealth of visual objects. Numerous others also moved along the coast: members of the Geological Survey of Canada, such as George Dawson; the Krause Expedition for the Bremen Geographical Society; J.A. Jacobsen for the Berlin Museum; Dr Sheldon Jackson in Sitka; the Jessup Expedition for the American Museum of Natural History; collectors working independently for the Smithsonian, or for museums in Princeton, Chicago, Philadelphia and New York; missionaries such as Thomas Crosby and Alfred Hall; local collectors such as C.F. Newcombe; and countless other traders, colonists, artists and photographers.[78] Many of these collectors and artists would have shared boats with Carr, used the same white contacts, and employed the same Native guides and informants.

Here we can locate another important element of Carr's early work. As already seen in her preoccupation with authenticity, Carr's desire to 'picture totem poles in their original setting' resonates with the rhetorics of 'salvage', a point conceded by Gerta Moray and Douglas Cole.[79] There was a sense of urgency to Carr's task. Along most of the coast the practice of erecting poles had declined, and many others showed signs of deterioration and decay. In the two decades prior to her project a large number had been removed and relocated to museums and fairs.[80] Franz Boas claimed that the Jessup North Pacific Expedition alone had between 1897 and 1900 secured 6600 ethnographic artifacts and nearly 2000 specimens of physical anthropology.[81] These salvage anxieties were further inflected by nationalism and regional sentiment. Northwest Coast Indians were often viewed by Canadians as distinctive elements of a national heritage; thus, the desire to preserve 'relics' of Native culture was akin to preserving 'ruins' in Europe. Already in 1892 an official in the BC government had expressed concern over the loss of what he considered the province's heritage.

> With regard to Indian curios, it is much to be regretted that so many are being exported from the Province. The funds at the disposal of the museum are quite inadequate to meet the high prices demanded by the dealers, so that it is to be hoped all who can will assist in securing rare specimens, and that persons visiting the outlying districts will avail themselves of their greater facilities for bartering with the Indians, and add to the, at present, somewhat meager collection now in the museum.[82]

In 1905 the province established what was to become the Royal British Columbia Museum, with the intent of keeping at least part of the region's heritage at home.[83] At the time of Carr's journeys, however, its collection was considered incomplete and hopelessly inferior to the extensive collections of Northwest Coast native artifacts held by museums outside the province.[84]

Carr would later write that in the face of these rapid changes, her project was to record what remained before they too suffered a similar fate. Carr's 'Lecture on totems' delivered at the end of her documentary project clearly lays out these anxieties.

> Soon, soon, the old villages will be a thing of the past. Even in the north it is the same. Settlers are pouring into the new country; silently, gradually the old is slipping away. We hardly know where or how long till we wake to the fact that they used to be and now are not . . . [The] object in making this collection of totem pole pictures has been to deposit there wonderful relics of a passing people in their own original setting. The identical spots where they were carved and placed by the Indians in honour of their chiefs.[85]

Likewise, as she presented her paintings to the public, she framed them in terms of the nation's 'heritage':

> I glory in our wonderful west and I hope to leave behind me some of the relics of its first primitive greatness. These things should be to us Canadians what the ancient Briton's relics are to the English. Only a few more years and they will be gone forever into silent nothingness and I would gather my collection together before they are forever past.[86]

Again, while this provides a view into a prevailing current of colonial culture, it is important not to collapse Carr's painting into her writing but instead to ask what *visual* form such imperial nostalgia took. We can see this by returning to Carr's work, but also by comparing it with that of others. As already noted, Carr's 'record' was focused primarily on artifacts, not people, although as Moray correctly notes, people were more present in these images than in Carr's later work. It is equally important to recognize that Carr's vision was widely shared. Nowhere is this more starkly evident than in the remarkable similarities in viewpoint and subject matter between Carr's painting of a stairway and longhouse in 1912 (Figure 4) and a photograph of the same scene by Edward Curtis taken around 1914 (Figure 8).[87] Curtis is today known widely for his remarkable 20-volume photographic/ethnographic catalogue of the 'North American Indian' in which he provided a visual record of Indian 'types' and tribal 'characteristics', both physical and cultural.[88] There is no evidence that Carr and Curtis ever met, or that either had seen the other's work. But in many ways this is beside the point, for what is remarkable is the degree to which both found the *same* objects 'ready-at-hand' (both materially and ideologically) and the manner in which both apprehended them from almost exactly the *same* viewing point.[89]

Like Curtis, Carr's acquaintance with, and careful attention to, 'Indians' was not centered on the ongoing life and adaptation of a people negotiating a modernizing world in which 'tradition', 'culture' and 'identity' were being continuously reworked (much as they were for Europeans too).[90] It is difficult to see this as anything other than an imperial nostalgia, albeit an ambivalent one. Carr's art may have rendered visible the continuing presence of Native people in the province, but it tended to frame them only as

FIGURE 8 Tenaktak House, Harbledown Island. Photograph by Edward Curtis (1914). Reproduced from the Collections of the Library of Congress

anachronistic relics, and thus may have done as much to reinforce as contest the widely held assumption that 'Indians' had no future. There is little sense of the massive social, economic and technological changes that were occurring along the coast, even if these are registered with some frequency in her writing. Brass bands and gasoline street lights had no place in Carr's imaginative landscapes.[91] Despite her 'long attachment' to the coast a great deal remained outside the bounds of her 'organic' vision. Ironically, Carr's defenders have appealed to the region's modernization to justify the absence of people in Carr's images, since the villages that Carr visited were often abandoned by able-bodied workers seeking employment in the province's growing resource sector.[92] Yet, Carr seemed strangely unable to reflect upon this as an essential element of what it meant to 'be Indian' during her documentary period. At least in her visual work, modernity belonged only to Europeans.

We can now see that Moray's painstaking reconstruction of Carr's viewing sites, by which she establishes the documentary nature of Carr's early work in contrast to her later aestheticizing of these artifacts, actually tells a far different story than the one she intended it to. In her study Moray painstakingly mapped the place from which Carr sketched each pole in order to show the care Carr took to 'accurately' represent each totem in their 'original settings'. As Moray explains: 'these maps, together with an extensive study of ethnographic photographs in museum archives, and of secondary literature on the sites, have been used to reconstruct what Carr would have seen on her visits and her activity on each site'.[93] Yet, what Carr would have seen was not merely a product of her presence (and diligence) but of her construction as an *observer*. Moray simply reproduces Carr's ideological positioning as a detached observer with a narrow focus on 'relics', an observer who remained quite uninterested in capturing the wider cultural and economic geographies of the region. Hence, these maps, and the historical photographs from which they are based, do not attest to an 'objective' scientific interest, and a deliberate anti-colonial political intent, so much as tell us about a visuality grounded firmly in the discursive and spatial organization of the region's postcolonial cultures. Carr's viewing sites were chosen specifically to capture *totems*. Her viewing position, often on beaches leading to or fronting the villages, reproduced the detached gaze of the sovereign European subject surveying the village from afar rather than attending to 'interiors'. As is today common knowledge, Edward Curtis also went to great lengths to 'frame' native scenes (Figure 9), something that required an elaborate act of 'distancing' at the same time as it sought to capture the essence of the culture. Carr's perspective was little different (Figure 10).

Going deeper: from primitive artifacts to primal nature

Most accounts of Carr's life and work posit a break between this early 'documentary' period and her later 'mature' works, as well as a second shift part way through the latter period when the rainforest replaced totem poles as her primary object of interest. By tracing Carr's early journeys I have suggested that there may be as many important continuities as discontinuities.[94] It is not insignificant, for instance, that for the remainder of her career Carr retraced many of the routes of this original journey, although the

FIGURE 9 Reconstructing pre-contact Native societies: Edward Curtis in Blunden Harbour, circa 1915. Photograph by Edmund A. Schwinke. Reprinted with permission of the Burke Museum of Natural History and Culture

specific details of her itinerary changed. Nor it is a minor point that she returned at times to material collected during her documentary project, reworking the canvasses and images in different ways as her artistic sensibilities and aesthetic influences changed. This occurred as early as 1912, while Carr was in France, even before her documentary project was complete.

For Carr's interpreters, her trip to France is usually seen as an omen of future changes.[95] Her work is understood as moving from naturalism to a more individual, subjective vision influenced by various modernisms. This, in turn, is seen to be enhanced

FIGURE 10 Emily Carr painting in Tanoo, 1913. Province of British Columbia Archives, F00254. Reprinted with permission of BC Archives

by her contacts with Group of Seven members after 1927.[96] Shadbolt interprets this as a sign of her artistic development; for Moray it is a deflation of her initial political impulse. But perhaps a different point can be made that departs from both interpretations: that this new aesthetic was less of a departure than commonly thought, and that it had as its precondition the 'artifactual vision' of Carr's early period, in which totem poles and other Native artifacts were already significantly decontextualized within a romantic tourist gaze and the salvage anxieties of European societies. If so, then her later work – the work that today circulates so widely, and which Crosby finds so deeply implicated in a colonialist visuality – may have as many continuities as discontinuities with her early work.

Indeed, given her initial concern with authenticity, it is curious that Carr appears to have had no serious misgivings about moving from a documentary objectivity to a more self-consciously subjective vision, to the point of painting over existing works. This should perhaps come as no surprise. The post-Impressionist movements that emerged in Europe during the first decade of the twentieth century – Expressionism, Fauvism and early forms of Modern Primitivism – shared one key assumption: that the move away from naturalism was in reality a move towards a *less normalized* vision and that this allowed the apprehension of a more fundamental form and spirit. In the first decades of the twentieth century, perhaps more than any other time in the history of Western art, artistic expression was seen as an attempt to express a reality that lay behind mere appearances, a search for informing principles that transcended culture, politics and history altogether.[97] Primitivism, in particular, sought after the elemental.

In other words, Carr's movement away from description – already evident in France – was likely done in the name of a purer vision.[98] Naturalism only reproduced surface to the disregard of depth; modernism, in its various forms, assumed greater depths beneath surface appearances. Carr's growing acquaintance with European modernisms – with their search for more elemental forms and their appeal to primitive sources – may simply have heightened a propensity for displacement and abstraction that was already present in Carr's early work. Only now, rather than seeking to return the tourist object to its original place – a peculiar project on its own – Carr turned to study the artifacts *themselves* for formalized rules of form and composition and as the expression of deeper spiritual forces.

Carr's appropriation of Native artifacts within modernist conventions became more pronounced after her inclusion in the National Gallery's 1927 show. From this point on, Carr would receive considerable acclaim, and would have much more frequent contact with Eastern artists and cultural elites. This in turn would shape her work in important ways. Lawren Harris acquainted Carr with theosophy and with the transcendentalism and vitalism of Walt Whitman. Other Group of Seven painters provided inspiration or competition.[99] After viewing A.Y. Jackson's Skeena River paintings, for instance, Carr began to realize the many ways BC landscapes could be approached through a far more aestheticized and stylized vision, and began to accept her designation as BC's member of the Group of Seven.

> I loved his things, particularly some snow things of Quebec and three canvasses up Skeena River. I felt a little as if beaten at my own game. His Indian pictures have something mine lacks – rhythm, poetry. Mine are so downright . . . I worked for history and cold fact. Next time I paint Indians I'm going off on a tangent tear. There is something bigger than fact: the underlying spirit, all it stands for, the mood, the vastness, the wildness, the Western breath of 'go-to-the-devil-if-you-don't-like-it', the eternal big spaciousness of it. Oh the West. I'm of it and I love it.[100]

The totem poles and other artifacts that preoccupied Carr in her earlier work remain present – at least until the early 1930s – but they are now seen in new ways. Influenced by Whitman's vitalism, she brings a new attention to movement, space, and spirituality. The totems lose almost all detail as they become monumentalized and metaphorical (Figure 1).

Between 1928, when Carr again traveled north to many of the sites that she had visited earlier, and 1932 when Carr shifted her gaze to the forest, Carr no longer conceived her work as a 'record'. As Moray, Shadbolt and others have noted, she instead found ***in*** the totems expressions of what she believed to be the insights and values of native culture, and also resituated these artifacts as symbols of the *spirit* of the Canadian west. Shadbolt's summary of this period brilliantly – albeit unwittingly – captures the significance of this shift, while reinforcing the conventional view that these works were the culmination of Carr's *intimate knowledge* of native art.

> Her understanding of Indian art is not in fact reflected in her work until after 1927, when she strips the poles of *excessive detail*, removes them from *distracting settings* and concentrates on their *sculptural strength* and *expressive energy*.[101]

In these last Indian paintings villages lose all specificity, often simply incorporated into the enclosing forests such that their totems are surrounded only by a threatening wilderness, stripped of any 'cultural' context whatsoever. Increasingly, Carr drew the totem and the forest into the same horizon of meaning. Native cultures – reduced to the expression of a vague primeval spirit – appear to emerge *from nature itself*. Shadbolt codifies this reading of Carr's work, but her canvasses bear it well.

> [Carr had] long since known and felt the art of the Indian . . . there comes a moment of insight when she sees its profound and inevitable relationship to the environment in which it developed. *The totem poles literally come out of the giant trees*. But as brooding silent watching presences they involve a more profound generic relation to the hostile world of the rainforest.[102]

Shadbolt claims that this was the moment when Carr recognized the essential and eternal bond between Native and nature, such that the artifact now stood metonymically for a more original *natural* identity, common to all humanity, but most fully expressed by primitive peoples. Thus, Shadbolt argues, Carr recognized in her last Indian works and subsequent forest paintings that 'the Indian carved into his pole the same silences, mysteries, orifices and excrescence as nature has in hers'.[103] Native life mirrored nature, and the totem poles, Shadbolt implies, were merely signposts leading Carr deeper into the primeval forest, revealing a domain 'vast and rich enough to provide [Carr] with pictorial metaphors for all the expression she wished to express'.

Shadbolt's view no doubt reflects what Neil Smith describes as the almost 'instinctive Romanticism' that pervades North America culture.[104] But similar views were prevalent in Carr's time too. Maurice Barbeau, an acquaintance of Carr and curator of the National Museum of Man, argued that Indian art found its source in 'nature itself'.[105] Indeed, although uncritical of the terms of her analysis, Shadbolt is likely not far off the mark. Already in Carr's work from the early 1930s, totem poles are depicted as an element internal to nature. The irony hardly needs mention: in her later works Carr placed the totem back in the forest, reiterating the decontextualization of the poles that she so stridently decried in Sitka. At times they are simply pictured at the moment of their reclamation by nature. In these paintings totems lose their verticality, overwhelmed by growth. They are, finally, vanquished, as much by Nature as by the cultural violence of colonialism. Eventually Indian themes disappear entirely from Carr's work, replaced by nature itself.[106]

Carr's forest images from the late 1930s are as popular today as her aestheticization of Native culture from the early 1930s, due in part to the growth of BC's environmental movement in the 1970s and since. Indeed, Carr's forest images *are* often provocative. Although the interiors of her forests appear 'dark', 'mysterious' and 'veiled' – familiar tropes of feminized nature – hers is not a nature that passively awaits its unveiling before a penetrating and possessive masculinist gaze. Few of Carr's forest images invite possession by the viewer. In cases like *Cedar* (Figure 2) they are viewed from inside, from a vantage point that precludes grasping them in their totality. More often than not her forests appear *impenetrable*, to the point where they threaten to engulf the viewer. Indeed, the images can effect a paranoid anxiety in viewers, undermining their ability to master space.

Viewed in light of the ecological – and social – catastrophe of much industrial forestry, it is perhaps justified that many find in Carr's images a utopian statement of hope.[107] Yet, such interpretations fail to attend to other important aspects of these works. Specifically, they ignore the histories of 'seeing' that brought Carr to the point where the forest could stand for primal energies and forces and, moreover, could *stand apart* from its cultural surrounds. Lawren Harris is often credited with prodding Carr to end her fascination with totems and move to the more elemental forces of wild nature of which the totems were only an expression. In one of most quoted letters in Canadian art history Harris wrote to Carr that,

> the totem pole is a work of art in its own right and is very difficult to use it in another form of art. But how about seeking an *equivalent for it* in the exotic landscape of the Island and coast, making your own form and forms within the greater form. Create new things from the landscape . . .[108]

That Lawren Harris and Emily Carr could find in nature an equivalent for the totem pole suggests that for both 'nature' belonged to the same visual economy as did 'Indians': one could be substituted for the other. Here it is important to recall my earlier discussion of Carr and the tourist/ethnological gaze. Once Native culture is staged as a natural culture – the artifacts no longer part of the life of a modernizing people but symbols instead of the Indian's unity with nature – it is a simple matter to conflate the two. By purifying Native culture – emptying it of any foreign or contaminating modern elements – the nature of which it is a mere element itself appears pure. Native culture (past) and nature (present) become the same, the former simply an expression of the latter's temporality. There is no need, therefore, to mark a contemporary native presence in a natural landscape.

Like her images of native artifacts, Carr's famous rainforest paintings retain aspects of a colonialist visuality. This differs from other accounts that understand Carr's shift to the forest as the result of either an 'inner journey' (whereby her journey into nature mirrors her increasing maturity as an artist) or as a result of external 'influences' (her immersion in Walt Whitman, her introduction to Theosophy, the prodding of Harris, and so on) that caused her to abruptly change course. To be sure, these influences were important. Yet, they did not lead Carr to abandon a 'love' of Natives for a 'love' of nature. The latter was already anticipated in the former. The effect of the developmental narrative often woven around Carr's work, in which ruptures are posited between her 'early' and 'later' periods, and within her later periods, has been that Carr's forest paintings have been treated in very different ways from her Indian works, and often discussed as if unrelated to them. Even those most critical of Carr's appropriation of Native culture fail to link her 'artifactual vision' with her 'rainforest' imagery. With the rhetoric of *rupture* written twice over it is easy to lose sight of those elements in her first Indian paintings that remain as irreducible traces in her celebrated rainforest images. If we lose sight of these traces, we lose sight of the relationship between our present-day environmental imaginaries and the colonial histories of seeing that remain embedded in them.

Painting land claims – the salvation art of Yuxweluptun[109]

I have focused extensively on Carr in order to consider what it might mean for her work to be installed in the present as the most developed and sophisticated 'BC seeing'. That the VAG, itself located at the proprioceptive center of Vancouver, places her work at the center of Northwest Coast visual cultures – both physically and metaphorically – raises important questions about the way in which the region's colonial pasts continue to infuse its postcolonial present.[110]

With this in mind, let me return to the exhibition with which I opened this paper. Canvasses such as *Scorched Earth, Clear-cut Logging on Native Sovereign Lands, Shaman Coming to Fix* (Figure 3) are undeniably challenges to the legacy bequeathed not only by Northwest Coast landscape artists who preceded Yuxweluptun, but to a broader colonialist visuality that has framed the region. In this sense, Yuxweluptun's work is *citational*; it recalls, even as it contests, colonial ways of seeing, and Carr's work in particular. In important ways the cultural legibility of his work depends on this. I will say more about his work momentarily. First, however, it should be noted that juxtaposing Yuxweluptun and Carr comes with certain dangers. To begin, it risks setting up Yuxweluptun's 'BC seeing' as a corrective to Carr's, as if it is somehow less ideological in character. If we understand visuality as always already historical, then we must understand Yuxweluptun's work also in terms of its enabling conditions (of which one is, undoubtedly, the experience of colonialism and its displacements). More problematic is that juxtaposing the two risks reducing Yuxweluptun's art to a reaction to colonialism, paradoxically recentering European colonialism as that which most defines the experience of contemporary First Nations peoples. This would be a grave mistake. While attending to 'postcoloniality' means recognizing the ways in which places and cultures have been 'worked over' by colonialism, it does not mean that the experience of the colonial subject is *reducible* to colonialism. Indeed, one of the troubling aspects of 'postcolonial studies' has been its tendency to privilege subjects and practices that best fit its theoretical interests ('hybridity', 'mimicry'), while other subjects and practices that do not so readily reflect the effects of colonial encounters are left unexamined. Certainly one of the appeals of Yuxweluptun's work – along with other Native and American Indian artists such as Carl Beam, Teresa Marshall, James Luna and Hachivi Edgar Heap of Birds – is that it fits the postcolonial 'slot' rather well (much like Carr's work fitted an anxious nationalism in the 1930s). This is not to discredit the work of these artists; rather, it is to keep in mind that there may be *other* ways of 'being Native' that tend to escape critical commentary.[111]

A final, related problem with comparing Yuxweluptun and Carr is the risk of reducing visuality on the coast to a simple Native–White binary. This not only sets up a questionable dichotomy, but erases important differences *among* Native and Euro-Canadian artists (not to mention other artists who, in a thoroughly 'multicultural' province, do not fit this binary frame).[112] Yuxweluptun's work can be made representative of 'Native art' only with considerable violence to the diversity of artistic production among First Nations artists; this repeats Carr's earlier tendency to collapse different Native groups into an undifferentiated 'Indianness'. 'Being Native' is not a fixed site; it consists of heterogenous and shifting locations that must be understood in their relationality and in their singularity. Likewise, it would be a grave mistake to understand Yuxweluptun's art as purely *oppositional*. This would recapitulate an old nostalgia for a place of 'pure' difference, as if it were possible to speak, write or paint from a location that was somehow external to the social, cultural, political and technological relations of postcolonial global capitalism. Yuxweluptun makes no such claims for his art, even as he insists on its political intent and on the unique perspective that he has as a First Nations person who lives in the long shadow of colonialism. Rather than oppositional, it is perhaps better to understand Yuxweluptun's work as *differential*, in the sense that it disrupts or refracts 'BC seeing', not in the name of returning to a 'pure' vision, or a mythical place prior to the toxic neo/post/colonial environments that he so brilliantly depicts, but in order to imagine a future postcoloniality in which 'nature', 'culture' and 'nation' are understood in different, more socially just, ways.[113]

With these cautionary statements in mind, let me turn to Yuxweluptun's work.[114] Like Carr, Yuxweluptun also seeks to depict the 'deeper truth' of West Coast landscapes. But, unlike Carr, who in her decontextualization of the artifact sought to escape history in order to capture the region's underlying essence (and thereby erased modern Native territorialities) Yuxweluptun narrates his 'BC seeing' as irrevocably historical, stating that 'if I am to talk about my work, I will have to start with the past'.[115] This statement offers several readings. In one respect, it points to biography and thus to Yuxweluptun's unique experience as a First Nations person. Born of an Okanagan mother and a Coast Salish father, Yuxweluptun grew up an 'urban Indian', living in Kamloops and in the Vancouver suburb of Richmond.[116] Both of his parents were heavily involved in Native politics across BC, with the result that Yuxweluptun was not only exposed to the complexities of Native–White relations in the region, but also to the diverse experiences of *different* First Nations communities.[117] Thus, in contrast to the sedentarist metaphysics that has so dominated Western representations of non-Western others (which assume that the identity of the 'other' is bound to place in contrast to the mobility of the European subject), Yuxweluptun's life has been lived *between* places, and *between* cultures.[118] This mobility is mirrored in his artistic training. Avoiding schools that taught Native artists 'traditional' techniques, Yuxweluptun chose to study at the province's premier art school: the Emily Carr College of Art and Design.[119] This helps us see more clearly the complex 'folds' within which postcolonial identities, art and politics occur. It also necessarily cautions us from seeing Yuxweluptun's art as *representative* of all Native art. Like Carr, his vision is historically constituted, yet in important ways, singular.

The matter of history can also be understood more broadly. Yuxweluptun's statement alerts us to that fact that the 'reality' he paints is *historical*: he paints the legacy of colonialist practices both in the past and in the present. To understand the present – and his experience of 'being Native' – one has to attend to the histories that displaced Native peoples, constituted this corner of the Americas as *Canadian*, incarcerated First Nations on postage-stamp-sized reserves, disrupted Native cultural practices, and mapped a web of disciplinary practices across the spaces and bodies of aboriginal peoples. Far from merely an alternative depiction of 'national' space within a pluralistic culture, every brush-stroke in Yuxweluptun's work is a measured and reflexive encounter with the changing configurations of Native lands and cultural identities as they are lived and experienced today – differentially – by aboriginal peoples along the Northwest Coast. Thus, when he speaks of his work he speaks not of color, line or style, but of genocide, despotism and concentration camps; not of an 'inner vision' or of 'surrealism' as a modernist aesthetic, but of a life that was *made surreal* by the displacements of colonialism and by the administrative apparatus of a colonialist state; not of primeval nature but of a toxicological colonialism.[120]

What is of interest to me, then, is the manner in which Yuxweluptun asks us to imagine other paths through the complex terrain of vision and landscape in postcolonial British Columbia that Carr's work forecloses. Evident immediately is his refusal to separate 'art' from 'politics', or contests over 'nature' from contests over 'nation'. This is clear in canvasses such as *Scorched Earth*, in which the landscape is clearly constituted through violence and political conflict. Perhaps for this reason he refers to his work as 'salvation art'.[121] 'Salvation', in this instance, is multivalent and paradoxical. It plays on a long history of efforts by Christian missionaries to drive a wedge between Native people (especially children) and their cultural and religious traditions. It also appropriates the teleology of salvation history, although with a final destination somewhat different from the one conventionally imagined. Finally, it also, not unlike Emily Carr, points to an enduring *spiritual* relation to the land among First Nations peoples, albeit a spirituality that does not collapse 'Natives' into 'nature', but instead exists as a resource for efforts to 'heal' an environment exploited by industrial capitalism.

At the heart of Yuxweluptun's art is the matter of land. As Charlotte Townsend-Gault has summarized in a pithy statement, Yuxweluptun does not paint land*scapes*, but land *claims*.[122] Key in this effort has been the rearticulation of nature as a cultural and political artifact, rather than a purely physical domain, and thus crisscrossed by prior political claims that are mobilized again in contemporary struggles over land, resources and nation. Here the distance between Carr and Yuxweluptun is most pronounced. If Carr purified coastal landscapes into two elements – a decaying native culture and a brooding, spiritualized, all-encompassing nature – Yuxweluptun does the opposite. Nature in his work does not overwhelm a vanquished culture, nor is Native culture collapsed *into* nature in such a way as to erase the violence of colonial dispossession. Rather, in his work so-called 'natural' landscapes are assembled from elements of Northwest Coast Native culture.

Yuxweluptun's *Scorched Earth* is illustrative of these multivalent aesthetic/political gestures. We can read it, for instance, as an ecological critique of the violence of industrial forestry. The tree-stumps in the foreground, tracing a line backward to denuded slopes, and the brown squares of clearcuts in the distant hills speak of the violence wreaked on the forest. Both the land and the sun weep in the face of this travesty, as does a figure resting on one of the distant hills. But this image articulates much more, for Yuxweluptun does not simply paint a 'green' critique, but asserts native *rights* to a land being destroyed.[123] The various components of the landscape – trees, mountains, the sun, even the tree-stumps – are constructed out of Coast Salish motifs. *There is no nature apart from these*. Whereas for Carr, especially in her later work, Native culture was merely a fleeting expression of nature, for Yuxweluptun nature is an artifact of culture.

We see this again in *Clayoquot* (Figure 11) where clearcuts are depicted as gaping wounds in a cultural landscape, the square spaces of the rationalized forest gouged into hills formed from Coast Salish ovoids.[124] Even the tree-stumps are constructed from these forms in what is a clear claim of sovereignty over territories already worked over by industrial colonialism. The presence of the shaman in *Scorched Earth* stakes out a similar position: any future healing of this wounded earth will hinge on the recognition of Native land rights and Native cultural/environmental traditions.

There are additional complexities to Yuxweluptun's work that deserve brief mention.

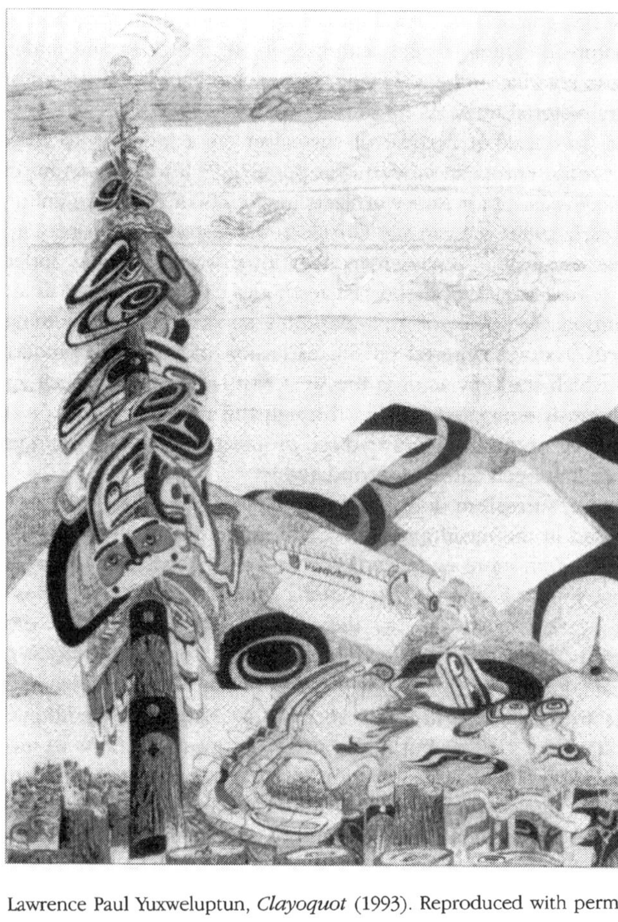

FIGURE 11 Lawrence Paul Yuxweluptun, *Clayoquot* (1993). Reproduced with permission of artist

Earlier I noted that Carr envisioned her early documentary project as an attempt to establish a permanent record. Yuxweluptun has claimed to do something entirely different: 'my work is to *record*'.[125] The shift is subtle but the implications immense. Whereas Carr's desire to produce a 'record' assumed a static culture threatened by the forces of modernity, Yuxweluptun sets out 'to record' ongoing struggles over identity and land that exceed the traditional/modern dichotomy. There is an important lesson in this for any progressive anti-racist, postcolonial environmentalism. For Yuxweluptun, Coast Salish culture is not something fixed and therefore under threat from a colonizing modernity. Modernity does not arrive from the outside like a stealthy intruder insinuating itself into the cultural traditions of the 'other'. Rather the 'traditional' is always already dynamic, a shifting constellation of ideas, forms and beliefs that are resources for building alter/native modernities at once continuous with, but not imprisoned by, the past.

This is also, I think, the way in which Yuxweluptun's complex 'borrowings' from surrealism must be approached. Yuxweluptun finds in surrealism a method that expresses the fantastic nature of life as a First Nations person, for, operating under the sign of Reason, European colonialism bent the coordinates and lines of Native culture, distorting the 'real' in very material ways. As he puts it, 'my reality was surreal'.[126] But this can be pushed further. To speak of his use of surrealism as a 'debt' is to forget the *prior* borrowings that made European modernisms possible.[127] If we reinsert into the story the circulation of Northwest Coast Native artifacts in circuits of European culture during the late 1800s and early 1900s, we can see European modernisms as enabled at least in part by the 'jolt' that these objects gave to conventional ways of seeing. Indeed, it is only through an act of forgetting that modernist aesthetics can be made out as a development *internal* to Europe. To speak of Yuxweluptun's surrealism as something 'borrowed' merely reinforces a story centered on the extension of European modernity across a global stage in which the only actor is the West writ large. Indeed, such an assumption can be surreptitiously smuggled back in through the much-hyped trope of 'hybridity'. Yuxweluptun makes a much more radical *proprietary* claim to modernism itself, repatriating what had been earlier appropriated.[128]

Crucially, though, surrealism does not return to the West Coast without remainder; it has been reworked in the meantime. To reclaim modernism by way of Dali is in itself a provocative move, even more so given fierce battles over what counts as, and who can speak for, Native 'tradition' among First Nations and non-Natives alike. This issue infuses Yuxweluptun's work, much as it did Carr's. His efforts to destablize the tradition/modernity distinction have led him to inscribe Coast Salish motifs onto hotdogs and automobiles, 'queering' conventional notions of Native culture. Again, this is serious play. It engages with demands in White society that Natives be traditional in order to preserve their status as 'Native', but it also places in question efforts to recover cultural traditions among First Nations. His solution is to recode 'change' as intrinsic to Native cultures, and thus a sign of vibrancy rather than decay: 'I work from the native perspective that all shapes and any elements can be changed to anything to present a totally native philosophy. It allows me to express my feelings freely and show you a different view'.[129]

Ultimately Yuxweluptun provides us with a differential optic, one that displaces rather than inverts colonialism's binaries, and thereby produces a crisis in discourses of national sovereignty. The Native 'reality' that he paints is traditional but not static, modern but still Native. West Coast landscapes are ecological but no less cultural for being so. His 'salvation art' does not seek an idyllic past, it points to an open future, one no longer constrained by a national(ist) landscape imaginary that at once incorporates Natives as the nation's 'origin' and at the same time relegates them to the margins by placing them 'in' nature.

Conclusion: fractured visualities, multiple temporalities

The juxtaposition of Carr and Yuxweluptun testifies to the fractured and political nature of visuality on the West Coast of North America, both in the past and the present. As I have attempted to show, neither Carr nor Yuxweluptun take us to the 'essence' of these landscapes. Despite the rhetoric of art critics who have installed Carr as the artist who best captured the unique character of the coast, 'seeing BC' is something contested, both in the past and in the present. Likewise, Yuxweluptun's work does not reveal the landscape's singular 'truth', even though it testifies to the particular historical form these landscapes take in the colonial present. More than Carr perhaps, Yuxweluptun recognizes the 'essence' of the coast to be its continuous *remaking* by historical and political forces. Ironically, his attention to the flux of postcolonial landscapes and identities may be the very reason for his success in a society that increasingly celebrates hybrid aboriginal identities, yet remains somewhat disturbed by claims of difference that cannot be easily incorporated within the 'multicultural' state.

For my purposes, the popularity of *both* artists today points to the paradoxical nature of postcolonial settler societies and the multiple temporalities and trajectories of decolonization found in them. Important political events, such as the recent Nisga'a land claim settlement in northern BC, can be taken as evidence of a progressive decolonization where the wrongs of the past are righted in the present. Yet, the continued circulation of 'Carr' – and the 'salvation art' of Yuxweluptun – disturbs this comforting story of inevitable decolonization. It reminds us that the 'post' in postcoloniality refers to continuities as much as discontinuities, and that decolonization is about politics, not history. It is worth noting that this does not mean that Carr's work remains unchanged by the context of its display, or that it retains its hold on Canadians' imaginations in the same way today as in the past. A more complete discussion would attend to how her art is viewed in a 'knowing' or 'ironic' manner. It would also have to account for the degree to which visitors to the VAG may simply bypass Carr as a throwback to an earlier era, and instead quickly move on to the gallery's 'temporary' exhibits. Yet, as I have shown elsewhere, echoes of Carr's vision are today found in widely disparate sites.[130] Even in its multiple registers, the continuous returns to Carr today suggest that the 'optics' of colonialism do not disappear entirely with the end of formal colonialism; they exist as colonialism's afterlife, the continued presence of the past in the present.

Notes

[1] Ruth Appelhof, *The expressionist landscape: North American landscape painting 1920–1947* (Seattle, University of Washington, 1988), p. 77.

[2] Jonathan Crary, 'Modernizing vision', in H. Foster, ed., *Vision and visuality* (Seattle, Bay Press, 1988), p. 48.

[3] There is now an extensive literature on these debates in the context of the Americas. Some of the best writings have emerged from the Latin American context, often using the specific nature of Latin American postcolonialities to rework key concepts and theories. For recent English language debates, see the journal *Nepantla*. See also Florencia Mallon, 'The promise and dilemma of Subaltern Studies: perspectives from Latin American history', *American Historical Review* **99** (1994) 1491–515; J. de Alva, 'The postcolonialization of the (Latin) American experience: a reconsideration of "colonialism", "postcolonialism", and "mestizaje" '. In Gyan Prakash, ed., *After colonialism: imperial histories and postcolonial displacements* (Princeton, Princeton University Press, 1995), pp. 241–75; and Walter Mignolo, *Local histories/global designs: coloniality, subaltern knowledges and border thinking* (Princeton, Princeton University Press, 2000). For some recent writings in geography on colonialism, postcoloniality and North America, see Dan Clayton, *Islands of truth: the imperial fashioning of Vancouver Island*. (Vancouver, UBC Press, 1999); Bruce Braun, 'Buried epistemologies: the politics of nature in (post)colonial British Columbia', *Annals of the Association of American Geographers* **87** (1997), pp. 3–31; Matthew Sparke, 'A map that roared and an original atlas: Canada, cartography, and the narration of nation', *Annals of the Association of American Geographers* **88** (1998), pp. 463–95; and Kay Anderson, 'Thinking "postnationally": dialogue across multicultural, indigenous, and settler states', *Annals of the Association of American Geographers* **90** (2000), pp. 381–91.

[4] This differentiation exists internal to nation-states such as Canada, and not just between different national contexts. The experience of the Cree in Quebec, the Inuit of Nunavut, and the Nisga'a of British Columbia cannot be collapsed into a monolithic 'indigeneity' or 'postcoloniality'.

[5] A call for attending to the singularity of any specific postcoloniality is not the same as claiming its autonomy from a wider set of colonial relations and practices. Rather, the specificity of any postcolonial situation is a direct result of the socio-spatial *networks* that connect places and actors to other places and actors.

[6] For discussion, see Clayton, *Islands of truth*.

[7] See Anderson, 'Thinking "postnationally" '.

[8] I borrow the phrase 'after colonialism' from Gyan Prakash, who argues that the very ambiguity of the phrase allows for a reading of the present as always already marked by histories of colonialism, something less clear in the term 'postcolonialism'. See Gyan Prakash, *After colonialism*.

[9] Between 1927 and 1986 the National Gallery made no purchases of 'Indian' art, preferring to look to Europe for its acquisitions and cultural roots. See Diana Nemiroff, 'Modernism, nationalism and beyond', in Diana Nemiroff, Robert Houle and Charlotte Townsend-Gault, eds, *Land, spirit, power: First Nations at the National Gallery of Canada* (Ottawa, National Gallery of Canada, 1992), p. 18).

[10] See Scott Watson, 'The modernist past of Lawrence Paul Yuxweluptun's language allegories', in Scott Watson, ed., *Lawrence Paul Yuxweluptun: born to live and die on your colonialist reservation* (Vancouver, Morris and Helen Belkin Art Gallery, 1995), and W. Jackson Rushing, *Native American art and the New York avant-garde* (Austin, University of Texas, 1995).

11 See also Matthew Sparke, 'Between demythologizing and deconstructing the map: Shawnadithit's New-Found-Land and the alienation of Canada', *Cartographica* **32** (1995), pp. 1–21.

12 For an historical account of the Indian Land Question, see Cole Harris, *Making Native space: colonialism, resistance, and reserves in British Columbia* (Vancouver, UBC Press, 2002).

13 I discuss this at greater length in *The intemperate rainforest: nature, culture and power on Canada's West Coast* (Minneapolis, University of Minnesota Press, 2002).

14 W.J.T. Mitchell, ed., *Landscape and power* (Chicago, University of Chicago Press, 1994).

15 Homi Bhabha, *The location of culture* (New York, Routledge, 1994).

16 Stephanie Walker, *This woman in particular: contexts for the biographical image of Emily Carr* (Waterloo, Wilfrid Laurier University Press, 1996); Timothy Brennan, 'The national longing for form', in Homi Bhabha, ed., *Nation and narration* (New York, Routledge, 1990), pp. 44–70.

17 Watson, 'The modernist past'; Rushing, *Native American art*. The members of the 'Group of Seven' included Lawren Harris, Tom Thomson, J.E.H. MacDonald, Arthus Lismer, Frederick Varley, Frank Johnston and Franklin Carmichael.

18 This view was sanctioned by none other than the Governor General of Canada, Vincent Massey:

> [This] group of gifted artists . . . turned their backs on Europe – quite deliberately – and surrendered themselves to their own environment, striving to uncover its secret. The inevitable happened. The Canadian landscape took possession of them. They abandoned the methods and techniques which were alien to Canada, and recorded its beauty faithfully in the clear lights, bold lines and strong colours which belong to it.

It was F.B. Houser, however, who first articulated this view, writing that the modern Canadian School 'was inspired as the result of a direct contract with Nature herself'. F.B. Houser, *The Canadian art movement: the story of the Group of Seven* (Toronto, Macmillan, 1926), p. 24. This had required 'a new type of artist; one who divests himself of the velvet coat and flowing tie of his caste, puts on the outfit of the bushwacker and prospector; closes with his environment; paddles, portages and makes camp; sleeps in the out-of-doors under the stars; climbs mountains with his sketch box on his back, p. 15.

19 Gerta Moray, 'Northwest Coast Native culture and the early Indian paintings of Emily Carr, 1899–1913' (PhD diss., University of Toronto, 1993). In a 1933 *Saturday Night* review Mortimor Lamb described Carr as a West Coast painter 'free from all influences than those created by her own responsiveness to a scenic environmental itself elemental, austere and yet scenically romantic'. In the same magazine, Ira Dilworth would later add to Carr's mystique, describing her as an artist who fought her way to the objects themselves. Carr, he explained, traveled 'with a courage that we at the present day can scarcely understand' traveling into 'the most solitary parts of the British Columbia coast', overcoming 'poor transportation' and 'often extremely dangerous stretches of sea'. Ira Dilworth, 'Emily Carr: Canadian Artist-Author', *Saturday Night* **57** (1 November 1941), p. 26.

20 Doris Shadbolt, *The art of Emily Carr* (Toronto, Clarke, Irwin, 1979).

21 *Ibid.*, p. 11.

22 Arjun Appadurai, 'Disjuncture and difference in the global cultural economy', *Public Culture* **2**(2) (1990), pp. 1–24; Akhil Gupta and James Ferguson, 'Culture, power, place: ethnography at the end of an era', in Akhil Gupta and James Ferguson, eds, *Culture, power, place: explorations in critical anthropology* (Durham, Duke University Press, 1997).

23 Crary, 'Modernizing vision'; Donna Haraway, *Simians, Cyborgs and women: the reinvention of nature* (New York, Routledge, 1991).

[24] Derek Gregory, *Geographical imaginations* (Oxford, Blackwell, 1994); John Rajchmann, 'Foucault's art of seeing', *October* **44** (1988), pp. 89–117.

[25] See Maria Tippett, *Emily Carr: a biography* (Toronto, Oxford University Press, 1979); Paula Blanchard, *The life of Emily Carr* (Seattle, University of Washington Press, 1987); Moray, 'Northwest Coast Native culture'.

[26] Shadbolt, *The art of Emily Carr*, pp. 194–95, emphasis added. Similar rhetorics are found in Shadbolt's earlier account of Carr's life, where, after tracing Carr's many contacts, she concludes that, 'despite all this, the popular view which sees her as simply the West Coast painter of Indian subjects and rainforests is, though incomplete, justified . . . She remained a non-intellectual who responded primarily to the spirit of things and people and had to learn intuitively'. Doris Shadbolt, *Emily Carr* (North Vancouver, J.J. Douglas, 1975), p. 4.

[27] Eric Newton, 'Canadian art through English eyes', *Canadian Forum* **28** (1939), pp. 344–45.

[28] Brennan, 'The national longing for form'.

[29] This story of a 'retreat' to Victoria is somewhat misleading. Carr remained in contact with a vibrant artistic community in Seattle, including the cubist painter Mark Tobey, continued to exhibit, and had contact with eastern curators and ethnologists.

[30] Shadbolt, *The art of Emily Carr*, p. 195, emphasis added.

[31] Moray makes this point explicitly.

[32] Stuart Hall, 'When was the post-colonial? Thinking at the limit', in Iain Chambers and Lidia Curti, eds, *The post-colonial question* (London, Routledge, 1996), pp. 242–60.

[33] Moray, 'Northwest Native Culture'.

[34] Marcia Crosby, 'Construction of the Imaginary Indian', in Stan Douglas, ed., *Vancouver anthology: the institutional politics of art* (Vancouver, Talonbooks, 1991).

[35] *Ibid.*, p. 270.

[36] *Ibid.*

[37] Robert Berkhofer, Jr., *The white man's Indian: images of the American Indian from Columbus to the present* (New York, Vintage, 1979); Edward Said, *Orientalism* (New York, Vintage, 1978). See also Daniel Francis, *The imaginary Indian: the image of the Indian in Canadian culture* (Vancouver, Arsenal Pulp, 1992).

[38] Crosby, 'Construction', p. 276.

[39] See Heather Dawkins, 'Paul Kane and the eye of power: racism in Canadian art history', *Vanguard*, 1986 (Sept), pp. 24–27; Maureen Ryan, 'Picturing Canada's native landscape: colonial expansion, national identity and the image of a dying race', *RACAR* **18** (1992), pp. 138–49; Loretta Todd, 'What more do they want? In Gerald McMaster and Lee-Ann Martin, eds, *Indigena: contemporary native perspectives* (Hull, Canadian Museum of Civilization, 1992); Gerald McMaster and Lee-Ann Martin, 'Introduction', in Gerald McMaster and Lee-Ann Martin, eds, *Indigena*; Robert Fulford, 'The trouble with Emily: how Canada's greatest woman painter ended up on the wrong side of the political correctness debate', *Canadian Art* **10** (1993), pp. 32–39. For discussion of these debates see Rosemary Coombe, *The cultural life of intellectual properties* (Durham, Duke University Press, 1998).

[40] A similar point is made by Ryan, 'Picturing'. Surveying a number of nineteenth-century representations of native peoples in North America, Ryan argues that these images claimed 'aesthetic proprietorship' over native lands, and 'could naturalize in aesthetic forms a racial theory that had served to marginalize a population of the country whose presence and title to the land had been at odds with Anglo-European expansion and settlement within the colony' (p. 45). More recently Rosemary Coombe has explored the trope of 'appropriation', a key issue in Crosby's critique of non-Native artists and writers. For Coombe, 'cultural appropriation' is something that must be evaluated *contextually* and in terms of relations of *power*. For

instance, Coombe explores the ways that 'subaltern' groups use mass-media texts, celebrity images, trademarks and so on, to forge identities and communities. Understood in this context *cultural appropriation* is a subversive act within 'media-saturated consumer societies' (Coombe, *Cultural life*, p. 209). As Coombe notes, however, this celebratory notion of appropriation is cast in doubt in the case of the appropriation of 'Native' artifacts. Although concerned that First Nations critiques of appropriation often rely on tropes of 'authenticity', and notions of 'property' borrowed from the West, she argues that First Nations claims of 'cultural authenticity' should be seen as effect to foreground 'the[ir] inability to name themselves and a continuous history of having their identities defined by others' (p. 227). Coombe concludes that by contextualizing debates over appropriation it becomes possible to 'enact and practice an ethics of appropriation that attends to the specificity of the historical circumstances in which certain claims are made' (p. 230).

41 Bhabha, *Location*; Leela Gandhi, *Poscolonial theory: a critical introduction* (New York: Columbia University Press, 1998).

42 See Harris, *Making native space*.

43 See also Cole Harris, *The resettlement of British Columbia* (Vancouver, UBC Press, 1996); Brett Christophers, *Positioning the missionary* (Vancouver, UBC Press, 1999)

44 Carr's departures from what was considered 'proper' objects for women to paint is one basis upon which she has been celebrated as an early feminist. For a discussion of limits to women's artistic endeavors see Griselda Pollock, *Vision and difference: femininity, feminism and histories of art* (London, Routledge, 1988).

45 Moray, 'Northwest coast Native culture', p. 9.

46 *Ibid.*, pp. 14, 16.

47 *Ibid.*, p. 107. We should hasten to note that this does not return us to something akin to Shadbolt's autonomous artist whose work is an expression of 'inner vision'. Rather, it brings into view the *singularity* of Carr's artistic vision and cultural production – or of the *event* of representation more generally – as an effect of the multiple, fractured social and psychic spaces of BC's postcolonial cultures.

48 Douglas Cole, 'The invented Indian/the imagined Emily', *BC Studies* **125/126** (2000), p. 152.

49 Emily Carr, *Growing pains: the autobiography of Emily Carr* (Toronto, Oxford University Press, 1946), p. 211.

50 See Robert Linsley, 'Painting and the social history of British Columbia', in Stan Douglas, ed., *Vancouver anthology: the institutional politics of art* (Vancouver, Talonbooks, 1991), pp. 225–46.

51 This narrative is frequently found in writings about Carr. See Dilworth, 'Emily Carr', and Shadbolt, *The art of Emily Carr*.

52 Maria Tippett's biography is perhaps the most thorough in its discussion of the social networks and material practices of travel. See Tippett, *Emily Carr*.

53 Crary, 'Moderning vision'. See also Rajchmann, 'Foucault's art of seeing'.

54 Jonathan Crary, *Techniques of the observer: vision and modernity* (Cambridge, MIT Press, 1990).

55 Nicholas Green, *The spectacle of nature: landscape and bourgeois culture in nineteenth-century France* (Manchester, Manchester University Press, 1990).

56 Harris, *The resettlement of British Columbia*.

57 Gilles Deleuze and Felix Guattari, *Anti-Oedipus: capitalism and schizophrenia*, trans. Robert Hurley, Mark Seem and Helen Lane (New York, Viking Press, 1977).

58 The growth of steamship service occurred for several reasons. First, by the close of the nineteenth century, advances in technology had made such travel relatively inexpensive, and

far quicker than before. Secondly, in conjunction with improving transportation links, lumber camps and fish plants grew in number along the coast, resulting in a growing network of towns and villages stitched together by new social actors – machines. There is a wealth of material on these early steamship routes, much in the form of memoirs, others in the form of histories of individual companies. See Wiggs O'Neill, *Steamboat days on the Skeena River* (Kitimat, Northern Sentinal Press, 1960); Robert Turner, *The Pacific Princesses: an illusrated history of the Canadian Pacific Railway's Princess fleet on the North Coast* (Victoria, Sono Nis Press, 1977); Gerald Rushton, 'Union Steamship Company of British Columbia, a short history', in Ruth Greene, ed., *Personality ships of British Columbia* (West Vancouver, Marine Tapestry, 1969); Gerald Rushton, *Whistle up the inlet: the Union Steamship story* (Vancouver, J.J. Douglas, 1974); Gerald Rushton, *Echoes of the whistle: an illustrated history of the Union Steamship Company* (Vancouver, Douglas & McIntyre, 1980); John Munro, 'Coastal shipping and the development of coastal British Columbia', unpublished ms, 1988; Norman Hacking, *The Princess story: a century and a half of West Coast shipping* (Vancouver, Mitchell Press, 1974).

59 See also Raymond Williams, *The country and the city* (New York, Oxford University Press, 1973); Patricia Jasen, *Wild things: nature, culture and tourism in Ontario, 1790–1914* (Toronto, University of Toronto Press, 1995).

60 Harris, *The resettlement of British Columbia*, p. 183.

61 This does not mean that these communities were now excluded from processes of modernization, only that it placed new demands on *Native* mobility.

62 Wolfgang Schivelbusch, *The railway journey: the industrialization of time and space in the nineteenth century* (Berkeley, University of California Press, 1986).

63 *Ibid.*, p. 60.

64 White communities were more likely to be described in terms of their level of development, pace of growth and services available; in other words, by the distance they had traveled toward an imagined destination – the metropolitan center. Native villages were appraised through an entirely different lens. Maturin Ballou's description was in many ways typical:

> From time to time small native villages are seen on the islands and the mainland, all typical of the people, and quite picturesque in their dirtiness and peculiar construction. Some of the cabins are built of boards, but mostly they are rude, bark-covered logs. In front of these dwellings stand totem poles, presenting hideous faces carved upon them in bold relief, together with uncouth figures of birds, beasts, and fishes. A portion of these tall posts are weather-beaten and neglected, significantly tottering on their foundations, green with mold, unconsciously foreshadowing the fate of the aboriginal race.

Maturin Ballau, *The new Eldorado: a summer trip to Alaska* (Boston, Houghton, Mifflin & Company, 1890), p. 194. Linda Nochlin notes that the picturesque in nineteenth-century art found its themes in the 'marginal' and the 'peripheral', the 'rough and ready' and the 'unrefined'. While potentially subversive of dominant culture (and classical painting), the picturesque could also be deeply nostalgic, focusing on that which was 'vanishing' or becoming 'obsolete'. See Linda Nochlin, *The politics of vision: essays on nineteenth-century art and society* (New York, Harper and Row, 1989).

65 Among the first was John Muir, who traveled the passage in 1879 and wrote 'half-booster, half-scholarly' articles in the *San Francisco Evening Bulletin*. For a good summary of the trade, see Ted C. Hinckley, 'The inside passage: a popular gilded age tour', *Pacific Northwest Quarterly* (April, 1965), pp. 67–74. These trips were popular with 'educated classes' and artists,

a point remarked upon in the memoirs of many travelers. Nor were women excluded from these paths. This is borne out in the description provided by Septime M. Collis who traveled the route in 1890.

> The pleasure is much enhanced to by the fact that those who are your fellow-passengers are apt to be ladies and gentlemen, by which I mean persons whose good breeding naturally tends to a regard for the comfort of their companions; and among them you will find men and women, young and old, of bright intelligence, who, devoting their time to travel, are full of fact and anecdote – scientists, savants, authors, and artists of renown from all part of the earth.

See Septima Collis, *Woman's trip to Alaska* (New York: Cassell Publishing Company, 1890), p. 193. It was the practice on some of these inland cruises to segregate social groups. While Native men and women were provided only with deck fare passage (available to non-Native men too), non-native *women* were permitted only to purchase cabin fares.

66 Turner, *Pacific Princesses*.

67 This does not mean that all travel experience is determined in advance. Indeed, travel can be, and often is, an *uncanny* experience that disturbs conventional views. See Derek Gregory, 'Between the book and the lamp: imaginative geographies of Egypt, 1849–60', *Transactions of the Institute of British Geographers* **20** (1995), pp. 29–57.

68 Alexander Badlam, *Wonders of Alaska* (San Francisco, Bancroft, 1891), p. iv.

69 *Ibid.*, p. 8, emphasis added. 'Kodaking' became popular along the Inside Passage during this period, not only because the technology had democratized photography, but also because the region now was set up for the viewer as a *visual resource*, framed to signify the 'primitive' and 'pristine'. See Collis, *Woman's trip*, for a description of 'kodaking'.

70 Emily Carr, *The heart of a peacock* (Toronto, Oxford University Press, 1953).

71 Rushing, *Native American art*.

72 See Tippett, *Emily Carr*, p. 289, no. 13. As important, the argument that Carr's travels and work challenged conventional roles for women needs to be approached with caution. Many of the paths Carr traveled were readily available to women, especially women from among the upper classes to which Carr belonged, even if she disowned the connection. Even when she departed from these paths, she was usually under the (paternal) 'care' of white traders, missionaries and relatives.

73 Knapp, 1980. Douglas Cole 1985, *Captured heritage: the scramble for the Northwest Coast artifacts* (Seattle, University of Washington Press, 1985).

74 Carr, *Heart of a peacock*, pp. 80, 82.

75 Carr returned to paint in Sitka on a later journey. It is also during her first journey that Carr met Theodore Richardson, an American painter from Minneapolis who lived part of the year in Sitka painting Native scenes.

76 Donna Haraway, *Primate visions* (New York, Routledge, 1989)

77 Ballou, *The new El Dorado*, p. 228.

78 Cole, *Captured heritage*; Aldona Janaitas, *From the land of the totem poles* (New York: American Museum of Natural History, 1988); Ira Jacknis, 'George Hunt, Kwakiutl Photographer', in Elizabeth Edwards, *Anthropology and photography, 1860–1920* (New Haven, Yale University Press, 1992).

79 See also James Clifford, *The predicament of culture* (Cambridge, Harvard University Press, 1988); Renato Rosaldo, *Culture and truth* (Boston, Beacon Press 1989).

80 See Cole, *Captured heritage*. It is worthy of note that the practice of erecting totem poles had expanded markedly *after* contact with Europeans and that what was for a time the object that

most signified 'authentic' Northwest Coast Indian culture in Western imaginations was actually a product of complex transcultural exchanges.

81 Rushing, *Native American art*.

82 Quoted in Chorley-Smith, *White bears and other curiosities: the first 100 years of the Royal British Columbia Museum* (Victoria: Royal British Columbia Museum, 1989), p. 142.

83 Established first as a museum of natural history, it very shortly began to add native artifacts. It was common at the time for ethnological collections to be treated as components of natural history collections.

84 C.F. Newcombe, *Guide to anthropological collection in the provincial museum* (Victoria, 1909).

85 Emily Carr, 'Lecture on totems', unpublished ms, 1913. Public Archives of British Columbia.

86 Carr, 'Lecture on totems', p. 53.

87 It is instructive that Carr identifies this site incorrectly, a sign of her lack of attention to *differences* among and between native groups, something that she shared with other tourists but not ethnologists for whom the differences were crucial.

88 Bill Holm and George Quimby, *Edward S. Curtis in the land of the war canoes: a pioneer cinematographer in the Pacific Northwest* (Vancouver, Douglas and McIntyre, 1980); Peter Jackson, 'Constructions of culture, representations of race: Edward Curtis's 'way of seeing', in Kay Anderson and Fay Gale, eds, *Inventing places: studies in cultural geography* (Melbourne, Longman Cheshire, 1992), pp. 89–106.

89 The striking similarities between their projects becomes evident in Curtis's 'General introduction':

> Months of arduous labour have been spent in accumulating the data necessary to form a *comprehensive and permanent* record of all the important tribes of the United States and Alaska that *still retain* to a considerable degree their primitive customs and traditions. The value of such a work, in great measure, will lie in the breadth of its treatment, in its wealth of illumination, and in the fact that it represents the result of personal study of a people who are *rapidly losing the traces of their aboriginal character and who are destined to become assimilated with the 'superior race'* (emphasis added).

Edward Curtis, *The North American Indian* (Seattle, 1907), I, p. 1. The preface to Curtis' collection was written by none other than Theodore Roosevelt, the American president most associated with the 'frontier' and its 'heroic' masculinity.

> In Mr. Curtis we have both an artist and trained observer, whose pictures are pictures, not merely photographs; whose work has far more than mere accuracy, because it is truthful. All serious students are to be congratulated because he is putting his work in permanent form; for our generation offers the last chance for doing what Mr. Curtis has done. The Indian as he has hitherto been is on the point of passing away. His life has been lived under conditions through which our own race passed so many ages ago that no vestige of their memory remains. It would be a veritable calamity if a vivid and truthful record of these conditions were not kept . . . [Mr. Curtis] is a close observer, whose qualities of mind and body fit him to make his observations out in the field, surrounded by the wild life he commemorates. He has lived on intimate terms with many different tribes of the mountains and the plains.

90 For an account that locates an ambivalent imperial discourse similar to Carr's and Curtis's some 80 years earlier, see Gareth E. John, 'Cultural nationalism, westward expansion and the

production of imperial landscape: George Catlin's Native American west', *Ecumene* **8**(2), pp. 175–203.

91 See Crosby, 'Construction', Scott Watson 'Disfigured nature: the origins of modern Canadian landscape', in Diana Augaitas and Helga Pakasaar, *Eye of nature* (Banff, Walter Phillips Gallery, 1991).

92 In other cases, Carr visited sites that had been abandoned as a result of depopulation. Yet, it is telling that she chose to picture these, rather than modern native villages.

93 Moray, 'Northwest Coast Native culture', p. 6, emphasis added.

94 My intention is not to resurrect the fiction of the authorial subject whose life can be read as a single, unbroken trajectory from beginning to end. But to bend the stick too far in the other direction, such that temporality and discontinuity become the same is, I think, equally problematic.

95 While in France Carr studied with a number of artists, including one – William Gibb – who became her mentor, and who was in turn a friend of Picasso and Matisse. During her stay Carr abandoned the documentary, naturalist and conservative realism that characterized her early work in favour of a newly acquired knowledge of postimpressionism, particularly Fauvism, and the early stages of modern primitivism. Carr's years in France have been documented by Tippett, *Emily Carr*; Shadbolt, *The art of Emily Carr*; Ian Thom, *Emily Carr in France* (Vancouver, Vancouver Art Gallery, 1991); and Moray, 'Northwest Coast Native culture'.

96 See Shadbolt, *The art of Emily Carr*; Moray, 'Northwest Coast Native Culture'.

97 See Clifford, *Predicament*. The German critic Kasimir Edschmid summarized this well in a 1918 discussion of Expressionism:

> We no longer have the concatenation of facts: factories, houses, illness, whores, outcry and hunger. Now we have the vision of them. *Facts have significance only so far as the artist's hand, reaching through them, grasps at what lies behind*. Art, which seeks only the essentials, excludes the incidental.

Quoted in Kristian Sotriffer, *Expressionism and fauvism* (Vienna, Anton Schroll and Company, 1972), p. 6.

98 Carr often portrayed herself – and has been portrayed by critics and historians – as nonintellectual, and therefore unconcerned with the changing currents of aesthetic theory. In the words of Shadbolt, '[Carr] responded primarily to the spirit of things and people and had to learn intuitively'. Doris Shadbolt, *Emily Carr*, p. 4. This 'nonintellectual' posture was in line with other post-Impressionist painters. See Sotriffer, *Expressionism*.

99 Shadbolt notes that Carr owned a 'well-thumbed' copy of Ralph Peason's *How to look at modern paintings*, which gave particular attention to Cubism, which Carr used to great effect in her depiction of both artifact and forest. The American painter, Mark Tobey, is widely considered to have been instrumental in Carr's Cubist turn.

100 Emily Carr, *Hundreds and thousands: the journals of Emily Carr* (Toronto, Clartke, Irwin & Company, 1996), p. 5.

101 Shadbolt, *The art of Emily Carr*, p. 30, emphasis added.

102 *Ibid*., p. 33, emphasis added.

103 *Ibid*., p. 36.

104 Neil Smith, 'The production of nature', in George Robertson, Melinda Mash, Lisa Tickner, Jon Bird, Barry Curtis, and Tim Putnam, eds, *Future natural: nature, science, culture* (London, Routledge, 1996.

105 Nemiroff, 'Modernism'.

106 Carr's advancing age and deteriorating health must also be seen as important factors that made

it difficult for her to continue her study of artifacts in the Native villages of the coast. Yet, this is not explanation enough for why her Native themes ended. The region surrounding Victoria contained ample Native material in the 1930s, only, just as when Carr began her career decades earlier, it was difficult to depict these Native groups in the same romantic and nostalgic registers that Carr used on her northern material. Tellingly, in the early 1940s, confined to her studio, Carr returned to native themes, drawing again on her already large archive of images from earlier trips north.

107 Linsley, 'Painting and social history'.

108 Quoted in Shadbolt, *The art of Emily Carr*, p. 76, emphasis added.

109 This section expands on brief comments made in Bruce Braun, Colonial vestiges: representing forest landscapes on Canada's west coast, *BC Studies* **112**, 5–37.

110 I owe this interpretation of the VAG's place in Vancouver's cultural geography to Stan Douglas, 'Introduction' in Stan Douglas, *Vancouver anthology*.

111 For a critique of the privileging of the 'hybrid', 'exile' and 'migrant' in contemporary cultural theory see Caren Kaplan, *Questions of travel: postmodern discourses of displacement* (Durham: Duke University Press, 1996), and James Clifford, *Route: travel and translation in the late twentieth century* (Cambridge: Harvard University Press, 1997). To criticize postcolonial theory for its emphasis on mobility, however, is not the same as dismissing postcolonial criticism because it is *written* by those whose mobility has enabled them to occupy privileged positions in the US academy.

112 The VAG's 1995 exhibition *Topographies: aspects of recent B.C. art*, went some way toward exploring this fractured visual field.

113 I borrow this distinction from Donna Haraway, *Simians*.

114 Yuxweluptun's work has received increased critical attention. Especially useful are the papers written for his solo show at the Morris and Helen Belkin Art Gallery, Vancouver, in 1995. See Yuxweluptun, *Lawrence Paul Yuxweluptun: born to live and die on your colonialist reservation* (Vancouver, Morris and Helen Belkin Art Gallery, 1995).

115 Yuxweluptun, 'Artist's statement', in *Born to live and die*, p. 1.

116 Importantly, Yuxweluptun has also maintained links to rural Native communities and currently lives in Fort St. James.

117 Both parents were involved with the North American Indian Brotherhood. For a time his father headed the Union of BC Chiefs and his mother worked with the Indian Home-makers Association of BC. See Charlotte Townsend-Gault, 'The salvation art of Yuxweluptun', in Yuxweluptun, *Born to live and die*.

118 Lisa Malkki, 'National geographic: the rooting of peoples and the territorialization of national identity among scholars and refugees', in Gupta and Ferguson, *Culture, power, place*.

119 The resurgence of 'traditional' Northwest Coast Indian art has been highly controversial. While individuals such as Bill Reid and Robert Davidson have been instrumental in reinvigorating traditional artistic styles and techniques, they have also been heavily criticized for placing Northwest Coast Indian art in an straightjacket of 'tradition'. Whether these criticism are fair is not entirely clear, but certainly one of the great ironies of this resurgence is that it has leaned heavily on the work of non-Native art historians such as Bill Holm, whose highly influential studies of Northwest Coast Indian art have served to 'codify' its forms and styles. See Bill Holm, *Northwest Coast Indian art: an analysis of form* (Seattle, University of Washington Press, 1965).

120 Yuxweluptun, 'Artist's statement'.

121 Townsend-Gault, 'Salvation art'.

122 Townsend-Gault, 'Salvation art'. Robert Linsley makes the point that Yuxweluptun's landscape

paintings recall multicoloured resource and land-use maps. What makes his paintings remarkable, then, is precisely that these resource landscapes are no longer seen to exist outside social relations as 'natural' resources, but are infused with social relations and colonial legacies. See Robert Linsley, 'Yuxweluptun and the West Coast landscape', in *Born to live and die*.

123 Townsend-Gault, 'Salvation art'.

124 Although this has not received much attention, Yuxweluptun's intervention in the politics of Clayoquot Sound – traditional *Nuu-chah-nulth* territories – is not without controversy.

125 Quoted in Townsend-Gault, 'Salvation art', p. 12, emphasis added.

126 Quoted in Townsend-Gault, 'Salvation art'. This mirrors Frida Kahlo's refusal to accept the surrealist label: 'I never painted dreams. I painted my own reality'. As quoted in Hayden Herrera, *Frida: A biography of Frida Kahlo* (New York, Harper and Row, 1983), p. xi–xii.

127 See Watson, 'Modernist past'; Townsend-Gault, 'Salvation art'; Charlotte Townsend-Gault, 'Hot dogs, a ball gown, adobe and words: the modes and materials of identity', in W. Jackson Rushing III, ed., *Native American art in the twentieth century* (New York, Routledge, 1999).

128 Townsend-Gault refers to this as 'reciprocal appropriation'. Townsend-Gault, 'Salvation art'.

129 Yuxweluptun, 'Artist's statement', p. 1. This statement should be read as in part a reaction to the way in which Native traditions were 'revived' in the 1960s as a catalogue of forms that appeared timeless and unchanging. See Holm, *Northwest Coast Indian art*.

130 See Braun, *The intemperate rainforest*.

Part V
Cultures of Nature and Science

[17]

SCIENCE, OBSERVATION AND ENTERTAINMENT: COMPETING VISIONS OF POSTWAR BRITISH NATURAL HISTORY TELEVISION, 1946–1967

Gail Davies

Popular culture is not the endpoint for the communication of fully developed scientific discourses; rather it constitutes a set of narratives, values and practices with which scientists have to engage in the heterogeneous processes of scientific work. In this paper I explore how one group of actors, involved in the development of both postwar natural history television and the professionalization of animal behaviour studies, manage this process. I draw inspiration from sociologists and historians of science, examining the boundary work involved in the definition and legitimation of scientific fields. Specifically, I chart the institution of animal ethology and natural history film-making in Britain through developing a relational account of the co-construction of this new science and its public form within the media. Substantively, the paper discusses the relationship between three genres of early natural history television, tracing their different associations with forms of public science, the spaces of the scientific field and the role of the camera as a tool of scientific observation. Through this analysis I account for the patterns of cooperation and divergence in the broadcasting and scientific visions of nature embedded in the first formations of the Natural History Unit of the British Broadcasting Corporation.

In spring 1957 the previously informal collaborations between naturalists, broadcasters and scientists directed at taking and broadcasting film of animals for British television were recognized in the establishment of the Natural History Unit (NHU) of the British Broadcasting Corporation (BBC). Based in Bristol and self-sufficient, with its own producers, library, film-editing facilities and eight permanent staff, the NHU was charged with supplying the majority of natural history output on the BBC.[1] In a paper to the BBC board of management, unit pioneer Desmond Hawkins identified the vision of natural history television framing their early development. In what has been called 'the most important and influential document in the unit's history',[2] he asserted:

> The spirit of scientific enquiry must have pride of place. In handling this subject we expose ourselves to the critical scrutiny of scientists, and their approval is an important endorsement. Moreover, it is their work that throws up the ideas and instances and controversies from which programmes are made. We look to them as contributors, as source material, as consultants, and as elite opinions on our efforts. In short we need their goodwill.[3]

The choice for head of unit, the ornithologist Bruce Campbell, reinforced an ethos in which scientific skills were valued above broadcasting experience. As Chris Parsons writes in his history of the unit: 'Although he had no experience as a producer, Bruce [Campbell] had shown himself to be truly interested in broadcasting and his impeccable ornithological background, together with his involvement with various naturalist's organisations provided just the right credentials for the leader of the NHU at the time.'[4] Certainly, there was competition from other styles of natural history television. Wildlife adventure programmes, previously successful in the cinema; broadcasts of the popular spectacles of the zoo; and a scientific vision of nature that emulated the naturalist's lecture all featured in early natural history broadcasts. However, only one genre was widely supported by film-makers and institutionalized within the first formations of the NHU: an expert vision of nature that stressed the importance of field observation and scientific interpretations of animal form and function.

Through the lens of the NHU, this paper explores how negotiations over the meaning of natural history films were implicated in the co-constitution of postwar broadcasting and natural history sciences. I argue that early natural history television was not the endpoint for the communication of fully developed scientific discourses on animal behaviour, but constituted a set of narratives, values and practices with which both scientists and broadcasters engaged. The paper traces relationships between broadcasters, film-makers and scientists simultaneously involved in the formalization of a unit for the production of natural history films and the establishment of animal behaviour studies as a biological science. From a loose association of individuals interested in the development and communication of natural history, patterns of cooperation and divergence emerge, which can be traced through discussions of the appropriate vehicle for the presentation of natural history on television. This paper examines these competing visions of nature as individuals are brought together by a belief in the restorative power of the natural, in the moral value of field observation and in the camera as a tool of science as well as of entertainment. Shared values in the public science of natural history and ethology, an emerging field science for the study of animal behaviour, establish natural history films as boundary objects, ensuring that cooperation is achieved, even though the films have different meanings to participants.

The research deals symmetrically with the natural science and cultural production of natural history film-making, positioning the development of natural history television within accounts of boundary-making practices between nature and culture, expert and lay, subject and object.[5] The work speaks theoretically to a growing dialogue between cultural geography and science studies through its treatment of science as a set of material and social practices located in particular places,[6] extending this to argue for the importance of the spaces of popular culture in the constitution of science. I have also drawn productively from recent work on animal geographies which explores the material and symbolic spaces occupied by animals in human culture.[7] Woven through accounts of early natural history film-making are reflections on the establishment of a proper location for understanding animal behaviour, whether that be the zoo, the safari or the field; the moral value attributed to interactions with animals in certain environments; and the formulation of an appropriate aesthetic regime to represent this in both scientific and public communications. Through its particular focus on the institutional formation of the NHU, a third literature to which this research alludes is that on the development of organizations.[8] This reminds us that processes of organization are necessarily heterogeneous, and thus, whilst suggestive of a common myth in the development of early British natural history film-making, this paper does not preclude the importance of other formats for the presentation of animals on television in the later history of the Unit.[9] The research draws on a range of interviews and documentary evidence within the NHU, including films, scripts and papers, as well as published accounts of the development of the Unit and biographies of key individuals involved in the multiple contexts of postwar British natural history.

The paper is presented as follows. First, I consider research on the role of popular culture in the construction of science, drawing on three modes of boundary work between science and non-science that have proved productive in historical accounts of natural history. Secondly, I discuss the context for the emergence of natural history film-making, and describe conceptions of natural history films from several different but related worlds, including those of the broadcaster, the ethologist and the natural history film-maker. Thirdly, I introduce three programme formats illustrating experiments in early natural history programming. I explore how wildlife adventure programmes, outside broadcasts based in the zoo, and the television format of the naturalist lecture are all responses to the challenges and contexts of early natural history film-making. Through their different technologies for filming animals and the narratives used to frame animals and environments, each type of programme suggests a particular model of human and animal interaction and presents a divergent view of nature. The patterns of centrality and marginality that emerge within these genres of natural history film-making are explored through their intersection with the changing practices of natural history sciences, broadcasting and natural history film-making. Concluding, I reflect empirically on the subsequent development of the unit and theoretically on the importance of considering the role of the media in the spatialization of differently situated scientific knowledges.

Bounding nature: institutional science and popular culture

Historians of science have persuasively argued against conceptions of the popularization of science that assume the existence of two independent and monolithic cultures – expert and lay – and construct science as a body of pre-established facts which are communicated to a passive public.[10] Shapin, for one, asserts that the professionalization and popularization of scientific knowledges are coterminous.[11] Publics are not involved only as audiences for non-controversial science constructed in the laboratory, museum or field, but are a constitutive part of the processes of producing and negotiating scientific methods and knowledges. On this basis popular culture is not simply the endpoint for a fully developed scientific discourse. Rather, it offers a set of identities, spaces and technologies which scientists have to negotiate in the establishment of scientific boundaries, the public legitimation of science and the differentiation of scientific and lay visions of nature.

Thomas Gieryn's concept of boundary work captures this process.[12] Gieryn explores the processes through which scientists assert the autonomy of scientific practices vis-à-vis applied science or technology, in order to lay claim to funding, whilst simultaneously claiming science as the foundation of technological progress. The notion of boundary work draws attention to the processes through which scientists, science journalists and the other allies define, delimit and defend certain fields; maintaining boundaries between science and non-science, science and technology and different paradigms of science. Scientific work is both heterogeneous and co-operative, and boundary work simultaneously connects and differentiates diverse interests within the locations of science, as well as linking core scientific debates to public science metaphors.[13] The task of demarcating science from non-science is a social process involving the attribution of selected characteristics to the practitioners, methods, knowledges, values and institutions of science.[14] It involves multiple actors engaged in negotiations between different worlds and, as Gieryn indicates, boundary work is complex, contested and contingent.[15] Three key modes of boundary work are valuable for understanding relations within the contexts of British postwar natural history: the public presentation of science, the spaces of field observation and the technical objects of scientific practice.

An important facet of boundary work is distinguishing science from other activities through advocacy in a cultural space, located a significant distance from the spaces of science.[16] Turner identifies public science as 'the body of rhetoric, argument and polemic produced . . . to persuade the public or influential sectors thereof, that science . . . is worthy of receiving public attention, encouragement and financing'.[17] The process of forging space for new forms of scientific understanding simultaneously involves creating a public form, or public science. This has historically been a literary form.[18] The nineteenth century saw an explosion in new journals and books about natural history, providing the means of communication between specialities and between men of science and the public.[19] Such communication contexts advocated public support for science and, through the religious, intellectual and utilitarian values claimed for it, distinguish the identity of amateur and professional. A similar role is played by the boundary work of public demonstration or experiment, which effectively legitimizes science through a smooth public presentation concealing the untidy craft of the scientist: 'caging Nature's caprices in thick walls of faultless display'.[20]

Public science functions as the front stage of scientific professionalisation. As Livingstone suggests, 'the Goffmanesque distinction between front and back regions, between public and private spaces snakes its way through the experimental programme more generally'.[21]

The rhetorical strategies through which public science is presented are important.[22] Following Myers, it is possible to identity the use of both 'narratives of nature' and 'narratives of science' in negotiating scientific value and identity.[23] The former are important in legitimating the relative positions of competing scientific paradigms through reference to wider discourses of nature that reinforce the social values of science.[24] Boundary work is also achieved through attributing unique characteristics to a scientific identity through narratives of science. In creating the role of natural history professional, the counter-category of amateur is simultaneously forged through a narrative of science that inscribes intellectual and social boundaries of class and gender.[25] This is often pursued in public presentations seeking to fashion a credible scientific identity through using rhetorical devices that displace other actors.[26]

Secondly, boundary work between science and other activities depends crucially on where practices are positioned, and on the spaces privileged for science. This builds explicitly on geographical interests in the location of science. As Thrift *et al.* observe, 'the study of science as a social construction has been pursued through a peculiarly spatial imaginary, which always attaches insight to the site. The locales in which scientific knowledge is produced are not seen as passive backdrops, but as vital links in the chain of production, validation and dissemination.'[27] Laboratories, field sites, scientific institutes and museums are socially and materially constructed as locations privileged for the practical activities that make and sustain scientific knowledge. Boundary work between scientists and publics are embedded in social and spatial relations, both in constructing legitimate locations for scientific observations and in the means of authoritative communication that transcend these sites.[28] From an initial focus on the laboratory, sociologists and historians of science are increasingly exploring other sites important in the construction of scientific knowledge including the natural history museum, botanical garden and the varied sites of fieldwork.[29]

The contested status of the field as a legitimate site for the generation of knowledge about nature is particularly important in the development of the natural history sciences. Field studies have generally had low status within the biological sciences.[30] Unlike the laboratory, zoo or museum, the spaces of the field cannot easily be restricted to sanctioned scientists or controlled for experimental conditions. Field studies suffer from accusations of subjectivity, and scientists struggle to enable an authoritative travel of knowledge between the undisciplined spaces of observation and the point of data analysis. Tensions between field naturalists and cabinet or museum naturalists can be traced from the eighteenth century onwards.[31] Burkhardt explores the diverse strategies pursued by naturalists from Cuvier, through Yerkes and Tinbergen, to establish the credibility of observation in the field for animal behaviour study. He introduces diverse boundary strategies for constructing the scientific field, including establishing the identity of a trustworthy observer, excluding other actors and using scientific instrumentation to assert the credentials of the field as a proper setting for the generation of scientific knowledge.[32]

As indicated above, boundary work also takes places through the negotiation of objects. In their study of the reorganization of ecological research at the Berkeley Museum of Vertebrate Zoology, Star and Griesemer discuss how cooperation is achieved through practices and objects that have varied meanings to diverse actors of science operating within different worlds. They define boundary objects as those 'both plastic enough to adapt to local needs and the constraints of the several parties employing them, yet robust enough to maintain a common identity across sites. . . . They have different meanings in different social worlds but their structure is common enough to make them recognizable, a means of translation.'[33] Boundary objects account for cooperation between heterogeneous viewpoints. In their study, such objects support the definition and maturing of the different worlds of museum administration, amateur collection and research science by standardizing the interfaces between them. 'The management of diversity proceeds with the use of objects which originate and continue to inhabit different worlds and which incorporate different meanings to make them coherent.'[34] Shared goals enable actors investing in the collaboration to work together despite their positions in different worlds.

In natural history work, boundary objects are produced when sponsors, theorists and amateurs collaborate to produce representations of nature.[35] Their boundary nature is indicated by their existence as concrete and abstract, particular and general, conventionalized and customized. In the representations of twentieth-century natural history, negotiations between actors on the status of visual observation, instrumentation and data recording are frequently articulated around the role of the camera. The camera as a boundary object is introduced in the work of Latour[36] and Winston,[37] and illustrated as an important point of negotiation within the complex matrix of science, art, hunting, education and entertainment in American natural history in the work of both Gregg Mitman and Donna Haraway.[38] The primate behaviour films of Clarence Ray Carpenter feature in Haraway's account of the discipline, as film becomes a key component in the development of field primatology. Similarly, Mitman suggests that the camera supports the credibility of field science through enabling observations in the field to transcend the point of observation.[39] The power granted to the camera as a scientific tool, able to move between the field site of natural history and the demands of scientific biology, was significant in the development of field animal behaviour studies. 'No longer dependent upon notebook and pencil, the biologist or anthropologist could record movement and behaviours on a medium that could then be transported to the laboratory, where movements could be slowed down and behaviour analysed, spliced, and edited.'[40] Film offered the opportunity to make the study of animal behaviour in the field more scientific.

However, film is also a technology of entertainment, and the use of hoaxing and set-ups to increase the spectacle of early wildlife photography in the United States was rife. Scientific film pioneers such as Carl Akeley at the American

Museum of Natural History set themselves against faking by encouraging associations with wildlife photographers Martin and Osa Johnson in attempts to secure and standardize film's scientific credentials.[41] A reading of the camera's ability to witness animal behaviour had to be sought and could not be assumed. As Haraway concludes, 'the camera, an eminently democratic machine has been crucial to crafting stories in biology. [However,] its control has eluded the professional and the moralist, the professional scientists.'[42] The apparent accessibility of this visual media challenges scientists' need to discipline access to means of representation and secure authority. In the interlinked worlds of field observation and natural history film-making, cameras and film are boundary objects whose stabilization is sought to enable the necessary collaboration between divergent worlds. I would assert that these negotiations around public science, the places of science and film as a boundary object could also be traced in the complex worlds of postwar British natural history.

Television, natural history and natural history television

Renewed transmissions of BBC television in 1946 introduced a new opportunity for the communication of ideas and images of animals within an already complexly configured space.[43] The worlds of broadcasting, biology and natural history in Britain underwent a number of transitions during this period. These contexts were shaped by an influential set of individuals involved in planning, ecology, ethology and broadcasting, moving across disciplinary boundaries more fluid than today. The worlds of professional biology and natural history were relatively close-knit, with amateurs and professionals mixing within clubs and societies, and working together on expeditions, at conferences, editing books and making television programmes.[44] These associations, centred on organizations like the British Trust for Ornithology, The Royal Society for the Protection of Birds, the British Ecological Society and the Royal Zoological Society, were highly influential: for example, in formulating shifts from preservation to conservation in planning for nature reserves,[45] and in rejuvenating field studies for the understanding of whole organisms within the biological sciences. The contexts for television were equally complex but less well established. Television was relatively experimental in the 1950s, with limited resources and few proven formats or experienced staff. Natural history film-makers operated across these worlds, building links within British regional contexts like London, Oxford and Bristol, through shared interests in developing and communicating a particular vision of nature. I now turn to chart these intersections between the visions of broadcasters, ethologists and natural history film-makers: three simultaneously moving targets, each changing shape to effect the position of the others.

The broadcaster's vision

The 1940s saw wide support for popular natural history, embracing conservation groups, nature photography and film, radio broadcasts and publishing ventures.[46] The resumption of television broadcasts after the war appeared to offer scope for furthering this. A small group of radio producers, broadcasters and scientists centred around the BBC in Bristol, including Desmond Hawkins, Peter Scott and James Fisher, had already established their reputation for natural history radio through home service broadcasts such as *The naturalist* (1946), *Bird song of the month* (1947) and *Birds in Britain* (1951). New ideas on nature and nature conservation were also popularised through publishing innovations like the Collins New Naturalist series.[47] The editors of this series, Dudley Stamp, James Fisher, Julian Huxley and John Gilmour, pioneered the publication of books combining the latest scientific research with a popular but unsentimental approach to natural history, based on field study. These books synthesized a British naturalist tradition with new developments in wildlife photography overseen by Eric Hosking; the artistic modernism of Clifford and Rosemary Ellis's dust jackets; and the developing fields of ecology and ethology. Nature was diverting and healing for a nation recovering from war, reasserting links with the past but also looking to the future. As James Fisher reputedly said to publisher William Collins in 1942, 'What this country needs is a good series of books on natural history to take its mind off the carnage.'[48] Natural history combined traditional values of empirical observation and the search for order with an emphasis upon scientific problem-solving, planning and educated citizenship held in high esteem in postwar British culture.

Television presented both opportunities and constraints for this public science of natural history. Broadcasting at the BBC was strongly shaped by the vision of the first director-general after the war, William Haley, who developed radio broadcasts in line with Reith's original intentions to 'inform, educate and entertain'.[49] He channelled resources into educational programming, introducing the Third Programme in 1946 and the Reith Lectures in 1948.[50] He did offer a supportive slot for natural history radio broadcasts from Bristol, scheduling them on Sundays after the one o'clock news.[51] However, he was less interested in developing the medium of television.[52] Television was still unproven in the late 1940s and early 1950s, with restricted signal transmission audiences and limited funding.[53] Compared to the large audiences and symbolic national presence achieved by radio during the war, television was a minority concern, with few specialist staff or programming strands. Television focused on emulating radio formats, with live outside broadcasts of national events such as the Olympics, the Proms and the Coronation; competing with cinema for immediacy, rather than attempting to emulate the quality of celluloid projections.[54] Live television, with its well-lit studios and large unwieldy electronic cameras, was a difficult environment for filming animals. Outside the studio, where developments in lightweight Bolex and Arriflex film cameras had made cine film more portable, there were still technological restrictions. Film stock required large quantities of light, limiting filming to animals active in daylight. Lens magnifications were constrained by technologies in glass lens manufacture, so film-makers had to approach close to the subjects being filmed. Both features in London and regional programming in Bristol endeavoured to overcome these problems, contributing a number of experimental strands and one off programmes. The

BBC also bought in existing film from film-makers previously popular in the cinema. However, in the early 1950s the BBC had no formalized policy for the production of natural history programming.

The BBC was, however, motivated by competition as this period progressed. The arrival of commercial Independent Television (ITV) in 1955 meant that British broadcasters sought to exert more direct control over film content and ownership, and to monitor and react to audiences. This was particularly important following the arrival of the ITV wildlife strand *Survival* in 1961. ITV broadcasts did not necessitate direct competition over scheduling, but did require the BBC to generate greater flexibility to audiences and better allocation of television resources. The ITV need to monitor and respond to commercially meaningful audience figures put pressure on the BBC to compete for respectable audience share. The BBC's audience research department was introduced in 1955, challenging the autonomy of its educational vision.[55] The development and maintenance of programme strands, balancing education and popular influences on programme content, were increasingly organized under the administration of a 'unit' within the BBC. Units were allotted a certain amount of broadcasting time, whilst producers within the unit were confined to proposals that conformed to the unit's general brief.[56] BBC broadcasters recognized the potential for natural history programming, but required a demonstrable track record before committing resources to a unit for this genre.

The ethologist's vision

The period before and after the Second World War saw development of the two most prominent approaches to the study of animal behaviour developed in the twentieth century – ethology and comparative psychology. Ethologists, such as Konrad Lorenz, Nikolaas Tinbergen and their associates, tended to study birds, fish and insects under field conditions. Their methods involved detailed observations and experiments in nature to reveal innate behavioural patterns related to evolutionary processes. Lorenz laid down many of the theoretical foundations of the new field of ethology through work on captive wild animals, reared in near-naturalistic conditions in a special experimentation station at his home in Austria where he could observe their behaviour closely.[57] The Dutch naturalist Tinbergen focused on field observation and experimentation, devising simple problems he could put to animals in the wild without disturbing them unduly.[58] The British naturalist Julian Huxley, himself a pioneer of the meticulous observation of birds in their environment, shared their ideals and was influential in bringing this research to the UK.[59] Ethologists channelled their work onto the visual sense, and made extensive use of film in research and public presentations of their work.[60] Competition for their interpretation of animal behaviour came from comparative psychologists, mainly based in the United States, who largely studied mammals under laboratory conditions and focused on learnt behaviour patterns.[61]

Both communities of researchers responded to challenges facing animal studies following decline in support for morphological and taxonomic studies and the rise of molecular biological experimental science.[62] However, they differed in their practices. Located in the field, classical ethologists worried about the distorting effects on natural patterns of behaviour created by moving animals from nature to the laboratory. Conversely, comparative psychologists were concerned about the lack of experimental control in the field, the absence of data analysis and ethologists' failure to consider environmental factors in development.[63] Ethologists had to fight to establish the idea that animal behaviour could best be understood by examining it in its 'natural' surroundings.[64] The field as a place of science, the status of the naturalist's vision and the moral value of empirical observation were central to boundaries between the concepts and practices of ethology and comparative ethology.

Early ethologists also differed from comparative psychologists in setting great store by their ability to communicate their vision of nature to the public, writing popular books and magazine articles for publications like *Country life*.[65] Tinbergen was 'keen to disseminate his ideas and discoveries about animal behaviour as widely as possible',[66] appearing in many films and television programmes. He emphasized this in his teaching: 'I try to impress on my students that half their work is communication. Science is a social effort, and scientists must adjust to the public.'[67] Ethologists recognized the importance of public science. However, it was not a predetermined scientific vision that was communicated. Dominant throughout their public presentations was an emphasis on the wonders of nature, the pleasures to be derived from its study and the importance of fieldwork as the proper location for the study of animal behaviour. Studying nature was also an exercise in accomplishing an ideal society, and field observation was a moral value before it became the hallmark of scientific ethology.[68]

In most accounts of this period of animal behaviour studies, the vision of ethologists dominates that of comparative psychologists. Dewsbury, for instance, identifies five dimensions in which ethologists could be said to have triumphed: through an evaluation of rhetoric used to describe their achievements; their presence in scientific histories of the period; their relative influence in cognate fields such as psychology; their role in training future biologists; and the 1973 Nobel prize for physiology or medicine awarded to Lorenz, Tinbergen and Karl Von Fisch.[69] However, there is often a contradiction within accounts of the dominance of the ethologist's vision of nature. Roell, for one, attributes the ethologists' success to their adoption of scientific method and experiment.[70] In comparison with the laboratory rigour of comparative psychologists, this seems difficult to substantiate. The power of ethology as a new set of practices and interpretative tools for understanding animal behaviour was not simply established by allying itself to the values of objective science. Ethology is a complex, contingent, historically situated and evolving set of practices and concepts forged by a disparate community of researchers and other allies. In forging this new vision for animal behaviour studies, a key role is played by scientific interactions with the media.

The natural history film-maker's vision

In the early 1950s a diverse set of individuals, including many mentioned above, began to experiment with natural history television broadcasts from Bristol. Desmond Hawkins, previously a writer, editor of literary criticism and radio producer in London, moved to Bristol after the war to take a course in television production and pursue interests in natural history.[71] His interests in nature and broadcasting inevitably brought him into contact with the naturalist, painter and broadcaster Peter Scott. Scott had radio and television experience, skills in filming and painting wildfowl, and was in the process of setting up the Wildfowl Trust at Slimbridge in 1946, just up the Severn estuary from Bristol. They were joined by the studio engineer Tony Soper, one production assistant paid for by television and the assistant film editor belonging to the West Region Film Unit, Chris Parsons.[72] These few individuals established links to the worlds of postwar natural history via other naturalists in publishing, conservation and scientific activities. An article on the *New Naturalist* photographic editor Eric Hosking illustrates the nature of these informal associations:

> A chance meeting with the well known ornithologist Guy Mountfort in a BBC studio led to an expedition to Spain in 1956, and every summer since then Hosking has been abroad adding to his already enormous library of bird photographs. The 1956 expedition was Las Marismas of the Coto Donana in Southern Spain, and was led by Guy Mountfort; party members included Lord Alanbrooke, Sir Julian Huxley, Max Nicholson, James Ferguson Lees and James Fisher.[73]

Early natural history film-makers were later to recruit within these networks; 'the people that were taken on here were great bird-watchers or members of the RSPB or the Entomological Society or Zoological Society.'[74]

Many of these naturalist film-makers looked to natural history television to

Figure 1 ~ Desmond Hawkins interviewing Peter Scott for the *Naturalist*, 1947 (copyright, BBC Natural History Unit)

provide something they had been unable to find within London broadcasting contexts or biological sciences. Hawkins's interests in literature embraced the study of landscape and nature in Hardy, Lawrence and Eliot, but he was frustrated with the London publishing world. Further, few British universities outside of Oxford, where Tinbergen established a new programme for animal ethology in 1949, then offered the opportunity to study animal behaviour in the field. As David Attenborough reflects, 'when I emerged from my degree the kind of research I saw going on in zoology in Cambridge was not the kind of behavioural ethological research I was interested in . . . I didn't want to go back to looking at dogfish in formalin. So I didn't go back to university at all.'[75] A shared desire to engage in and communicate a specific form of nature contemplation, stressing the primacy of experience and a moral value for fieldwork, motivated these individuals. However, they struggled with limited staff, experience and resources to commission new material and realise this vision on television. 'Because West Regions' bids were made from premises which were essentially radio-orientated, the means by which it approached the new medium were necessarily basic, [and] invariably inventive.'[76] Further resources and control could most easily be obtained through the establishment of a centre for the production of natural history television. A unit status would ensure that commissioning editors in London were committed to a quota of natural history, allowing scope to define their vision for natural history. However, unable to shoot specially new celluloid footage of animals, they were initially reliant on the use of mobile outside broadcast units and studios at Bristol, or existing sources of film from which to build their reputation. Drawing upon disparate modes of communication including the zoo, wildlife safari cinema and naturalist lecture, they nevertheless sought to establish a new unit for the production of natural history programmes which met broadcasting requirements and supported an appropriate aesthetic, moral and scientific vision for presenting and interpreting animal behaviour.

Competing visions of natural history television

The first natural history television broadcast after the war was an outside broadcast from Slimbridge in May 1953 presented by Scott and produced by Hawkins.[77] This marked the start of a rich period of natural history programming dominated by three genres of television. The wildlife series of Armand and Michaela Denis were transmitted from 1954; a variety of programmes featuring animals associated with zoos developed from 1954; and the programme strand *Look* was initiated from Bristol in 1955. This section introduces these three film genres and their relationships within the diverse social, spatial and technical worlds of postwar natural history.

Adventures in natural history

Nature films, travelogues, scenics and expedition films had been an important part of non-fiction cinema entertainment since the 1920s. From the incipient days of cinema when Cherry Keaton accompanied Roosevelt on hunting trips to East Africa, wildlife safari films, featuring explorers rather than naturalists, had become popular forms of animal entertainment. These early films marked a shift from big-game safaris to 'hunting with a camera',[78] emulating and communicating the experience of the hunt, but replacing the gun with the camera as the means through which trophies are captured and nature mastered.[79] The film-maker is still manly adventurer and hero of empire; indeed, as Carl Akeley put it, 'camera hunting takes twice the man that gun hunting takes.'[80] In these films nature is constructed as a primitive wilderness, 'implicated in broader movements to create and preserve a vision of nature as timeless domain for white Euro-American men'.[81]

The Belgians Armand and Michaela Denis, who started making films for cinema with the Renault-sponsored *Wheels across Africa* in 1930, brought this for-

Figure 2 ~ Armand and Michaela Denis on the front cover of the *Radio Times*, advertising their series *Safari to Asia*, 3 December 1960 (copyright, BBC Natural History Unit)

mat to BBC television in 1954 with the series *Filming wild animals*. This husband and wife team focused on the pursuit of animals, composing their films as a quest in which Michaela pursued animals with a still camera and Armand shot their progress using a 16mm portable film camera. The films were located in the national parks and game reserves of East Africa, places forged out of the African landscapes through the preservationist impulses and images of empty wildernesses held by European colonial administration.[82] Their carefully scripted narratives revisited these symbolic geographies, traversing bush and hunting-track, culminating in visits to the game warden's garden. The films are constructed using codes of fiction cinema, allowing the audience to experience the hunt through camera techniques such as point-of-view shots, and with edited sequences of pursuit adding drama to their adventures. They engage viewers through amusing dialogue, reflecting on the excitement and danger of their expeditions and including the non-expert, as the older Armand introduces the younger Michaela to the wildlife of East Africa. Their interpretations of animals are unashamedly anthropomorphic and include the search for such named individuals as the idiosyncratically long-horned rhinoceros, star of the 1958 film *Search for Gertie*.[83] The following extract, spoken by Armand, accompanied by visuals cutting between the couple filming and close-ups of the rhino, captures something of their style:

> I never feel at ease on foot approaching a rhino in open country with no protection whatever in case of a charge. A lot of people have got killed that way. I signal to Michaela, 'This is not the long-horn Gertie. Let us go.' Oh yes, Michaela wants at least one more picture. At last, to my relief, we're on our way back and when we least expect it a long pointed incredibly sharp forward-pointing horn emerges from behind a bush. We've no time to hide, we are at work taking pictures. This is Gertie all right – at last, the legendary Gertie, the most abnormal and freakish rhino horn we have ever seen. She sighs. She is aware of us, but she continues undisturbed.

The Denises' films were bought in by the BBC and edited for transmission by producers at Bristol. Over 89 programmes were transmitted, with series like *Filming wild animals* and *On safari* running until 1968. Records suggest they were popular with the public: audience shares ranged from 5 to 13%, and audience indices spaned from 65 to 80, attaining levels still considered good for natural history films.[84] Their broadcast hours were included in total programme output advancing the productivity of producers in Bristol with controllers in London.[85] For the BBC producers, buying in safari films provided access to an established genre of programme-making able to overcome the technological restrictions for filming animals. Animals were relatively easy to film in the open spaces of the Serengeti; many African mammals are active during the day, so filming could take place in good light and game wardens provided access to known haunts of key charismatic animals. These were clearly important films within the broadcasting context of the 1950s.

However, they were also criticized. Chris Parsons suggests that the series were 'embarrassingly expensive' and required 'a lot of post-production support' to edit into appropriate length and style for broadcast on the BBC.[86] He also records concern about their content.

> Some of us were uneasy about the Unit being responsible for the Denis series . . . It was undeniable that Armand and Michaela Denis were very popular with viewers and their films frequently contained excellent animal behaviour footage from Africa; but their personality double act – with alternating commentary – together with their 'pets' in the warden's garden routine made some of us uneasy.[87]

Despite their 'undeniable' popularity and their financial value, they were never central to the way members of the NHU defined their identity. Their programme output and audience ratings supported the productivity required for a unit, but unease with their vision of nature means that they are marginalized in histories of the unit's formation.

Explanation can be drawn from the narratives, locations and style of filming in these programmes. The Denises' films were located in environments closely associated with the Victorian pursuit of zoological specimens through hunting[88] and the interwar focus on reserving nature for the preservation of game animals.[89] The flamboyant and ritualistic hunting with a camera they present was difficult to reconcile with capturing animals on film for the purposes of supporting a modern science of animal behaviour. In their films the camera was positioned within a technological matrix that includes other accoutrements of the safari such as jeep and gun; and the conduct of Armand and Michaela Denis asserts the association of filming with the invasive practices of hunting. The camera's location within a particular hunting nature-culture devalued its status as a scientific instrument, and destabilized the associations with a scientific vision programme-makers were attempting to forge.

The view from the zoo

The 1950s also saw a proliferation of television programmes showing animals in zoos and studios. The zoo was an obvious site for early wildlife programmes through which captive animals could be incorporated into television programmes through outside broadcast technology. Access to all manner of animals could be guaranteed within the zoo, and 'you could get a signal either into a Post Office or telephone wires or by the radio dish'[90] to transmit the images live. This opportunity was widely exploited. The BBC unit for the production of television features in London produced *Looking at animals* and *All about animals* with George Cansdale. David Attenborough presented a series of *Zoo quests* (1954–61) from London studios, interspersing film footage of zoo collecting trips with studio footage, which introduced the animals close up. In Bristol *News from the zoos* (1959) was presented by James Fisher from a series of European zoos, with *World zoos* (1961) later extending this format. When ITV transmissions began in 1955, Granada built a studio within London Zoo where they presented *Zoo time* with Desmond Morris. Perhaps the best-known programme of this type was *Animal magic*, presented by Johnny Morris from Bristol Zoo in Clifton. This was to be one of the longest-running series for the NHU, transmitted for 21

Figure 3 ~ James Fisher with Bob Veeraswamy and a South American porcupine at Paignton Zoo, filming with an outside broadcast electronic camera for *News from the zoos*, 22 July 1959 (copyright, BBC Natural History Unit)

years, from 1962 to 1983.

The format of these programmes varied, but all featured zoo staff and broadcasters presenting animals within zoo enclosures and studios, and most did make 'good' television. As David Attenborough reflects,

> each week the curator of London Zoo brought a different animal into the studio and put it on a mat. It wasn't much fun for the animals who were plonked under the arc lights, but it was fairly apparent it was good television. The creatures would bite the curator, or relieve themselves down his jacket or escape and need to be chased around the studio.[91]

Broadcasting from the zoo or studio meant that programme-makers were able to share the visual appeal of exotic animals at the zoo.[92] Television programmes also innovated on the spectacle offered by zoos by showing presenters actually interacting with the animals. This tactile encounter with animals is promised at the zoo, where the real event is the physical presence of the animals themselves,[93] and zoos do offer opportunities for direct human animal interaction, through animal rides, demonstrations, children's zoos and public feeding. However, presenters, such as Johnny Morris, a radio comedian rather than a naturalist, were able to narrow further the gap between audience and animals. In his performance as zookeeper, Morris touched and interacted with all manner of animals, shifting back and forth between his own commentary and comic voiceovers for the animal's thoughts and conversations, giving voice, character and identity to the animals themselves.

Zoo programmes did become popular staples of wildlife television throughout the 1950s and 1960s, but their use became increasingly restricted to the lower budget and production values of children's television.[94] The presence of the studio, cameras and people at the zoo, and the light and heat of filming, meant 'natural' animal behaviour could not be captured. The interpretations of animal behaviour chosen by presenters like Johnny Morris again created unease at the NHU; at worst they were seen as sentimental and unscientific anthropomorphism, at best populist. As Chris Parsons recalls, 'His anthropomorphism offended some people, but I considered this a fair price to pay if we were able to reach viewers who did not normally watch *Look*.'[95] There was also increasing reluctance from zoos to be involved with this form of television. George Cansdale, presenter of *Looking at animals* and *All about animals* and superintendent of London Zoo from 1948 to 1953, was one of the first casualties of this shift. In 1953 he was sacked by the council of the London Zoological Society and his job divided between separate departments, reputedly because academic experts and officials at the zoological society disapproved of his television appearances in which he was seen playing with and cuddling all sorts of animals.[96]

Natural history programmes from the zoo were located within a complex and ideologically ordered space associated with Victorian developments in taxonomy.[97] Television had innovated on this view of nature through taking outside broadcast cameras backstage in these places of science, featuring footage of animals out of zoo opening hours, exhibiting unfamiliar behaviour in the studio, or interacting with people and celebrating a tactile experience of animals. The tactile interaction and subjective interpretation of animals presented by Johnny Morris and George Cansdale clashed with the scientific ambitions of men at the Royal Zoological Society and caused discomfort to film-makers in Bristol. As Birke suggests, 'Objective detachment is . . . stereotypically masculine in our culture . . . to identify with animals (a more 'feminine' position) is to cease to be objective.'[98] This presentation of animals and people contradicted the modernization of animal studies sought by zoo administrators and broadcasters, positioning the camera as a means of extending the spectacle of the least scientific parts of the zoo experience. This was acceptable for informing children about nature – with its long history of anthropomorphism – but inappropriate for a mature vision of modern natural history television. Emphasis was placed instead on the third strand of wildlife broadcasting, involving a different form of interaction with animals and drawing upon associations between naturalists and film-makers.

A programme of science and observation

The third strand of natural history film-making incorporated film footage of animal behaviour into the conventions of the naturalist's lecture. Starting with a number of experimental broadcasts from May 1953, *Look* was shaped into a series in August 1955 and ran for 15 years until 1969. *Look* was distinctively subtitled 'a programme of science and observation',[99] which differentiated it from entertaining antics at the zoo or romantic wildlife adventures. This was an approach pioneered by the Bristol producer Desmond Hawkins, as he describes:

> I suppose the aims I had were to break away from the old sort of anthropomorphic 'Dear Mr and Mrs Blackbird' kind of journalism, you know, that we had before. What I flew at my masthead were the words 'science and observation'. A programme of science and observation was what we tried to do. If you recall how much of television had been zoo animals, animals in studios, there was no doubt that you were being taken into the wild, that nothing was rigged or cheated and this was a tremendous advantage of course.[100]

Look aimed to take viewers out of the studio, shunning the showmanship of 'hunting with a camera', to show people a vision of nature untouched by humans emerging from the work of amateur naturalist film-makers and animal behaviour studies.

Look was broadcast live from the studios in Bristol and presented by Peter Scott. At the start of each programme Scott appeared seated behind his desk in a mock-up of the naturalist's study, surrounded by maps, charts, microscopes, field glasses and drawing-boards – props through which he would guide the viewer into new scientific perspectives on animal worlds. In the early programmes he would then move over to the sofa to meet his guest and maker of the film they were to introduce, watch and discuss. He would remove the film from the can, placing it in the projector and cranking the handle; the screen would flicker into life, introducing the latest footage of previously unseen animal behaviour. This programme format drew upon existing film footage, combining it with the demands of a live broadcast in the studio, where Scott held the programme together through his discussions with invited experts and drawings in the manner of a television lecture. Scott explains the evolution of this format in his memoirs:

> The pattern of the lecture was a short preliminary talk with drawings and a map followed by the first reel of film. There was then a short pause while the film reels were changed, during which I progressed the story with more drawings; finally at the end of the second reel of the film there was a brief conclusion. This technique was obviously capable of extension; so that instead of two reels one might have three or four sections of film with spaces for discussion in between them. This would then become more like a lantern slide lecture, but with film inserts instead of slides. Thus was the framework of the television lecture with film established.[101]

There were few resources to commission these short films. Producers at Bristol worked around these constraints by relying on the skills of amateur naturalist film-makers and scientists already filming animal behaviour. They used personal and professional networks to find people, often of private means, who had the time, skill and enthusiasm to extend the medium of wildlife photography from still to moving pictures.[102] As Tony Soper explains, 'we had to find film for these programmes. And we used to telephone round to Peter Scott's friends, to people like Eric Hosking, Lord Alanbrooke, people like that. Anyone who'd got an amateur film camera and did bird films in their holidays mostly.'[103] Pioneering film-makers like Heinz Sielmann followed the work of Lorenz, rearing wild animals in near-naturalistic conditions to observe and film their behaviour. Ernest Neal and Eric Ashby specialized in habituating mammal populations to obtain images of badgers and other British mammals. Other film-makers used field skills developed in the pursuit of animals and birds in the practices of European wildfowling to approach animals unseen. Further film was taken directly from ethological studies: Huxley's material of gannets, Fisher's of fulmars and Tinbergen's of herring gulls. Their aim was to inscribe a view of animal behaviour in the

Figure 4 ~ Peter Scott on location with the Kodak camera he used to shoot his lecture films, which featured in many of the early editions of *Look* (copyright, BBC Natural History Unit)

field apparently undisturbed by human intervention.[104] These early films were rarely made with the primary purpose of transmission on television, and many of them made no concessions to the demands of their new medium of broadcast. Sequences would be shot from one perspective, often featuring birds on the nest where their appearance could be assured, and transmitted with little editing. The emphasis of the film-makers was 'taking' images of animals rather than 'making' complete films. However, it was these films featuring natural animal behaviour captured in the field, their narratives, locations and film-making practices on the boundaries between television and science, which were most actively supported by pioneering natural history film-makers at the BBC.

The public science of *Look* supported the construction of the field as a site for the study of animal behaviour and the camera as a scientific method of recording animal movement. The presentation of film footage within the naturalist study affirms the authority of the camera as a scientific device with a status similar to the charts, maps and field glasses which surrounded Scott. The presentations of Scott differ from the anthropomorphic humour of Armand and Michaela Denis or the close identification with animals of Johnny Morris. His presentation of the trustworthy, gentlemanly identity of the naturalist reinforced his position as a credible spokesperson for science.[105] His introductions to the films structured the programme in the manner of a scientific text, 'a visual set of inscriptions produced by the instrument and a verbal commentary uttered by the scientist.'[106] Television programmes were able to make links between field, film and the practices of science that were difficult to achieve within the textual form of scientific articles. Their techniques of film montage cut between the empirical and abstract; 'the world of visible organisms and the unseen structures of information proposed in the theoretical model are seamlessly linked, literally in one line.'[107]

The legitimacy of these recordings of animal behaviour is further underlined by the processes and locations of filming. The animal's image is captured, but since the presenter stands outside the film no power is seen exercised over the animal. The naturalist film-maker looks but, unlike the hunter or the zookeeper, does not touch, and the images produced are thus able to transcend the moment of recording, universalized as natural behaviour. The camera constructs a landscape in which animal behaviour can be witnessed, but which is empty of human disruption. Without visible evidence of the complex connections between humans and animals in the urban zoo or the construction of exotic wilderness of east Africa, such sites could be coded as places for the professionalization of a modern biological science and as appropriate spaces for forging a new, forward-looking language for natural history television. Scientists and film-makers were brought together by the shared practices of detached observation, reinforcing the scientific status of ethology and realizing film-makers' vision for public natural history. However, this was not a popular science, for this public vision of nature divides expert and lay access to knowledge. The public presentations of ethological science in *Look* reinforced 'the critical boundary between watching and witnessing, between who is a scientist and who is not, and between popular culture and scientific fact'.[108] The move away from images of animals at the zoo, where animals could be seen interacting with presenters, meant that the spaces where viewers could see themselves within natural history programmes were devalued. With the subsequent disappearance of Armand and Michaela Denis from the screen, it would be many years before women were again fronting natural history programmes as credible spokespersons for animal behaviour. The stunning sequences of natural history behaviour in *Look* presented a view of nature only available to the specialist naturalist, ornithologist or film-maker. While natural history audiences could enjoy these, they were not experiences that many could empathize with or would seek for themselves.

The filming techniques and format pioneered by *Look* provided many successful films in the 1950s and 1960s for the NHU. Natural history footage from further afield was shown in *Look* and *Faraway look*, often featuring specially commissioned material from increasingly professional wildlife cameramen like Sielmann, Ron Eastman and Roger Jackmann. The academic format of natural history was developed in the series *Life*, presented by Desmond Morris, a former student of Tinbergen. There were collaborations with organizations like Oxford Scientific Films, linked to Tinbergen's Oxford programme of ethology, who were influential in developing high-magnification wildlife films for the study of insect behaviour. During the late 1950s producers at Bristol decided they could no longer use material filmed by naturalists or scientists with no concessions to television, concluding: 'We should tailor our films to suit the medium.'[109] As Chris Parsons observed, 'I was saddened by the realisation that this was probably the last time we would be able to use this great man's work [Viscount Alanbrooke], for his interest was in *taking* film of birds rather than *making* films about birds.'[110] However, the ethos of detached observation, field studies and scientific vision in programmes like *Look* and the associations between film-making and science forged were taken forward into core definitions of the NHU for natural history film-making at the BBC. They still endure in discussions at the unit today, supporting the lingering public service broadcasting remit of the BBC, and upholding a commitment to naturalism, the key to authority of this natural history form. As one of the longest-serving members of the Unit suggested:

> We've always had a good relationship with the scientific world. . . . And therefore, there was a sense of integrity, about our films which hopefully, you know, is not too badly strained even today. We haven't been making fairy stories. We haven't been making Disney films.[111]

Conclusion

These three programme formats, transmitted concurrently, represent different points of intersection between the worlds of postwar television and natural history, entertainment and science. All are innovative responses to the challenge of finding places for filming animals in early television, and all play an important role in establishing multiple resources for the institution of a designated unit for natural history presentations. Yet not all are privileged as places for constructing new knowledge about animals in postwar natural history. Their narratives of nature, science and adventure bisect attempts publicly to warrant the science of ethology and the moral values of field observation. Their location within the national parks of East Africa, the zoo and field are sites shared in the search for appropriate ways of filming animals for television programmes and studying animal behaviour. Each film location has a distinct geography involving different configurations of film-maker, broadcaster, scientist and animal, and constitutes a particular 'culture of nature'.[112] Different techniques of filming are used to incorporate images of animals into the worlds of television, only one of which constructs a credible spokesperson for nature in the practices of early ethology. The camera and film act as boundary objects between these practices, tying together diverse interests within these sites. Both the representational practices of television and science are important in understanding how certain films became both research tool and structuring metaphor in the production of animal behaviour research and the formations of early natural history television. Only the practices and presentations of public science in programmes like *Look* are capable of bringing together the diverse desires of film-makers and ethologists.

Through appeal to new forms of communication in the mass media, a vision of nature could be simultaneously professionalised and popularized. Early ethologists look to the media to legitimate and support their science of animal behaviour study. By positioning animal films as a way of moving between detached observation in the field and the public realms of science, a scientific view of animal behaviour could be universalized: the media 'publicly warranting that the knowledge produced in such places was reliable and authentic'.[113] The status of the field as scientific site and the use of film to move from field to study are mutually constituted – to establish a scientific identity for the field and the camera as an instrument of field science requires publicly attributing objective ideals to both at the same time.[114] The need to legitimate the scientific credentials of these concurrently is a driving motivation for the ethologist's appeal to the public, outside the normal channels of science. The BBC had established a reputation for educational programming, and with prior experience of natural history radio offered institutional veracity to its presentations of animal life: 'institutions play a crucial role in formulating boundary objects which allow for stable translations.'[115] At Bristol they found a group of film-makers who shared their ethic of close detachment and observation for nature, their view of the moral value of field experience, and a scientific and educational vision for the television camera. Producers at Bristol had come into natural history film-making, marginalized from the cultures of London broadcasting or seeking refuge from the increasingly mechanical objectivity demanded by molecular biology. In their desires to institute a unit for natural history film-making they were prepared to draw upon a range of television genres to fill schedules and support audience ratings, but consistently sought to formalize an ethos around ethological metaphors for understanding animal behaviour. The collaboration was mutually beneficial to the personal and professional satisfaction of individuals inhabiting both worlds.

The meanings attributed to these films and the purposes intended in their transmission differed, and were progressively to diverge from the late 1950s onwards. However, their narratives of nature form the basis of the productive collaboration between scientists and film-makers that has permeated the development of natural history film-making in Britain. The analysis offered here suggests this association supported the nature of the field as a scientific research site. It furthers recognition that the field is best understood as a heterogeneous and porous domain of scientific activity, involving multiple negotiations of many kinds between professionals and amateurs, by extending this analysis to include role of television natural history films. Analysis of these programmes as boundary objects deepens our understanding of how film simultaneously manages to advocate the primacy of experience whilst successfully excluding popular experiences and other knowledges of nature. Through exploring these relations from the perspective of the media, this paper has suggested that visualizations of nature are not simply pioneered in an expert realm to be picked up and popularized in the public sphere. The technologies, locations, rhetorics and institutional associations of competing visions of nature both define and re-create boundaries between expert and popular understandings of the natural world. Non-scientific institutions play an important role in the practices of knowledge-making, shaping the geographies of both scientific and popular knowledge production. Geographers have much to contribute to exploring the plural and contested sites involved in the making of science, tracing how diverse contexts and situations traditionally located 'outside' scientific work are articulated within it. Whilst historians and sociologists of science are increasingly attending to the articulation of science with its public dimension, less well observed are the struggles over these forms of expertise within the contested spaces of popular culture. This paper has been concerned to chart the dynamics between scientific and popular forms of knowledge within early natural history film-making, but many of the cultural spaces of postwar television and science remain to be explored.

Acknowledgements

I would like to acknowledge the support of the ESRC, without whose financial support this research would not have been possible. I am deeply indebted to everyone at the BBC's Natural History Unit who contributed their time. In particular thanks to Alan Baker, the head of the NHU film library, for leading me through the NHU archives, and to Helen Gilks from the NHU picture library for locating the images of early natural history film-makers reproduced here. I would like to thank Jacquie Burgess for formative guidance in the development of this research, Brian Balmer for helpful conversations on the nature of boundary work and Phil Crang for valuable comment in shaping this paper. Thanks finally to two anonymous reviewers for their very useful suggestions for improvement.

Notes

1 C. Parsons, *True to nature: Christopher Parsons looks back on 25 years of wildlife filming*

with the BBC Natural History Unit (London, Patrick Stephens, 1982), p. 58.

2 *Ibid.*, p. 173.

3 *Ibid.*, pp. 173–4.

4 *Ibid.*, p. 90.

5 J. Murdoch, 'Inhuman/nonhuman/human: actor-network theory and the prospects for a nondualistic and symmetrical perspective on nature and society', *Environment and planning D: Society and space* **15** (1997), pp. 731–56; S. Whatmore 'Hybrid geographies: rethinking the "human" in human geography' in D. Massey, J. Allen and P. Sarre, eds, *Human geography today* (Oxford, Polity Press, 1999), pp. 22–39.

6 See e.g. J. Murdoch, 'Towards a geography of heterogeneous associations', *Progress in human geography* **21** (1997), pp. 321–37; D. Livingstone, 'The spaces of knowledge: contributions towards a historical geography of science', *Environment and planning D: Society and space* **13** (1995), pp. 5–34; N. Thrift, F. Driver and D. Livingstone, 'The geography of truth', *Environment and planning D: Society and space* **13** (1995), pp. 1–3; S. Shapin, 'Placing the view from nowhere: historical and sociological problems in the location of science', *Transactions of the Institute of British Geographers* **23** (1998), pp. 5–19.

7 J. Wolch and J. Emel, *Animal geographies: place, politics and identity in the nature–culture borderlands* (London, Verso, 1998); C. Philo, 'Animals, geography and the city: notes on inclusions and exclusions', *Environment and planning D: Society and space* **13** (1995), pp. 655–82; C. Philo and J. Wolch, 'Through the geographical looking glass: space, place and human–animal relations', *Society and animals* **6** (1998), pp. 103–18.

8 J. Law, *Organizing modernity* (Oxford, Blackwell, 1994); W. Kaghan and N. Phillips, 'Building the tower of Babel: communities of practice and paradigmatic pluralism in organization studies', *Organization* **5** (1998), pp. 191–215.

9 G. Davies, 'Narrating the Natural History Unit: institutional orderings and spatial strategies', *Geoforum* (forthcoming).

10 S. Hilgartner, 'The dominant view of popularization: conceptual problems, political uses', *Social studies of science* **20** (1990), pp. 519–39; R. Cooter and S. Pumfrey, 'Separate spheres and public places: reflections on the history of science popularisation and science in popular culture', *History of science* **32** (1994), pp. 237–67; M. Bucchi, *Science and the media: alternatives routes in science communication* (London, Routledge, 1998).

11 S. Shapin, 'Science and the public', in R. C. Olby, ed., *Companion to the history of modern science* (London, Routledge, 1990), p. 1001.

12 T. Gieryn, 'Boundary-work and the demarcation of science from non-science: strains and interests in professional ideologies of scientists,' *American sociological review* **48** (1983), pp. 781–95; T. Gieryn, *Cultural boundaries of science* (Chicago, Chicago University Press, 1999).

13 S. L. Star and J. R. Griesemer, 'Institutional ecology, "translations" and boundary objects: amateurs and professionals in Berkeley's museum of vertebrate zoology, 1907–1939', *Social studies of science* **19** (1989), pp. 387–420.

14 Gieryn, 'Boundary-work and the demarcation of science from non-science', p. 782.

15 For example, whilst publics may be important to the constitution of a field of enquiry as a scientific realm, once established the public legitimation of science begins to rest precisely on its autonomy vis-à-vis public questions.

16 F. Turner, 'Public science in Britain 1880-191' *Isis* **71** (1980), pp. 589–608; T. Gieryn, G. Bevins and S. Zehr, 'Professionalisation of American scientists: public science in the creation evolution trials', *American sociological review* **50** (1985), pp. 392–409; in geography, see C. Withers, 'Towards a history of geography in the public sphere',

History of Science **37**, 115 (1999), pp. 45–78.

17 Turner, 'Public science in Britain', pp. 589–90.

18 Shapin argues that 'the differentiation and specialisation of science . . . created an opportunity for the explicit "popularisation" of science, and thus, for literary forms designed to convey otherwise inaccessible or impenetrable scientific knowledge to sectors of the public: 'Science and the public', p. 1001.

19 R. Barton, 'Just before nature: the purposes of science and the purposes of popularization in some English popular science journals of the 1860s', *Annals of Science* **55** (1998), pp. 1–33.

20 H.M. Collins, 1988, quoted in Livingstone, 'Spaces of knowledge', p. 23.

21 *Ibid.*, pp. 22–3.

22 A. Holmquest, 'The rhetorical strategy of boundary work' *Argumentation* **4** (1990), pp. 235–58.

23 G. Myers, *Writing biology: texts in the social construction of scientific knowledge* (Madison, University of Wisconsin Press, 1990).

24 Exploring the narratives of nature in popular writings on animal behaviour studies in the 1940s, Dewsbury explains the prominence European ethologists, such as Nikolaas Tinbergen, achieved over the more diffuse field of comparative psychology through the content of their well-targeted promotional activities: D. Dewsbury, 'Rhetorical strategies in the presentation of ethology and comparative psychology in magazines after World War II', *Science in context* **10** (1997), pp. 367–86.

25 A. Secord, 'Science in the pub: artisan botanists in early 19th century Lancashire', *History of science* **32** (1994), pp. 269–315.

26 The work of Tucker on scientific claims in the context of Victorian ballooning considers popular culture as a competitive field where 'scientists had to work against – or redirect – stereotypes of discoverers and explorers to fashion a credible identity for themselves': J. Tucker, 'Voyages of discovery on oceans of air: scientific observation and the image of science in an age of "balloonacy" ', *Osiris* **11** (1996), p. 175.

27 Thrift *et al.*, 'The geography of truth', p. 2.

28 'A trust relationship is central to the very idea of empirical science [and] that relationship is inscribed in space: those who have not seen these things know them by trusting those who have': Shapin, 'Placing the view from nowhere', p. 8.

29 H. Kuklik and R. Kohler, 'Science in the field', *Osiris* **11** (1996); N. Jardine, J. A. Secord and E. C. Spary, eds, *Cultures of natural history* (Cambridge, Cambridge University Press, 1996).

30 H. Kuklik and R. Kohler, 'Introduction', *Osiris* **11** (1996), pp. 1–14.

31 R. W. Burkhardt, 'Ethology, natural history, the life sciences, and the problem of place', *Journal of the history of biology* **32** (1999), pp. 489–508; see also D. Outram, 'New spaces in natural history', in Jardine *et al.*, *Cultures of natural history*, pp. 249–65.

32 See also Secord, 'Science in the pub'.

33 Star and Griesemer, 'Institutional ecology', p. 393.

34 *Ibid.*

35 *Ibid.*, p. 408.

36 B. Latour, *Science in action: how to follow scientists and engineers through society* (Cambridge, Cambridge University Press, 1996).

37 B. Winston, 'The documentary film as scientific inscription', in M. Renov, ed., *Theorizing documentary* (London, Routledge, 1993), pp. 37-57.

38 D. Haraway, *Primate visions: gender, race and nature in the world of modern science* (London, Routledge, 1989); G. Mitman, 'Hollywood technology, popular culture and the American Museum of Natural History', *Isis* **84** (1993), pp. 637–61.

[39] G. Mitman, 'When nature is the zoo: vision and power in the art and science of natural history', *Osiris* **11** (1996), pp. 117–43.

[40] Mitman, 'Hollywood technology', p. 639.

[41] Haraway, *Primate visions*, p. 44.

[42] *Ibid.*

[43] Television transmission in the UK had begun in 1922; however, television broadcasting was abandoned during the war since it was thought the transmitter at Alexandra Palace would act as a target for the bombing of London.

[44] P. Marren, *The new naturalists* (London, Collins, 1995).

[45] J. Sheail, *Nature in trust: the history of nature conservation in Britain* (Glasgow, Blackie, 1976); J. Sheail, 'War and the development of nature conservation in Britain', *Journal of environmental management* **44** (1995), pp. 267–83.

[46] D. Matless, 'Visual culture and citizenship: England in the 1940s', *Journal of historical geography* **22** (1996), p. 433.

[47] See Marren, *The new naturalists.*

[48] *Ibid.*, p. 21.

[49] Haley liked to use the image of British society as a cultural pyramid slowly aspiring upward, with the BBC acting as the lever that activated this progressive upward movement. Quoted in K. Kumar, 'Holding the middle ground: the BBC, the public and the professional broadcaster', in J. Curran, M. Gurevitch and J. Woollacott, eds, *Mass communication and society* (London, Edward Arnold, 1977), p. 246.

[50] J. Cain, The BBC: *70 years of broadcasting* (London, BBC Publications, 1992).

[51] Parsons, *True to nature*, p. 27.

[52] *Ibid.*

[53] Television had only one-tenth of the funding of radio. In 1950 the television signal only covered 33–50% of the British population, focused on London, the south-east and six other major cities, and broadcasting was limited to 30 hours a week. By 1953 only around 1 million television licences had been sold at a cost of £1 each: Cain, *The BBC*, p. 63.

[54] Parsons, *True to nature*, p. 27. For example, television news began simply as a reading of the news with no presenter in vision; pictures did not accompany news bulletins until July 1954: Cain, *The BBC*, p. 64.

[55] For further information on the role of audience measures in shaping broadcasting output, see I. Ang, *Desperately seeking the audience* (London, Routledge, 1991); J. Ettema and D. Whitney, *Audiencemaking: how the media create the audience* (London, Sage, 1994).

[56] M. Allaby, 'The Natural History Unit', *Vole* **12** (1978), p. 3.

[57] P. Klopfer, *Introduction to animal behaviour* (Englewood Cliffs, NJ, Prentice-Hall, 1974), p. 34.; Burkhardt, 'Ethology, natural history, the life sciences and the problem of place'.

[58] W.H. Thorpe, *The origins and rise of ethology* (London, Heinemann, 1979).

[59] A. Bramwell, *Ecology in the twentieth century* (London, Yale University Press, 1989).

[60] On the importance of the visual in the study of ethology, Mitman writes: 'Lorenz and Tinbergen developed a theoretical framework for animal behaviour that centred on animal posture and visual display as signs of stimuli that triggered innate releasing mechanisms in organisms': 'Hollywood technology', p. 652.

[61] Dewsbury suggests that ethologists and comparative psychologists came into direct conflict in 1953 through an exchange over Konrad Lorenz's theory of instinctive behaviour. Dewsbury, 'Rhetorical strategies in the presentation of ethology'.

[62] K. Vernon, 'Desperately seeking status: evolutionary systematics and the taxonomists' search for respectability 1940–60', *British journal for the history of science* **26** (1993),

pp. 207–27.

63 Dewsbury, ''Rhetorical strategies in the presentation of ethology'.

64 As Bramwell observes 'the idea that animal behaviour could be understood by close observation in the wild found opposition. It was difficult to carry out controlled experiments in these conditions': *Ecology in the twentieth century,* p. 41.

65 See e.g. K. Lorenz, *King Solomon's ring: new light on animal ways* (London, Methuen, 1952); N. Tinbergen *Curious naturalists* (London, *Country Life,* 1958); N. Tinbergen, 'The field is our laboratory', *Country life* **110** (1951), pp. 70–72.

66 Parsons, *True to nature,* p. 259.

67 E. Hall, 'A conversation with Nobel Prize winner Niko Tinbergen' *Psychology today* **7** (1974), p. 66.

68 H. van Rappard, 'Book review – The world of instinct: Niko Tinbergen and the origin of ethology in the Netherlands 1920–1950 by D. R. Roell', *Journal of the history of the behavioural sciences* **34** (1998), pp. 63–65.

69 Dewsbury, 'Rhetorical strategies in the presentation of ethology'. See also D. Dewsbury, 'Comparative psychology and ethology: a reassessment', *American Psychologist* **47** (1992), pp. 208–15; D. Dewsbury, 'Americans in Europe: the role of travel in the spread of European ethology after World War II', *Animal behaviour* **49** (1995), pp. 1649–63.

70 For a discussion of this book in English, see Rappard, 'Book review'.

71 D. Hawkins, *When I was: a memoir of the years between the wars* (London, Macmillan, 1989).

72 Parsons, *True to nature,* p. 52.

73 This extract is from an undated, unsourced article on Eric Hosking in the Natural History Unit archives.

74 John Sparks, series producer, Natural History Unit, interviewed 13 June 1995.

75 David Attenborough, quoted in J. Burgess and D. Unwin, 'Exploring *The Living Planet* with David Attenborough', *Journal of geography in higher education* **8** (1984), p. 112.

76 Parsons, *True to nature,* p. 26.

77 For a full filmography of BBC natural history films from 1953 to 1995, see G. Davies, 'Networks of nature: stories of natural history film-making from the BBC' (PhD thesis, University of London, 1998).

78 J. Ryan, *Picturing empire : photography and the visualization of the British Empire* (London, Reaktion, 1997); J. MacKenzie, *The empire of nature: hunting, conservation and British imperialism* (Manchester, Manchester University Press, 1988).

79 This form of film-making retains the characteristics of a hunt without harm to the animal; as Roosevelt suggests, 'the chief attractions lie in the physical hardihood for which the life calls, the sense of limitless freedom which it brings, and the remoteness and wild charm and beauty of primitive nature. All of this we get exactly as much in hunting with the camera as in hunting with the rifle': quoted in P. Schmitt, *Back to nature: the arcadian myth in urban America* (New York, Oxford University Press, 1969), p. 146.

80 Quoted in Ryan, *Picturing empire,* p. 137

81 *Ibid.*

82 R. Neumann, 'Ways of seeing Africa: colonial recasting of African society and landscape in Serengeti National Park', *Ecumene* **2** (1995), pp. 149–69; R. Neumann, 'Dukes, earls and ersatz Edens: aristocratic nature preservationists in colonial Africa', *Environment and planning D: Society and space* (1996), pp. 79–98.

83 This extract is taken from *Search for Gertie,* post-production script (NHU library, transmission date 1958).

[84] Audience measures were introduced following the advent of ITV, which put pressure on the BBC to compete with the ITV companies for a respectable audience share to support its statutory fee. The BBC's audience research department was introduced after 1955 to monitor the number of people who had seen a programme, counting in total those who had watched more than half of a given programme. They also established a numerical code or percentage indices for measuring audience appreciation, using structured questionnaires to gauge how educational and interesting the audience found each programme.

[85] Parsons, *True to nature*, p. 70.

[86] *Ibid.*, p. 72.

[87] *Ibid.*

[88] Ryan, *Picturing empire*, p. 99.

[89] Neumann, 'Dukes, earls and ersatz Edens'.

[90] John Sparks, series producer, Natural History Unit, interviewed 13 June 1995.

[91] Quoted in an interview with David Attenborough in the *Independent Magazine* (29 Sept. 1990), p. 48.

[92] J. Berger, 'Why look at animals?', in *About looking* (London, Writers & Readers, 1980), pp. 1–26.

[93] S. Montgomery, 'The zoo: theatre of the animals', *Science as culture* **4** (1995), pp. 565–600.

[94] From the end of *Animal magic* in 1982 until the animal hospital and zookeeper documentaries in the mid-1990s, zoos ceased to feature on television as places of entertainment or amusement, with programmes on animals in the zoo more typically taking a campaigning or critical stance. See e.g. *Zoo 2000*, transmitted in 1984, and *The state of the ark*, produced by Features in London in 1986. One exception to this was a 1987 live broadcast from the zoo, *Zooweek*, produced at the peak of the BBC's experiments with the short regular visits to outside broadcast locations in the live watch format.

[95] Parsons, *True to nature*, p. 103.

[96] *Guardian* 'The man from the Zoo: an obituary of George Cansdale' (26 Aug. 1993).

[97] K. Anderson, 'Culture and nature at the Adelaide Zoo: new frontiers in "human" geography', *Transactions of the Institute of British Geographers* **20** (1995), pp. 275–95; R. J. Hoage and W. A. Deiss, eds, *New worlds, new animals: from menagerie to zoological park in the nineteenth century* (London, Johns Hopkins University Press, 1996).

[98] Quoted in B. Crowther, 'Towards a feminist critique of television natural history programmes', in P. Florence and D. Reynolds, eds, *Feminist subjects: multi media* (Manchester, Manchester University Press, 1995), pp. 136–7.

[99] D. Hawkins, quoted in post-production script, *Wildlife jubilee*, produced by George Inger and Peter Bale (BBC Natural History Unit Library, transmission date 1982).

[100] Desmond Hawkins, quoted in post-production script *TV50 natural history*, written and presented by Desmond Morris, produced by Sue Bourne (BBC Natural History Unit library, transmission date 1986).

[101] P. Scott, *The eye of the wind* (London, Hodder & Stoughton, 1961), p. 601.

[102] An anniversary programme made to celebrate the 21st edition of the series introduces the people and animals involved in these early film sequences. Transmitted at 7.50 p.m. on 23 May 1956 with the title *The twenty first*, the programme opens with scenes from Peter Scott's footage of Iceland and his film of birds including phalaropes and limpkins; then more familiar seals; Heinz Sielmann's pioneering footage of woodpeckers; and swans, flamingos and storks. Footage of mammals revealed hamsters, pine martens, polecats and koala bears. Further film of birds

showed penguins, gannets and golden eagles. Other families of animals were represented by frogs and swallowtail butterflies. These short film extracts were counterpoised by cuts to discussions with Peter Scott in the studio. The closing credits introduce the early filming sources, listing film sequences by: Field Marshal Viscount Alanbrooke, Australian News and Information Bureau, Dick Bagnall-Oakley, BBC Film Unit, Sir Geoffrey de Havilland, Walter Higham, H. G. Hurrell, G. T. Kay, Captain C. W. R. Knight, Bernard Kunicki, Roger Tory Peterson, RSPB Film Unit, Peter Scott, Heinz Sielmann, William Sladen and G. K. Whitehead. The programme's film editor was Christopher Parsons and the programme was produced by Brandon Acton-Bond. Post-production script, *Look no. 21: the twenty first,* introduced by Peter Scott and Desmond Hawkins, produced by Brandon Acton Bond (BBC Natural History Unit Library, transmission date 1956).

[103] Tony Soper, quoted in post-production script, *Wildlife jubilee.*

[104] A hunting history was clearly still important to some individuals: 'This involved achieving the power of death over the quarry without exercising it . . . This advance was no sudden change in my outlook. I was still too much captivated by the ritual aspects of wildfowling to give it up altogether. So many things went with the actual shooting – the beauty, the natural history, the exercise, the memories, and particularly the technical skills which in the course of years I had acquired': Scott, *Eye of the wind,* p. 173.

[105] Shapin, 'Placing the view from nowhere', p. 8.

[106] Latour, *Science in action,* p. 71.

[107] G. Myers, 'Every picture tells a story: illustrations in E.O. Wilson's sociobiology', in M. Lynch and S. Woolgar, eds, *Representation in scientific practice* (London, MIT Press, 1990), p. 261.

[108] D. Haraway, *Modest_witness@second_millenium.femaleman©_meets_oncomouse™: feminism and technoscience* (London, Routledge, 1997), p. 33.

[109] Parsons, *True to nature,* p. 46.

[110] *Ibid.,* p. 65 (emphasis original).

[111] John Sparks, series producer, Natural History Unit, interviewed 13 June 1995.

[112] D. Matless, 'Moral geography in Broadland', *Ecumene* 1 (1994), pp. 127–55.

[113] Shapin, 1988, quoted in Livingstone, 'The spaces of knowledge', p. 21.

[114] Anecdotal evidence suggests this achievement cannot be taken for granted. The Collins New Naturalist series had pioneered the extensive use of modern photographic techniques in the production of these popular field guides to natural history. However, other natural history field guides were very slow to adopt such new techniques, suggesting that the camera took a long time to be accepted as the main form of scientific form of animal representation. The first Audubon field guide to use photographs of birds rather than paintings and drawings was not issued until 1977. B. Winston, 'The documentary film', p. 39.

[115] F. Harvey and N. Chrisman, 'Boundary objects and the social construction of GIS', *Environment and planning A* **30** (1998), p. 1687.

[18]

"ILLUSORY RICHES": REPRESENTATIONS OF THE TROPICAL WORLD, 1840-1950

David Arnold
Department of History, School of Oriental and African Studies, University of London, UK

ABSTRACT

This article traces the emergence of a negative strain in European and North American representations of the tropics, as earlier images of natural abundance were supplemented, and partly supplanted, by the fears and frustrations of would-be colonisers and the growing realisation of the technical difficulties of tropical "development". Taking Pierre Gourou's *The Tropical World* (1947) as an exemplary text that embodies attitudes accumulated over the previous century of scientific activity and colonial administration in the tropics, it is possible to see how, despite recognition of wide regional variations, the tropics as a whole were seen as constituting an impoverished and pestilential region, largely unsuited to white settlement and agriculture, and yet reliant upon outside agency for prospects of development. Without entirely ceasing to be landscapes of desire, the tropics represented a more primitive world than the northern temperate zone, a domain of largely untamed nature that served, by contrast, to demonstrate the moral and material "superiority" of northern climates, races and civilisations.

INTRODUCTION

In 1947 the French geographer Pierre Gourou published his magisterial survey, *Les pays tropicaux* (Gourou, 1947). Translated into English six years later (1953a), and twice revised and republished by 1961, *The Tropical World* was both a summation of Gourou's earlier research in pre-war Indo-China (Gourou, 1931, 1940) and a source of inspiration for a new generation of studies in tropical geography (*Études*, 1972). *The Tropical World* rapidly became not only a widely cited and respected work of modern geography; it was also a classic example of a "geography of the mind" (Lowenthal & Bowden, 1976), a high point in the discursive representation of the tropics (which will be referred to here as "tropicality") that had formed over the preceding century of imperial expansion and control. Written when Gourou was in his late forties, and published at a time when he had already made the giant leap from being a lycée teacher in Hanoi to holding a prestigious chair in "tropical studies" at the Collège de France in Paris, *Les pays tropicaux* carried forward into the metropolitan geography of the 1950s and 1960s many ideas and preconceptions that had their origins in the pre-war world, especially in colonial Southeast Asia, but which also sought to grapple with ideas of environmental determinism and "possibilism" that had been circulating in European and North American scientific and academic thought for decades. The text thus tells us as much about a collective (but by the 1950s already rather dated) northern world-view of the intra-tropical zone as it does about the "tropical world" it seeks to depict. While tropicality embraced a variety of cultural tropes, *The Tropical World* predominantly stands for one particular strand of this geographical and political discourse – a belief in the intrinsic "inferiority" of tropical as opposed to temperate environments and hence in the primitivism of the social and cultural systems to which the tropics gave rise.

TWO SIDES OF THE TROPICS

Before elaborating on Gourou's work and its significance, it is first necessary to understand the historical emergence of the tropics as a conceptual, and not merely physical, space. European ideas about the tropics, as they had developed as a result of exploration, conquest and colonisation since the late fifteenth century, were necessarily diverse, but fall broadly into two categories. Once Europeans had overcome the ancient belief that the "torrid zone" was too hot to be habitable (Oviedo, 1959:29-30), there emerged a powerful perception of the tropics as paradisiacal. As is evident from the Caribbean voyages of Columbus (Cohen, 1988), the tropics were seen as lands of great natural abundance, alive with luxuriant vegetation and exotic birds and animals, and blessed with perennially warm climates. Spared the cold and hunger of northern winters, humans in the tropics could enjoy easy, year-round, subsistence in return for minimal labour. The tropics, especially the islands of the Caribbean, the Indian Ocean and latterly the Pacific, conjured up visions of an earthly paradise, a veritable Garden of Eden. Owing much to early voyages of discovery and subsequently to travellers' tales and novels like *Paul and Virginia* (Bernardin de Saint-Pierre, 1982), the representation of the tropics as landscapes of desire was heightened by an intensified "feeling for nature" in eighteenth-century Europe (Charlton, 1984:109). By that time "a full-fledged myth of tropical exuberance" had become established in Europe (Curtin, 1964:60).

The discourse of tropicality parallels the idea of Orientalism as a cultural and political construction of the west discussed by Said (1978). The tropics, too, represented an enduring alterity, but one which qualifies and extends the Orientalist paradigm, not least by demonstrating that historically Europe possessed more than one sense of "otherness". Ideas of the tropics did not easily conform to a single stereotype. As Duncan (1990:182) remarks, "landscapes never have a single meaning; there always exists the possibility of different readings". The symbolism of the tropics was deeply ambivalent, for a landscape of seeming natural abundance and great fertility was also paradoxically a landscape of poverty and disease, and debate raged about the tropics and how they might most effectively be brought into productive subordination to the north (Livingstone, 1999). Broadly speaking, the tropics were represented as a landscape in which the power of nature dominated human existence and to no small degree determined its characteristics and quality. Partly because of this perceived dominance of nature and the role that naturalists, physicians and geographers played in constructing ideas about the tropics and in shaping colonial practice in the tropics, the emergence of tropicality was intimately associated not just with the visual arts and literature of Europe and North America, but also with its scientific disciplines and technical specialities. Representations of the tropics evolved symbiotically with the natural and human sciences: botany, zoology, anthropology and geography in particular bore the imprint of Europe's sustained encounter with the tropical world.

While it is beyond the scope of this discussion to explore the tropics' role in the scientific disciplines of Europe, it needs to be stressed how important scientific ideas and academic authority were to the construction of the northern idea of the tropics. Particularly influential here was the German naturalist Alexander von Humboldt. Through his

Personal Narrative of Travels to the Equinoctial Regions of the Americas (1814-1825) and his pioneering studies in biogeography, Humboldt helped invent the tropics both as a field for systematic scientific enquiry and a realm of aesthetic appreciation (Nicolson, 1990). Humboldt, who saw art and literature as vital aids to a holistic understanding of the science of nature, embodied the Romantic belief in the fundamental unity of the natural world. In his "ecstatic" response to the tropics, he discovered a world of "organic richness" and "abundant fertility" so different from temperate Europe and yielding irrefutable proof of the Cosmos as a "harmoniously ordered whole" (Humboldt, 1901, I: 2, 13). He inspired a number of North American painters to visit tropical America to try to capture on canvas the "romance of the tropics" and the dramatic landscapes Humboldt had described (Manthorne, 1989). He also tempted a generation of British naturalists, led by Charles Darwin, Joseph Dalton Hooker and Alfred Russel Wallace, to investigate the tropics for themselves.

There already existed a cult of the tropical picturesque in European travel accounts of the period, but the legacy of Humboldt added to the rich fusion of scientific and aesthetic appreciation in naturalist writings. Darwin's descriptions of Brazil during the voyage of *The Beagle* (1831-1836) are full of Humboldtian delight in tropical nature. Other visitors to Brazil, like H. W. Bates, whose sojourn on the Amazon lasted from 1848 to 1859, gave further authoritative evidence of the abundant natural riches and "teeming profusion" of the tropics. On the coast at Para, Bates found the climate "one of the most enjoyable on the face of the earth" and the soil "fertile in the extreme even for a tropical country". Inland at Barra "every inch" of the soil was of "the most exuberant fertility", while at Ega it was "of marvellous fertility, even for Brazil" (Bates, 1873:4, 19, 177, 275). Not all naturalists shared Humboldt's aesthetic appreciation of the tropics, and in an age in which professional scientists were replacing enthusiastic amateurs, many strove for greater objectivity of language and to distance themselves from impassioned Romanticism. Wallace, Bates's companion in Brazil from 1848 to 1852, was depressed by the monotony of tropical forests and homesick for an English spring (Wallace, 1889). Despite a further period of residence in tropical Southeast Asia, Wallace (1878:vii) remained convinced that travellers and naturalists had combined "to praise, and not infrequently to exaggerate, the charms of tropical life". Yet, as late as the 1880s, the German naturalist Ernst Haeckel (1883:102-03) was still discovering in Sri Lanka an "Eden", a "paradise", filled with all the "endless marvels" of tropical nature.

But Europe's engagement with the tropics contained, almost from the outset, a duality that made the tropics appear as much pestilential as paradisiacal. Powerfully negative representations of the tropics centred on images of primitiveness, violence and destruction – the speed and fatality of tropical diseases, the destructiveness of tropical storms, the ferocity of tigers and other carnivorous beasts prowling in malarious jungles (Oviedo, 1959:37) – as well as the detrimental effect of tropical abundance on the moral and physical well-being of human inhabitants, whose easy subsistence bred indolence and provided no stimulus to technological innovation and the arts of civilisation. While early encounters with the Caribbean evoked visions of a terrestrial paradise, subsequent European attempts to colonise the region spawned more negative images, centring on excessive heat, high mortality and the failure of white settlement schemes (Kupperman, 1984). For centuries "the tropics" signified for Anglo-Americans the coasts and islands of the New World from Jamestown to Bahia, a region that produced such valued crops as cotton, sugar and tobacco, over which the tropics enjoyed a near monopoly. But since these lands were, despite their riches, seemingly unsuited for white labour and settlement, recourse was had to African slaves, thus accentuating the paradox of a region of apparently great natural wealth being exploited through a primitive and inhumane system of labour. Early European writers on the West Indies saw little that was intrinsically unhealthy about the region, but by the second half of the eighteenth century, as plantation economies expanded and imperial rivalries intensified (bringing unprecedented numbers of white soldiers and sailors to the islands), a growing body of English and French medical treatises described the many dangerous diseases prevailing there and their greater virulence compared to those of Europe. There was a significant duality to this literature. In dwelling on the exceptional vulnerability of European constitutions, it reflected on their status as frail exotics exposed to tropical heat, humidity and disease, and the concurrent dangers of "intemperate" conduct in a hot climate. But it also described diseases suffered by African slaves that were detrimental to productive labour as well as being a danger to white slave-owners and their families (Sheridan, 1985:17-40). The perils of the tropics came increasingly to be located in such diseases as yellow fever, dysentery and malaria, and in ideas of race as well as place.

The fact that many diseases originated outside the region, especially in West Africa, prompted physicians to look beyond the Caribbean and to generalise about the climate and diseases of the tropics as a whole. A parallel development was the interest, first pursued by the Portuguese (Russell-Wood, 1992) but later systematically developed by the Dutch, French and English, in the dissemination of "useful" plants that would not flourish in northern climes but could be moved from one tropical or sub-tropical region to another, thus strengthening the visual impression as well as the economic utility of a single "tropical world". Although the complex of ideas associated with the tropics was first assembled in the Caribbean, it did not remain confined there. Even before the abolition of slavery in 1834 and the subsequent decline in the region's economic and political importance, the concept of the tropics was being exported, particularly through medical and botanical texts, to other intra-tropical regions, such as India and Southeast Asia, despite their lack of plantation slavery and their very different environmental and cultural conditions (Arnold, 1998). As Anderson (1992:140-41) has remarked, by the 1830s "whether a tropical region was defined most accurately by vegetation, parallels of latitude, isotherms, humidity, or a relative discomfort index may have remained open to dispute, but that such a place did exist, and was peculiar in some definable way, was no longer in doubt".

Gourou, writing in the 1940s, exemplified this relatively recent, but by then well established, idea of a "tropical world", endowed (despite its evident diversity) with a large measure of environmental homogeneity, viewed from an external, European perspective, and judged by the exacting standards of the northern temperate zone. It is striking that while Gourou, like many earlier authors, recognised considerable diversity within the tropics, as in his account of Tonkin (1931), with its sustained contrast between the malarious, mountainous interior and the populous Red River delta, he nonetheless found it meaningful to generalise about "tropical" Indo-China and subsequently the entire "tropical world". While naturalists and geographers commonly distinguished between hot, wet lowlands, dry savannas, and cold alpine areas, the first of these was generally taken to typify the tropics as a whole. This view was partly encouraged by a tendency to see upland regions as semi-European in their temperate flora and fauna (and thus not part of the true tropics at all), and to regard savannas as "unnatural" landscapes caused by deforestation and the subsequent degrading of vegetation and soils (Belt, 1888:185-87; cf. Gourou, 1940:381-91). It is, by contrast, indicative of the power of the ruling idea of the tropics that when Europeans encountered a landscape that did not conform to expectations of tropical fecundity and exuberance, they were disappointed or felt obliged to apologise for its deficiency. In 1880, a British geologist thought it necessary to remind his readers that many parts of eastern India did not look like the imagined tropics at all. Around Raniganj, he wrote, "There is generally a harshness and dryness about the vegetation of these jungles, together with want of varied colours, which I did not expect to meet with ... I soon found that the tropical luxuriance and density of undergrowth in the forests which I did expect, need not be looked for in these parts of India" (Ball, 1880:23). He was not alone in feeling acute disappointment that the tropics failed to match the conventional image of tropical abundance inspired by travel accounts, the writings of naturalists and displays of tropical vegetation in greenhouses at Kew or in botanic gardens in the tropics themselves. Having relished the luxuriant vegetation at Bombay, his first "tropical" port of call in the East, Haeckel (1883:69) was dismayed to find, when he travelled into the Western Ghats, "the total absence of the gorgeous vegetation of the tropics". Delighted by the botanic garden at Peradeniya near Kandy, he was equally perplexed by his first experience of unruly jungle, growing "without any kind of order" and "in wild confusion" (Haeckel, 1883:120-26).

In Gourou's account of the tropics there is scarcely a trace of the Edenic: poverty and pathogenicity are all pervading. The tropical world is full of "horrors", a region where climate and disease are "terrible foes" to mankind (Gourou, 1953a:12). Domesticated animals (so prominent a feature of the European scene) are rare, resulting in deficient human diets and poor transportation. The tropics, unaided, can nurture nothing higher than a "vegetable civilisation" (Gourou, 1953a:66). The famed fertility of the tropics is dismissed as a myth. The soils are "infertile"; laterite, for Gourou one of hallmarks of tropical poverty, is represented as a kind of "pedological leprosy" (Gourou, 1953a:21; Gourou, 1940:87). The overwhelming impression is of the "inferiority" (Gourou, 1953a:97, 141) not just of soils, but of the entire tropical region compared to the temperate zone. Situating himself firmly in the latter, Gourou (1953a:6) observes: "We who live in temperate lands find it difficult to realise how baneful Nature can be to man or to understand that in unreclaimed regions water may swarm with dangerous germs, myriads of blood-sucking insects may inject deadly microbes into the human body, and the very soil may be harmful to the touch". But this negative representation of the tropics was nothing new: indeed, to a large extent Gourou was simply recycling and dramatising pessimistic judgements that had long circulated in scientific and geographical discourse. It is striking that even when Gourou cites a recent source, *Nutrition in the Colonial Empire* (1938-39), that does not subscribe to earlier environmental and racial ideas to explain tropical backwardness but deploys in the colonies, as for the metropole, the "new knowledge" of nutrition, Gourou still incorporates this into his gloomy essentialising about the physical constraints that condemn humanity to poor health and low productivity in the tropics.

THE "PRIMITIVE" TROPICS

Created as much as discovered, the tropics were made to bear a moral message that flattered Europe's sense of superiority while denigrating its alien "other". Although describing the tropics as "nature's garden" might seem to suggest unqualified approval, in an age obsessed with improvement and progress, with racial origins and competitive evolution, there were definite disadvantages to being the denizens of an earthly paradise. This was a primitive world, a land that time forgot and civilisation had shunned. The "primordial" character of tropical landscapes was most forcefully suggested by the vast and seemingly inexhaustible forests. Whether naturalists found the forests of Amazonia grand or gloomy, they understood them to be an obstacle to progress and the advance of civilisation: Wallace (1889:32) was not alone in anticipating that the "woodman's axe" would be the "pioneer of civilisation" in Brazil as it had been in Canada.

To northern visitors, tropical landscapes were landscapes of regression, at times pleasurably so, for they evoked a child-like wonderment and excitement. To arrive on the Sri Lankan coast as dawn broke recalled for James Tennent (1859, II:99) "in an intensified degree, the emotions excited in childhood by the slow rising of the curtain in a darkened theatre to disclose some magical triumph of the painter's fancy, in all the luxury of colouring and the glory of light". Landing at Bombay, Haeckel (1883:66) "rushed like a child" to "seize all these new wonders". To naturalists the vegetation and wildlife of the tropics were even more suggestively windows onto an ancient world, providing (as for Darwin on the Galapagos Islands) clues to the origins and evolution of life itself. Wallace (1878:vii), who identified many "remarkable forms of life" now confined to the tropics, believed that "the relations of these to extinct types which once inhabited the temperate zones" opened up "many interesting questions as to the past history of the earth". At the very least, the tropics enabled northern naturalists to visualise a world that could only otherwise be found in the fossil record. In Sri Lanka, Haeckel was intrigued to find ferns that closely resembled those of the Carboniferous era. Many of the plant groups at Peradeniya reminded him of "the fossil flora of earlier geological ages, as represented by Meyer in his ingenious reconstructions of the scenery of the primaeval world" (Haeckel, 1883:139).

Perceptions of "primitiveness" in nature inevitably extended to the human inhabitants of the tropics as well. While in the temperate north, the races of Europe had, by their own estimation, gained an unprecedented mastery over nature, in the tropics nature's rule still seemed absolute. If the tropics were an Eden, it was one in which indigenes remained lost in the childhood of Man. Dwarfed by the landscape around them, tropical people (even in a society like Sri Lanka's, with the still visible relics of an ancient civilisation) were the "simple children of nature", giving to the outsider "some conception of that...primaeval paradise into which the human race was born" (Haeckel, 1883:154). Darwin, like many of his scientific contemporaries, was struck by what appeared to be the dominance of northern life forms over those of the tropics and southern latitudes, and readily extended this concept of biological superiority from plants to people (Browne, 1983:130). Wallace, with Darwin, co-founder of the theory of natural selection, saw evidence in tropical forests of a wider struggle for survival between living organisms, but in general he believed that the "genial heat and ample moisture" of tropics, which had "probably changed little even throughout geological periods", was more conducive to a greater diversity of species than the harsher conditions of the temperate world, which had been subjected to repeated climatic fluctuations and periodic glaciation. It was natural to assume that temperate races had needed to acquire similar qualities: in the tropics, by contrast, the "struggle for existence as against the forces of nature was ...always less severe" (Wallace, 1878:66, 122).

It was widely believed that humans had originated in the tropics, close to their primate cousins (Browne, 1983:170). But as the American geographer Ellen Churchill Semple put it, though born in the tropics, mankind only grew to maturity in the temperate zone: "Where man has remained in the tropics, with few exceptions, he has suffered arrested development. His nursery has kept him a child" (Semple, 1911:635). Alternatively, in the other belittling stereotype of the colonial age, tropical men resembled women, whether from their unfamiliar appearance or because tropical heat, humidity and disease had sapped their virility. Tennent (1859, II:107) noted of the

Singhalese that, "With their delicate features and slender limbs, their frequent want of beards, their use of earrings and their practice of wearing a cloth round the waist ..., which has all the appearance of a petticoat, the men have an air of effeminacy very striking to the eye of a stranger". Evidently, "masculine" races were born of harsher climes, a presumption elevated to a principle of governance in the "martial races" theory of late nineteenth-century India, with Sikhs, Pathans and Gurkhas favoured for military recruitment over the "effeminate" and "non-martial" races of tropical Bengal and Madras.

Describing the eastern regions of the Indonesian archipelago, Wallace reversed cherished notions of tropical idylls by arguing that the islanders suffered by the "excessive cheapness" of their staple, sago. Its sheer abundance was a "curse rather than a blessing", a cause of "great laziness" and "extreme...misery". The "habit of industry" not being acquired through "stern necessity", all labour was "distasteful" to them: sago-eaters, in consequence, lived in squalid huts and wore scant clothing. This led Wallace (1862:136) to reflect on the "singular fact that no civilised nation has risen within the tropics. That rigour of nature which some may have thought a defect of our northern climes has...been one of the acting causes of high civilisation". From Wallace in the 1860s to Gourou eighty years later it was virtually axiomatic among northern writers that there could be no indigenous civilisation in the tropics, for civilisation was the product of humans moving from their original home in the tropics, where the climate was kind and the living easy, to the more physically and mentally challenging temperate zone. Wallace (1864:164) argued that harsh soils and inclement seasons were more a stimulus to the emergence of a "hardier, more provident and social race" of men than year-round plenty: "Is it not the fact that in all ages, and in every quarter of the globe, the inhabitants of temperate have been superior to those of tropical countries? All the great invasions and displacements of races have been from North to South, rather than the reverse; and we have no record of there ever having existed, any more than there exists today, a solitary instance of an indigenous inter-tropical civilisation".

Gourou (1953a:32) likewise reasoned that the tropics, by and large, had supported only "primitive" modes of cultivation and civilisation. Even the civilisations of Indo-China were essentially "vegetable" ones, trapped in an ancient primitiveness, not far removed from the neolithic, and dependent on the products of nature, rather than industry, for food, clothing, utensils, and housing (Gourou, 1940:192-95). This was a strikingly negative conclusion for him to arrive at given his own account of intensive rice cultivation in the Red River delta and its capacity to support a high density of population; but even here Gourou (1931:82-90, 109, 241), operating within the colonial developmental logic of the time, saw a society that was still in its "infancy", dependent on the paternalistic care of French officials for maintaining river dykes, improving communications and transport, curbing disease, and ushering in the benefits of the modern age. Gourou acknowledged wide cultural differences between Asia and the tropical regions of Africa and America, but nonetheless concluded that Southeast Asia was no exception to the general rule that the tropics were incapable of creating their own civilisations, for that of Tonkin was, he believed, essentially derived from the ancient Chinese, just as India was indebted, not to any indigenous race like the Tamils, but to Aryan invaders from the north (cf. Gourou, 1953b:78-81). He likewise sought to explain the apparent paradox of Mayan civilisation in Central America through traces of a more temperate origin and saw in the decline of the Maya evidence from the past, still pertinent to the present, of how intensive agriculture and demographic growth could rapidly outstrip the productive capabilities of poor tropical soils (Gourou, 1953a:43-52). Civilisation could only arrive and be sustained in the tropics through outsiders' energy and agency.

In an age of imperial expansion and colonial rule, the fate of past civilisations carried stern warnings for the present. For many late nineteenth- and early twentieth-century thinkers, entering the tropics threatened a loss of masculine vigour to temperate races from the north. The tropics thus represented an impenetrable barrier to the worldwide expansion of the white races and sapped them of the racial superiority derived from their northern homelands. These themes were most forcefully articulated by geographers like Semple and Ellsworth Huntington, who deployed a mixture of environmental determinism and Social Darwinism to demonstrate that northern climates bred vigour and intelligence while the tropics were conducive to racial decay. According to Semple (1911:10), "The debilitating effects of heat and humidity aided by tropical diseases, soon reduce intruding peoples to the dead level of economic inefficiency characteristic of the native races". Thus, the history of Aryan civilisation in India was a tale of the moral and physical degeneration of a once-white race succumbing as a victim to the enervating, disease-sodden environment of tropical South Asia. "Everywhere in the Tropics the enervating effects of heat, moisture, and abundance make not only the natives averse to steady work, but start the energetic European immigrant down the same easy descent to Avernus" (Semple, 1911:627).

Although critics denounced Semple's arguments as crude, self-serving and simplistic (Febvre, 1925:95, 172), Huntington's "moral discourse of climate" (Livingstone, 1991, 1994) gained wide approval, though he too proposed, and sought to demonstrate historically and scientifically, that the temperate regions of northwestern Europe and the northeastern United States represented the "optimum conditions" for the development of human intelligence and physical vigour in contrast to the torpor and degeneracy of the tropics. In *Civilization and Climate*, first published in 1915 and intended as a contribution to "the new science of geography", Huntington (1924:56) observed that, whatever the cause, "it is generally agreed that the native races within the tropics are dull in thought and slow in action. This is true not only of the African Negroes, the South American Indians, and the people of the East Indies, but of the inhabitants of southern India and the Malay peninsula". Native lassitude was more than an obstacle to material progress in the tropics; it was also a threat to the white races that ventured there. "Experience shows that the presence of an inferior race in large numbers tends constantly to lower the standards of the dominant race". The tropics dragged even "superior" races down to their own "primitive" levels.

TRANSFORMING THE TROPICS

Huntington's views were remarkable for their crudeness and sweeping pan-tropical generalisations. It was more common among those who travelled and worked extensively in tropical areas to distinguish between races or to offer a variety of explanations for the alleged backwardness of tropical peoples – to reason, for instance, that it was oppressive Spanish rule that had reduced the native Indians of Nicaragua to poverty and indolence (Belt, 1888:170), that the African "Negro" ("the true child of tropical climes") was better adapted to the hot, wet tropics than the natives of Amazonia (Bates, 1873:283), or that Indian or Chinese immigrants were essential for a productive work-force (Willis, 1909:12). For Gourou, as for many of his immediate predecessors, the concern was not with a distant vision or transient experience of the tropics as a site of pleasure or repugnance, but as a practical locale for economic development. At the heart of the paradox of the tropics and the problem of their exploitation lay nagging doubts about the suitability of the tropics for the white settlement that had displaced more "primitive" races and transformed the "wildernesses" of North America and Australasia into regions of productive agriculture.

The arguments for and against white settlement in the tropics were well rehearsed even by the mid-nineteenth century. They drew not just on the fate of ancient civilisations or the collapse of Caribbean ventures, but also on what was seen by Anglo-Saxon writers as the failure of Iberian attempts at colonisation and the spread of civilisation. Indeed, much of the British and North American discourse on tropicality in the late nineteenth and early twentieth centuries reads as a critique of Spanish and Portuguese "miscegenation" and tropical degeneration, and hence as an argument for maintaining, by contrast, the vigour and purity of the Anglo-Saxon race. In 1843, Arthur Thomson, a Madras Army surgeon, surveyed the entire tropical world, but could find no region where Europeans had ever successfully established themselves. Where traces of European settlers were to be found, as in Latin America, it was only, he believed, as a result of racial intermixing and the consequent degeneration of their origin racial stock. He concluded that there was "little doubt" that "the tropical parts of the world are not suited by nature for the settlement of the natives of the temperate zone. European life in those parts is with difficulty prolonged, much sickness is suffered and their offspring become degenerate and cease to propagate their species in a few generations" (Thomson, 1843:132). A decade later Frederic Mouat of the Bengal Medical Service noted how all attempts to establish white colonies in the East Indies had failed: so adverse were climatic conditions that "the progeny of the third generation [of whites] was almost entirely female" and the majority of these were "sickly and barren". Only on the salubrious island of Reunion did he find Creoles who had escaped descent into "effeminate forms" (Mouat, 1852:63, 72).

And yet it was hard for naturalists and geographers to envisage the lasting transformation of the tropics without white settlers. Thomas Belt, a mining engineer and amateur naturalist, anticipated that Mexico and all of Central America would one day succumb to colonisation by the United States. Railways from the north would penetrate into the tropics and a "constant stream of immigration" would "change the face of the country", filling it with "farms and gardens, orange groves, and coffee, sugar, cacao and indigo plantations". "No progress", he added, "need be expected from the present inhabitants" (Belt,1888:178). If , as critics maintained, white settlement was not practicable (or achievable only at terrible cost), how could tropical agriculture ever be improved? How could the tropical world be made more productive and efficient without the stimulus and example of white labour? Were the tropics inherently unsuitable for Europeans, or could they be "made safe" by modern medical and sanitary science?

The debate raged back and forth, gaining renewed vigour from the expansion of the British, French, German and American empires in the 1880s and 1890s and by their direct engagement with the practical difficulties of ruling tropical lands and seeking to tap their native wealth. As late as 1939 Grenfell Price revisited the question why it was that the northern races had failed to colonise the tropics. He reviewed the work of those, like Huntington and Andrew Balfour of the London School of Tropical Medicine, who appeared pessimistic about the possibility of white settlement even with the advantages of modern medicine and sanitary science. But Price was more inclined to the views of those, like William C. Gorgas in Cuba and the Panama Canal Zone or Raphael Cilento in Queensland, who through their part in the "scientific invasion of the tropics", had shown that whites could live safely and labour efficiently in the tropics (Price, 1939). But, with recent developments in India and the Philippines in view, Price (1939:38) was forced to admit that the "progress of science" had not improved the political position of whites in the tropics. In the 1930s, Gourou (1931:42), too, saw it as a valid test of Tonkin's climate to ask whether it suited the white man, though he had in mind colonial administrators rather than settlers. By the 1940s, the issue was virtually dead: the prospects for white settlement anywhere in the tropics (Queensland apart) were increasingly remote as the Second World War gave way to insurgent nationalism and imminent decolonisation. Interest shifted from white colonisation to settling Vietnamese, Indonesian and Filipino migrants in the less populated portions of their own regions (Gourou, 1940:183-8; Pelzer, 1945; 1957).

The emergence in the 1890s and 1900s of tropical medicine and tropical agriculture exemplified the intensified search for practical solutions to the paradox of the rich-but-poor tropics. But these new specialties brought not only a new technical specificity to the investigation of the tropics, but also reinforced many negative representations. This was most evident in tropical medicine. Patrick Manson, the "father" of modern tropical medicine, investigated filaria in Amoy before retiring to London: his work inspired others, notably Ronald Ross of the Indian Medical Service in his quest for the cause and mode of transmission of malaria in the 1890s. Manson confessed he had difficulty in defining the boundaries of the tropics as far as disease was concerned, but based his call for Britain to introduce "special education in tropical medicine", firstly, on the fact that, since Britain had become the centre of "a great and growing tropical empire", it needed to develop a corresponding medical specialism, and, secondly, that tropical disease was "in many respects ... widely different from the diseases of temperate climates". There were "dozens" of diseases, "more or less special to the tropics", which needed "special knowledge" for their diagnosis and successful treatment. (Manson, 1897:842). Where earlier writers had blamed the climate, or lethal "miasmas" generated by tropical heat, humidity and decomposing vegetation, Manson could identify specific causes – parasites, insect vectors and "germs" – many of which were unknown or uncommon in temperate regions. With the experience of British India (rather than America and yellow fever) in mind, Manson identified the malaria parasite as "by far the most important disease agency in tropical pathology". Not only did it cause fatal fevers; it also undermined the health of millions, exposing them to other ailments, sapping their powers of resistance and recovery. Directly or indirectly, malaria was "the principal cause of morbidity and death in the tropics and sub-tropics" (Manson, 1898:1).

Ross similarly argued that malaria was responsible for between a quarter and a half of all sickness in tropical regions, but, more than that, malaria was the great obstacle to civilisation in the tropics. Malaria was "the great enemy of the explorer, the missionary, the planter, the merchant, the farmer, the soldier, the administrator, the villager, and the poor". It was responsible for rendering "the whole of the tropics comparatively unsuitable for the full development of civilization" (Ross, 1910:vii-viii). The identification of malaria with tropical backwardness and torpor became a recurring theme, and nowhere more so than in Gourou's writing between the 1930s and 1950s. For Gourou (1953a:6) malaria was "the most widespread of tropical diseases", and though (on the basis of French colonial research) he ingeniously explained the role of human activity in suppressing the disease in the Red River delta, he still saw it as one of the most important factors in causing low productivity and in inhibiting development throughout tropical Asia (Gourou, 1953b:30-3). To some extent he shared Ross's belief that through modern medicine and sanitary science, and the eradication of the anopheles mosquito, malaria could be overcome and the tropical world developed for the benefit of whites and indigenes alike. Even before the advent of DDT, the "conquest" of malaria seemed synonymous with the "conquest" of the tropics. This underscored the more general message of the time – that modern science had the capacity to make the tropics truly productive, to meet the world's (i.e., the north's) expanding needs for raw materials from timber to palm oil, and to open up the tropics as the last remaining "new lands" left on earth.

And yet Gourou seemed to remain somewhat sceptical as to whether natural conditions would ever make this ideal of tropical development possible.

The effect of tropical medicine, like tropical agriculture, veterinary and soil science, was as much to publicise the perils of the tropics as to advance practical solutions. The image of the tropics as a world set apart by nature, a world characterised by poverty, disease and backwardness, thus acquired a new scientific authority and specificity: the foundations had been laid for a reconceptualisation of the "backward" tropics as the Third World (much as package tourism authored a revised version of the tropical Eden). Thus, at a time when environmental determinism seemed less and less appropriate for northern temperate lands, it retained its influence in accounts of the "tropical world". Gourou (1953a:151) anticipated that, aided by modern science, change was now possible in the tropics, but solely along lines allowed by nature's "directions" and "vetoes". Any attempt, for instance, to impose temperate agriculture on the tropics would simply fail: "Tropical conditions" could "only be commanded by being obeyed" (Gourou, 1953a:119).

Gourou denied that the physical world simply and crudely determined human activity and social forms. He sought in his later works to distance himself from the climatic and environmental "determinism" of Ratzel, Semple and Huntington and also from the "possibilism" of Paul Vidal de la Blache (Gourou, 1953b:47). But in the eyes of many of his contemporaries, including the *Annales* historian Fernand Braudel, whose own work sought a union of history with a deterministic geography, Gourou remained broadly within the latter tradition (Braudel, 1971:7). In the 1920s, Lucian Febvre, another founder of the *Annales* School, had also sought to eschew crude determinism in favour of the Vidal de la Blache's "possibilism" (Livingstone, 1992:266-9), but when he looked at the tropics he could not fail to see a clear example of how powerfully nature could inform the human condition. In a manner that anticipated at least some of Gourou's later findings, Febvre denounced the "illusory riches" and supposed natural abundance of the tropics as the product of an "over-idealistic geography". The tropics had been misguidedly represented as "lands of promise", where nature, "generously heaped her treasures on man". Experience had, conversely, shown that the tropical rainforest (which Febvre took to be the archetypal tropics) was "a difficult place to live in". Despite superficial impressions of "inexhaustible fecundity", the fertility of the soil was "more apparent than real": the tropical forest was, in fact, "a desert, covered with verdure". Even if scientific agriculture were introduced, "too many enemies" lay in wait for it fully to succeed. Viewed objectively, he argued, "the balance sheet" of the tropics, was "poor" (Febvre, 1925:183-88). It is hard not to see the roots of Gourou's own pessimism about the tropics in this broadly deterministic geographical tradition, accentuated by the ambitions and frustrations generated by European and American tropical colonies, and articulated over more than a century through recurrent northern notions of tropical alterity.

REFERENCES

Anderson, W. (1992) 'Climates of opinion: acclimatization in nineteenth-century France and England', *Victorian Studies*, 35, 135-58.

Arnold, D. (1998) 'India's place in the tropical world', *Journal of Imperial and Commonwealth History*, 26, 1-21.

Ball, V. (1880) *Jungle Life in India: Or the Journeys and Journals of an Indian Geologist*, London: de la Rue & Co.

Bates, H.W. (1873) *The Naturalist on the River Amazons*, 3rd ed., reprinted London: J. M. Dent & Sons, 1910.

Belt, T. (1888) *The Naturalist in Nicaragua* (1st edition, 1874), 2nd ed., London: E. Bumpus.

Bernardin de Saint-Pierre, J.-H. (1982) *Paul and Virginia* (1st published 1788) London: Peter Owen.

Braudel, F. (1971) 'Préface' to P. Gourou, *Leçons de géographie tropicale*, Paris/Hague: Mouton.

Browne, J. (1983) *The Secular Ark: Studies in the History of Biogeography*, New Haven: Yale University Press.

Charlton, D.G. (1984) *New Images of the Natural in France: A Study in European Cultural History, 1750-1800*, Cambridge: Cambridge University Press.

Cohen, J.M. (ed.) (1988) *The Four Voyages of Christopher Columbus*, London: Cresset Library.

Curtin, P.D. (1964) *The Image of Africa: British Ideas and Action, 1780-1850*, Madison: University of Wisconsin Press.

Duncan, J.S. (1990) *The City as Text: The Politics of Landscape Interpretation in the Kandyan Kingdom*, Cambridge: Cambridge University Press.

Études de géographie tropicale offertes à Pierre Gourou (1972) Paris/Hague: Mouton.

Febvre, L. (1925) *A Geographical Introduction to History*, London: Kegan Paul, Trench, Trubner & Co.

Gourou, P. (1931) *Le Tonkin*, Paris: Exposition Coloniale Internationale.

Gourou, P. (1940) *L'utilisation du sol en Indochine française*, Paris: Centre d'Études de Politique Étrangère.

Gourou, P. (1947) *Les pays tropicaux: Principes d'une géographie humaine et économique*, Paris: Presses Universitaires de France.

Gourou, P. (1953a) *The Tropical World: Its Social and Economic Conditions and Its Future Status*, London: Longmans, Green & Co.

Gourou, P. (1953b) *L'Asie*, Paris: Libraire Hachette.

Haeckel, E. (1883) *A Visit to Ceylon*, London: Kegan Paul, Trench & Co.

Humboldt, A. von (1901) *Cosmos: A Sketch of a Physical Description of the Universe*, London: George Bell & Sons.

Huntington, E. (1924) *Civilization and Climate* (1st edition 1915), 3rd edition, New Haven: Yale University Press.

Kupperman, K.O. (1984) 'Fear of hot climates in the Anglo-American colonial experience', *William and Mary Quarterly*, 61, 213-40.

Livingstone, D.N (1991) 'The moral discourse of climate: historical considerations of race, place and virtue', *Journal of Historical Geography*, 17, 413-34.

Livingstone, D.N. (1992) *The Geographical Tradition: Episodes in the History of a Contested Enterprise*, Oxford: Blackwell.

Livingstone, D.N. (1994) 'Climate's moral economy: science, race and place in post-Darwinian British and American geography', in A. Godlewska & N. Smith (eds), *Geography and Empire*, Oxford: Blackwell, 132-54.

Livingstone, D.N. (1999) 'Tropical climate and moral hygiene: the anatomy of a Victorian debate', *British Journal for the History of Science*, 32, 93-110.

Lowenthal, D. & Bowden, M.J. (eds.) (1976) *Geographies of the Mind: Essays in Historical Geography in Honor of John Kirtland Wright*, New York: Oxford University Press.

Manson, P. (1897), 'The necessity for special education in tropical medicine', *Lancet*, 2, 842-5.

Manson, P. (1898) *Tropical Diseases: A Manual of the Diseases of Warm Climates*, London: Cassell.

Manthorne, K.E. (1989) *Tropical Renaissance: North American Artists Exploring Latin America, 1839-1879*, Washington: Smithsonian Institution Press.

Mouat, F.J (1852) *Rough Notes on a Trip to Reunion, the Mauritius and Ceylon.* Calcutta: Thacker, Spink & Co.

Nicolson, M. (1990) 'Alexander von Humboldt and the geography of vegetation', in A. Cunningham & N. Jardine (eds.) *Romanticism and the Sciences*, Cambridge: Cambridge University Press, 169-85.

Nutrition in the Colonial Empire (1938-39), *First Report of the Committee on Nutrition in the Colonial Empire* (Cmd 6050), *Parliamentary Papers*, X.

Oviedo, G.F. (1959) *Natural History of the West Indies*, Chapel Hill: University of North Carolina Press.

Pelzer, K.J. (1945) *Pioneer Settlement in the Asiatic Tropics: Studies in Land Utilization and Agricultural Colonization in Southeastern Asia*, New York: American Geographical Society.

Pelzer, K.J. (1957) 'Geography and the tropics', in G. Taylor (ed.) *Geography in the Twentieth Century*, 3rd ed., London: Methuen, 311-44.

Price, A.G. (1939) *White Settlers in the Tropics*, New York: American Geographical Society

Ross, R. (1910) *The Prevention of Malaria*, London: Murray.

Russell-Wood, A.J.R. (1992) *A World on the Move: The Portuguese in Africa, Asia and America, 1415-1808*, Manchester: Carcanet.

Said, E.W. (1978) *Orientalism*, London: Routledge & Kegan Paul.

Semple, E.C. (1911) *Influences of Geographic Environment on the Basis of Ratzel's System of Anthropo-Geography*, London: Constable.

Sheridan, R.B. (1985) *Doctors and Slaves: A Medical and Demographic History of Slavery in the British West Indies, 1680-1834*, Cambridge: Cambridge University Press.

Tennent, J.E (1859) *Ceylon*, 3rd edition, London: Longman, Green, Longman & Roberts.

Thomson, A.S. (1843) 'Could the natives of a temperate climate colonize and increase in a tropical country and vice versa?', *Transactions of the Medical and Physical Society of Bombay*, 6, 112-38.

Wallace, A.R (1862) 'On the trade of the Eastern Archipelago with New Guinea and its islands', *Journal of the Royal Geographical Society*, 32, 127-37.

Wallace, A.R. (1864) 'The origin of human races and the antiquity of man deduced from the theory of "natural selection"', *Journal of the Anthropological Society*, 2, 158-70.

Wallace, A.R. (1878) *Tropical Nature and Other Essays*, London: Macmillan.

Wallace, A.R. (1889) *A Narrative of Travels on the Amazon and Rio Negro*, 2nd edition, reprinted New York: Haskell House Publishers (1969).

Willis, J.C. (1909) *Agriculture in the Tropics: An Elementary Treatise*, Cambridge: Cambridge University Press.

[19]

Race, space and moral climatology: notes toward a genealogy

David N. Livingstone

This paper is a contribution to the genealogy of the presupposition that climatic zones and their inhabitants may legitimately be portrayed in the vocabulary of moral judgement. Focusing on four different scholarly domains at different points in time—Enlightenment philosophy, early Victorian anthropology, late nineteenth century tropical medicine, and early twentieth century climatology—it discloses the extent to which intellectual exchange among a Western elite was concerned with the mobilization of 'moral climatology' for racial purposes. The persistent and widespread inclination to use climate as the vehicle for moralistic construals of global space is thus demonstrated.

Introducing moral climatology

"It is not surprising that tropical lands should as a whole be thinly populated and that they should be backward in civilization. The contribution of these regions to the fund of world civilization is small, whether in the material, intellectual or moral sphere".[1] With these words the distinguished French geographer, Pierre Gourou of the Collège de France, reached towards the conclusion of his 1947 volume on *The Tropical World*. Recently described as a "magisterial survey",[2] this epitome of his wide field experience in Indo-China during the 1930s delivered to a generation of geographers a zonal imaginary out of which a suite of tropical geographies could be hewn.[3] And its influence spiralled well beyond the geographical academy. Its translator, E. D. Laborde, fervently hoped that the book would find its way into "the hands of every colonial administrator and planter", and it was welcomed as such by *The Economist*, *The Times Literary Supplement* and *International Affairs*. Courtesy of Pierre Gourou, the tropical world—that imagined singularity routinely presented as pessimistic, pathological and pestilential—was kept in discursive circulation despite customary concessions to its subregional variations. The following extracts from the first three chapters, are emblematic of the judgemental vocabulary in which the whole was cast. "Hitherto lands with a hot, wet climate", chapter one begins, "have proved less favourable than the temperate belt as an environment for man". "In physical and mental activity", chapter two tells us, ". . . man is restricted in the tropics by serious maladies whose existence is entirely due to the hot, damp climate. We who live in temperate lands find it difficult to realize how baneful Nature can be to man". "On the whole", chapter three concludes, "tropical soils are less favourable to man than temperate soils, which are richer and more stable. . . . It is therefore not surprising that tropical peoples are often few in number and have not built up higher civilizations of their own". [4]

If these few excerpts make clear the censorial tenor of Gourou's tropical geography, they no less bear witness to the temperate world as the mensural standard against which its tropical 'other' is again and again unfavourably compared. Temperate normalcy remained inviolate, its features valorized as the touchstone of mental and moral excellence alike. My main objective in what follows is to work towards a genealogy of an inclination, hinted at here, to think of climate in moral categories. This involves both a widespread tendency to deploy moralistic language in depicting climatic conditions *and* a conviction that it is entirely reasonable to read moral order straight off patterns of global climate. Before I begin this task, however, we need to recollect that a good deal of Gorou's intellectual purchase was derived from the simple fact that there were many before him.

Of course, the categories to which Gourou had recall had already received re-inforcement in the wake of European colonialism, but it would be mistaken to consider them simply a product of Europe's imperial moment.[5] The roots of this entire predilection can be traced back at least to Aristotle who, in the seventh book of the *Politics*, portrayed the peoples of cold climates, especially Europe, as full of spirit while those from Asia as deficient in spirit but endowed with skill and intelligence. Again, during the Renaissance, Bodin elaborated a geography of suspicion, treachery, cruelty and drunkenness, grounded in a synthesis of Hippocratic medicine and astrological theory.[6] And then, in the early seventeenth century Saluste Du Bartas, described by John Prest as a French prototype of Milton, imagined the Garden of Eden as blessed with:

> A climate temp'rate both for cold and heat,
> Which dainty *Flora* paveth sumptuouslie
> With flowrie Ver's innameld tapistrie.[7]

Such veneration of the temperate—and corresponding contempt for the tropics—found repeated expression in the succeeding centuries. During the first decades of the eighteenth century, Fontenelle urged that neither the torrid nor polar zones were suited for science, Chardin considered that hot climates "enervate the Mind" and the abbé du Bos—who incidentally commented on the geographical distribution of poetry and painting—found in climatic factors the source of annual differences in crime rate.[8] A half century or so later William Falconer—antiquarian, natural historian, and associate of Joseph Banks—drew analogies between an Anglo-Saxon present and a Pacific past in his discussion of the links between climatic circumstance and forms of punishment. In his 1781 *Remarks on the Influence of Climate* he urged that the "intermediate climates have always been esteemed, both in ancient and modern times, to be the most favourable to human nature". Even the "fickleness" and "uncertainty of temper" that from time to time attended temperate civilizations were veiled virtues: "highly favourable to the advancement of science", such attributes were crucial to constitutions "dissatisfied with the present, and of course inquisitive after what is better in future". To Falconer it was precisely such a disposition that induced "a spirit of examination and enquiry".[9] Inquiries of this stripe, as John Gascoigne has observed, provided a climatic hypothesis which "was one of the first, and one of the most fertile, organising principles for a 'Science of Man' ".[10]

A century further on and we come to Henry Drummond, evangelical exponent of Spencerian evolution and professor of natural science at the Free Church of Scotland College in Glasgow.[11] Having moved into "the heart of this mysterious Africa" and having endured a "climate ... so pestilential that when two go, you and your friend are simply nursing each other time about", Drummond told the readers of his 1891 *Tropical Africa* that it was "a wonderful thing to look at this weird world of human beings—half animal half children, wholly savage and wholly heathen". To have done so was "to have watched the dawn of evolution ... to have the great moral and social problems of life, of anthropology, of ethnology, and even of theology, brought home to the imagination in the most new and startling light".[12] Part travelogue, part diary, and part scientific communiqué,[13] Drummond's *Tropical Africa* traded liberally in the currency of adjectival derision: the rivers were "sullen", the mangroves "foetid", the crocodiles "revolting", the climate "pestilential", the land "unstirred by civilisation", and the people "primeval"—children "of nature, thoughtless, careless, and contented".[14] So when Pierre Gourou insisted in 1947 that tropical soils delivered a yield unexpectedly "inferior to that of temperate soils", that in "temperate lands limestone never gives rise to forms so hostile to man", that "by their unhealthiness tropical lands are much less favourable than temperate regions to the formation and perpetuation of ... high civilisation", he was simply updating a long tradition of tropical censure and temperate celebration.[15] Mapping tropical geography was a performance in delineating the boundaries of cultural possibility.

We are, by now, well aware that works of the kind represented by this rather randomly selected suite of testimonials materially contributed to what has come to be called 'Tropicality'.[16] Now, let me be perfectly clear about this. Nothing I have said so far should be construed as implying that the tropics were *invariably* cast in a negative light. Various scientific travellers from before Humboldt to after Darwin found the tropics exhilarating and rhapsodized about them with intoxicated rapture; early conservationists participated in what has been called an 'Edenic discourse' projecting the possibility of rebuilding a fruitful Paradise in tropical island environments; fiction writers could find in them a rich source of romantic inspiration; the experiences of artists in the tropics challenged the representational conventions of their own day; and philosophers, of course, from time to time celebrated the qualities of 'the noble savage'.[17] So we stand witness here to anything but a discursive singularity; contestation and contradiction attended tropical nature writings in a host of ways. As Nancy Stepan has shown, for example, Alfred Russel Wallace's narrative of tropical nature tellingly challenged the kind of heroic romanticism given voice in Philip Henry Gosse's *The Romance of Natural History*.[18] Or again, Marion Newbigin's intriguing little volume on *Animal Geography* presented the tropical rainforest *both* as barren of foodstuffs to support anything other than the most meagre forms of human subsistence *and* as enjoying an unparalleled wealth of species, abundance of life and luxuriance of undergrowth.[19] In not dissimilar terms, Patrick Manson reconceptualized species abundance as parasitic fecundity when he told the readers of his *Tropical Diseases* that "just as the fauna and flora of the tropical world are infinitely richer in species than those of colder climates, so there is a corresponding distribution in the wealth and poverty of pathogenic organisms".[20] My intention is not to pursue tropical paradox, however. What instead excites my interest is the propensity to resort to moralistic idioms in the evocation of climatic conditions.[21] Exercises in regional climatology, to put it another way, were frequently adventures in moral cartographics. That the languages of climatic description and moral judgement have been much more frequently elided in western thought than is commonly supposed is the proposition I seek to demonstrate here. What I am hoping to inaugurate, then, is the genealogy of a presupposition—the presupposition that climate may coherently be construed in the language of morality. To be sure, this discursive assumption has been articulated differently in different times and places; it has displayed local inflections and delivered diverging verdicts. But the self-same inclination has persistently reasserted itself and it is the outline form of this trajectory that I want to trace here.

My method of working is simply put; it is, as it were, to peel back four overlapping layers of western thought. My claim is not that each of the intellectual communities I discuss explicitly relied on their forebears or were filially connected with those who had gone before. Rather I want to show a marked persistence over time (from the European Enlightenment to the early twentieth century) and across scholarly domains (from philosophical reflection to anthropological narrative, and from tropical medicine to modern climatology) of the tendency to use climatic conditions as a vehicle for the transportation of western moral freight. Having said this, it is important that a number of caveats are registered at our point of departure. First, the term 'race' frequently surfaces in the writings of those on whom I call to give witness. No essentialist definition is implied by my use of the term; the concept was doubtless understood differently at different points in time and in different circumstances. Its deployment here is in full recognition of its historical contingency, but I do not see it as my task, in this inquiry at least, to elucidate its multiple significations.[22] Second, an analysis of reports—agricultural, administrative, medical—produced by those who had first-hand experience of tropical environments and whose findings were intended for local consumption might well disclose a different construal of moral climatology than that depicted in the literature I have chosen to dwell on. My reason for focusing on the sources I do is that they circulated widely among Western intellectual elites and uncover something of the degree to which scholarly discourse was implicated in moralistic readings of global climate. Third, my textual focus should not be taken to imply that I deny the importance of other cultural forms; what I have called elsewhere 'tropical hermeneutics' is an enterprise of much wider dimensions than textual interrogation and, in particular, embraces those performative enactments embedded in material practices and concrete operations.[23] Fourth, while my exegesis rotates around a *western* moral ecology of tropicality, it would be mistaken to assume that such manoeuvres were unique to the West. The moral stigmatizing of people who inhabited certain regions or zones, I suspect, was every bit as much a part of, say, non-European Hindu and Islamic cultural formations. Whether these constructions reinforced or challenged western conceptions remains an open question. Finally, any attempt to take catch the spirit of three centuries of intellectual history is bound to incur charges of historiographical naivete or monumental arrogance. Both may be justified. But my narrative is put forward in the conviction that if the cost of scholarly safety is over-specialization, narrowness and fragmentation then the price is, simply, too high.

Philosophical weather

In recent years the extent to which the western philosophical tradition has been implicated in a variety of racial projects has begun to be uncovered. A number of works scrutinizing what has been called the 'racist culture' of philosophy or the 'racial economy' of science have made their appearance.[24] The implications of this work have yet to be fully registered, but they are likely to be of considerable dimensions. It suggests, for instance, that classical philosophical quandaries like the mind-body problem—routinely read as decorporealized metaphysics—should not be uncoupled from how philosophers thought about particular minds and particular bodies. The 'body' in question, as it were, was rarely the non-western racial body or the artisan's body, and the 'mind' never the black servant's mind or the female mind. Yet if racial obsessions were a far more dominant strain in western philosophizing than is regularly acknowledged, just as arresting in these self-same philosophical accounts of racial difference and 'the nature of human nature' is the presence of geographical factors, in particular climate, as the declarations of Enlightenment philosophers from Hume to Hegel attest. Thus the philosophers' claim to have transcended the sixteenth century humanists' interest in ethnography, geography and history and to have substituted 'abstract axioms' for 'concrete diversity' cannot be taken as an unproblematic achievement. Descartes may have felt that he had overcome what Toulmin calls "a youthful fascination with ethnography and history" because to him history was "like foreign travel. It broadens the mind, but it does not deepen it".[25] In fact the *universals* in which rationalists traded were only secured by a privileging of certain—often geographically-referenced—*particulars*.

Take then David Hume's 1748 essay "Of National Characters".[26] Here Hume engaged in a detailed rebuttal of the view that national tempers were influenced in any material way by physical causes such as climate drawing on the support of Strabo's *Geography*. And yet he did feel constrained to concede that "there is some reason to think, that all the nations, which live beyond the polar circles or between the tropics, are inferior to the rest of the species, and are incapable of all the higher attainments of the human mind". Moreover, he approvingly called upon the testimony of Lord Bacon to the effect that "where the native of a cold climate has genius, he rises to a higher pitch than can be reached by the southern wits". Thereby, without succumbing to climatic determinism, Hume gestured towards a geography of mental and moral attainment. And nowhere was this suspicion more clearly intimated than in his comments on the distribution of certain 'passions':

> The only observation, with regard to the difference of men in different climates, on which we can rest any weight, is the vulgar one, that people in the northern regions have a greater inclination to strong liquors, and those in the southern to love and women. One can assign a very probable *physical* cause of this difference. Wine and distilled waters warm the frozen blood in the colder climates . . . [a]s the genial heat of the sun, in the countries exposed to his beams, inflames the blood, and exalts the passion between the sexes.

What this admission discloses is Hume's preparedness to countenance—as others had done before him—the influence of climate on certain 'passions' connected with what he called "the grosser and more bodily organs of our frame", even while exempting "the operations of the mind and understanding" from the self-same influences.[27]

That same year, 1748, Montesquieu's *The Spirit of Laws* made its first appearance.[28] The distinctive role that Montesquieu allocated to climate—the first and most powerful of all empires, as he put it—in his controversial analysis of human behaviour and cultural forms, which called forth censure from Jansenist and Jesuit sources, is well known, as is his reliance on some of the earlier writers advertised above.[29] So too is the diversity of opinion on the extent to which he may be characterized as a geographical determinist—a matter on which Collingwood and Cassirer have taken markedly different viewpoints.[30] I do not intend to adjudicate on this matter. Instead I want simply to readvertise the extensive deployment of moralistic vocabulary in the famous Part Three of *The Spirit of Laws*. What sustained Montesquieu's analyses were two discrete bodies of Enlightenment literature: first, the travel writings of a range of authors reporting on overseas voyages and second, the physiological theories of figures like Boerhaave that enabled Montesquieu to conceive of the nervous system as consisting of tiny tubules that carried nerve juice, or animal spirits, around the body. It was this conviction that allowed him to claim that inhabitants of cold climates were more vigorous since the fibres in their cardiovascular systems contracted in cold air and this, in turn, allowed for faster flowing of the blood.[31] Irrespective of the mechanisms involved, what is particularly conspicuous in Montesquieu's portrait is his moral mapping of the climata, as the following gobbets plainly attest: "the passions of the heart are extremely different in different climates" (Book XIV, 1); the "inhabitants of warm countries are, like old men, timorous; the people in cold countries are, like young men, brave" (XIV, 2 [2]); "As climates are distinguished by degrees of latitude, we might distinguish them also, in some measure, by degrees of sensibility" (XIV, 2 [8]); in "northern countries, we meet people who have few vices, many virtues, a great share of frankness and sincerity. If we draw near the south, we fancy ourselves removed from all morality; the strongest passions multiply all manner of crimes" (XIV, 2 [12]).

The case of Immanuel Kant, who, even while defending monogenism, looked at least in part to climate as the source of racial differentiation, is also worthy of brief perusal. In his 1775 essay "On the Different Races of Man", for example, Kant identified four distinct varieties of the human species each with its own natural dispositions. Here he made reference to what he called the "stem genus" of each race and specified "that portion of the earth between the 31st and 52nd parallels in the Old World" as the zone "in which the most happy mixture of influences of the colder and hotter regions and also the greatest wealth of earthly creatures is encountered".[32] Conversely, as he made clear in an exposition of national characteristics as reflected in the aesthetic and moral "feeling of the beautiful and sublime", inhabitants of certain other zones enjoyed "no feeling that rises about the trifling".[33] Given these sentiments, it comes as no surprise to find Kant, in his posthumously published lectures on *Physical Geography*, elaborating a global cartographic in which the temperate world was exalted to the apogee of human excellence. Here he observed that in "the hot countries the human being ... does not ... reach the perfection of those in the temperate zones. Humanity is at its greatest perfection in the race of the whites"; that "all inhabitants of the hottest zones are exceptionally lethargic"; and that:

> The inhabitant of the temperate parts of the world, above all the central part, has a more beautiful body, works harder, is more jocular, more controlled in his passions, more intelligent than any other race of people in the world. That is why at all points in time these peoples have educated the others and controlled them with weapons.[34]

Similar obsessions were also prominent in Hegel's *Lectures on the Philosophy of World History* given between 1822 and 1828. In the introductory volume dealing with "Reason in History", Hegel appended a lengthy discussion of what he called "the geographical basis of world history". Given his philosophical preoccupations with "the Idea of the spirit", Hegel's interest in world history predictably stemmed from his concern to discern how that self-same spirit displayed "itself in reality as a series of external forms". The way this played itself out for Hegel was in what he termed the "various national spirits", and he considered it geography's task to elucidate how the shape of specific spiritual forms was conditioned by "a particular external nature".[35] The whole undertaking was what might be called an exercise in the universal spiritualization of space, a stance entirely in keeping with his approving resort to the geographical writings of Carl Ritter.[36] "The spiritual character of the three continents" he wrote for instance, "varies in accordance with ... natural differences. In Africa proper, man has not progressed beyond a merely sensuous existence ... Asia is the land of antithesis, division, and expansion ... Europe is the land of spiritual unity". For, as he subsequently put it, the "natural world and the spiritual world together form the living totality of *history*".[37]

My concern here, however, is not to dilate upon the Hegelian philosophy of geo-history. Instead it is again to underscore the moral climatology that from time to time obtruded into the analysis. While Hegel, like Hume, was chary of admitting too much climatic influence on human affairs he freely conceded certain global generalities in his elucidation of spiritual space. "Climate does have a certain influence", he wrote, "... in that neither the *torrid* nor the *cold regions* can provide a basis for human freedom or for world-historical nations". Such extremities were simply "not conducive to spiritual development". Neither "the torrid nor the frigid zone" permitted the human race to "acquire sufficient resources to allow [it] to participate in higher spiritual interests". "The torrid and frigid regions, as such," he reaffirmed, "are not the theatre on which world history is enacted. In this respect, such extremes are incompatible with spiritual freedom." Having thereby edged much of global geography to the margins of historical significance, Hegel could conclude: "All in all, it is therefore the *temperate zone* which must furnish the theatre of world history. And more specifically, the northern part of the temperate regions is particularly suited to this purpose".[38]

Even such a brief excursion as this into the climate of European philosophical opinion serves to uncover an Enlightenment textual community revolving around the moralization of global space. And even allowing for the suspicion that the concept of 'race' in, say, Scottish philosophy and German idealism was far from identical, I believe we can still detect a racialization of global space as well. The considerable dependence of such writers on each other, and on a suite of treatises produced by scientific travellers, highlights what Emmanuel Chukwudi Eze dubs "an astonishing level of ... intertextuality. There is, so to speak, quite a promiscuous theoretical as well as stylistic dependence of one writer on another".[39] And yet it would be mistaken to conclude that the moralistic deployment of climatic imagining was always prosecuted in the same way. Consider now, for example, the transatlantic case of Samuel Stanhope Smith, moral philosopher, advocate of the Scottish philosophy of common sense, and forebear of American anthropology.[40] In his *Essay on the Causes of the Variety of Complexion and Figure in the Human Species*, which made its first appearance in 1787, Smith sought to refute the polygenist anthropology emanating in particular from the pen of Kames who insisted that the "different species of men [were] fitted for the different climates" rather than being a product of those self-same regimes.[41] For Smith, by contrast, climate was, very largely, adequate to account for human variability and the diagnostic features of human societies were to be understood as responses to such environmental conditions. But Smith's project cannot simply be read as a Montesquieu-like endorsement of climatic determinism; the mainsprings of his anthropological engagement were, in a profound sense, a matter of moral philosophy. As he confirmed in the 1810 edition of his *Essay*, "the denial of the unity of the human species tends to impair, if not entirely destroy, the foundations of duty and morals, and, in a word, of the whole science of human nature. No general principles of conduct, or religion, or even of civil policy could be derived from natures originally and essentially different from one another, and, afterwards, in the perpetual changes of the world, infinitely mixed and compounded".[42] What bothered Smith, at base, was that Kames-style polygenism subverted the unity of the human constitution and therefore any moral basis for political intercourse—a crucial consideration in the immediate aftermath of the American Revolution whose apologists were busily repudiating "the props upon which virtue had traditionally rested—tradition itself, divine revelation, history, social hierarchy, an inherited government, and the authority of religious denominations".[43] Smith's resort to climate as a determinant of racial variation was therefore with the purpose of preserving a universal *moral* human nature—a common moral sense—in the face of seeming anthropological disruption. Here then was another facet of the 'moral climatology' so characteristic of Enlightenment thinking.

Doubtless other cases might be called upon to illustrate the general tendency to elide moral commentary and climatic description. The writings of Buffon and John Locke, for example, might profitably be read with an eye to precisely this tendency. But comprehensive survey is not my quarry here. Rather it is to disclose a widespread presumption among Enlightenment *philosophes* that it was entirely appropriate to speak of global climate in the language of moral axiom.

The atmosphere of Victorian anthropology

The treatises of these Enlightenment thinkers, it is clear, contributed in various ways to the cultivation of a philosophical anthropology congregating around such enigmas as the 'nature of human nature' and the political implications of anthropological diversity. In this environment what I have called moral climatology was often called upon to bear the weight of the suspicion that human nature was neither geographically universal nor historically constant.[44] Indeed the conjectural human history that flourished among the Enlightenment *philosophes* was precisely designed to explain "how our nature would have undergone change and become what it is".[45] Now I want to move beyond Enlightenment philosophy to early nineteenth century empirical anthropology—a discipline that took shape in the interface between colonial expansionism and missionary endeavour.[46] My claim here is that the sorts of moralistic construal of regional climate that so captivated the arm-chair philosophers were reproduced and expanded by writers who spent lengthy periods of time in colonial settings and who published detailed ethnological accounts of the peoples among whom they were stationed. First-hand anthropological experience in tropical climates served to reinforce the heritage of racialized philosophical anthropology in which they were wittingly or unwittingly domiciled. Of course caricature is all too easy here and Gillian Beer tellingly points to the ways in which ethnographic writing could challenge taken-for-granted domestic assumptions. The need to "discriminate the different registers across which such terms", now offensive, as "savages" or "natives" could function, is well taken.[47]

Nevertheless, as illustrative of how moralizing manoeuvres could manifest themselves in Victorian anthropology, I want to dwell on the case of John Crawfurd (1783–1868), Fellow of the Royal Society, ethnologist, colonial envoy and civil surgeon, who spent several years in various capacities in Southeast Asia during the first quarter of the nineteenth century, and who, on his return to England, played a crucial role in the early history of the Ethnological Society of London of which he held the vice-presidency and presidency during the final decade of his life.[48] Bearing in mind that the Ethnological Society was decidedly humanitarian at least compared with its more racialist successor, the Anthropological Society of London, Crawfurd's declarations are thus all the more remarkable.[49] Of course I am not suggesting that Crawfurd spoke in any unique or representative way for Victorian anthropology. My claim, rather, is that from a point of strategic institutional significance his voice records at least some the currents of thought coursing through the anthropology of his day.

In 1820 Crawfurd published a three volume *History of the Indian Archipelago* in which he elaborated *in extenso* on the spectrum of cultural forms from anarchy to despotism he had discerned during his time in tropical Asia.[50] Under the former conditions, mere savagery was to be found, while the latter regime produced citizens who were, as Victor Savage puts it, "deprived the liberty of benefiting from the comforts of civil life surrounding them".[51] What is particularly notable in Crawfurd's analysis, however, is his attempt to find in climatic conditions the mainspring of such bi-polar cultural extremism. No people, he insisted, "inhabiting a warm climate has ever known how to reconcile freedom and civilization". The root cause, of course, was that "the softness and fruitfulness of the climate" failed to induce "that wholesome discipline by which man, in severer regions, is bred to habits of hardihood, enterprise, and independence".[52] Climatic luxuriance, environmental abundance and moral jeopardy apparently went hand in hand. Remarkable here, moreover, is the way that the very conditions which had suggested to earlier writers that the tropics were likely to be the site of Paradise, had become the basis for the vilification of warm climates. Such had long been the case. Sir Walter Raleigh, for instance, had suspected that the seemingly paradisial character of the tropical world, with its botanical riches, masked its true character; and he thus found himself pressed to the pathological conclusion that, "nature being liberall to all without labour, necessitie imposing no industrie", the tropics necessarily corrupted men and women by inducing "idlenesse [which] bringeth forth no other fruites than vaine thoughts and licentious pleasures".[53] Two centuries later, Crawfurd was re-telling that self-same tale as he depicted the deadening monotony of a tropical climate and its anaesthetizing effects on mind and will.

Moreover, the longitudinal consequences of climate's moral imperatives were immensely significant. If modes of Asiatic government were reflective of a soft and warm climate, then legal structures, religious practices and other cultural forms no less took shape under the influence of climatic conditions. It was, for example, India's own "forms of government" that made it impossible "to believe that a rational Christianity either was, or ever can be, under such circumstances, the character of religion among them".[54] In consequence the "barbarians of the Archipelago" were condemned to remain in a morally darkened space; for it was simply impossible for such peoples to "ever adopt a material and beneficial portion of the humanity, improvement, and morality of Europe, without, at the same time, adopting the religion with which these concomitants of civilization are so closely interwoven".[55] If what he called "a perfect freedom of colonization and settlement" was a precondition for the dissemination of *"civilization and Christianity"*, these processes were ultimately rooted in climatic realities. The "feebleness of intellect" that he discerned in the Indian islands, arose from "such a state of society, and such a climate" as to render "the greatest powers of the *native* mind" incomparably weaker than even "the ordinary standard of the human understanding in the highest stages of civilisation".[56] In this way Asiatic mental and moral temperaments were declared to be the naturalized products of climate. Take, for instance, his typification of Java (from a passage in which the moralistic adjectival modifiers are palpable) where a 'soft' climate had produced, as it were, a 'soft' people:

> The shepherd state, the offspring of the cold and immeasurable plains of Tartary, and the school both of war and government, could have no existence among the woods, the narrow valleys, and soft climate of Java. The Javanese are *naturally* an unwarlike people, and it is the necessary consequence of their luxurious climate, that they should want the hardihood and manly virtues of the semibarbarians of severer regions. The fertility of their soil, and the benignity of their climate, are a sort of hot-bed, in which has sprung up a sickly civilization, wanting the vigour and hardihood of the plant of a rougher clime, and more stubborn soil.[57]

These fixations were to remain a life-long passion with Crawfurd, and he further refined his moral climatology in imaginative ways as a glance at some of his pronouncements during the extraordinarily active final half-decade of his life will illustrate. During these years, when the *Transactions of the Ethnological Society* established a new series, Crawfurd's name surfaced as, by far, its most frequent contributor, dealing with such subjects as the connections between racial history and language, ethnic intermixture, domesticated animals, cannibalism and many more.[58] To be sure, he re-confirmed such well-worn sentiments as the view that "mere intemperance of climate ... is sufficient to prevent man from making any advance towards civilization, and to hold him permanently in the savage state" and that the "races of man which inhabit Africa correspond with the disadvantages of its physical geography".[59] But he also now deployed moral climatology as an explanans of civilization mediated through the role of domesticated animals. Since, to him, the extent of civilization was "ever in proportion to the number and variety of the animals which man has subdued to his service", and since climatic conditions directly influenced the geography of domesticated animal distribution, it was obvious that nature governed, to some degree at least, the global pattern of civilization.[60] Again, as a staunch defender of polygenism, he used geographical arguments to dismiss the "imagined distant migrations" that monogenists had to assume to sustain their theories of the unity of the human race and biological evolution.[61]

All in all, Crawfurd mobilized his convictions about innate racial quality and the moral order of climate to deliver a textual globe whose narrative legend was replete with such descriptors as "feeble", "effeminate", "despotic",[62] "rude", "unambitious", and "unenterprising".[63] Further testimonial confirmation of the vitality of moral climatology from the writings of other Victorian anthropologists could certainly be supplied. James Hunt, for example, in an explicit discussion of what he called 'ethno-climatology' was sure that tropical climates induced "a low sense of morality" and produced inhabitants "essentially sensual"; temperate regimes, by contrast, delivered "increased activity of the brain".[64] But enough has been said to give some sense of how moral climatology surfaced in the empirical writings of anthropologists in the period.

Medicalizing moral meteorology

Even while early nineteenth century ethnologists were extending the climatic imperatives of Enlightenment philosophical anthropology, moves towards the medicalization of the project were already well under way. What is noteworthy here, I think, is the fact that throughout the nineteenth century, both before *and* after the rise of the new parasitology (which effectively discredited the older environmental tradition of medicine) moral climatology persisted in one form or another. It surfaced either in what might be called a derivative or distributive form, namely, that mental and moral worth were *derived from* climatic circumstances, or that these self-same qualities were *distributed* according to climatic zone. James Ranald Martin, for example, built a moral component into the very fabric of medical topography when he elaborated, as a foundational axiom of the discipline, the principle that "that a slothful, squalid-looking population invariably characterises an unhealthy country". This, he believed, confirmed his judgement that when "a body of men who are idle and intemperate" move into an area,

> the immediate result will be that the soil will deterioriate for want of proper care, the weeds will reappear, the drains will become obstructed, the edible products of the earth will lessen in quantity, and diminish in their nutritive quality; the inhabitants will become unhealthy from the bad state of their grounds; and the diminution of their

physical powers thus produced will disable them progressively more and more from remedying the causes of the evil. [65]

Thus while it is true to say, as David Arnold puts it, that the "interest in medical topography served to strengthen the concept of disease as being environmentally determined", it is equally clear that the enterprise was often an exercise in the plotting of regional moralities.[66]

Consider first, then, the case of James Johnson (1777–1845), an Irish navy surgeon who spent lengthy periods of time in both Egypt and India before establishing himself in a distinguished London practice.[67] In 1813, he published an influential treatise on the effects of tropical climates on European constitutions. The pessimistic tincture of Johnson's work is well known to medical historians[68] as he insisted that "the tender frame of man is incapable of sustaining that degree of exposure to the whole range of causes and effects incident to, or arising from vicissitudes of climate." Degeneracy, disease, drooping and death were the inevitable consequences of a failure to "study well the climate, and mould our obsequious frames to the nature of the skies under which we sojourn".[69] Johnson's medical interventions were part and parcel of his moral universe. Thus when he appended a section on 'Tropical Hygiene' to his text, he concluded with the following observations on the conduct of the passions:

> The removal of religious and moral restraint—the temptations to vice—the facility of the means and the force of example, are the real causes of this 'bias to pleasure'; and in respect to the *effects* of licentious indulgences between the tropics, I can assure my reader that he will find, probably when it is too late, how much more dangerous and destructive they are than in Europe . . . The monotony of life, and the apathy of mind, so conspicuous among Europeans in hot climates, together with the obstacles to matrimony, too often lead to vicious and immoral connexions with Native females, which speedily sap the foundation of principles imbibed in early youth, and involve a train of consequences, not seldom embarrassing, if not embittering every subsequent period of life! It is here that a taste for some of the more refined and elegant species of literature, will prove an invaluable acquisition for dispelling *ennui*, the moth of mind and body.[70]

That these judgements were to prove long lasting is nowhere more clearly revealed than in the fact that some 40 years later, when James Ranald Martin "ventured to re-cast and re-write the entire work", these paragraphs remained materially unchanged.[71]

Johnson's predilection, here disclosed, towards the painting of moral geographies did not lie buried within the covers of what was to become Victorian England's Bible of tropical medicine.[72] Some years later, in 1831, when he published his reflections on an autumnal excursion through Europe taken in 1829, he added "Observations and Reflections on the Moral, Physical, and Medicinal Influence of Travelling-Exercise, Change of Scene, Foreign Skies, and Voluntary Expatriation". To be sure, his focus here was on the temperate world, rather than the tropics, but his suspicions about the effects of Italy's climate were precisely because "thermometrically speaking . . . the Summer heat of the Italian valleys approaches the temperature of the tropics".[73] The "seductive" nature of the climate, however, could not gainsay the "statistical facts" he adduced to substantiate his conviction that the "climate of Italy . . . is much less favourable to the health and longevity of man than in England".[74] My reason for advertising these judgements, however, is not to do with exposing British haughtiness; rather it is to redraw attention to the ease with which climate was read through moralistic lenses and translated into the language of diagnostic medicine. For having reviewed what he called the 'Medicinal Influence of an Italian Climate', Johnson moved seamlessly on to the "Moral Influence of an Italian Climate". Certainly he confessed the latter was "not quite so tangible or obvious" as the former, but he nonetheless itemized "Cleanliness and Delicacy—Industry—Patriotism—Morality— Religion" under the general rubric. The following excerpt serves to convey some sense of the entire undertaking:

> Numerous moral and political causes ... have combined with the burning skies of an Italian Summer, and the mild atmosphere of Winter to relax the corporeal energy and intellectual vigour of the permanent inhabitants—and, that a portion of this inactivity—this *vis inertiae*, is communicated, partly by the force of example, partly by the operation of climate, to all who reside for any length of time beneath the skies of Italy, there cannot be a shadow of doubt. This effect is not peculiar to that country alone, but is seen in all hot climates, operating with more or less force, on those who migrate thither from northern regions.[75]

The medical and moral 'pathologization of space' that flourished under the environmentalist paradigm, however, did not disappear with the triumph of the new Tropical Medicine that surfaced in 1898 with the publication of Patrick Manson's *Tropical Diseases*,[76] the first edition of which was adorned with a miniature icon of "La Belle Sauvage" on the frontispiece.[77] Allow me to illustrate this by calling attention to Luigi Westenra Sambon who joined Manson at the London School of Tropical Medicine soon after its establishment in 1899, and emerged as a dedicated advocate of the germ theory.[78] It was the evidence that he accumulated (along with George Carmichael Low) during work in a highly malarious region of Italy—the Roman campagna—that provided crucial evidence in support of Manson's mosquito-transmission theory of malaria. The matter of human acclimatization to the tropical world was a major concern of Sambon and he addressed several audiences on the topic, publishing his own ideas in several different outlets.[79] Throughout he vigorously supported the view that Europeans could rapidly and entirely adjust in a short space of time to the tropical world, that tropical climate had been the victim of prejudice and unwarranted medical stereotyping, that "the difficulties in the way of colonisation are not due to climate, but to parasitism" and that acclimatization "is, to a great extent, a mere question of hygiene".[80] Nevertheless, it was clear that, to Sambon, *medical* hygiene encompassed *moral* hygiene as well:

> personal habits are of the utmost importance; temperance and morality are powerful weapons in the struggle for life ... Sexual immorality under the influence of a tropical climate, and in the presence of a native servile and morally undeveloped population, raises to a climax unknown amid the restraints of home life, and becomes one of the most potent causes of physical prostration.[81]

For Sambon, the Darwinian struggle against tropical parasites called for the self-same moral circumspection that earlier advocates of environmental medicine had demanded. Parasitology did not exorcise tropical medicine of its climatic obsessions; rather it reworked climatic categories but continued to privilege such factors "as predisposing, conducive, or even essential prerequisites, either for the existence of the germ, or for its transference, or for it to thrive".[82] In this discursive environment, plenty of space was left for moral prophylaxis.

The climate of moral opinion

Although periodic moral commentary routinely punctuated the writings of both anthropologists and students of tropical medicine, we might be justified in thinking that the writings of climatologists and meteorologists would remain immune to such speculative readings of global weather. Again, however, what we find is a persistent inclination—right into the mid-twentieth century—to supplement scientific accounts of atmospheric conditions with moralistic pronouncements. Three cases will serve to sketch in the main features of the landscape.

Seas and Skies in Many Latitudes was a special kind of travel book; reminiscent of the immensely popular *Chasing the Monsoon* which was appeared in 1990,[83] it was a kind of meteorological travelogue neatly encapsulated in its subtitle, *Wanderings in Search of Weather.*[84] The book came out in 1888 and its author was Ralph Abercromby, a Fellow of the Royal Meteorological Society, author of *Principles of Forecasting by Means of Weather Charts*, and *Weather*, and contributor to the *Philosophical Magazine* on such matters as the velocity of ocean waves. While the seeking out of strange weather was the aim of the project, in particular the hunting of the hurricane, he devoted much time to interspersing his account with periodic asides on heat exhaustion, frostbite and other weather-related conditions. But beside this, the moral economy of climate obtruded regularly. Take, for example, his reminiscences on his time in Egypt. Here, he observed, the "traveller from Europe enters . . . another world of scenery and climate, of man and of thought". And most prominent in what might be called the directory of difference was the "manifestation in religion" of Oriental "character and temperament". These features were directly mappable on to climatic circumstance, and he embarked on the project of showing "how far certain peculiarities of climate have favoured the development of various religions in the neighbourhood of Syria and Arabia, and how the extensions of Islam is almost conterminous with certain very dry regions of the earth".[85] The basic argument ran something like this. Because the development of "that metaphysical temperament which is indispensable to the founder of a new religion requires solitude and comparative idleness", it was clear that in "any climate among the lowest races men are too much occupied with procuring food to have time or inclination for anything else", and, therefore, that "people in that stage of civilisation have little or no religion, among other good reasons because the necessary conditions for thinking are impossible among them". No genuine religion, for example, could even grow in "many tropical climates, like the South Sea Islands" because "subsistence is so easy that men need not work".[86] By contrast, echoing Ernest Renan's dictum that "le désert est monothéiste",[87] Abercromby insisted that it was:

> . . . in the dry deserts of Syria and Arabia we find all the requirements for the development of religious thought. The struggle with the forces of nature, compared with either that against the rank forest and savage beasts of the tropics, or the cold winter of northern climates, is so easy, that nations in those dry regions appear to have been the first to establish a tolerably organised state of society, where men have leisure to think, and to consider how far the welfare of the community can be improved.[88]

In Abercromby's telling, then, the geographical distribution of Islam mapped pretty well onto the rainless areas of Africa and Asia. Regional climatology was thus, to a large degree, an exercise in the naturalizing of regional religious systems. It was also a project in the medicalization of space in which, recycling standard medical opinion, he casually discriminated sickly and salubrious spaces. When writing on Manila under the title 'Health in the Tropics', he expanded on "the debilitating effect of mere residence in a tropical climate. The moment an Englishman enters the steamy air of the Tropics he begins to lose his superfluous energy, and sooner or later, according to his constitution and habits of life, his nervous and physical strength is impaired." To Abercromby, then, it was certain that a

> great deal of the failure to open up such countries as Africa, is due to the climatic demoralisation of the agents employed, rather than to the intrinsic difficulties of the enterprise. The best protector of the African savage from European aggressors is the deadly climate of that dark continent. When trying to discover the origin of this peculiar debilitating effect of tropical climate, one must certainly lay the blame on the steamy nature of the heat.[89]

These dangers of demoralization, degeneracy, depravity and debility however, were not just climatologically distributed; they were also profoundly gendered as the following not a little quaintly attests:

> The languor of the Tropics is only a sensation, and one which increases with indulgence. It is impossible to conceive a more unhealthy life than that of many European ladies in India, who lie on their couches all day and do absolutely nothing but think how hot it is and how ill they are. Lawn-tennis has had a most beneficial influence on the habits of Englishwomen in the Tropics; for the pleasure and excitement of the game compels them to take exercise, and to forget their ailments for a longer or shorter period.[90]

Sentiments of this general class continued to manifest themselves in standard climatology textbooks until well into the twentieth century as the following two cases reveal. The first suite of ideas and practices congregates around the climatologist, Robert De Courcy Ward, who taught at Harvard for the first three decades of the century.[91] During a distinguished career he edited the *American Meteorological Journal*, authored a variety of books including *The Climates of the United States* (1925) and collaborated with Köppen on a new edition of the prestigious five-volume *Handbuch der Klimatologie*. Here, however, I want initially to refer to his immensely popular text on *Climate* which appeared in 1908 as a companion volume to the Hann *Handbuch*. Having disposed of the classification of climates and the meteorological characteristics of the three major bands, Ward turned to what he called "The Hygiene of the Zones" and to the impact of tropical, temperate and polar regimes on the human race. Ward was fully aware of the medical research carried out by Manson and Ross on malaria, Reed and Carroll on yellow fever, Koch on cholera, and Kitasato on plague, and he reviewed their findings for his readers. Even while affecting a cautious espousal of the 'broad generalizations' of the likes of Montesquieu, Buffon and Hume, it was absolutely clear that he believed that everything from mental characteristics to governmental systems were "affected to a greater or less degree" by climate, and happily cited the deterministic construals of Oriental religion by the likes of Renan and Abercromby.

Any detailed recitation of yet another portrayal of tropical climate as "debilitating and ennervating" and of the "ignorance, superstitutious . . . and sexual passion" of its inhabitants, would amount to little more than multiplying testimonials. What is conspicuous here, however is the degree to which such flaws provoked political imperatives. In tropical zones, he insisted, "voluntary progress toward a higher civilisation is not reasonably to be expected" and accordingly the "tropics must be developed under other auspices than their own".[92] What made these sentiments particularly ominous were the contextual circumstances that allowed Ward to mobilize applied climatology in the cause of political demagogy. For even as *Climate* was making its way onto the bookshelves in 1908, Ward was winding up his chairmanship of the Immigation Restriction League of Boston—a pressure group composed of Boston Brahmins pushing for legislative mechanisms to secure selective immigration.[93] As a Harvard man, he occupied a position of strategic institutional power from which he could mobilize scientific authority in the cause of racial politics. In fact much of his own research effort rotated around what he himself labelled 'anthropoclimatology'—a mélange of geography, history, medicine, meteorology, hygiene and ethnology. Using the vocabulary of science, like other nativists, he easily constructed with remorseless zeal a racial hierarchy to which genteel Bostonians could willingly give their allegiance. This is amply illustrated by his contribution to a 1930 book entitled *The Alien in Our Midst*, or *"Selling our Birthright for a Mess of Industrial Potage"*—a collaborative volume of essays, written by "a number of Americans (present and former) on immigration and its results" and drawn together by Madison Grant and Charles Stewart Davison:

> "Never shall ye make the crab walk straight. Never shall ye make the sea-urchin smooth." Thus many centuries ago, Aristophanes disposed of the fallacy of the Melting Pot. ... What goes into the Melting Pot determines what will come out of it. Put in sound, sturdy stock akin to the pioneer breed ... then we shall develop a race here worthy to carry on the ideals and traditions of the founders of our country. But if the material is a polyglot assortment of nationalities, physically, mentally, morally below par, then there can be no hope of producing anything but an inferior race.[94]

Within climatology, such ideas did not go away. The following year, 1931, Austin Miller's standard textbook on *Climatology* made its first appearance. By the time of its fifth edition in 1957, it was still declaring that "Psychologically, each climate tends to have its own mentality, innate in its inhabitants and grafted on its immigrants" and that direct behavioural correlates of climatic governance were readily discernible across the globe. "The enervating monotonous climates of much of the tropical zone, together with the abundant and easily obtained food-supply," Miller told his readers, "produce a lazy and indolent people, indisposed to labour for hire and therefore in the past subjected to coercion culminating in slavery."[95] Two things are remarkable here, namely, the way in which moralistic descriptions of climate—enervating, monotonous, lazy, indolent—were still being presented as settled scientific maxims, and the ease with which intellectual potential and moral behaviour were thoroughly naturalized courtesy of the inexorable operation of climate's moral imperatives.

Conclusion: mapping moral space

The theme that runs like a high voltage current through the moral climatology of Enlightenment philosophers and polygenist anthropologists, medical topographers and regional climatologists, is the idea that—in one way or the other—global space can be parcelled into a series of bi-polar antipathies: the salubrious and insalubrious; the superior and the subordinate; the moderate and the intemperate. What such fixations delivered was a textual cartographic that undergirded the mental map of the modern West and taught us, to use John Willinsky's phrase, how "to divide the world".[96] Indeed there were many occasions on which these metaphorical mappings materialized into what Martin Rudwick has called the "visual language" of cartography.[97]

In the Hippocratic tradition, the mapping of diseases, for example, had long been what Michael Osborne tellingly characterizes as "a medical analogue of cartography".[98] Thus during the French expeditions into the Orient, military leaders regarded cartography as the most valuable of the various scientific skills at their disposal. That medical topography in this vein was an exercise in the mapping of moral space is clear from Osborne's further comment that Bory de Saint-Vincent depicted Greece "as a degraded place with degraded and wretched people in need of restoration and the benefits of French civilization".[99] Here, disease ecology and moral cartography were much closer than distant cousins. Within anthropology,[100] George Gliddon's scurrilous map of the "Geographical Distribution of the *simiæ* in relation to that of some inferior types of men" stands out as a particularly offensive case. Produced to confirm his claim that "within the black circumvalling line which surrounds the zone occupied by the *simiæ*, no 'civilization' . . . has ever been spontaneously developed since historical times", the map's message was simply that "the most superior types of Monkeys are found to be indigenous exactly where we encounter races of some of the most inferior types of Men".[101] As for geography, we might refer to Ellsworth Huntington's charts of civilization, genius, health, and what he called 'climatic energy', constructed to give visual form to his racist convictions about the superiority of temperate Europe.[102] Or yet again, we might advert to Griffith Taylor's extraordinarily imaginative maps, drawing on geological analogy, to portray linguistic "shatter-belts", "zones and strata" of racial distribution, and "lava-flow" diagrams of evolution, formulated to support the contention of his mentor William Diller Matthew that "the higher races of man are adapted to a cool-temperate climate [where] they reach their highest physical, mental and social attainments".[103]

These endeavours were but the latest expression of a moral ordering of space that had its roots in the encounter with difference that attended European voyages of reconnaissance. I have no doubt that Peter Gay is correct when he observes that the "prehistory of the social sciences had its beginning in the emergence of cultural relativism, the bittersweet fruit of travel. Whether realistic, embroidered, or imaginary, whether on ships or in libraries, travel was the school of comparison, and travelers' reports were the ancestors of treatises on cultural anthropology and political sociology".[104] Nor is there any doubt that Paul Hazard is right when he observes that travel meant "comparing manners and customs, rules of life, philosophies, religions; arriving at some notion of the relative; discussing; doubting".[105] But I am also convinced that the West persistently devised mechanisms to ensure that cultural relativism rarely implied cultural equivalence and that the temperate world would exercise dominion over tropical realms. Among these, moral climatology, I believe, had a key role to play. This is not to say that the moral economy of climate was always construed in the same way. To the contrary. Rather what we have witnessed here is what might be called a *common condition of discursive possibility*; by this I mean that it was routinely believed that climate could reasonably be represented in the language of moral judgement. The genealogy I have tried to uncover here may be seen as little more than a single drama performed again and again by different actors in different settings—philosophers, ethnologists, medical practitioners, climatologists, geographers. I hope my narrative amounts to more than just this of course; but even if not, it still attests to the remarkable capacity of a set of ideas to flourish in different temporal and cognitive niches.

School of Geography
The Queen's University of Belfast
Belfast BT7 1NN
Northern Ireland

Acknowledgements

I am most grateful to David Arnold, Felix Driver, Michael Heffernan, Nuala Johnson, Luciana Martins, James Ryan, Stephen Williams, Charles Withers and three anonymous readers for careful readings of an earlier version of this paper and for many insightful observations on it. I am very pleased to acknowledge the support of the British Academy, in the form of a two-year Research Readership, which facilitated the completion of this paper.

Notes

[1] P. Gourou, *The Tropical World: Its Social and Economic Conditions and its Future Status* (London 1953) 141. First published in French in 1947.

[2] D. Arnold, 'Illusory riches': northern representations of the tropical landscape, 1850–1950, *Singapore Journal of Tropical Geography*, **21** (2000) 6–18, quotation on p. 6.

[3] Though, interestingly, R. W. Steel's survey of 'Geographers and the Tropics' in R. W. Steel and R. Mansell Prothero (Eds), *Geographers and the Tropics: Liverpool Essays* (London 1964) 1–29, made only passing reference to Gourou's contributions and insisted that (p. 2) "there is no special branch of the subject to be recognized as 'tropical geography' ".

[4] Gourou, *Tropical World*, 1, 6, 24.

[5] Derek Gregory, for instance, speaks of the way in which European colonization "emphasised the virtues, and by implication, the normalization of temperate regimes". See D. Gregory, *Explorations in Critical Human Geography*, (The Hettner Lecture for 1997, Department of Geography, University of Heidelberg 1998) 64.

[6] See the discussion in C. J. Glacken, *Traces on the Rhodian Shore: Nature and Culture in Western Thought From Ancient Times to the End of the Eighteenth Century* (Los Angeles and Berkeley 1967) 434–441.

[7] S. Du Bartas, *His Divine Weekes and Workes* (translated by J. Sylvester) (1605) cited in J. Prest, *The Garden of Eden: the Botanic Garden and the Re-creation of Paradise* (New Haven 1981) 66. The reference to Du Bartas as a French Milton occurs on p. 34.

[8] Glacken, *Traces on the Rhodian Shore*, 551–622.

[9] W. Falconer, *Remarks on the Influence of Climate, Situation, Nature of Country, Population, Nature of Food, and Way of Life on the Disposition and Temper, Manners and Behaviour, Intellects, Laws and Customs, Form of Government, and Religion, of Mankind* (London 1781) 20, 73.

[10] J. Gascoigne, *Joseph Banks and the English Enlightenment* (Cambridge 1994) 142.

[11] Drummond's version of Spencerian evolution is manifest in his *Natural Law in the Spiritual World and The Ascent of Man.* A telling analysis of his outlook is to be found in J. R. Moore, Evangelicals and evolution: Henry Drummond, Herbert Spencer, and the naturalisation of the spiritual world, *Scottish Journal of Theology* **38** (1985) 383–417.

[12] H. Drummond, *Tropical Africa* (London 1891) 3, 10, 4.

[13] The book included research on the significance of mimicry in insect evolution, a theory of the African white ant, together with geological and meteorological notes.

[14] Drummond, *Tropical Africa*, 11, 12, 4, 90, 55.

[15] Gourou, *Tropical World*, 14, 23, 49.

[16] David Arnold describes the tropics as a conceptual as well as physical space. See D. Arnold, *The Problem of Nature: Environment, Culture and European Expansion* (Oxford 1996) chapter 8 'Inventing tropicality'. See also the forthcoming special issue of the *Singapore Journal of Tropical Geography* edited by F. Driver and B. Yeoh on 'Constructing the tropics' **21** (2000).

[17] See, for instance, M. L. Pratt, Alexander von Humboldt and the reinvention of América, in *Imperial Eyes: Travel Writing and Transculturation* (London 1992); J. Krasner, *The Entangled Eye: Visual Perception and Representation of Nature in Post-Darwinian Narrative* (Oxford 1992); L. Martins, A naturalist's vision of the tropics: Charles Darwin and the Brazilian landscape, *Singapore Journal of Tropical Geography*, **21** (2000) 19–33; R. H. Grove, *Green Imperialism: Colonial Expansion: Tropical Island Edens and the Origins of Environmentalism, 1600–1860* (Cambridge 1995); N. Rennie, *Far-Fetched Facts: The Literature of Travel and the Idea of the South Seas* (Oxford 1995); B. Smith, *European Vision and the South Pacific* (New Haven 1985); B. M. Stafford, *Voyage Into Substance: Science, Nature and the Illustrated Travel Account*, 1770–1840 (Cambridge, MA 1984); K. E. Manthorne, *Tropical Renaissance: North American Artists Exploring Latin America, 1839–1879* (Washington 1989).

[18] N. L. Stepan, Tropical nature as a way of writing, in A. Lafuente, A. Elena and M. L. Ortega (Eds), *Mundialización de la Ciencia y Cultura Nacional* (Madrid 1993) 495–510.

[19] M. I. Newbigin, *Animal Geography: The Faunas of the Natural Regions of the Globe* (Oxford 1913) 93, 96.

[20] P. Manson, *Tropical Diseases: A Manual of the Diseases of Warm Climates* (London 1898) xvi.

[21] Elsewhere I have referred to this species of endeavour as 'the moral economy of climate'. See D. N. Livingstone, The moral discourse of climate: historical considerations on race, place, and virtue, *Journal of Historical Geography* **17** (1991) 413–434, and Idem., Climate's

moral economy: science, race and place in post-Darwinian British and American geography in N. Smith and A. Godlewska (Eds), *Geography and Empire* (Oxford 1994) 132–154.
[22] For a relevant survey see I. Hannaford, *Race: The History of an Idea in the West* (Baltimore 1996).
[23] See the discussion in D. N. Livingstone, Tropical hermeneutics: fragments for a historical narrative, *Singapore Journal of Tropical Geography* **21** (2000) 93–100.
[24] See, for example, R. H. Popkin, The philosophical bases of modern racism, in C. Walton and J. P. Anton (Eds), *Philosophy and the Civilizing Arts: Essays Presented to Herbert W. Schneider on his Eightieth Birthday* (Athens, Ohio 1974) 126–65; H. M. Bracken, Philosophy and racism in H. M. Bracken (Ed.), *Mind and Language, Essays on Descartes and Chomsky* (Dordrecht 1984) 51–66; R. H. Popkin, Medicine, Racism and anti-semitism: a dimension of Enlightenment culture, in G. S. Rousseau (Ed.), *The Languages of Psyche: Mind and Body in Enlightenment Thought* (Berkeley 1990); D. Goldberg, *Racist Culture: Philosophy and the Politics of Meaning* (Oxford 1993); S. Harding (Ed.), *The "Racial" Economy of Science* (Bloomington 1993); E. C. Eze, *Race and the Enlightenment: A Reader* (Oxford 1997); R. H. Popkin, Eighteenth century racism, in R. H. Popkin (Ed.), *The Pimlico History of Philosophy* (London 1999) 508–515.
[25] See S. Toulmin, *Cosmopolis: The Hidden Agenda of Modernity* (Chicago 1990) 32, 33.
[26] This essay included the infamous footnote supporting polygenism and affirming the inferiority of blacks. See the discussion in R. H. Popkin, Hume's racism, *The Philosophical Forum* **9** (1977–78) 211–26, reprinted in R. H. Popkin, *The High Road to Pyrrhonism* (Indianapolis 1993) 251–66.
[27] D. Hume, Of national characters, in E. F. Miller (Ed.), *Essays: Moral, Political and Literary* (Indianapolis 1985) 197–215, on 207, 209, 213, 215.
[28] Montesquieu, *De L'Esprit des Lois* (Geneva 1748). The first English translation of this original edition by Thomas Nuget was published in London in 1750, and the citations that follow are drawn from this version.
[29] See the recent observations in Arnold, *Problem of Nature*, 18–25.
[30] See R. G. Collingwood, *The Idea of History* (Oxford 1946) and E. Cassirer, *The Philosophy of the Enlightenment*, translated by F. C. A. Koelln and J. P. Pettegrove (Boston 1955).
[31] D. W. Carrithers, Introduction, in D. W. Carrithers (Ed.), *The Spirit of Laws by Montesquieu. A Compendium of the First English Edition* (Berkeley 1977) 1–88, 46–47. The importance of this physiological context is clear in Montesquieu's "An essay on the causes affecting minds and characters", appended to the above edition of *The Spirit of Laws*. Paragraphs nine to 15 deal with "How the condition of the fibres of the nervous system can affect minds and characters". More generally, see also D. W. Carrithers The Enlightenment science of society, in C. Fox, R. Porter and R. Wokler (Eds), *Inventing Human Science: Eighteenth-century Domains* (Berkeley 1995) 232–70.
[32] I. Kant, Von den verschiedenen Racen der Menschen, reprinted as On the different races of man, in E. W. Count (Ed.), *This is Race: An Anthology Selected From the International Literature on the Races of Man* (New York 1950) 16–24, 23. Kant's racial thinking is discussed in K. W. Scheidt, The concept of race in anthropology and the divisions into human races from Linneus to Deniker, in Count, *This is Race*, 354–91. Scheidt makes the telling observation that "Kant reached raciology via a concern with geography", p. 372. See also D. Lloyd, Kant's examples, *Representations* **28** (1989) 34–54.
[33] I. Kant, Of national characteristics, in *Observations on the Feeling of the Beautiful and Sublime*, translated by J. T. Goldthwait (Berkeley and Los Angeles 1960), and reprinted in Eze, *Race and the Enlightenment*, 49–57.
[34] I. Kant, On countries that are known and unknown to Europeans, in *Physical Geography*, translated by K. M. Faull and E. C. Eze, in Eze, *Race and Enlightenment*, 58–64 on 63, 64.
[35] G. W. Friedrich Hegel, *Lectures on the Philosophy of World History. Introduction: Reason in History*, translated by H. B. Nisbet (Cambridge 1975 [1830]) 152.
[36] Ritter is cited on 173, 176, 233 of Hegel's *Lectures*.
[37] Hegel, *Lectures*, 172–73, 191.
[38] Hegel, *Lectures*, 154–55.
[39] Eze, Introduction, in *Race and Enlightenment*, 6.
[40] Aspects of Smith's climatic thinking are discussed in D. N. Livingstone, Geographical inquiry, rational religion, and moral philosophy: Enlightenment discourses on the human condition, in D. N. Livingstone and C. W. J. Withers (Eds), *Geography and Enlightenment*

(Chicago 1999) 93–119. Smith's anthropology has been the subject of numerous analyses, among them W. Stanton, *The Leopard's Spots: Scientific Attitudes Toward Race in America, 1815–59* (Chicago 1960) and T. F. Gossett, *Race: the History of an Idea in America* (Dallas 1963).

[41] Quoted in I. S. Ross, *Lord Kames and the Scotland of His Day* (Oxford 1972) 334.

[42] S. S. Smith, *Essay on the Causes of the Variety of Complexion and Figure in the Human Species* (Cambridge, Mass. 1965 reprint of 1810 edition) 149.

[43] M. A. Noll, The rise and long life of the Protestant Enlightenment in America, in W. M. Shea and P. A. Huff (Eds), *Knowledge and Belief in America. Enlightenment Traditions and Modern Religious Thought* (New York 1995) 88–124 on 100. See also the discussion in M. A. Noll, *Princeton and the Republic, 1768–1822. The Search for a Christian Enlightenment in the Era of Samuel Stanhope Smith* (Princeton 1989).

[44] See the discussions in R. Smith, The language of human nature, in C. Fox, R. Porter, and R. Wokler (Eds), *Inventing Human Science. Eighteenth-century Domains* (Berkeley 1995) 88–111; R. Smith, *The Fontana History of the Human Sciences* (London 1997); and P. B. Wood, The science of man, in N. Jardine, J. A. Secord and E. C. Spary (Eds), *Cultures of Natural History* (Cambridge 1996) 197–210.

[45] R. Wokler, Anthropology and conjectural history in the Enlightenment, in Fox *et al.*, *Inventing Human Science* 31–52 on 33.

[46] See, for example, G. W. Stocking Jr., *Victorian Anthropology* (New York 1987).

[47] See G. Beer, Speaking for the others: relativism and authority in Victorian anthropological literature, in *Open Fields: Science in Cultural Encounter* (Oxford 1996) 71–94, on 78.

[48] Something of Crawfurd's outlook is discussed in Livingstone, Moral Discourse of Climate, *op cit.*, and Climate's moral economy, *op. cit.* See also the discussion in V. R. Savage, *Western Impressions of Nature and Landscape in Southeast Asia* (Singapore 1984) 179–82.

[49] See R. Rainger, Race, politics, and science: The Anthropological Society of London in the 1860s, *Victorian Studies* **22** (1978) 51–70 and G. W. Stocking, Jr., What's in a name? The origins of the Royal Anthropological Institute: 1837–1871, *Man* **6** (1971) 369–90.

[50] On the history of the idea of Oriental despotism both before and after Montesquieu, see F. Venturi, Oriental despotism, *Journal of the History of Ideas* **24** (1963) 133–142; M. Richter, Despotism, in P. P. Wiener (Ed.), *Dictionary of the History of Ideas* (New York 1973), Vol. 2, 1–18; M. Curtis, The oriental despotic universe of Montesquieu, *Princeton Papers in Near Eastern Studies* **3** (1994) 1–38.

[51] Savage, *Western Impressions*, 180.

[52] J. Crawfurd, *History of the Indian Archipelago: Containing an Account of the Manners, Arts, Languages, Religions, Institutions, and Commerce of its Inhabitants* (Edinburgh 1820, 3 volumes), Vol. III, 4.

[53] Sir Walter Raleigh, *The History of the World* (London 1614), Book I, chapter III, para 15. On the issue of the geography of paradise more generally, see C. W. J. Withers, Geography, Enlightenment, and the paradise question in Livingstone and Withers (Eds), *Geography and Enlightenment*, pp. 67–92.

[54] Crawfurd, *History of the Indian Archipelago*, vol. II, 276.

[55] Crawfurd, *History of the Indian Archipelago*, vol. II, 279.

[56] Crawfurd, *History of the Indian Archipelago*, vol. II, 287.

[57] Crawfurd, *History of the Indian Archipelago*, vol. II, 296, emphasis in original.

[58] See, for example, J. Crawfurd, On the Antiquity of man from the evidence of language, *Transactions of the Ethnological Society of London* **2** (1863) 170–81; On the commixture of the races of man as affecting the progress of civilization (Europe), *Transactions of the Ethnological Society of London* **2** (1863) 201–13; On language as a test of the races of man, *Transactions of the Ethnological Society of London* **3** (1865) 1–9; On the so-called Celtic languages in reference to the question of race, *Transactions of the Ethnological Society of London* **4** (1866) 71–100; On cannibalism in relation to ethnology, *Transactions of the Ethnological Society of London* **4** (1866) 105–24.

[59] J. Crawfurd, On the connexion between ethnology and physical geography, *Transactions of the Ethnological Society of London* **2** (1863) 4–23, 4, 8.

[60] J. Crawfurd, On the relation of the domesticated animals to civilization, *Transactions of the Ethnological Society of London* **2** (1863) 387–468 on 387.

[61] J. Crawfurd, On the early migrations of man, *Transactions of the Ethnological Society of London* **3** (1865) 335–56 on 335. His repudiation of monogenism is clear in On Sir Charles

Lyell's 'Antiquity of man' and on Professor Huxley's 'Evidence as to man's place in nature', *Transactions of the Ethnological Society of London* **3** (1865) 58–70; On the plurality of the races of man, *Transactions of the Ethnological Society of London* **6** (1868) 49–58.

[62] J. Crawfurd, On the commixture of the races of man as affecting the progress of civilisation, *Transactions of the Ethnological Society of London* **3** (1865) 98–122 on 100.

[63] J. Crawfurd, On the physical and mental characteristics of the negro, *Transactions of the Ethnological Society of London* **4** (1866) 212–39 on 217.

[64] J. Hunt, On ethno-climatology; or the acclimatization of man, *Transactions of the Ethnological Society of London* **2** (1863) 53.

[65] J. R. Martin, *The Influence of Tropical Climates on European Constitutions, Including Practical Observations on the Nature and Treatment of the Diseases of Europeans on Their Return from Tropical Climates* (London 1856) 138. Martin was already the author of *Notes on the Medical Topography of Calcutta* (Calcutta 1837).

[66] D. Arnold, Introduction: tropical medicine before Manson, in D. Arnold (Ed.), *Warm Climates and Western Medicine: the Emergence of Tropical Medicine, 1500–1900*, [Clio Medica 35, Wellcome Institute Series in the History of Medicine] (Amsterdam 1996) 1–19, on 9. The wide availability of medical topographies of local regions meant that these were from time to time brought together into single texts, such as Sir James Clark *The Sanative Influence of Climate* (London 1846) which was really an inventory of regional climates, characterized as either salubrious or insalubrious.

[67] *DNB*, qv.

[68] M. Harrison, *Public Health in British India: Anglo–Indian Preventive Medicine 1859–1914* (Cambridge 1994) 43–49. See also M. Harrison, 'The tender frame of man': disease, climate, and racial difference in India and the West Indies, 1760–1860, *Bulletin of the History of Medicine* **70** (1996) 68–93.

[69] J. Johnson, *The Influence of Tropical Climates on European Constitutions: Being a Treatise on the Principal Diseases Incidental to Europeans in the East and West Indies, Mediterranean, and Coast of Africa* (London 1821) 2, 3.

[70] Johnson, *Influence of Tropical Climates*, 531.

[71] Martin, *The Influence of Tropical Climates on European Constitutions*, v. The relevant paragraphs are to be found on 139.

[72] On the idea of moral geographies, see F. Driver, Moral geographies: social science and the urban environment in mid-nineteenth century England, *Transactions of the Institute of British Geographers* **13** (1988) 275–87.

[73] It would be mistaken to generalize from this case since Italy's climate was often positively regarded—medically and aesthetically—by those taking the Grand European tour. The point is that when conceived in tropical categories, a climate was routinely regarded in a negative light.

[74] J. Johnson, *Change of Air, or the Pursuit of Health; an Autumnal Excursion Through France, Switzerland, & Italy, in the Year 1829; With Observations and Reflections on the Moral, Physical, and Medicinal Influence of Travelling-exercise, Change of Scene, Foreign Skies, and Voluntary Expatriation* (London 1831) 260, 262, 264. A very different medical judgement of Italy's climate is to be found in T. K. Chambers, *Some Effects of the Climate of Italy* (London 1865). He personally acknowledged his "debt of gratitude for the sweet influences of Italian skies" in the preface.

[75] Johnson, *Change of Air*, 279, 284. It should be noted that, coming from Ulster, Johnson's observations on Italian Catholicism are far from unpredictable: "In respect to the pompous formalities, the gorgeous imagery, the superstitious rites, the solemn mockeries, and the sickening delusions of Italian WORSHIP, whatever influence they may have on people immersed in ignorance and trammelled in priestcraft—they can have but one of two effects upon Englishmen—that of turning the Romish religion into ridicule, in strong minds—or that of overpowering and *converting* minds that are weak!", 291.

[76] H. Naraindas, Poisons, putresence and the weather: a genealogy of the advent of tropical medicine, *Contributions to Indian Sociology n.s.* **30** (1996) 1–35 on 3. For general background to Mansonian tropical medicine see Arnold, Introduction: Tropical medicine before Manson, *op cit*; M. Worboys, Manson, Ross and colonial medical policy: tropical medicine in London and Liverpool, 1899–1914, in R. Macleod and M. Lewis (Eds), *Disease, Medicine, and Empire: Perspectives on Western Medicine and the Experience of European Expansion* (London 1988) 21–37; M. Worboys, Germs, malaria and the invention of mansonian tropical medicine:

from 'diseases in the tropics' to 'tropical diseases' in Arnold, *Warm Climate* 181–207; and W. Anderson, Immunities of empire: race, disease, and the new tropical medicine, 1900–1920, *Bulletin of the History of Medicine* **70** (1996) 94–118. On Manson, see P. H. Manson-Bahr, *Patrick Manson: The Father of Tropical Medicine* (London 1962).

[77] Here Manson defined 'tropical' in climatic, rather than geographical, terms by insisting: "I employ the term 'tropical' in a meteorological rather than in a geographical sense, meaning by it sustained high atmospheric temperature". Manson, *Tropical Diseases*, xi. The contributions made by Sambon were explicitly noted in the preface.

[78] What follows draws on D. N. Livingstone, Tropical climate and moral hygiene: the anatomy of a Victorian debate, *British Journal for the History of Science* **32** (1999) 93–110.

[79] For example, the following pieces by Sambon contain very substantially the same material: Acclimatization of Europeans in tropical lands, *Geographical Journal* **12** (1898) 589–99; The possibility of the acclimatisation of Europeans in tropical regions, *British Medical Journal* **1** (1897) 61–6; L'acclimatation des Européens dans les régions tropicales, *Bulletin de la Société de Géographie, Lyon* **17** (1901) 44–66. On the issue of human acclimatiation in the period more generally, see D. N. Livingstone, Human acclimatization: perspectives on a contested field of inquiry in science, medicine and geography, *History of Science* **25** (1987) 359–94.

[80] Sambon, The possibility of the acclimatisation of Europeans, 66.

[81] Sambon, The possibility of the acclimatisation of Europeans, 66. Two years later remarkably similar language was used by William Z. Ripley when he wrote: "One of the most subtle physiological effects of a tropical climate is a surexcitation of the sexual organs, which *in the presence of a native servile and morally underdeveloped population* often leads to excesses even at a tender age" (emphasis added). See W. Z. Ripley, *The Races of Europe: A Sociological Study* (London 1900) 562. Robert Young takes this extract as emblematic of the sexual obsessions that characterized colonial encounters. See R. J. C. Young, *Colonial Desire: Hybridity in Theory, Culture and Race* (London 1995) 150f.

[82] Naraindas, Poisons, 30.

[83] A. Frater, *Chasing the Monsoon* (New York, 1990).

[84] I am grateful to one of my research students, Gary McKee, for drawing this volume to my attention. The fact that we turn here to a popular, rather than scientific treatise, may reflect the change in vocabulary that was effected through the professionalization of science at the end of the nineteenth century. Scientists were likely to disclose their cultural inclinations more directly in works of popular science, rather than in academic treatises.

[85] R. Abercromby, *Seas and Skies in Many Latitudes or Wanderings in Search of Weather* (London 1888) 36, 38.

[86] Abercromby, *Seas and Skies*, 38, 39.

[87] Cited in Pierre Deffontaines, *Géographie et Religions* (Paris 1948) 130.

[88] Abercromby, *Seas and Skies*, 40.

[89] Abercromby, *Seas and Skies* 365, 366.

[90] Abercromby, *Seas and Skies*, 271. In similar vein he also commented on nervous disorders: "nervous seizures are not uncommon on board ship in the tropics and sometimes in an emigrant ship they become quite an epidemic. When one woman breaks out, perhaps a dozen others will follow suit in twenty-four hours. These mental epidemics can be readily checked and cured by suitable moral treatment, but they are very interesting illustrations of mental and nervous health. The heat..., the close association of a large number of women, together with other causes, have a prejudicial effect on the portions of the nervous system which are disturbed in these attacks", 167, 168.

[91] Some biographical details are available in R. E. Dickinson, *Regional Concept: the Anglo–American Leaders* (London 1976) 277–79.

[92] R. DeC. Ward, *Climate, Considered Especially in Relation to Man* (London 1908) 220, 222, 227, 228.

[93] Something of this episode is discussed in D. N. Livingstone, 'Never shall ye make the crab walk straight': an inquiry into the scientific sources of racial geography, in F. Driver and G. Rose (Eds), *Nature and Science: Essays in the History of Geographical Knowledge* (Historical Geography Research Series, No. 28, 1992), 37–48.

[94] R. DeC. Ward, Fallacies of the melting pot idea and America's traditional immigration policy, in M. Grant and C. S. Davison (Eds), *The Alien in Our Midst or 'Selling Our Birthright for a Mess of Industrial Potage'* (New York 1930) 230–31. See also R. DeC. Ward,

Can the white race become acclimatized in the tropics?, *Gerlands Beiträge zur Geopolitik* **32** (1931) 149–157.

[95] A. A. Miller, *Climatology* (5th ed., London, 1947. First pub. 1931) 2. Such prescriptions could surface in less censorious ways. Thus, Vance observed that "the manners of leisure, a languid courtesy, perfection in the nice arts of life, and a contempt for manual labor" which characterized the Southern States of the U.S.A., "reflect some of the workings of climate". See also R. B. Vance, *Human Geography of the South: a Study in Regional Resources and Human Adequacy* (Chapel Hill 1932) 370.

[96] J. Willinsky, *Learning to Divide the World: Education at Empire's End* (Minneapolis 1998).

[97] M. J. S. Rudwick, The emergence of a visual language for geological science, 1760–1840, *History of Science* **14** (1976) 149–195. More generally on this theme see J. B. Harley, Deconstructing the map, *Cartographica* **26** (1989) 1–19, and J. B. Harley, Cartography, ethics and social theory, *Cartographica* **27** (1990) 1–23.

[98] On revivals of the Hippocratic tradition, see J. C. Riley, *The Eighteenth-century Campaign to Avoid Diseases* (Basingstoke 1987).

[99] M. Osborne, Resurrecting Hippocrates: hygienic science and the French scientific expeditions to Egypt, Morea and Algeria, in Arnold, *Warm Climate*, 80–98 on 80, 90.

[100] On a related theme see J. R. Camerini, The physical atlas of Heinrich Berghaus: distribution maps as scientific knowledge, in R. G. Mazzolini (Ed.), *Non-verbal Communication in Science Prior to 1900* (Florence 1993) 479–512.

[101] G. R. Gliddon, The monogenists and the polygenists: being an exposition of the doctrines of schools professing to sustain dogmatically the unity or the diversity of human races; with an inquiry into the antiquity of mankind on earth, viewed chronologically, historically, and palaeontologically, in Nott and Gliddon, *Indigenous Races of the Earth*, 650.

[102] E. Huntington, *The Character of Races, as Influenced by Physical Environment, Natural Selection and Historical Development* (New York 1924) 232.

[103] W. D. Matthew, *Climate and Evolution* (2nd ed. Special Publications of the New York Academy of Science 1939) 43; G. Taylor, The ecological basis of anthropology, *Ecology* **15** (1934) 223–242; G. Taylor, The zones and strata theory: a biological classification of race, *Human Biology* **8** (1936) 348–367. Taylor's cartographic ventures are discussed in H. Winlow, *Cartographic representations of race, c. 1850–1930*, PhD Thesis, The Queen's University of Belfast 1999.

[104] P. Gay, *The Enlightenment: an Interpretation* (New York 1970), vol. II, 319.

[105] P. Hazard, *The European Mind: 1680–1715* (New York 1969) 27–28. Originally published as *La Crise de la Conscience Europeenne* (Paris 1935).

[20]

Cultures of Knowledge: Investigating Intellectual Property Rights and Relations in the Pacific

Bronwyn Parry

Department of Geography, University of Cambridge, Cambridge, UK;
bcp20@cus.cam.ac.uk

Although often presented as inherently normative, Euro-American systems of intellectual property rights (IPRs) law are, like earlier systems of biological classification, best understood as particular, culturally defined systems for codifying knowledge employed to discipline objects, phenomena and social relations. Despite their partiality, such systems have successfully colonised new domains, recently underpinning a new uniform global regime for the protection of IPRs (GATT/WTO TRIPs). In this paper I reveal the central role that global institutions now play in accelerating the universalisation of specific "cultures of regulation": acting as powerful vectors for the transmission of particular types of knowledge and arbiters of the "normative" bases of global regulatory regimes. Recent empirical evidence from the Pacific illustrates how the TRIPs regime facilitates the commodification and appropriation of intellectual, cultural and biological resources in that region and highlights the development of alternative sui generis systems of IPR protection that challenge the normativity and hegemony of this regime. The article serves as an entry point for further research into the geography of knowledge systems—the way in which the colonisation of certain regulatory systems and forms facilitates the pursuit of particular interests or sustains relations of domination.

Introduction

As Peter Drahos (4) argues in his 1996 work *A Philosophy of Intellectual Property Rights*, "[T]he view that property is (necessarily) a thing is nowadays seen as quaint and false, or at least, not helpful". Rather, he (1996:4) suggests, property might now be more usefully understood as "the embodiment of a rights relationship between one person and another, or one person and many others … a contest for the control of objects that people need or want and sometimes upon which their very survival, either individually or as a group, depends". Although all forms of property rights can be thought of as linked by this shared teleology, complications arise in instances where the object or phenomenon warranting protection is itself "intangible". Intellectual property rights (IPRs) are then brought into play.

IPRs prevent the unlicensed reproduction of ideas, information or expressions that have commercial value. These may include literary, artistic or scientific works, performances, inventions in all fields of human endeavour, industrial designs, trademarks or designations.

Patents—one form of IPR—are monopolies granted to inventors as a reward and incentive to create. They afford the inventor exclusive rights to use the invention for a limited period of time, the power to prevent unlicensed use by others or the power to secure royalties from licensed use. A cursory review of the daily papers reveals the intensity of current debates surrounding the patentability of particular objects and phenomena, the rights of individuals, groups or institutions to monopolise these objects or phenomena through systems of IPR law, and countries' or communities' rights or ability to resist the imposition of IPR regimes.

These debates have been exacerbated by decisions to extend IPRs to a range of physical phenomena previously unrecognised within this system—engineered genetic materials, for example. Such developments have illustrated that intellectual property rights now provide an increasingly important mechanism for controlling not only intellectual and cultural expressions but also biological materials,[1] revealing the absolute fragility of distinctions between intellectual and physical property, tangible and intangible resources, nature and culture.

Although the tradition of positing rights in intangibles had its genesis in Roman law, it has since progressively colonised new domains informing the development of contemporary systems of IPR law in Europe and, more recently, in the US. The decision to base a new global system of trade-related IPR law (World Trade Organisation [WTO] legislation on trade-related intellectual property rights [TRIPs])[2] on precedents established within this particular juridical tradition has revealed the critical role that global institutions play in creating new cultures of regulation: acting as powerful vectors for the transmission of specific, culturally determined systems for codifying knowledge and as self-appointed arbiters of the "normative" bases of global regulatory regimes.

Whilst a considerable amount of research has been undertaken in geography on the relationship between geography, the law and systems and cultures of regulation, much of this research has been restricted to investigating the impacts of particular regulations on social or spatial form at a local, regional or national scale—the impact of federal or local regulations on urban housing markets, labour relations or union elections, for example (Clarke 1989; Dowall 1984; Jarman and Kouzmin 1993; Peck 1990).

Developments such as the introduction of the WTO TRIPs agreement suggest however, that it is a timely moment to investigate the geography of regulation at quite a different scale. Regulation, following a Foucauldian model, not only refers to legislative practice but encompasses, when more broadly conceived, particular systems for codifying knowledge (such as the Linnaean system of biological classification or the Euro-American system of IPR law) employed to discipline objects, phenomena and social relations. During the 1980s and 1990s, Thrift

(1985, 1996) and Blomley (1994) argued for greater research into "geographies of knowledge": the ways in which the spatial availability of knowledge[3] affects the production and transformation of social life. Such an agenda might be profitably extended to include studies of the geography of *knowledge systems*: the way in which the colonisation and, more recently, globalisation of certain regulatory systems and forms facilitate the pursuit of particular interests or sustain relations of domination. I agree with Barnes' (1995) suggestion that political and economic geographers ought to pay closer attention to the cultural aspects of global economic change, particularly as they are evidenced in power structures, the operation of discourse and the role of global institutions. Studies that investigate how regulatory systems (broadly conceived here as systems for codifying knowledge) are devised, enacted, disseminated and received may provide a useful entry point into the somewhat opaque and as yet largely neglected geography of the knowledge/regulation agenda.

Two recent contributions (Castree 1999; Whatmore 1999) have also questioned geography's relative silence on issues relating to the rise of biotechnology and the contested nature of knowledge claims that play such a central role in its governance. Studies of intellectual property rights and relations can prove useful within both contexts, as they provide insights into the machinations of the life-sciences industry as well as broader debates about the nature and ascendancy of particular knowledge claims and systems and their role in the regulation of intellectual, cultural and biological property.

This paper, then, has three aims. The first of these is to explore how the category of "incorporeal rights" came to be invented and incorporated into English and American systems of IPR law, how critical (but arbitrary) judgments extending the scope of patentability to life forms were arrived at in US courts, and how these precedents came to be adopted as the basis of a uniform, global system of IPR law. Here I concentrate on revealing the cultural determinacy—as opposed to inherent normativity—of the judgments that underpin the Euro-American system of IPR law[4] and the conflicts that arise when such systems are universalised.

My second aim is to reveal the impact that the universalisation of this system of IPR law is having on one region of the world—the South and West Pacific. The reader may consider this an unlikely case study, having reason to assume that this region remains far removed from such developments. However, recent research (Parry 2000, 2003; ten Kate and Laird 1999) has revealed that Pacific countries have been increasingly targeted by corporations, organisations and institutions seeking materials that might form the basis of "patentable" products and processes, an escalation induced in large part by the globalisation of a uniform system of IPR law. In the second section of this paper,

I draw upon my recent empirical research in the Pacific to outline the nature and extent of these collecting activities and the responses (cultural, political and legal) that the peoples of the South and West Pacific are making to them.

My third aim is to highlight current initiatives (academic and political) directed at challenging the normativity and hegemony of the Euro-American system of IPR law in the Pacific region. I discuss alternative frameworks for the protection of intellectual and cultural property that are derived from and responsive to local and regional contexts, as well as a series of further, as yet unexplored, issues and challenges. By investigating the progressive universalisation of Euro-American systems of IPR, detailing their reception in the Pacific and outlining proposed strategies of resistance, I hope to reveal something of the complexity of the emerging geographies of IPR law and to illuminate methods of engagement with them.

IPR Law: Genesis

Intellectual property rights law had its genesis in classical Roman law, most notably in the writings of Gaius. Gaius created a taxonomy of law, dividing it into categories relating to either persons, actions or things. Significantly, in relation to the latter, he established a distinction between corporeal and incorporeal things. Within Gaius' paradigm, corporeal entities were existent and tangible whilst incorporeal entities were not existent, but subsistent: that is, "They subsisted by virtue of human mental life ... they were things super-imposed by the mind onto the corporeal world" (Drahos 1996:17). The generation of a new category of incorporeal things had serious implications for the law, as it implied a requisite need to take account of the existence of these unobservable entities. This, in turn, paved the way for the creation of new legal conceptions of what might be encompassed by the term "property". Following the rediscovery and revival of classical Roman law by Inerius and his followers in 11th-century Bologna, the category of *res incorporales*[5] was imported into various European common-law and civil legal systems.

At a time in England (1680–1700) when debates about property centred upon what seemed an intractable philosophical dilemma—how, "if God gave the Earth to 'Mankind in common' it was then possible for any one person to have property in any one thing?"—it fell to Locke to construct an argument that equality and the commons could indeed coexist with individual appropriation and individual property rights. Proceeding from the assumption that every man (sic) already has "*A Property* in his own *Person*", Locke argued that an individual's labour power also belongs to that individual, leading him to conclude that "Whatsoever then he removes out of the State that nature hath provided, and ... hath mixed his *Labour* with, and joyned to it something that is his own, thereby makes it his *Property*" (Locke treatise II:27; emphasis in original).

In keeping with Locke's interpretation of natural law, the right to a reward for labour became unassailable, even in instances where the product of those labours was an immaterial thing such as an idea, invention, design or literary work. In the burgeoning capitalist economy of 17th- and 18th-century Europe, it became essential for those in possession of intangible assets—authors, artists and inventors—to be able to exploit them commercially. This could only occur if the law recognised property in immaterial or abstract expressions. Recent work by anthropologists and critical legal theorists (Aoki 1996; Boyle 1996; Coombe 1996) has pointed up the troubling centrality that is given to the figure of the romantic individual author and his "original creations" in Western intellectual property law, illustrating the inordinate emphasis that this law places on possessive individualism—what Gordon (1993:1533) refers to as the "strong Lockean 'I made it—it's mine' justificatory pattern". As Breyer (1970:281) notes, the assignation of intellectual property rights to individual authors or inventors is exceptional both in law and political economy given that "we do not ordinarily create or modify property rights, nor even award compensation, solely on the basis of labour expended".

Cornish (1999:111) has suggested that the 1624 English Statute of Monopolies, which created patents of fourteen years duration for "any manner of new manufacture", reflected broader economic objectives—specifically industrial growth and employment, rather than simply a "just" reward for expended labour. Certainly, by the time of the Industrial Revolution Britain had discovered that substantial economic returns could be derived from the exportation of protected technologies and inventions. Patent applications increased accordingly, rising from less than ten per year in 1750 to 110 in 1810 and 458 in 1840 (Boehm 1967:22–30).

A concomitant desire to protect fledgling industries and emergent technologies prompted America's founding fathers to enshrine the ability of "*authors and inventors*" to secure limited but exclusive rights to "*their respective Writings and Discoveries*" in the American Constitution (emphasis in original). The ratification of America's first Patent Act in 1790 created an upsurge in patent claims, many of which were extremely dubious. The task of weeding out fraudulent and duplicative claims proved contentious and was not resolved until the Patent Act of 1836 introduced a requirement that patent applications be examined for novelty, utility and patentability. Responsibility for adjudging what objects and phenomena might conceivably be disciplined or controlled within this particular regulatory system was vested initially with patent examiners, but ultimately with the judiciary.

Although all forms of law are cultural constructs, law-making and legal interpretation are rarely understood as subjective processes, but rather as normative ones informed and structured by a "deep" logic and rationality and underpinned by a set of underlying principles that are universally shared (see Unger 1983). However, a critical analysis of judgments made under Euro-American law reveals the partiality and cultural determinacy of interpretations and determinations of patentability and the central role that they have played in constructing biological materials as a form of intellectual property.

Evolving Geographies of IPR

Within Euro-American systems of IPR law, ideas and innovations are protected using different mechanisms. Copyright protects creative or inventive "works"; however, it does so by protecting the form, rather than the content, of a work. Patents, alternatively, protect the application of an idea. This is what makes it possible to argue that it is not modified animals or plants per se that are being patented in composition of matter claims, but rather the application of an idea or innovative process that results in this new composition of matter (the genetically engineered plant or animal). "Inventors" are thus granted rights to patent modified biological materials on the understanding that this will provide the means to control access to the ideas and innovations that inhere in them and that are the products of their intellectual labours.

However, inventions must meet particular criteria—they must be "novel", involve an inventive step, have industrial application and be able to be adequately disclosed by description. Certain categories of objects and phenomena are statutorily excluded from patent protection. These technically include "products of nature", "discoveries" and "biological processes for the production of plants and animals". A closer analysis of judgments relating to the filing of patents for both engineered biological products and processes within European and US jurisdictions reveals the extent to which determinations of patentability have, over the last decade, come to rest on selective, culturally derived interpretations of, for example, what constitutes an "invention" rather than a "discovery", of what is or is not a product of nature, and of what is meant by "new" or "novel" (new or novel to whom?).

The way in which the legal distinction between "discovery" and "invention" has been drawn in US patent law provides a salient case in point. To the extent that genetic research uncovers information about genes that already exist in nature, the question arises as to whether a gene so identified, is a "discovery" or an "invention" entitled to patent protection. Whilst the US Patent Act denies protection to discoveries, it limits that exclusion to "discoveries *as such*" (see US

Patent Act, US Code: Title 35 § 100–103; emphasis in original). Interpretations within US patent law have determined that an invention that makes *some practical use* of a discovery can be patented, as this is decreed to be more than a discovery "as such". This practical use may simply involve "purifying" the material by isolating it from its "natural" context.[6] Member states of the European Patent Convention will grant patents to applicants on the basis of their ability to establish the chemical structure of natural substances. Similarly, under US patent law it is possible to make a successful composition-of-matter claim for chemical compounds by simply "naming the compound or ingredients" (Halpern, Nard and Port 1999:237). As Correa (1992:155) concludes, "[A] very thin line separates invention from discovery in those countries".

With the development of sophisticated genomic software packages, it has become possible to search databases of DNA sequences for genes with specific protein functions that may have pharmaceutical applicability. Under US law, it is possible to apply for a patent on these genes and proteins by isolating the relevant cDNA and inserting it into a vector, as this is construed as an inventive act. The US Patent and Trademark Office has issued some 1000 patents on human genes on this basis (Steinberg 2000:1). Although James Watson, the former director of the Human Genome Project for the National Institutes of Health (NIH), once famously remarked that automated gene-hunting "could be done by monkeys" (quoted in Beardsley 1998:31), patents are nonetheless granted, as there is no requirement that these inventive acts be either complex or difficult to undertake.

Advocates of patenting argue that to treat the isolation of a gene or compound as a mere discovery would fail to provide sufficient reward for the work entailed in achieving this result. Conversely, opponents argue that to treat isolated genes or compounds as patentable inventions when the sole contribution of the "inventor" has been to reveal or discover the structure or composition of an existing natural substance is to stretch the concept of "invention" beyond reason.

Similar issues arise in relation to determinations of what constitute "naturally" and "nonnaturally" occurring materials. Until 1980, the distinction between "naturally" and "nonnaturally" occurring phenomena had remained fairly secure within US patent law. A long history of legal precedent had established that products of nature and physical phenomena would remain outside the scope of patentability, on the basis that they constituted "manifestations of nature, [that must remain] free to all men [sic] and reserved exclusively to none".[7] It was on this basis that the US Patent and Trademark Office first rejected a patent application for a genetically engineered bacterium produced by the microbiologist Ananda Chakrabarty on behalf of the General Electric Corporation. In a landmark 1980 judgment,[8] the US

Supreme Court later overturned this decision on appeal, ruling that although comprised of natural materials, the bacterium was not "a hitherto unknown natural phenomenon, but a non-naturally occurring 'manufacture' or 'composition of matter'—a product of human ingenuity".[9] Chakrabarty was deemed to have "invented" the micro-organism by successfully combining four plasmids and inserting them into a bacterium—a complex process at that time. In a dissenting opinion, four out of nine of the justices argued against Chakrabarty's claim, noting that Congress, in drafting the patent legislation, had "chosen carefully limited language granting protection to some kinds of discoveries but explicitly excluding others, including living material".[10] The intention of the legislation, they argued, was clear: to ensure that individuals or industries were prevented from "securing a monopoly on living organisms, no matter how produced or how used".[11]

The decision gave rise to an intense debate over the question of the *degree* of inventiveness that must be involved in order to make good a claim for patent protection. The debate was fuelled by a comment that Chakrabarty made at the time. When asked to describe how he had produced the new bacterium, Chakrabarty responded: "I simply shuffled the genes—and changed bacteria that already existed" (quoted in Shiva 1993b:33). His comments gave rise to serious concerns that monopoly rights would eventually be given for minor modifications to micro-organisms, plants or animals that, as Juma (1989:165) argued, "may turn out to be no more than the embodiment of cosmetic improvements".

Equally contentious judgments have been made in relation to determinations of who might qualify as "an inventor". Euro-American systems of IPR law have limited capacity to accommodate patent applications from collectives, preferring, in the Lockean tradition, to grant rights to individual authors or inventors. Consequently, varieties of seeds and crops or medicinal remedies that are products of generations of communal innovation often prove ineligible for protection, as no single "inventor" can be identified. Modifications to plants and crops, varieties of seeds and production techniques developed for use in subsistence communities may also be excluded from patentability, as they are intended for domestic, communal or agrarian use, rather than industrial use—a further requirement for patentability under the Euro-American system of IPR law.

These judgments have contributed to the creation of a new cosmology—a new system for classifying and controlling natural materials based on particular culturally defined interpretations and determinations. The judgments have been largely self-serving and protectionist, extending patentablity to those intangible objects and phenomena (ideas, inventions, codified knowledge) that form the basis of productivity in the post-industrial information age but that only a select number of countries have the technology to develop or exploit.

This might have proven of little consequence had these determinations remained confined to Euro-American jurisdictions. However, they have since been promoted and adopted as the basis of other national systems of IPR law[12] and, more recently, of a new global system of trade-related IPR law (GATT/WTO TRIPs). During negotiations for the Uruguay round of General Agreement on Tariff and Trade (GATT) talks in 1994, the US, the EU and Japan lobbied intensively for an international guarantee of respect for intellectual property, demanding that the 134 member states of the GATT (now the WTO) formally acknowledge the need to "make patents available for any inventions, whether products[13] or processes, in all fields of technology, provided that they are new, involve an inventive step and are capable of industrial application" (GATT TRIPs Article XXVII:1). Members were subsequently required to introduce "appropriate" (defined therein as internationally recognisable and enforceable) systems of IPR law based on established legal precedents. These proved to be the comparatively liberal precedents established within the world's two largest patent offices, the US and the Japanese.

Whilst many developing nations objected vociferously to this attack on their sovereignty, others acquiesced as they feared retaliatory actions or exclusion from access to the sophisticated, but protected, technology so essential for further economic development. Adherence to the WTO TRIPs agreement has since become nonnegotiable for member states, with any breach of its "detailed and onerous requirements" (Cornish 1999:119) leading to the invocation of the Special Omnibus 301 trade-related dispute resolution procedures, which include suspension of concessions, imposition of import quotas and exclusion of exports. As Cornish (1999:119) concludes, "[T]he U.S government clearly intends to continue pursuing individual countries with import sanctions wherever and whenever they offend its sometimes myopic sense of property in intellectual property matters".[14]

A Protouniversal Culture of Regulation?

It has been argued that the homogenising force of globalisation is so powerful that it has the capacity to create what Featherstone (1990:2) refers to as a "protouniversal culture". Whilst such terms are usually used in relation to media and consumer culture, it is also possible to situate an analysis of regulatory cultures within this context. Although classificatory and regulatory regimes begin as localised, culturally determined systems for codifying knowledge, they may diffuse rapidly and through adoption or imposition, quickly attaining ascendancy at a variety of scales. The Linnaean system of classification—another system for disciplining biological materials—became internationalised in this way. However, it was a selective internationalisation, as the system was adopted or imposed—that is to say, it colonised—along fairly narrow networks. The Euro-American system of IPR law is also spreading geographically, but by virtue of the omnipotence of global institutions, it appears destined to become comparatively more pervasive and hegemonic in a much shorter space of time.

The decision to base a new global regulatory framework on selective, arbitrarily determined judgments made under the Euro-American system sparked a vitriolic debate about the legitimacy of the GATT TRIPs agreement, giving rise to charges of cultural hegemony, "bio-piracy" and "bio-colonialism" (see Gibbs 2000; McAfee 1999; Shiva 1993a, b, 1997) Trenchant critics such as Shiva argue that regulatory regimes such as the WTO TRIPs agreement are not normative. Rather, they impose upon other countries and cultures "the globalised version of a very local and parochial tradition"—a tradition that legitimates the appropriation of biological and intellectual resources by providing protection "only when knowledge and innovation generates profits and not when it meets social needs" (Shiva 1993b:32). The universalisation of such systems threatens cultural diversity, Shiva (1993b:33) argues, by creating "a reductionist monoculture of knowledge which displaces other ways of knowing, other objectives for knowledge creation and other modes of knowledge sharing".

Alternatively, Aoki (1996:1344) cautions against the reification of globalisation and global regulatory frameworks, arguing that to conceive of them as inexorably advancing, homogeneous or unitary phenomena is to risk obscuring the constitutive role that nation-states (and, indeed, resistance movements) play in influencing supranational law and policy-making. Certainly, it would be naïve to assume that all developing countries see the introduction of new IPR regimes as a threat; on the contrary, many positively embrace them as useful mechanisms for securing adequate compensation for ideas, innovations, designs and inventions that have been all too readily appropriated in the past (Bhat 1996; Maskus 2000; Thompson and Rushing 1996).[15]

However, such analyses are perhaps overly optimistic about the transformative capabilities of governmental or nongovernmental actors, remaining, as Coombe (1996:1363) suggests: "far too sanguine, too untroubled by Uruguay Rounds, GATT systems, Dunkel Drafts and Special 301s ... those legal instruments and institutions that so effectively enforce borders by putting 'NO TRESPASSING' signs around [them]". Whilst well-publicised countervailing political and legal struggles, ranging from the Dunkel riots of the early 1990s to the more recent "Battle of Seattle", imply robust strategies of resistance (Wainwright, Prudham and Glassman 2000), over the past decade, many developing countries have quietly succumbed to sustained pressure to introduce and recognise systems of IPR.

The progressive universalisation of formalised systems of intellectual-property protection has facilitated the creation of new markets for previously unprotected products and processes, intensifying demand for materials—such as genetic resources, images or specialised knowledge—that might form the bases of these products and processes. One significant consequence of this has been an exponential increase in new collecting activities (see Parry 2003 for empirical evidence of these trends). One region of the world has proven of particular interest to collectors: the South and West Pacific. The reason for this, whilst apparently obscure, may be encapsulated in one word: *novelty.*

One of the principal tenets of the intellectual property rights law is that the material to be subject to protection be original or "novel"—that is to say, as yet uncharacterised within the panoply of inventions and objects already "claimed" by systems of IPR law. In an increasingly globalised, postmodern world in which a comparatively limited number of images, objects and forms have been endlessly picked over, manipulated and reworked, the task of finding "novel" materials not already classified within, and protected by, systems of IPR law proves an unenviable one. Collectors have consequently begun to extend the scope of their activities to include formerly untapped regions—in this case, the South and West Pacific.

Collecting Activities in the Pacific

In the early 1990s, the ratification of the GATT TRIPs agreement and the introduction of new Biodiversity Convention protocols establishing "just and equitable" mechanisms for the exploitation of genetic resources combined to create a new regulatory and economic space within which genetic materials could be traded on normative terms. The effect on collecting practice was immediate and marked. A series of new global collection programmes was instituted by US-based research organisations, multinational pharmaceutical companies and other public and private companies seeking patentable genetic materials (Parry 2000, 2003; Reid, Laird and Meyer 1993; ten Kate and Laird 1999). The scale of these collecting activities was unprecedented, with institutes such as America's NIH collecting in over 40 different countries simultaneously (Parry 2003). Biological collections were undertaken in several countries in the Pacific, including Australia, the Philippines, Micronesia, Papua New Guinea, Indonesia, Palau, Samoa and Fiji. The Fijian case provides an illustrative example of the effect that these developments have had on the exploitation of genetic resources in the region.

Convinced by arguments that the conservation of the natural environment can be effected through its commodification, researchers working at the University of the South Pacific's (USP) chemistry department, which has long-standing interests in the pharmacological properties and medicinal uses of plants and animals, declared in 1995 that bioprospecting[16] might provide a useful means of "enabling local communities concerned about the loss of their biological wealth to link up with the private sector wishing to locate unique chemical compounds for medical and other purposes"(Aalbersberg 1998:1). Several large pharmaceutical corporations were already seeking access to, or collecting in, the Pacific. The USP quickly entered into a commercial partnership with the pharmaceutical corporation Smith Kline Beecham, which was then bioprospecting in Micronesia. Having secured the negotiated consent of two community groups in Fiji, one coastal (the Verata) and one rainforest (the Namosi), workers began collecting samples in 1996. However, although the collection process had begun, a bioprospecting agreement (establishing terms and conditions of collection and use, IPR rights and compensation) had yet to be agreed to, primarily because the Fijian government was still finalising a new Sustainable Development Bill hastily redrafted to include regulations governing access to and use of genetic resources.

Without established legislation, the government and the university found themselves in a poor negotiating position. Offered a standard industry contract that unilaterally set "appropriate" conditions and terms of use (including the assignation of exclusive IPRs to the collecting organisation), they had only two immediate options: accept or withdraw. They accepted. For unrelated reasons, Smith Kline Beecham later withdrew from the project, forcing the USP to begin an immediate search for another commercial partner. The Strathclyde Institute of Drug Research (SIDR) at Strathclyde University, Glasgow, which acts as a broker supplying genetic and biochemical samples to corporate and commercial concerns, willingly agreed, as they had recently signed a contract to supply a Japanese pharmaceutical company with 5000 samples of previously unscreened genetic and biochemical material.

The collapse of the initial partnership provided a rare opportunity for the Fijian state to reassess its position. Accepting their limited expertise in relation to the commodification and patenting of genetic resources, the Fijian government and the USP together sought the assistance of the nongovernmental organisation (NGO) the Rainforest Alliance. The Alliance approached a number of environmental advisers, NGO advocates, ministerial officers and international IPR specialists—of whom I was one—to create an international panel of advocates who could collectively review the terms of any new contract negotiated with the SIDR. The panel conferred on a number of occasions, exchanging information relating to best practice, benefit-sharing provisions and contractual omissions.

These assessments revealed considerable inadequacies in the proposed USP/SIDR contract. Under the terms of this contract, USP, rather than SIDR, underwrote the costs of collection and preparation of samples using conservation aid grants. USP was also required to relinquish sovereignty over the material by allowing SIDR to retain exclusive control over future use and transfer of the collected samples. USP were not assured of receiving royalties on income that SIDR generated from licensing out the samples to third parties (eg other pharmaceutical companies) and neither the Fijian communities nor the USP retained any rights to be named on patent applications. Following the network's revision of the contract, the USP and the communities now receive 60% of any *fees*[17] obtained from companies who license out samples from the SIDR for evaluation purposes. The SIDR retains 40% of these fees. Collection costs are still met by the USP, but conservation-grant money is now dispersed directly to the affected communities. The SIDR retains the right to license out collected samples to third parties in perpetuity, although they may not do so without the express agreement of the USP. Similarly, they retain the right to file patent applications on any resulting products or processes; however, the USP may also be named as a joint partner. The USP has also secured the right to a small percentage *royalty* on any income produced from products derived from the collected material.

The question of how royalties would be divided between the USP and the source communities was to be determined within two years of the signing of the agreement but has not, to my knowledge, been publicly disclosed. Significantly, the contract was drawn up between the SIDR and the USP, as the SIDR preferred not to contract directly with the Fijian community groups. The resulting agreement is considered to provide some of the more generous terms of compensation to be found in recent bioprospecting agreements.[18] However, the advocates involved in the negotiation process remain cognisant of the fact that it is an agreement devised within—and very much constrained by—industry norms and the formal strictures of existing intellectual property rights law. Our broader concerns regarding issues of patent rights, sovereignty, enforceability, monitoring and disbursement of royalties remained unresolved and will perhaps never be adequately addressed within this paradigm. As one network member remarked, "[E]ach of these agreements [the standard industry contract and our revised version] really only serve the bare-bones function of transferring samples from one party to another with little suggestion that any larger objectives are being served" (unpublished memo/correspondence 1997).

Nonetheless, collection activities in other countries have not been monitored at all, and many are not subject to any contractual agreement. For example, a Japanese biotechnological research institute recently undertook an extensive—but completely uncompensated—collection of marine organisms in Micronesia.[19] The geography of the Pacific region—in which are found EEZs (Exclusive Economic Zones—maritime zones surrounding island states) in excess of one million square kilometres—militates against the effective monitoring of these unlicensed activities.

Collections of plant, animal, marine, microbial and fungal organisms are not the only resources that have proven of interest to those searching for potential sources of novel and hence "patentable" materials. Indigenous knowledge—for example, collected information pertaining to the identification, usage and efficacy of particular medicinal plants—has formed the basis of patent applications for novel *processes*, such as previously unrecognised treatments for disease. However, since Western systems of IPR law are predicated on recognition of an individual inventor, few attempts have been made to include indigenous communities in patent applications. As a consequence, a disturbing degree of paternalism has crept into negotiations for the granting of co-inventorship rights. For example, investigations undertaken for my doctoral research revealed that a US-based ethnobotanist involved in studying the pharmacopoeia of the Samoan people has since been named as the co-inventor on a patent for Prostratin,[20] rather than the healers who supplied the materials, on the basis that it was he who contributed (but that is to say *divulged* rather than *generated*) the ethnobotanical information on uses of the plant on which the pharmaceutical is based (Cragg interview 1996).

Similar collections of ethnobotanical knowledge about plant and animal species have also been undertaken in other countries in the region, including Australia, Papua New Guinea, Indonesia and the Philippines. Much of this collected information has since been transferred onto the very large NAPRALERT database of ethnobotanical information located at the University of Illinois at Chicago.[21] Although such databases are composed of information collected from various communities, the intellectual property rights of the contributors are seldom recognised, and only rarely do they receive any benefit from the publication of their ideas and innovations. However, the databases themselves are usually copyrighted, allowing the holder of the copyright to charge prospective users a fee to access the information stored on them.[22]

Other abstract expressions of indigenous knowledge and culture have also acquired a certain cachet, providing exotic new motifs for an image-weary marketplace. Photographs of indigenous people, their homes or their villages increasingly form the basis of trademarks or copyrighted advertisements, whilst the unauthorised reproduction of more abstract objects—including various types of spiritually significant cultural property such as artefacts and works of art—continues apace. As Blakeney (1995) notes, for example,

> [A] matter of increasing concern to Aboriginal peoples [in Australia] has been the utilisation of reproductions of traditional Aboriginal designs as a means of decorating a host of mundane products developed primarily for the tourist trade such as tea-towels, pencil cases, key rings, tee-shirts, drink coasters [with] at the higher end of the market, the current vogue for folkloric ornamentation resulting in the use of Aboriginal designs on wall hangings, carpets and posters.

In the mid-1990s, concern about the appropriation of indigenous materials in the Pacific reached new heights, as indigenous people themselves became the object of new collection programmes. In 1995, the US Department of Health announced that it was applying for a patent on a T-cell line developed from a blood sample collected from a member of Papua New Guinea's Hagahai tribe[23] in 1989 (Jenkins et al 1989a, b). The patent was granted in 1995 despite strong objections and appeals by the Papua New Guinea government, but has since been abandoned.[24] Rockefeller University researchers based in New York City have also announced that they will be undertaking a study of the genetic make-up of the inhabitants of Kosrae in Micronesia (where obesity is very high) as part of their wider search for an "obesity gene". Although a patent has yet to be granted on the gene, Amgen (a California-based pharmaceutical company) has already agreed to pay the university US$20 million for the right to license out the gene for use in pharmaceutical and gene-therapy development. Genomic information collected under such projects, including the Human Genome Diversity Project, is becoming an increasingly valuable resource, and a bitter contest is currently being waged over who should control genomic databases (public or privately held companies)[25] and on what terms and conditions (Arthur 2000; Radford 2000).

Responses

These developments raise the question of how the peoples of the South and Western Pacific have responded to these activities. Predictably for such a diverse area, responses have been varied and complex. Whilst political reaction and the development of policy are inherently linked, important distinctions between them remain. In the following section, I thus review each in turn before outlining new indigenous approaches to the protection of intellectual and cultural property that challenge the inherent normativity of existing IPR regimes.

Policy Responses

The speed with which new collection programmes were implemented caught several countries unaware, making them slow to recognise and regulate these activities. Heavily targeted countries were amongst the first to respond. These included Australia, which developed and published a detailed, formal policy on access to its genetic and biochemical resources in 1994,[26] and the Philippines, which drafted its Presidential Executive Order on The Prospecting of Biological and Genetic Resources in 1994, ratifying it in 1995. The Philippines government elected to take a proactive stance in relation to the exploitation of its genetic resources by enforcing stringent access conditions and demanding generous benefit-sharing provisions. The provisions were designed to safeguard the Philippines' sovereignty over their genetic resources whilst ensuring adequate compensation for their exploitation. Ironically, it has since been argued (Lesser 1996; Reid 1996) that these provisions were excessively restrictive, stifling commercial interest in bioprospecting in that country. In fact, the legislation was later redrafted and several of these requirements were relaxed.

In 1996, Fiji amended existing legislation to include appropriate mechanisms to protect genetic resources and indigenous information. However, the amendments[27] all draw on prior established legislation in relation to the use of genetic resources, most of which has evolved in response to the needs of the biotechnology industry. For example, the amendments again enshrine the seemingly incontrovertible right of bioprospectors to secure intellectual property protection for collected materials, including forms of indigenous knowledge, subject only to the requirement that they provide "a fair return" for the use of those resources (Sustainable Development Bill, section 249, subsection 249 1 [a]).

By 1997, several Pacific states had still not implemented any effective legislation or policies and remained vulnerable to inequitable exploitation. The South Pacific Regional Environmental Programme (SPREP)[28] hosted a workshop for 14 of its member states in 1998 with the aim of addressing "the top priority concerns of our region: intellectual property rights and their role in the conservation and use of biodiversity ... and regimes needed in the Pacific to make sure countries have control over their plant and animal genetic resources" (SPREP 1998a:1). However, whilst organisations such as SPREP search for strategies to resist the further monopolisation of their biological resources, Asia Pacific Economic Cooperation member states are simultaneously being encouraged to attend symposiums promoting the further harmonisation and enforcement of IPR rights in the region.

Political Responses

To concentrate only on policy outcomes would be to overlook entirely the disputatious political process that has surrounded the reception of new trade-related systems of IPR law in the Pacific and the development of a regional position on their adoption. The indigenous peoples of the Pacific have found themselves in a paradoxical position in relationship to the recognition of existing forms of IPR. As Brush (1999) suggests, nowhere are these paradoxes more clear and pressing than in the issue of the rights that indigenous peoples have over genetic resources and their knowledge of these resources. Several of the earliest soft-law declarations designed to promote and protect indigenous rights[29] explicitly assert indigenous peoples' rights to intellectual and cultural property as part of a broader claim to self-determination and social, economic and cultural autonomy. The Mataatua Declaration,[30] for example, declared that indigenous peoples of the world "must be recognised as the exclusive owners of their intellectual and cultural property" (cited in Posey and Dutfield 1996:205).

However, these strident claims for intellectual property rights clearly conflicted with indigenous groups' assertions about the cultural imperialism implicit in the progressive universalisation of Western systems of IPR law. By 1995, opposition to the introduction of western systems of IPR had greatly increased in the region, hardened partly through "consciousness-raising" by NGOs philosophically opposed to the patenting of life-forms and partly through increased first-hand experience of bioprospecting. At the United Nations Development Programme (UNDP)-funded Regional Consultation on Indigenous People's Knowledge and Intellectual Property Rights, held in Suva, Fiji in April 1995, participants sought to "reaffirm that imperialism is perpetuated through intellectual property rights systems ... and to commit ourselves to raising public awareness of the dangers of expropriation of indigenous knowledge and resources". They elected to do so by declaring the Pacific Region "a lifeforms patent free zone", calling for a moratorium on bioprospecting in the Pacific and urging indigenous peoples "not to co-operate in bioprospecting activities until appropriate mechanisms are in place" (UNDP 1995:3–4) Although demands for a moratorium were ultimately rejected, disquiet prevails in many quarters.

Nevertheless, with the expiration of the grace period for implementation of GATT/WTO TRIPs requirements imminent, most Pacific countries are now obliged to focus on immediate implementation of "appropriate" systems of IPR law. The universalisation of this system is now all but complete, with the Association of South-East Asian Nations, the Mekong River Basin Countries, and the Asia Pacific Economic Co-operation Forum drawn into compliance with WTO TRIPs (Blakeney 1998). As we enter the new millennium, policy initiatives edge their way forward, progress characterised by this uneasy rapprochement between the need to engage with—and, ideally, profitably exploit—a system that becomes every day more catholic, whilst minimising the risk of being subsumed to it with no beneficial result.

Member states are conscious that they must now conform to *all* of the GATT TRIPs resolutions; they may not—as they have done in the past—rely on selective adoption or exemptions to shape supranational legislation to suit national conditions. Although opportunities to effectively counter the normativity of this system remain severely restricted, a constellation of actors from nongovernmental, indigenous and academic communities are currently engaged in developing alternative models of property protection in the Pacific that seek to creatively reinterpret these norms. It is to a discussion of these alternatives that I now turn.

Strategies of Resistance

In their recent work, *Global Business Regulation*, Braithwaite and Drahos (1999) articulate a plan of action for recovering citizen sovereignty eroded by processes of globalisation, one that involves the implementation of a number of strategies of resistance. Although they acknowledge the formidable challenge of devising effective challenges to the TRIPs agreement, they remain optimistic that microstrategies —those derived from, and responsive to, local or regional contexts— may provide workable alternative frameworks. The creation of what are termed "sui generis" (of its own kind, unique or particular) systems of IPR law in the Pacific provides an example of the viability of this strategy. A significant but often overlooked feature of the TRIPs legislation is that it only establishes minimum (albeit stringent) standards of IPR protection. Individual nation-states retain the right to devise sui generis systems of IPR protection as long as they meet the requirements for international enforceability. Therein lies an opportunity to challenge the normativity of the existing regime. It is through the creation of what Knorr-Cetina and Mulkey (1983:132–133) refer to as "transepistemic communities" of activists, academics and specialists (such as that employed in the Fijian case) that this opportunity might best be realised. Two examples provide cases in point.

In 1997, the Papua New Guinean government began to investigate the possibility of developing a sui generis system for protecting intellectual, biological and cultural property rights in PNG based on a combination of customary practice and existing IPR law. Actors with expertise in this area were drawn together to discuss objectives and strategies. A collaborative project involving researchers from the PNG National Cultural Commission, the University of Papua New Guinea law faculty and the faculties of social anthropology at Cambridge and Brunel universities has since been formed to address this issue. The aim of the project, which focuses on PNG, is to employ anthropological methods of enquiry to investigate new forms of economic transaction emerging in the Pacific as a consequence of both increased international interest in intellectual property protection and domestic interest in the protection of intellectual, biological and cultural property. As the investigators suggest, the concept of "IPR" has itself become a powerful metaphor for the idea of controlling all sorts of potentially valuable intangible assets, as the "runaway" character of recent claims made in the name of cultural "property" attests. In PNG, as in many other countries, an ever-greater number of objects and phenomena are thus being swept into the category of things requiring "compensation". This, in turn, has led to unrealistic expectations for recompense, whilst concealing other forms of reward for human endeavour, creativity and inventiveness (Kallinoe and Leach 2000; Strathern 1999a, b, 2000)

As Strathern (2000:7) argues, ideas derived from Euro-American culture have provided "both the language for the notion of intellectual property *and* a language for resistance". Whilst they underpin the development of IPR, they also underpin the way in which opposition to this paradigm has been constructed—notably through the development of a counterdiscourse of "traditional resource rights" that appeals to "community" and "collective rights". The investigators hope to challenge this long-standing but unreflexive and stereotypical "individual"-vs-"community" dichotomy by revealing the inventiveness of indigenous approaches to the regulation of transactions that provide important alternative models of proprietorship. Their preliminary findings suggest, for example, that a model of "multiple" rather than either "collective" or "individual" rights may provide an alternative regime for protecting information or materials transacted between groups, especially where the gaining and disposal of the resources does not extinguish the parties' mutual interest in the thing—where the transmission of knowledge is involved, for instance.

By focussing on the social and economic realities of current ownership claims and transactional behaviour in PNG, the study aims to contextualise the language of claims within the contemporary workings of indigenous economies by providing data on current modes of transmission of knowledge and related assets. It is hoped that the research will elucidate different modes of thinking about tangible and intangible resources and provide insights into alternative understandings of what constitutes "property" and of how it might be transacted. This, in turn, might inform the creation of an alternative framework and language for understanding IPRs and relations with applicability within the wider field of resource rights.

Australia's Aboriginal and Torres Strait Islander Commission (ATSIC) is also establishing a taskforce comprised of lawyers, indigenous activists and NGOs to implement new guidelines and recommendations (Janke 1998) on the development of sui generis legislation to protect indigenous cultural and intellectual property[31] and intangible cultural heritage. Research conducted by ATSIC has documented a long history of misuse and unauthorised reproduction of indigenous cultural property, including works of art, designs and other intangible cultural expressions (such as languages, oral traditions, songs and dances) and has identified a need to establish more effective mechanisms for their protection.

The authors of this report argue that the development of a sui generis system is particularly important, as the effective protection of indigenous cultural knowledge, expressions and products is beyond the scope of existing intellectual property laws. Such laws, they maintain, rarely allow for the protection of intangible expressions of culture, generally argue that such expressions do not meet the requirement for "originality" as defined by the Australian Copyright Act, and only provide protection for a fixed period, limiting protection for intergenerational use of indigenous cultural knowledge, artefacts and expressions. ATSIC is committed to the development of options for reform of the Copyright Act to provide more effective protection for indigenous intellectual and cultural property. New frameworks would draw on the development of international standards formulated by UN agencies such as the World Intellectual Property Organisation (WIPO) and the United Nations Educational, Scientific and Cultural Organisation (UNESCO).

Conclusions and Future Challenges

Although geographers have long expressed interest in the nexus between geography and the law, little attention has yet been paid to exploring the relationship between geography and regulation as might be enacted through a study of the geography of knowledge systems. By analysing the progressive universalisation of one particular, culturally defined system for codifying knowledge—in this case, Western systems of IPR law as exemplified by the GATT/WTO TRIPs agreement—I hope to have provided an illustration of the way in which the colonisation and, more recently, globalisation of certain regulatory regimes facilitate the pursuit of particular interests and sustain relations of domination.

Recent contributions have sought to critique and challenge what is now considered by many to be a simple but pervasive narrative: that of globalisation as a monolithic, inexorably advancing phenomenon that overwhelms local experience and leads inevitably to the destruction of regional cultures (Appadurai 1996; Jackson 2000; Miller 1997). Commentators such as J K Gibson-Graham (1997:5) have taken particular exception to what they see as a thoroughly overdetermined conception of global capitalism that portrays it as colonising, disciplining, penetrative, appropriating and ultimately triumphant, arguing that it must be possible to "theorise capitalism without representing dominance as a natural and central feature of its being". Whilst I agree in principle, my first-hand experience of the progressive colonisation of the GATT TRIPs agreement suggests that this particular regulatory regime has proven a most effective instrument for disciplining and monopolising objects and phenomena that have, until now, remained outside the grasp of global capital. It is difficult to underestimate the dominance and degree of geographic penetration of this particular regulatory form, both of which are clearly evidenced in rates of adoption and conformity. In circumstances such as these, capitalist hegemony does not seem at all "fictive". Rather, it seems alive and well and still characterised, as Jameson (1991:412) once suggested, "by the relentless saturation of any remaining voids and empty places".

However, I do agree with Gibson-Graham (1997:3) that such depictions of capitalist hegemony can serve to obscure "the presence and vitality of noncapitalist economic forms". Echoing Strathern, they suggest that the discourse of capitalist hegemony provides not only a conception of what is to be resisted or changed, but also the parameters within which opposition to this paradigm might be constructed. In drawing attention to some of the alternative frameworks for the protection of intellectual, cultural and biological resources that are currently emerging in the Pacific, I hope to have presented some evidence to support the existence of vibrant, alternative approaches to economic relations—in this case, property protection.

Whilst the development of new alternative regimes is encouraging, it should not invoke complacency. Although the projects outlined in the previous section are significant, they only provide an entry point for future research in this domain. Several important epistemological issues clearly await more detailed attention, providing fertile ground for interdisciplinary ventures. Anthropologists, geographers, historians of science and others interested in the geography of knowledge systems might together undertake studies that detail how the Euro-American system of IPR law has infiltrated specific countries, usurping existing systems of property protection, and in so doing might illustrate how particular knowledge systems gain ascendancy and colonise new cultural terrains. Economic and cultural geographers might examine how particular materials (forms of cultural property, for example) become constituted as commodities able to be traded within new resource economies and subordinated to systems of IPR law. They might also detail expectations in relation to compensation for uses of biological and cultural property and assess the degree to which they are linked to the introduction of systems of IPR, investigate within this context the complex relations between "property claims" and sociocultural factors, provide realistic and critical appraisals of the effectiveness of IPR as an instrument in the regulation of access to resources involving indigenous knowledge or cultural property and explore the nexus of relations between customary law and international law in the development of viable alternative sui generis systems for the protection of biological, cultural and/or intellectual property.

The scope for future research investigating issues relating to intellectual property rights and relations, not only in the Pacific but in a variety of other countries as well, is considerable. Such research will play an important role in generating more reflexive conceptions of property, alternative frameworks for its protection and, ultimately, effective challenges to the normativity of this particular culturally defined system for codifying knowledge. I encourage geographers to claim their place at the forefront of this research.

Acknowledgments

I am indebted to Dr. Richard Howitt of Macquarie University, Sydney for suggesting that this might prove a timely moment to write such an article. I would like to thank King's College, Cambridge for supporting this research through their Research Fellowship programme. Thanks also to Marilyn Strathern, Phil Howell, Bill Adams, Laura Cameron and Cathy Gere for their valuable comments and criticisms on an earlier draft. I would also like to thank the referees for their very helpful comments. Any omissions or errors remain my own. Finally, my thanks to Juliet for making fieldwork in the Pacific even more enjoyable than it might otherwise have been.

Endnotes

[1] In cases where property rights such as patents are used to control access to modified or engineered biological materials.

[2] The World Trade Organisation (WTO) formerly the General Agreement on Tariff and Trade (GATT) legislation on Trade Related Intellectual Property Rights (TRIPs).

[3] For example, how stocks and types of knowledge are accumulated and disseminated. See Thrift (1996:96–97).

[4] I realise that it is contentious to cast the Euro-American system of IPR law as a monolithic or undifferentiated legislative canon. To do so is to risk obscuring the deep historical divisions that have existed within EU courts and between EU and US courts over determinations of patentability, authorship, and so on. Nonetheless, it is clear that they share the same genealogy and conform in many respects. The introduction of the 1998 European Directive on Biotechnology, which effectively creates a single market for biotechnological products, reflects a further harmonisation of EU and US law. For example, the directive requires EU member states to provide IPR protection for "biotechnological inventions", with this now liberally interpreted to include "elements" isolated from the human body or otherwise produced by means of a technical process—an interpretation clearly based on precedents set under US law. See Crespi (2000).

[5] Literally, "incorporal things".

[6] The key case in which the concept of purification is established is *In re Bergstrom*, 1970 427 F. 2d 1394, 166 USPQ 256.

[7] This was established in *Funk Brothers Seed Co. vs Kalo Inoculant Co.*, 1948 333 US 127, 68 S. Ct. 440, 92 L. Ed. 588 at 130.

[8] 35 U.S.C.A. § 101—*Diamond vs Chakrabarty* 100 S. Ct. 2204, 447 US 303, 65 L. Ed. 2nd. 144.
[9] Cited in *Diamond vs Chakrabarty*, 447 US 303, at 309.
[10] *Ibid.* at 320.
[11] *Ibid.*
[12] For example, within a decade of the *Diamond vs Chakrabarty* decision, the European Patent Office guidelines determined that whilst "to find a substance freely occurring in nature is mere discovery and therefore not patentable—if a substance found in nature has first to be *isolated* from its surroundings [rescued from the chaos] ... if it is 'new' in the absolute sense of having no 'previously recognised existence' [within Western systems of knowledge classification] then the substance *per se* may be patentable" (*Diamond v Chakrabarty*, 447 US 303 [1980]).

The Japanese patent office later extended this definition further to allow claims on chemical compounds contained in natural substances "*in any instance* where 'taking possession' and confirmation of the structure of the product could not be effected without artificial aids". For a more detailed discussion, see Bent et al (1987).
[13] Including genetically engineered organisms and purified biochemical compounds
[14] In 1997, the Office of the US Trade Representative reported that the WTO had initiated actions against ten countries and was placing a further 46 trading-partner countries on its "watch lists" (see Office of the US Trade Representative Web site).
[15] My thanks to Timo Goeschl for directing me to these readings.
[16] Defined by Reid et al (1993:v) define bioprospecting as "the exploration of biodiversity for commercially valuable genetic and biochemical resources".
[17] These are one-off payments for securing access to the material—typically UK£15 per month per sample. Royalties, on the other hand, are a negotiated percentage of the net income produced from pharmaceuticals derived from the collected materials (typically 1–5%). They are paid successively over a specified period of time.
[18] For example, Aalbersberg (1998:3) reported that "[T]he 60:40 split compares favourably with that offered by many other collectors/brokers which may be as low as 10:90 [respectively]".
[19] This was reported directly to me by international lawyers working for a global environmental law organisation in America. They requested that the information remain unattributed, as identification may compromise their work in the region.
[20] Prostratin is a treatment for yellow fever/hepatitis developed by America's National Cancer Institute. The other named inventors were technicians within the NIH.
[21] For a detailed discussion of the establishment of this database, see Farnsworth (1990).
[22] In the case of the NAPRALERT database, information is given free of charge to the World Health Organisaton and to scientists or institutions in developing countries who request it. The database is now online worldwide. Prospective users pay a licence fee of US$100.00 per year. Commercial users are charged higher (undisclosed) fees (Farnsworth 1990:18).
[23] The Hagahai are a 260 strong hunter-gatherer community that reportedly had its first contact with missionary workers in 1984.
[24] Following a sustained protest campaign, no recent attempts have been made to work the patent. See United States Patent and Trademark Office Archive Pat. No. 5, 397, 696. For a more detailed review of the Hagahai case see Cunningham (1998), Frow (1997:153–154), *Cultural Survival Quarterly* (1996) vol. 20 Hagahai special issue.
[25] Although I would argue that the distinction between public and privately funded enterprises is far from secure.
[26] A working paper produced by the Coordination Committee on Science and Technology (CCST) for the Department of Prime Minister and Cabinet in 1994 takes as one of its principal stated objectives the need to generate "consistent and comprehensive arrangements for dealing particularly with foreign companies and governments wishing

to access our biological resources ... and to create and use mechanisms to protect intellectual property rights associated with Australia's biological resources" (CCST 1994:1–2).

[27] Contained in Fiji's Sustainable Development Bill Section 249 Subsections 1–18.

[28] SPREP is an autonomous regional organisation comprising 26 member states, including all 22 Pacific Island nation-states. Its stated role is "to promote co-operation in the South Pacific region and to provide assistance in order to protect and improve its environment and to ensure sustainable development for present and future generations" (see Web site). The workshop was hosted in association with the Foundation for International Environmental Law and Development (FIELD), the World-Wide Fund for Nature (WWF) and the London-based School of Oriental and South Asian Studies (SOAS).

[29] These include the Charter of the Indigenous Tribal Peoples of Malaysia, the Kari-Oca Declaration, the Indigenous People's Earth Charter and the U.N. Draft Declaration of the Rights of Indigenous Peoples (Part IV. 29).

[30] The Mataatua Declaration came out of the First International Conference on the Cultural and Intellectual Property Rights of Indigenous Peoples, held in Aotearoa (New Zealand) in 1993 and attended by a number of Pacific Rim countries, including Australia, the Cook Islands, Fiji and the Philippines.

[31] This is defined in Janke (1998:2) as:

> the expressions, products, knowledge and practices that make up the collective heritage and culture of indigenous people. It embraces the intangible ideas and knowledge associated with artistic works and designs and other forms of cultural expression such as music, dance, song and story. It also includes traditional knowledge, innovations and practices relevant to the conservation and sustainable use of biological diversity.

References

Aalbersberg W (1998) *The development of bioprospecting agreements in Fiji.* BCNET position paper. Fiji: University of The South Pacific Publishers

Anderson R (1995) Oceans plundered in the name of medicine. *New Scientist* 148 (2005):5

Aoki K (1996) Intellectual property and sovereignty: Notes towards a cultural geography of authorship. *Stanford Law Review* 48:1293–1355

Appadurai A (1996) *Modernity at Large: Cultural Dimensions of Globalization.* Minneapolis: University of Minnesota Press

Arthur C (2000) The race is on to decode "God's software", but the debate rages over who should benefit from the results. *The Independent* 16 March:3

Barnes T (1995) Political economy 1: "The culture, stupid". *Progress in Human Geography* 19(3):423–431

Beardsley T (1998) An express route to the genome? *Scientific American* 279(2):30–32

Bent S et al (1987) *Intellectual Property Rights in Biotechnology World-Wide.* New York: Stockton Press

Bhat M (1996) Trade-related intellectual property rights to biological resources: Socioeconomic implications for developing countries. *Ecological Economics* 19(3):205–217

Blakeney M (1995) *Milpurrurru & Ors v Indofurn & Ors*: Protecting expressions of Aboriginal folklore under copyright law. *Murdoch University Electronic Journal of Law* 2(1). Available at http://www.murdoch.edu.au/elaw/indices/title/blakeney_abstract.html (last accessed 14 May 2002)

Blakeney M (1998) The role of regional commercial unions in Europe and Asia. *Prometheus* 16(3):41–50

Blomley N (1994) *Law, Space and the Geographies of Power.* London: Guilford Press
Boehm K (1967) *The British Patent System.* Vol 1, *Administration.* Cambridge, UK: Cambridge University Press
Boyle J (1996) *Shamans, Software and Spleens: Law and the Construction of the Information Society.* Cambridge, MA: Harvard University Press
Braithwaite J and Drahos P (1999) *Global Business Regulation.* Cambridge, UK: Cambridge University Press
Breyer S (1970) The uneasy case for copyright: A study of copyright in books, photocopies and computer programmes. *Harvard Law Review* 84:281–289
Brush S (1994) A non-market approach to protecting biological resources. In T Greaves (ed) *Intellectual Property Rights for Indigenous People: A Sourcebook* (pp 131–145). Oklahoma City: The Society for Applied Anthropology Publishers
Brush S (1999) Bioprospecting the public domain. *Cultural Anthropology* 14(4): 535–555
Castells M (1989) *The Informational City: Information Technology, Economic Restructuring and the Urban Regional Process.* Oxford: Blackwell
Castells M (1996) *The Information Age: Economy Society and Culture.* Vol 1, *The Rise of the Network Society.* London: Blackwell
Castree N (1999) Synthesis and engagement: Critical geography and the biotechnology century. *Environment and Planning A* 31(5):761–763
Clarke G L (1989) The context of federal regulation: Propaganda in US union elections. *Transactions of the Institute of British Geographers* NS 14:59–73
Coombe R (1996) Authorial cartographies: Mapping proprietary borders in a less-than-brave new world. *Stanford Law Review* 48:1357–1366
Coordination Committee on Science and Technology (CCST) (1994) *Access to Australia's Biological Resources.* Canberra: Australian Government Publishing Service
Cornish W R C (1999) *Intellectual Property: Patents, Copyright, Trademarks and Allied Rights.* London: Sweet and Maxwell
Correa C (1992) Biological resources and intellectual property rights. *European Intellectual Property Review* 14(5):154–157
Cragg, G (1996) Interview with Dr. Gordon Cragg, director of the Natural Products Division, The National Cancer Institute, Bethesda, Maryland, 19 March
Crespi S (2000) An analysis of moral issues affecting patenting inventions in the life sciences: A European perspective. *Science and Engineering Ethics* 6(2):157–180
Cultural Survival Quarterly (1996) Hagahai special issue. Vol 20
Cunningham H (1998) Colonial encounters in postcolonial contexts: Patenting indigenous DNA and the Human Genome Diversity Project. *Critique of Anthropology* 18:205–233
Dowall D (1984) *The Suburban Land Squeeze: Land Conversion and Regulation in the San Francisco Bay Area.* Berkeley: University of California Press
Drahos P (1996) *A Philosophy of Intellectual Property Rights.* Aldershot: Dartmouth Press
Farnsworth N (ed) (1990) The role of ethnopharmacology in drug development. In *Proceedings of the Ciba Foundation Symposium Bioactive Compounds From Plants* (pp 2–21). Chichester: Wiley
Featherstone M (ed) (1990) *Global Culture: Globalisation and Modernity.* London: Sage Publications
Frow J (1997) *Time and Commodity Culture: Essays in Cultural Theory and Postmodernity.* Oxford: Clarendon Press
General Agreement on Tariff and Trade (GATT) TRIPs Article XXVII:1 http://www.wto.org/english/docs_e/legal_e/legal_e.htm (last accessed 5 June 2002)
Gibbs D (2000) Globalisation, the bioscience industry and local environmental responses. *Global Environmental Change* 10:245–257
Gibson-Graham J K (1997) *The End of Capitalism (As We Knew It).* London: Blackwell

Gordon W (1993) A property right in self-expression: Equality and individualism in the natural law of intellectual property. *Yale Law Journal* 102:1533–1557
Government of the Philippines (1995) Presidential Executive Order on the Prospecting of Biological and Genetic Resources. Order no. 247. Adopted on 18 May
Greaves T (ed) (1994) *Intellectual Property Rights for Indigenous Peoples: A Sourcebook.* Oklahoma City: The Society for Applied Anthropology Publishers
Halpern S, Nard C and Port K (1999) *Fundamentals of US IPR Law: Copyright, Patents and Trademarks.* The Hague: Kluwer Law International
Jackson P (2000) Commodity cultures: The traffic in things. *Transactions of the Institute of British Geographers* NS 24:95–108
Jameson F (1991) *Postmodernism, or the Cultural Logic of Late Capitalism.* Durham, NC: Duke University Press
Janke T (1998) *Our Culture, Our Future—A Report on Australian Indigenous Cultural and Intellectual Property Rights.* Commissioned and published by the Australian Institute of Aboriginal and Torres Strait Islander Studies and the Aboriginal and Torres Strait Islander Commission (ATSIC). Canberra: Australian Institute of Aboriginal and Torres Strait Islander Studies and ATSIC
Jarman A and Kouzmin A (1993) Australian metropolitan development: Local government reform and urban growth in the 1990s. *Environment and Planning C: Government and Policy* 11:143–160
Jenkins C, Dimitrakakis M, Cook I, Sanders R and Stallman N (1989a) Culture change and epidemiological patterns among the Hagahai, Papua New Guinea. *Human Ecology* 17(1):27–57
Jenkins C, Bhatia K, Prasad M, Koki G and Lombange J (1989b) Immunogenetic studies of two recently contacted populations from Papua New Guinea. *Human Biology* 61(1):45–64
Juma C (1989) *The Gene Hunters.* London: Zed Books
Kallinoe L and Leach J (eds) (2000) *Rationales of Ownership.* New Dehli: UBS Publishers' Distributors Ltd
Knorr-Cetina K and Mulkey M (eds) (1983) *Science observed: perspectives on the social study of science* London: Sage
Leff D (1994) Just discovered obesity gene points to distant slimming therapy. *BioWorld Today.* 5:232
Lesser W (1996) International treaties and other legal and economic issues relating to the ownership and use of genetic materials. In K Hoagland and A Rossman (eds) *Global Genetic Resources: Access, Ownership, and Intellectual Property Rights* (pp 31–50). The Beltsville Symposia in Agricultural Research. Washington, DC: The Association of Systematics Research Publishers
Locke J (1690) *Two Treatises of Government.* Edited by P Laslett. Cambridge, UK: Cambridge University Press
Maskus K (2000) *Regulatory Standards in the WTO: Comparing Intellectual Property Rights with Competition Policy, Environmental Protection and Core Labour Standards.* Washington, DC Institute for International Economics working paper. http://www.iie.com/CATALOG/WP/2000/00-1.htm (last accessed 14 May 2002)
McAfee K (1999) Selling nature to save it? Biodiversity and the rise of green developmentalism. *Environment and Planning D: Society and Space* 17:133–154
Miller D (1997) *Capitalism: An Ethnographic Approach.* Oxford: Berg
Office of the US Trade Representative. http://www.ustr.gov/enforcement/dispute.shtml (last accessed 5 June 2002)
Parry B C (2000) The fate of the collections: Social justice and the annexation of plant genetic resources. In C Zerner (ed) *People, Plants and Justice* (pp 374–402). New York City: Columbia University Press

Parry B C (2003) *Bio-commodities: Revealing the Social and Spatial Dynamics of Trade in Genetic Resources.* New York City: Columbia University Press
Peck J (1990) The state and the regulation of local-labour markets—Observations on the geography of the youth training scheme. *Area* 22(1):17–27
Posey D and Dutfield G (1996) *Beyond Intellectual Property: Towards Traditional Resource Rights for Indigenous Peoples and Local Communities.* Ottawa: International Development Research Centre
Radford T (2000) Blair and Clinton in plea over gene code. *The Guardian* 15 March:3
Reid W (1996) Regulating the biotrade to promote the conservation and sustainable use of biodiversity. In K Hoagland and A Rossman (eds) *Global Genetic Resources: Access, Ownership, and Intellectual Property Rights* (pp 303–315). The Beltsville Symposia in Agricultural Research. Washington, DC: The Association of Systematics Research Publishers
Reid W, Laird S and Meyer C (1993) *Biodiversity Prospecting: Using Genetic Resources for Sustainable Development.* Baltimore, MD: World Resource Institute Publishers
Rockefeller University researchers. Genetic basis of obesity in humans. http://www.rockefeller.edu/research/heads.htm (last accessed 14 May 2002)
Rural Advancement Foundation International (RAFI) (1995) Gene hunters in search of disease genes collect human DNA from remote island populations. RAFI Communiqué May/June 1995. http:/www.rafi.org/article.asp?newsid=207 (last accessed 5 June 2002)
Shiva V (1993a) *Monocultures of the Mind: Perspectives on Biodiversity and Biotechnology.* Penang: Third World Network Publishers
Shiva V (1993b) Why we should say "no" to GATT TRIPs. *Third World Resurgence* 39(Nov):32–39
Shiva V (1997) *Biopiracy: The Plunder of Nature and Knowledge*. Boston: South End Press
South Pacific Regional Environmental Programme (SPREP) (1998a) Press release: Pacific countries focus on using international law to protect their biodiversity. http://www.sprep.org.ws/PressRelease/1998/pr9810_.htm (last accessed 14 May 2002)
South Pacific Regional Environmental Programme (SPREP)(1998b) Who profits from the Pacific's natural treasures? *SPREP Environmental Newsletter* 52 (June):2
South Pacific Regional Environmental Programme (SPREP) http://www.sprep.org.ws (last accessed 5 June 2002)
Steinberg D (2000) Will genomics spoil gene ownership? *The Scientist* 14(17):1–5
Strathern M (1999a) Property, transactions and creations—New economic forms in the Pacific. Unpublished manuscript, Cambridge, UK
Strathern M (1999b) *Property, Substance and Effect: Anthropological Essays on Persons and Things*. London: Athlone
Strathern M (2000) Multiple perspectives on intellectual property. In K Whimp and M Busse (eds) *Protection of intellectual, biological and cultural property in Papua New Guinea.* Canberra: Asia Pacific Press and Conservation Melanesia Publishers
ten Kate K and Laird S (1999) *The Commercial Use of Biodiversity: Access to Genetic Resources and Benefit-Sharing.* London: Earthscan
Thompson M and Rushing F (1996) An empirical analysis of the impact of patent protection on economic growth. *Journal of Economic Development* 21(2):61–79
Thrift N (1985) Flies and germs: A geography of knowledge. In D Gregory and J Urry (eds) *Social Relations and Spatial Structures* (pp 366–403). Basingstoke: Macmillan
Thrift N (1996) *Spatial Formations.* London: Sage
Unger R (1983) The critical legal studies movement *The Harvard Law Review* 96(3):320–432

United Nations Development Programme (UNDP) (1995) *Statements of the Regional Meetings of Indigenous Representatives on the Conservation and Protection of Indigenous Knowledge.* Nairobi: UNDP Publications

Wainwright J, Prudham S and Glassman J (2000) The Battle in Seattle: Microgeographies of resistance and the challenge of locally constructed futures. *Environment and Planning D* 18(1):5–13

Whatmore S (1999) Geography's place in the life-science era? *Transactions of the Institute of British Geographers* NS 24:259–260

[21]

Materialist returns: practising cultural geography in and for a more-than-human world

Sarah Whatmore

Oxford University Centre for the Environment, School of Geography

This paper surveys the return to materialist concerns in the work of a new generation of cultural geographers informed by their engagements with science and technology studies and performance studies, on the one hand, and by their worldly involvements in the politically charged climate of relations between science and society on the other. It argues that these efforts centre on new ways of approaching the vital nexus between the bio (life) and the geo (earth), or the 'livingness' of the world, in a context in which the modality of life is politically and technologically molten. It identifies some of the major innovations in theory, style and application associated with this work and some of the key challenges that it poses for the practice of cultural geography.

Thinking is neither a line drawn between subject and object nor a revolving of one around the other. Rather thinking takes place in the relationship of territory and earth... involving a gradual but thorough displacement from text to territory.[1]

Something/happening

It seems pertinent, even unavoidable, to begin by confessing that I still feel something of an outsider in the 'cultural geography' camp – at least as it came to be configured in the formative years of my research career in the late 1980s as the rise of 'cultural studies' in the UK gained disciplinary purchase in the guise of 'the new cultural geography'.[2] That project's signature concerns with the politics of representation and identity cast my obdurately earthy interests in cultivation and property, growing and eating, in a very unfashionable light. At that time such interests found a more convenient if not very permissive home in political economy – where the 'matter of nature', as Margaret Fitzsimmons so memorably put it, was marginally less marginalized.[3] So, in a small but not insignificant way, my being invited to present the *cultural geographies* annual lecture[4] is testament to some kind of realignment of intellectual energies underway; that moment of *fabulation* that Deleuze conjures[5] in which cultural forces regroup and start to generate their own stories: stories which enter the world as envoys of 'something happening' – giving that something/happening both shape and momentum.

This paper might best be thought of as just such a self-conscious act of storying – an envoy of the recuperation of 'materiality' that is gathering force in this something/happening through energies as diverse as postcolonial, feminist, landscape, urban, legal and performance studies.[6] Through these diverse currents, cultural geographers have found their way (back) to the material in very different ways that variously resonate with what I take to be amongst the most enduring of geographical concerns – the vital connections between the *geo* (earth) and the *bio* (life).[7] The durability of these concerns bears the hallmark of geography's history, which (like anthropology and archaeology) took shape before the division of academic labours into social and natural sciences became entrenched. It is a division with which these disciplines have never been entirely comfortable, and with which they continue to wrestle more self-consciously, and sometimes productively, than others. With the advent of the 'new cultural geography', this *earthlife* nexus was written out of, or more accurately, into the ancestral past of cultural geography – at least in the Anglophone research community.[8]

I argue here that this nexus is currently being recharged and taken in unfamiliar directions by a new generation of cultural geographers, not least through multiple engagements with the 'geo/bio-philosophy' of Deleuze and Guattari[9] from which this intervention pushes off. Such engagements have been direct, through close readings of their work and the philosophical industry that it has spawned, and indirect, through the twin intermediaries of science and technology studies and performance studies in which it is differently, and variously, inflected.[10] A common commitment in such work is a view of science and philosophy as projects in which theory does not take on a representational function, but rather an active and practical one, such that every theory acts as a 'mechanics' – simultaneously a technology of practice and an intervention in the world.[11] But this storying of cultural geography's recuperation of the material works against forging 'it' into the latest in a weary and wearying succession of 'new turns' that have been written into the intellectual history of cultural geography, still less one that is uniformly or exclusively Deleuzian.

Instead, I want to emphasize that this recuperation manifests a rich variety of analytical impulses; philosophical resources and political projects that don't 'add up' to a singular 'new' approach, let alone one that has a monopoly of insight or value. To this end, I use the language of *re*turns to suggest that what is new (as in different) about the something/happening in cultural geography is a product of repetition – turning seemingly familiar matters over and over, like the pebbles on a beach – rather than a product of sudden encounter or violent rupture. Just as importantly, what is different or innovatory about these materialist returns is generated as much by the technologically and politically molten climate that informs cultural geographers' intellectual investments and worldly involvements as by any academic repositioning. In this case, I think there can be little doubt that the materialist returns of cultural geography today are bound up with the proliferation of what Bruno Latour calls 'matters of concern'[12] and Michel Callon calls 'hot situations'[13] associated with the intensification of the interface between 'life' and 'informatic' sciences and politics. This intensification has been witnessed in serial public controversies since the 1990s, from GM to nanotechnology,[14] in which the practices of social, as well as natural, scientists have been caught up.

These involvements are reflected in cultural geography's reworking of substantive topics (like domestication and sensory perception) in unfamiliar, and sometimes unsettling, directions such as the burgeoning interest in animal cultures,[15] 'post-humanism'[16] and geographies of 'affect'.[17]

I want to start by articulating the broadest sense of this claim that a new generation of cultural geographers is returning to the rich conjunction of the *bio* and the *geo* – or, for want of a more felicitous expression, to what the writer Jeanette Winterson calls the 'livingness' of the world.[18] My argument is that the most important difference (a big claim, I know) in the 'something/happening' in cultural geography's materialist recuperations is that this return to the livingness of the world shifts the register of materiality from the indifferent stuff of a world 'out there', articulated through notions of 'land', 'nature' or 'environment', to the intimate fabric of corporeality that includes and redistributes the 'in here' of human being.[19] In this it shares the same impulse as Derrida's frequently articulated insistence on addressing philosophical enquiry to 'the entire field of the living, or rather to the life/death relation, beyond the anthropological limits of "spoken language"'.[20] Importantly, however, this redirection of materialist concerns through the bodily enjoins the technologies of life and ecology, on the one hand, and of prehension and feeling, on the other, in refiguring the ontological disposition of research – drawing cultural geographers into new conversational associations; research practices and modes of address that collectively mark what I have called 'more-than-human' approaches to the world.[21] I will tease out this rather bold claim by sketching some of the key facets of the materialist returns now in play, as I see them, and the kinds of challenges that they pose.

Materialist recuperations

In the last edition of the *Dictionary of human geography*, Denis Cosgrove in his entry on cultural geography distinguishes 'classical' from 'new' styles of cultural geography by reference to their approaches to the study of landscape.[22] The former, associated with the work of Carl Sauer and the Berkeley school that he inspired, has as its reference point his iconic essay 'The morphology of landscape' in which

> cultural landscape is fashioned from a natural landscape by a cultural group. Culture is the agent, the natural area the medium, the cultural landscape is the result.[23]

By contrast, 'new' cultural geography is associated with the flowering of cultural studies in Britain, as signalled by the no less totemic essay of Daniels and Cosgrove introducing their book *Iconography and landscape*, in which landscape is defined as

> a cultural image, a pictorial way of representing... surroundings. This is not to say that landscapes are immaterial. They may be represented in a variety of materials... in paint on canvas, in writing on paper, in earth, stone, water and vegetation on the ground.[24]

The point I want to draw from these exemplary quotations is rather different from that for which they have come to stand in demarcating a 'new' from a 'classical' regime.

Despite the significant differences they articulate, what I find most striking about them is that they share an overriding common currency, namely that they both cast the making of landscapes (whether worked or represented) as an exclusively human achievement in which the stuff of the world is so much putty in our hands. On these accounts, as I have suggested elsewhere, 'the world remains untroubled and untroubling, waiting impassively for us to make up our minds and making no difference' to the landscape (or knowledge, or environment...) in the making.[25] By the same token, cultural geography's investments in questions of identity and culture have remained largely wedded to that most vociferously silent and self-evident subject of the social sciences, the 'in-here' of human being. So it is that recent contributions have sought to do (at least) three things. The first has been to re-animate the missing 'matter' of landscape, focusing attention on bodily involvements in the world in which landscapes are co-fabricated between more-than-human bodies and a lively earth.[26] The second has been to interrogate 'the human' as no less a subject of ongoing co-fabrication than any other socio-material assemblage.[27] The third in my list has been the redistribution of subjectivity as something that 'does not live inside, in the cellar of the soul, but outside in the dappled world'.[28]

This redistribution of energies puts the onus on 'livingness' as a modality of connection between bodies (including human bodies) and (geo-physical) worlds. In turn, that acts as a rallying point for geographers (and others) working against the lexical cast of the 'new' cultural geography and the humanist commitments of cultural geography more broadly, bringing all manner of philosophical resources to bear on their efforts. These include the corporeal materialisms inspired by Foucault's bio-cultures, Merleau-Ponty's ontology of the flesh and the feminist corporeal ethics of Diprose[29] and others; and the energetic materialisms inspired by the relational ontologies of Spinoza, Whitehead and Deleuze (amongst others), such as Stengers' co-fabrication or 'working together'.[30] In conjunction with the molten question of what 'livingness' means in a life science era,[31] such resources and energies redirect materialist concerns in ways that have profound ethical and political, as well as analytical, consequences. As the political theorist Jane Bennett recently put it, they attempt to hold onto the relational and emergent imperatives of material force in which the 'thing-ness of things' – bodies, objects, arrangements – are always in-the-making and 'humans are always in composition with nonhumanity, never outside of a sticky web of connections or an *ecology* [of matter]'.[32]

If these are some of the lineaments of the differences/innovations wrought by the materialist returns of a cultural geography attentive to the livingness of the world, how is this attentiveness playing out in terms of more specific research directions and impulses? I want to outline four commitments being taken forward in diverse ways in such work that strike me as being of particular importance.

The first is a shift in analytic focus from *discourse* to *practice*. Inspired by numerous and non-additive efforts to work against the grain of the logocentric conception of social agency – 'I think therefore I act' – that is a familiar mantra of orthodox social science. This shift is associated by some with the so-called 'practice turn'[33] and a variety of approaches which relocates social agency in practice or performance rather than discourse – thinking and acting through the body – and reworks discourse itself as a specific kind of practice.

The second is a shift from an onus on *meaning* to an onus on *affect*. The bodily register of current work reopens the interval between sense and sense-making, and multiplies the sensory dimensions of acting in the world and the milieux of inter-corporeal movement. Affect refers to the force of intensive relationality – intensities that are felt but not personal; visceral but not confined to an individuated body. This shift of concern from what things mean to what they do has methodological consequences for how we train our apprehensions of 'what subjects us, what affects and effects us' or 'learn to be affected'.[34]

The third redirection of effort is towards *more-than-human* modes of enquiry. Such modes of enquiry neither presume that socio-material change is an exclusively human achievement nor exclude the 'human' from the stuff of fabrication. Animals and technological devices have variously been used as 'agents provocateurs' in tackling the question of difference and rigorously working it through the specific materialities and multiplicities of subjectivity and agency.[35] Such modes of enquiry attend closely to the rich array of the senses, dispositions, capabilities and potentialities of all manner of social objects and forces assembled through, and involved in, the co-fabrication of socio-material worlds.

The fourth shift is from a focus on the *politics of identity* to *the politics of knowledge*. Here two currents come together in addressing concerns with the ways in which knowledge is produced, hardwired into the social fabric and contested in a variety of public forums. One of these concerns the redistribution of expertise attendant on the recognition of multiple knowledge practices and communities that bear on the framing of inherently uncertain socio-technical problems.[36] The other concerns the practice of science (including social science) in constituting the phenomena that it studies as 'reliable witnesses' where that reliability is guaranteed by allowing phenomena to work against, or to exceed, our experimental expectations.[37]

Practising more-than-human geographies

> For if the look purchases the transcendence of the human only at the expense of repressing the other senses (and more broadly the material and the bodily with which they are traditionally associated), then one way to recast the figure of vision (and therefore that with which it is ineluctably associated) is to resituate it as only one sense among many in a more general – and not necessarily human – bodily sensorium.[38]

I have sought to argue that the creativity of cultural geography is generated not by a succession of 'new' turns but by the gathering force of constant re-turns to enduring preoccupations with the processes and excesses of 'livingness' in a more-than-human world. Trying not to solidify the heterogeneity of ideas and practices at work in the recuperation of materiality in cultural geography into the latest such 'turn', I have outlined some of what I see as the most important aspects of an ongoing realignment of intellectual energies. It is a realignment that promises much in terms of equipping geography in the life science era, but one that brings real and pressing methodological and political challenges in its wake. Before one gets carried away with their claims to novelty, it is worth recalling earlier efforts to marry the 'bio' and 'geo' in cultural geography. Thus, for example, buried in his 'morphology of landscape' essay is an appeal by Carl Sauer (following Vidal de la Blache) that

> Geographers should avoid considering the earth as the scene on which the activity of man (sic) unfolds itself, without reflecting that this scene is itself living.[39]

In similar vein, current interests in performativity are anticipated in J.B. Jackson's concerns with the 'vernacular landscapes' generated by what he saw as the inexhaustible capacity for improvisation in people's everyday ways of making themselves at home in the world with and against the grain of 'aristrocratic' or political designs.[40] That said, the differences are profound, as these preoccupations with the intersection between the 'bio' and the 'geo' become charged with the socio-technological possibilities, political registers, cultural sensibilities, and intellectual enthusiasms of a new generation of geographers. The next edition of the *Dictionary of human geography* is a few years off, but some glimpse of the ways in which the treatment of landscape in the updated entry on cultural geography might be recharged by this work are signalled in Cary Wolfe's observations on the bodily sensorium, quoted at the start of this final section.

So what difference do the materialist recuperations and research directions in cultural geography that I have sketched above make to the question of how, as Karen Barad puts it, 'matter comes to matter'?[41] On the one hand, such work is clearly marked by the distinctive axes of academic exchange that inform it, notably the burgeoning fields of study emerging between disciplines – such as science and technology studies, performance studies, and feminist studies. On the other, it is their engagement in matters of public controversy and everyday concern taking place in a proliferation of other cultural and political forums that most stands out. Here, what makes the question of materiality matter is the molten climate of relations between science and society, technology and democracy[42] in which the knowledge practices of social and natural scientists, civil servants and corporate lawyers, NGOs and direct action groups, citizens and consumers rub up against one another in the event of all manner of knowledge controversies. Such controversies (around genetic engineering, MMR injections, pharmaceutical patenting, stem-cell harvesting, reproductive cloning, for example) are at once about the most mundane and intimate aspects of social life – food, health and kinship – *and* the sites of prolific inventiveness in the life sciences. Taking the working example of what we might call 'bio-geographies', a field set to explode in the cultural geography literature, it seems,[43] I want finally to illustrate some of the differences in the ways in which cultural geographers are currently recuperating the material, and to signal some of the major challenges that they pose.

The first major difference is that 'life' itself has changed, to become the latest addition to that peculiar socio-material assemblage called 'natural resources'. The treatment of animals and plants as biological resources is hardly new, but with the rise of genetic and, more particularly, post-genomic biotechnologies any vestige of difference between the ability to manipulate and commodify their bodies and human bodies has been removed. This disturbing levelling of biological differences, reinforced by the re-materialization of biological entities in the guise of machine-readable informatic codes,[44] has profound effects on what bodies count and what counts as bodily in the work of cultural geographers today.[45] Not least are the considerable additional skills required to study the detailed knowledge practices involved in the production and circulation of such bio-technological artefacts, if cultural geographers are to get to grips with the *specificity* (as against the originality) of knowledge objects like artificial life forms. The cultural potency of 'artificial life' suggests that it might be possible to learn from the repertoire of techniques employed in artistic work that engages science and/or scientists to stage public experiments in the possibilities of reworking hum/ani/machine interfaces through robotic, neurological and genomic amplifications or extensions of bodily competences and temporalities.[46] For example, the Australian performance artist Sterlarc, who has worked with robotics scientists at Sussex University in devising an 'exoskeleton', seeks to produce a choreography of movements in which

> instead of seeing the human body as the choreographer and the robot as the instrument, I really see the two working together. That is how it becomes an artistic performance. I have no desire to control the machine.... I am open to its doing the unexpected. In this sense the human body has always been a kind of cyborg. .... I am not satisfied with just theorising about it. I want to experience what actually happens and then try to articulate what that means.[47]

A second major difference is the changed relationship between science and society in which new scientific knowledge claims and/or artefacts, particularly in those fields that touch the visceral vernacular of social anxiety relating to food or health, have become routinely controversial matters. Such controversies take cultural geographers to unfamiliar forums. At one end of the spectrum stand the law courts in which the artefacts themselves are called upon as material witnesses[48] in the determination of competing claims to the 'intellectual property' in new biological artefacts.[49] At the other are the proliferation of impromptu 'hybrid' forums that swell in the face of new technologies – like GM or mobile phone masts – gathering to them all manner of concerned citizens and/or consumers; seasoned advocacy groups; scientific dissidents and the like that can change the commercial and regulatory fabric of such technologies in unpredictable ways. How do social scientists, including cultural geographers, position themselves in these forums? As the clamour grows for greater 'interdisciplinarity' as a way of addressing such knowledge controversies, cultural geography's rich tradition of experimentation provides a valuable resource for resisting the pressures on us (from within and outside the discipline) to assume the position of 'interpreters' between concerned publics and natural scientists.

As I see it, perhaps the greatest challenge presented by these 'more-than-human' styles of working is the onus they place on experimentation and, by implication, on taking (and being allowed to take) risks. Let me dwell momentarily on just two aspects of this experimental imperative. First is the urgent need to supplement the familiar repertoire of humanist methods that rely on generating talk and text with experimental practices that amplify other sensory, bodily and affective registers and extend the company and modality of what constitutes a research subject.[50] Second, the experimental demands of 'more-than-human' styles of working place an onus on actively redistributing expertise beyond engaging with other disciplines or research fields to engaging knowledge practices and vernaculars beyond the academy in experimental research/politics such as the 'deliberative mapping' exercise pioneered by Gail Davies and her collaborators in relation to xeno-transplantation.[51] I hope and trust that *cultural geographies* will continue to play its part as a leading journal in which scholars can take risks and experiment; in which the worldliness that has been the hallmark of geographical endeavours is reinvigorated; and in which conversations and politics proliferate in generative ways rather than hardening into orthodoxy.

Acknowledgements

I should like to acknowledge the valuable feedback I received from audiences on versions/parts of this paper presented at the 'Envisioning geographies' symposium at UCL; the Lennart Andersson Lecture at Karlstadt University; and the 'Exhibition of British geography' plenary session at the IGU conference in Glasgow.

Notes

1 G. Deleuze and F. Guattari, *What is Philosophy?* (French original 1991), trans. Graham Burchell and Hugh Tomlinson (London, Verso, 1994).

2 See C. Philo, ed., *New words, new worlds: reconceptualising social and cultural geography* (Lampeter, Dept of Geography, Saint David's University College, Lampeter, 1991).

3 M. Fitzsimmons, 'The matter of nature', *Antipode* 21 (1989), pp. 106–20.

4 This is a version of the *Cultural geographies* annual lecture given at the centennial conference of the Association of American Geographers (Philadelphia, March 2004). My thanks to the journal's editors, Mona Domosh and Philip Crang, for the challenging invitation to present the lecture, and to Hodder Arnold for sponsoring the event.

5 Deleuze is here following Bergson. See G. Deleuze, *Negotiations* 125 (New York, Columbia University Press, 1995), p. 125.

6 Exemplary of these recuperations of the material in various fields of cultural geography are the following: in relation to the postcolonial, I. Cook and M. Harrison, 'Cross over food: re-materialising postcolonial geographies', *Transactions of the Institute of British Geographers* 28 (2003), pp. 296–317; in relation to feminist scholarship, C. Nash, 'Genetic Kinship', *Cultural studies* 18 (2004), pp. 1–34; in landscape studies, J. Wylie, 'On ascending Glastonbury Tor', *Geoforum* 33 (2002), pp. 441–54; in urban studies, A. Latham and D. McCormack, 'Moving cities: rethinking the materialities of urban geographies', *Progress in human geography* 28 (2004), pp. 701–24; in relation to legal geographies, D. Delaney, 'Making nature/marking humans: law as a site of (cultural) production', *Annals of the Association of American Geographers* 91 (2001), pp. 487–503; and in relation to performance studies, J.D. Dewsbury, P. Harrison, M. Rose and J. Wylie, eds, 'Enacting geographies', special issue of *Geoforum* 33 (2003).

Sarah Whatmore

7 There are clearly other ways of conceiving of the re-materialization of cultural geography that owe more to anthropological traditions in the study of material culture; see e.g. P. Jackson, 'Rematerialising social and cultural geography', *Social and cultural geography* 1 (2000), pp. 9–14.

8 Notable exceptions include the institutional hold of cultural ecology in the Nordic countries and its persistence as an active research grouping in the Association of American Geographers. It should also be noted that reservations about the 'linguistic' turn in British cultural geography were articulated even at its height (see esp. Philo, *New words, new worlds*).

9 Both 'geo-philosophy' (see J. Bonta and J. Protevi, *Deleuze and geo-philosophy: a guide and glossary* (Edinburgh, Edinburgh University Press, 2004)) and 'bio-philosophy' (see K. Ansell-Pearson, *Germinal life: the difference and repetition of Deleuze* (London, Routledge, 1999)) provide useful ways into the work of Deleuze and Guattari.

10 Hence e.g. Latour's famous quip that the acronym ANT (Actant Network Theory) could just as easily have been ART (Actant Rhizome Theory). See T. Crawford, 'An interview with Bruno Latour', *Configurations* 1 (1993), pp. 247–68.

11 T. Murphy, 'Quantum ontology: a virtual mechanics of becoming', in E. Kaufman and K. Heller, eds, *Deleuze and Guattari: new mappings in politics, philosophy and culture* (Minneapolis, University of Minnesota Press, 1998), p. 213.

12 A term he uses in contrast to 'matters of fact' and as shorthand for refusing the distinction between what is controvertible (e.g. values) and what is not (e.g. observational data). B. Latour, *Politics of nature: how to bring the sciences into democracy* (Cambridge, MA, Harvard University Press, 2004).

13 A term he uses to describe situations in which 'everything becomes controversial [in] the absence of a stabilised knowledge base'; M. Callon, 'An essay on framing and overflowing', in *The laws of markets* (Oxford, Blackwell, 1998), p. 260.

14 See H. Nowotny, P. Scott and M. Gibbons, *Rethinking science: knowledge, and the public in an age of uncertainty* (Oxford, Polity, 2001).

15 C. Philo and C. Wilbert, eds, *Animal spaces: beastly places* (London, Routledge, 2000).

16 N. Castree and C. Nash, eds, 'Mapping posthumanism', theme issue of *Environment and planning A* 36 (2004).

17 D. McCormack, 'An event of geographical ethics in spaces of affect', *Transactions of the Institute of British Geographers* 28 (2003), pp. 488–507.

18 J. Winterson, *Gut symmetries* (New York, Knopf, 1997), p. 85.

19 See P. Sheehan, ed, *Becoming human: new perspectives on the inhuman condition* (Westport, CT, Praeger, 2003).

20 J. Derrida and E. Roudinesco, 'Violence against animals', in *For what tomorrow ... a dialogue*, trans. J. Fort (Stanford, CA, Stanford University Press, 2001), p. 63.

21 S. Whatmore, *Hybrid geographies: natures cultures spaces* (London, Sage, 2002).

22 D. Cosgrove, 'Cultural geography', in R.J. Johnston, D. Gregory, G. Pratt and M. Watts, eds, *The dictionary of human geography*, 4th edn (Oxford, Blackwell, 2000), pp. 134–38.

23 C. Sauer, 'The morphology of landscape' [1925], in J. Leighley, ed., *Land and life: selections from the writings of Carl Ortwin Sauer* (Berkeley, University of California Press, 1963), p. 343.

24 S. Daniels and D. Cosgrove, 'Iconography and landscape', in S. Daniels and D. Cosgrove, eds, *The iconography of landscape* (Cambridge, Cambridge University Press, 1988), p. 1.

25 S. Whatmore, 'Generating materials', in M. Pryke, G. Rose and S.Whatmore, eds, *Using social theory: thinking through research* (London, Sage, 2003), p. 92.

26 For instance, S. Hinchliffe, '"Inhabiting" – landscapes and natures', in K. Anderson, M. Domosh, S. Pile and N. Thrift, eds, *Handbook of cultural geography* (London, Sage, 2003), pp. 207–26.

27 See e.g. K. Anderson, 'White natures: Sydney's Royal Agricultural Show in post-humanist perspective', *Transactions of the Institute of British Geographers* 28 (2003), pp. 422–41.

28 B. Latour, 'Body, cyborgs and the politics of incarnation', in S. Sweeney and I. Hodder, eds, *The body* (Cambridge, Cambridge University Press, 2002), p. 140. See e.g. Wylie, 'On ascending'.

29 R. Diprose, *Corporeal generosity* (London, Routledge, 2002).

30 I. Stengers, *Power and invention: situating science* (Minneapolis, University of Minnesota Press, 1997).

31 See R. Doyle, *Wetwares: experiments in postvital living* (Minneapolis, University of Minnesota Press, 2003).

32 J. Bennett, 'The force of things: steps to an ecology of matter', *Political theory* 32 (2004), p. 365 (emphasis original).

33 T. Schatzki, K. Knorr-Cetina and E. Von Savigny, eds, *The practice turn in contemporary theory* (London, Routledge, 2001).

34 Latour, 'Bodies, cyborgs and the politics of incarnation', p. 140.

35 E.g. C. Wolfe, *Animal rites: American culture, the discourse of species and posthumanist theory* (Chicago, University of Chicago Press, 2003).

36 See Nowotny *et al.*, *Re-thinking science*.

37 Stengers, *Power and invention*, p. 85.

38 Wolfe, *Animal rites*, p. 3.

39 Sauer, 'The morphology of landscape', p. 321.

40 J.B. Jackson, *Discovering the vernacular landscape* (New Haven, CT, Yale University Press, 1984), pp. 7–8.

41 K. Barad, 'Posthumanist performativity: toward an understanding of how matter comes to matter', *Signs* 28 (2003), pp. 801–32.

42 M. Callon, P. Lascoumes and Y. Barthe, *Agir dans un monde incertain: essai sur la dêmocratie technique* (Paris, Seuil, 2001).

43 In the context of giving the *cultural geographies* lecture, I was responding here to a number of sessions at the Centennial Annual Meeting of the Association of American Geographers at which I was speaking, and thinking in particular of the 'Geographies of biotechnology' sessions organized by Beth Greenhough and Emma Roe.

44 See Doyle, *Wetwares*.

45 See P. Thurtle and R. Mitchell, eds, *Semiotic flesh: information and the human body* (Seattle, Walter Chapin Simpson Centre for the Humanities, 2002).

46 O. Dyens, *Metal and flesh* (Cambridge, MA, MIT Press, 2001).

47 Quotation from www.stelarc.va.com.au.

48 See M. Strathern, *Property, substance and effect: anthropological essays on persons and things* (London, Athlone Press, 1999).

49 See Delaney, 'Making nature/marking humans'.

50 For an example, see S. Whatmore and S. Hinchliffe, 'Living cities: making space for urban nature', *Soundings: journal of politics and culture* 22 (2003), pp. 37–50.

51 See G. Davies, J. Burgess, J. Eames, M. Mayer, K. Staley, A. Stirling and S. Williamson, *Deliberative mapping: appraising options for addressing the kidney gap* (Wellcome Trust Final Report Grant 064492, 2003). See also: http://www.deliberative-mapping.org

Name Index